The Conquest of the Incas

John Hemming was born in Canada in 1935 and was educated at Eton College, McGill University, and Magdalen College, Oxford, where he read history. Mr. Hemming is an enthusiastic traveler and explorer. In the early sixties he took part in the Iriri River expedition into unexplored areas of central Brazil. He is on the Council of the Royal Geographical Society and of the Anglo-Brazilian Society, and is a Trustee of Primitive Peoples Fund. This book took him six years to write, and during that time he consulted more than a thousand books and documents and traveled to Peru and Spain to investigate sources.

John Hemming

The Conquest of the Incas

A Harvest Book • Harcourt, Inc.

San Diego New York London

Library of Congress Cataloging-in-Publication Data

Hemming, John, 1935–

The conquest of the Incas.

(A Harvest book)

Bibliography: p.

1. Peru—History—Conquest, 1522-1548.

2. Incas. I. Title.

F3442.H47 1973 985′.02 73-4616

ISBN 0-15-602826-3

Printed in the United States of America

B D F H J I G E C A

FOR MY PARENTS

AND GODPARENTS

CONTENTS

LIST OF ILLUSTRATIONS

Between pages 418 and 419

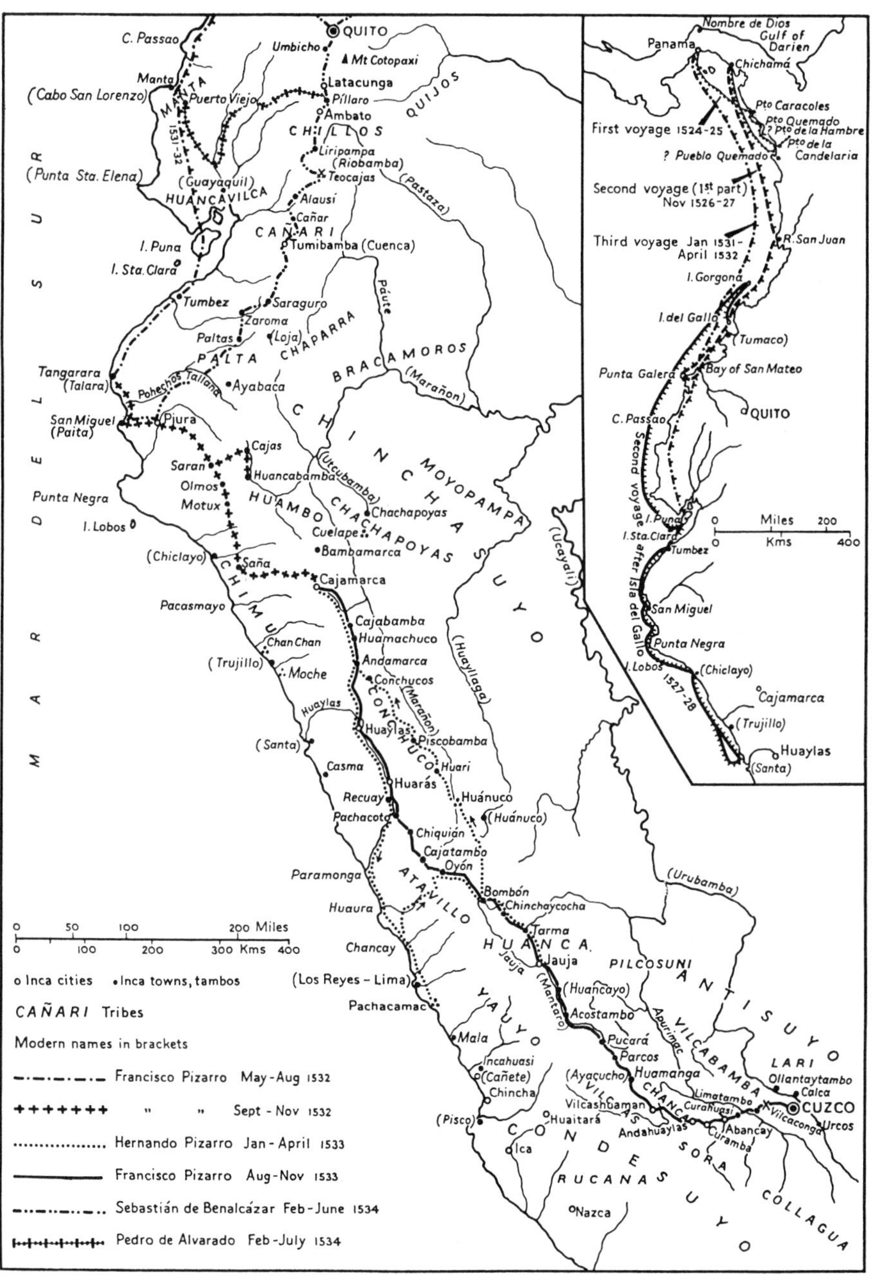

1. The early voyages and the march on Cuzco

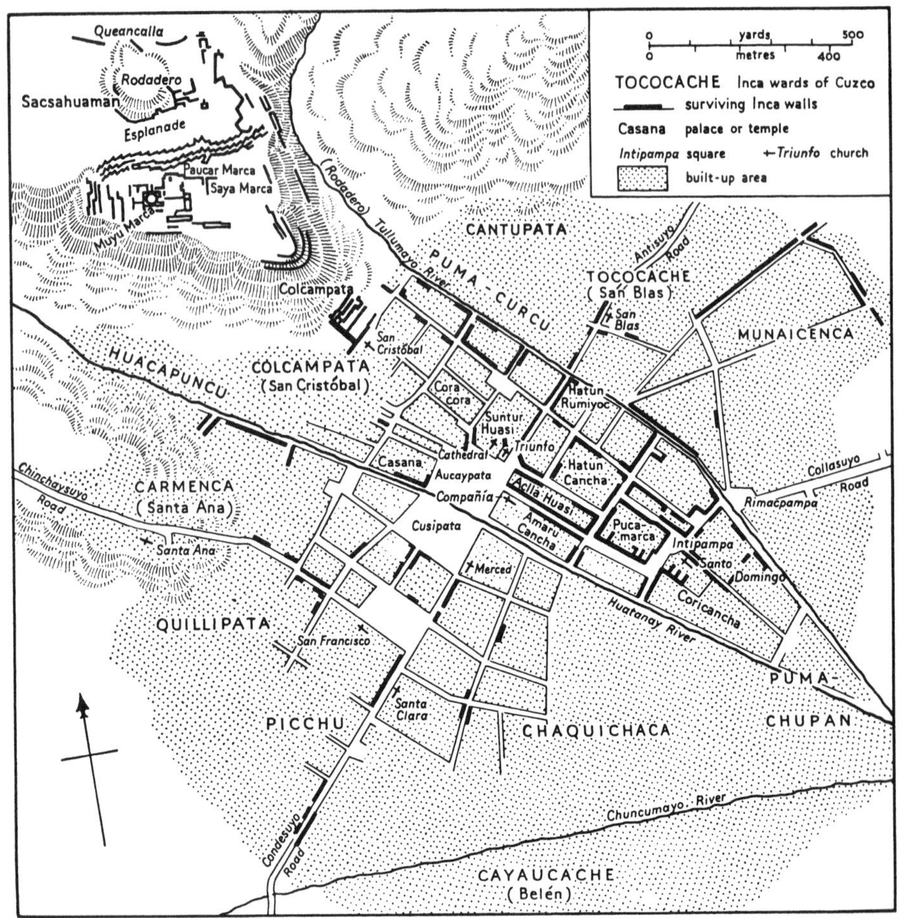

2. Cuzco at the time of the Conquest

3. The Quitan Campaign

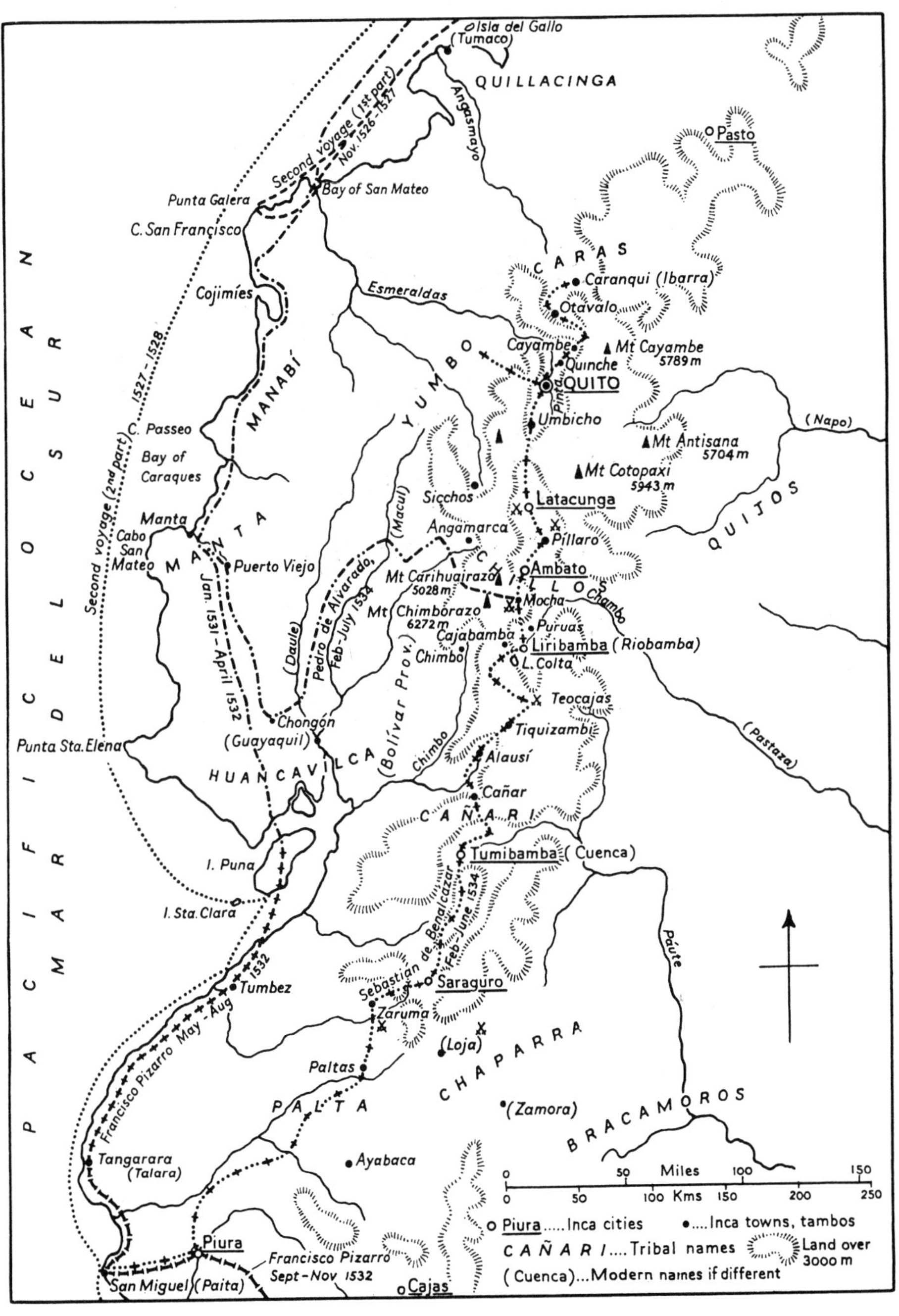
Isla del Gallo
(Tumaco)
QUILLACINGA
Angasmayo
Pasto
Second voyage (1st part) Nov. 1526-1527
Bay of San Mateo
Punta Galera
C. San Francisco
CARAS
Caranqui (Ibarra)
Otavalo
Cojimíes
Esmeraldas
Cayambe
Mt Cayambe 5789 m
Quinche
QUITO
Pinta
YUMBO
MANABÍ
Umbicho
(Napo)
Mt Antisana 5704 m
Mt Cotopaxi 5943 m
C. Passeo
Bay of Caraques
Sicchos
Latacunga
QUIJOS
Manta
Cabo San Mateo
Angamarca
Píllaro
(Macul)
Pedro de Alvarado, Feb-July 1534
MANTA
Puerto Viejo
Ambato
Mt Carihuairazo 5028 m
CHILLOS
Chambo
Mocha
Mt Chimborazo 6272 m
Puruas
Cajabamba
Liribamba (Riobamba)
Chimbo
L. Colta
Jan. 1531 - April 1532
(Daule)
(Bolívar Prov.)
Teocajas
Chongón
Tiquizambi
(Guayaquil)
(Pastaza)
Punta Sta. Elena
Alausí
Chimbo
HUANCAVILCA
Cañar
CAÑARI
Tumibamba (Cuenca)
I. Puna
I. Sta. Clara
Sebastián de Benalcázar Feb-June 1534
Páute
1532
Tumbez
Saraguro
Zaruma
(Loja)
CHAPARRA
Paltas
Francisco Pizarro May-Aug
PALTA
(Zamora)
BRACAMOROS
Second voyage (2nd part) 1527-1528
PACIFIC OCEAN
MAR DEL SUR
Tangarara (Talara)
Ayabaca
0 50 Miles 100 150
0 50 100 Kms 150 200 250
Piura ..... Inca cities
.... Inca towns, tambos
CAÑARI .... Tribal names
Land over 3000 m
(Cuenca)... Modern names if different
Piura
Francisco Pizarro Sept-Nov 1532
San Miguel (Paita)
Cajas

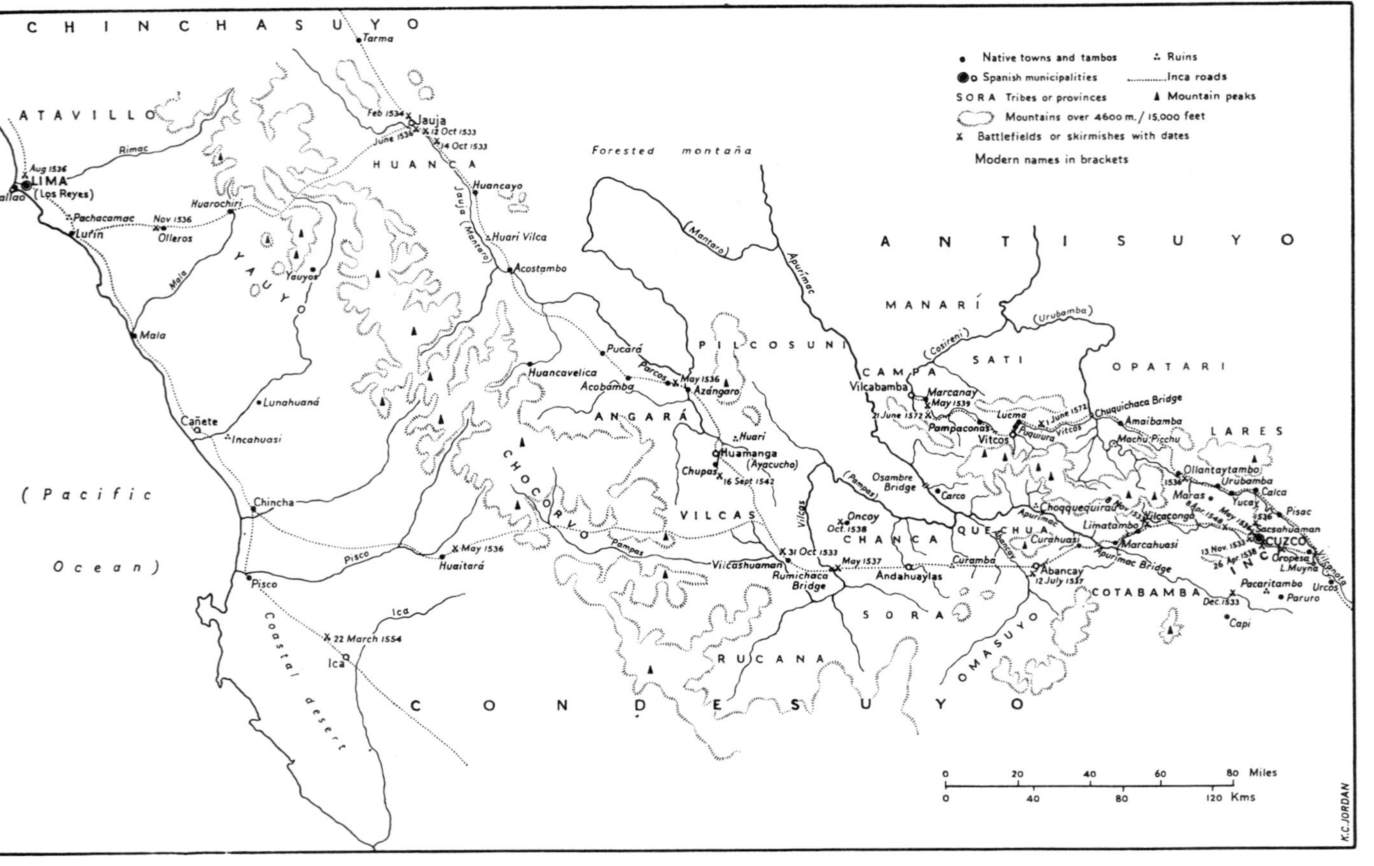

4. Southern Peru during Manco's Rebellion and the Civil Wars

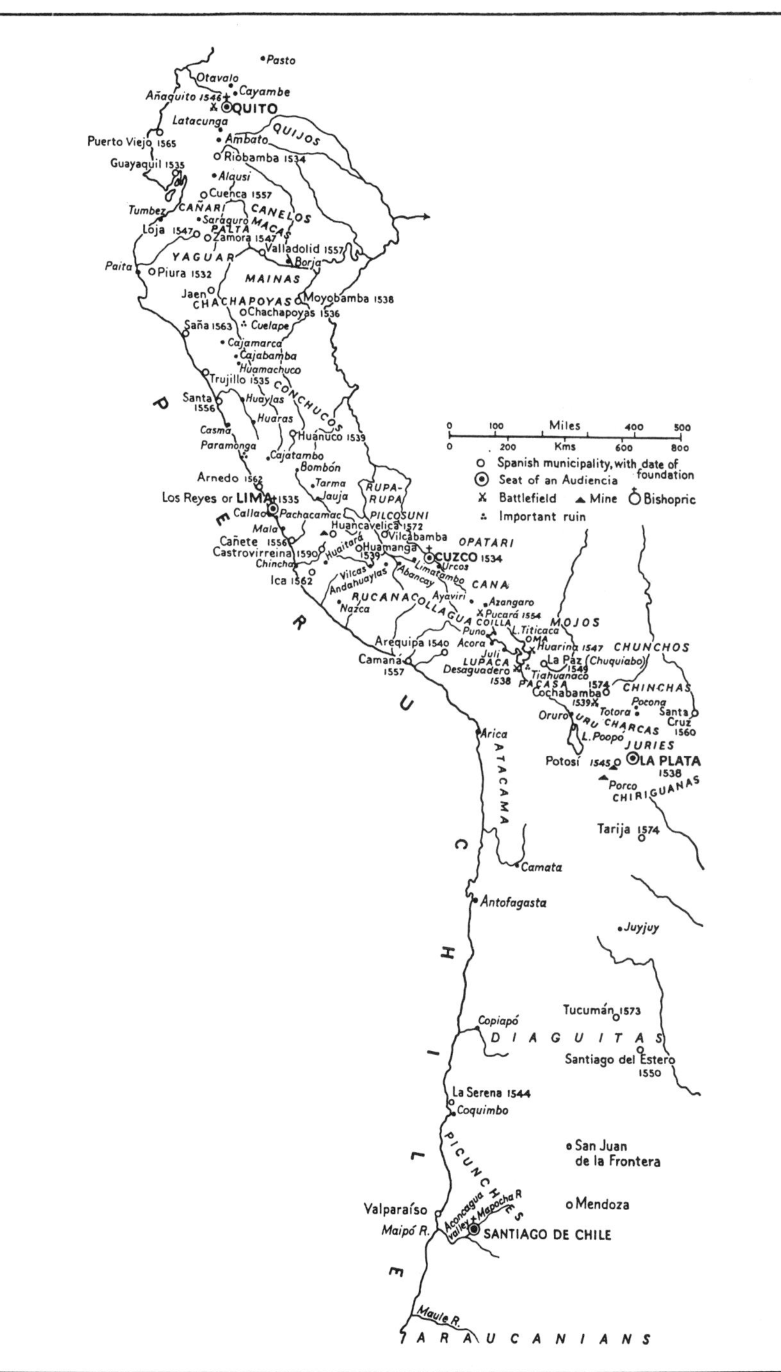

5. The Spanish occupation of the Inca Empire

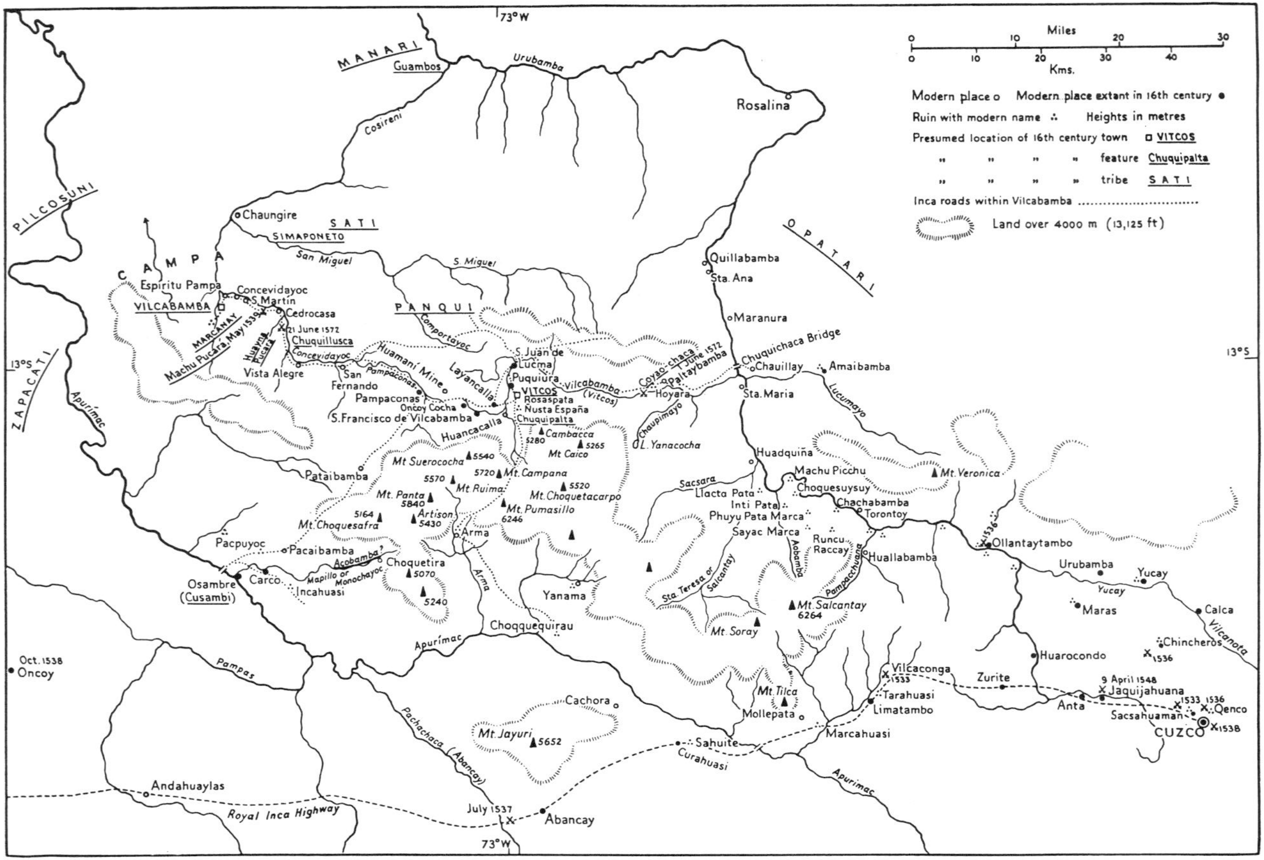

Miles 0 10 20 30
Kms. 0 10 20 30 40

Modern place o Modern place extant in 16th century •
Ruin with modern name ∴ Heights in metres
Presumed location of 16th century town □ VITCOS
" " " " feature Chuquipalta
" " " " tribe SATI
Inca roads within Vilcabamba ....
Land over 4000 m (13,125 ft)

6. Vilcabamba

PREFACE

This book was written largely to satisfy my own curiosity. I had spent nearly a year travelling to all parts of Peru and visiting most of its known ruins. This led to a fascination by the Conquest, and particularly by relations between Spaniards and Indians. This book therefore concentrates on contacts between the two races, at all levels from the Inca royal family down to humble peasants and conscripted miners.

Because of this emphasis I have not attempted a survey of Inca society: there are many excellent books on the subject, and little new material to justify another. I describe the Inca way of life where it is relevant to the narrative of the Conquest or it impressed the conquistadores. Similarly, I describe Spanish society and the passionate complexities of the civil wars between the conquerors primarily to show the effect that these had on the Indians.

This book has given me an opportunity to refute some misconceptions and to reconstruct gaps in the chroniclers' narratives. It has been said that the Incas succumbed without a struggle. To restore their honour I have described in detail the fighting of the Conquest and native rebellions. The Conquest of Quito and Manco Inca's second rebellion are subjects that have not received due attention before now. Nor has the small native state of Vilcabamba, the sanctuary in which Manco Inca and his sons tried to confront or co-exist with Spanish-occupied Peru.

There have also been misconceptions about Spanish conduct in post-Conquest Peru. The 'leyenda negra', the legend of Spanish atrocities, is still being hotly debated. Here I have tried to penetrate the clouds of conflicting hyperbole in contemporary reports and treatises. This is an area where much fine historical research is being done. An accurate picture of life in Peru during the period of imposition of colonial rule is emerging from careful studies of legal, governmental and municipal records.

The fate of the last Incas is another question that aroused my curiosity. I have therefore attempted to record the lives of the various survivors of the Inca royal house, and have traced their descent in a series of family trees. I have also tried to resolve the mystery surrounding the location of the 'lost city' of Vilcabamba.

The Peru that was conquered was the last advanced civilisation completely isolated from the rest of mankind. But although the story of the Conquest is fantastic, the participants were real men. I have sought to remove the Incas from the realm of prehistory and legend, and to show them as men struggling against a terrible invasion. It is difficult to see events from their side, for the Spaniards alone had writing, and our record of events is almost always witnessed through Spanish eyes. This imbalance has been somewhat righted by three sources rediscovered during the present century. The most remarkable is a report dictated by the Inca Titu Cusi Yupanqui for the benefit of the King of Spain. It not only provides an excellent historical summary of the Conquest through Inca eyes, but is also the only autobiographical memoir by a leading figure of *either* side. Another excellent source that consistently viewed events from the Indian side is the *Historia general del Perú* by the friar Martín de Murúa. This neglected work gave me great pleasure. It frequently corroborated names or events that were otherwise in doubt, and it filled tantalising gaps in the other chroniclers' narratives. The third important discovery is the *Nueva corónica y buen gobierno* by the eccentric mestizo Felipe Guaman Poma de Ayala. It is a wild and incoherent book, a mixture of delightful drawings and a text in which Spanish and Quechua are freely interspersed. But although the author was naïve and dotty, his illustrations and descriptions of Inca and sixteenth-century life are authentic. As a half-caste he had intimate experience of the good and evil of life under the Spaniards.

The other great mestizo author was the Inca Garcilaso de la Vega. His *Comentarios reales* has dominated modern knowledge about the Incas, because it is a powerful classic, entertaining and of considerable literary merit. Garcilaso left Peru when he was twenty, and his history was written decades later, largely from other sources. He told many delightful stories about his childhood in Cuzco. But as a historian Garcilaso has forfeited my confidence: he meanders, forgets, romanticises or blatantly distorts too often to remain authoritative.

In writing this book I have been working in the shadow of William Hickling Prescott, the blind Bostonian lawyer whose classic *History of the Conquest of Peru* was published in 1847. Prescott was a masterly critical historian, and he had access to the manuscripts of all the best eyewitnesses of the Conquest. He wrote an immortal narrative of the

Conquest itself and of the Spaniards' civil wars until 1548. But he did not attempt to explore social conditions in post-Conquest Peru in any way, and did not describe the Inca state of Vilcabamba or the integration of the Incas into Spanish society.

Since Prescott's time the archives have yielded their treasures. The Spaniards had a passion for keeping records and notarising every aspect of their lives. Countless thousands of documents have been published in modern collections that sometimes run to over a hundred volumes but often have no sequence or index. Historical journals have also proliferated, and there have been many fine specialised studies by professional historians. Almost none of these sources was available to Prescott. They form an immense body of material, from which I have assembled this book.

I have quoted wherever possible from contemporary authors, to capture the excitement of eyewitness reports. The reference of every quotation is given at the back of the book. References to other passages are indicated by an asterisk (*) in the main text, but where the reference note provides additional information, a dagger (†) is used. The reader may also be helped by the glossary of Quechua and Spanish words appearing in the text, the Table of Weights and Measures, and the Chronology that appear at the back of the book.

ACKNOWLEDGMENTS

MY voyages to and from Peru were given to me by Mr Knud Lauritzen, shipowner of J. Lauritzen Lines, Copenhagen: without his generosity this book would never have been written. My travels about Peru were helped by many kind friends, particularly Mr Enrique Novak of the Banco de Crédito del Perú, who gave me introductions in many towns.

I am grateful to various people who have helped me acquire illustrations, particularly Mr Alfred M. Bingham, the son of Senator Hiram Bingham, Mr J. H. K. Harriman of Lima and Miss Virginia Fass, who developed my own photographs. I am also most indebted to Ann Kendall, Francis Greene and William Oddie for taking photographs for me in Cuzco and Trujillo. Having spent many hours in libraries, I am also grateful for the assistance of the staffs of the British Museum Reading Room, and of other libraries in England, Spain, Peru, France and the United States.

The Conquest of the Incas

I

CAJAMARCA

ON 25 September 1513 a force of weary Spanish explorers cut through the forests of Panama and were confronted by an ocean: the Mar del Sur, the South Sea or Pacific Ocean. This expedition was led by Vasco Núñez de Balboa, and one of its senior officers was a thirty-five-year-old captain called Francisco Pizarro. Six years after the first discovery the Spaniards established the town of Panama on the Pacific shore of the isthmus. Panama became a base in which to build ships to explore and exploit this unknown sea. It was the threshold of a vast expansion.

Spain was developing with explosive force during these years. Throughout the Middle Ages the crusading knights of Castile had been driving the Mohammedans out of the Iberian peninsula. The final victory of this reconquest came in January 1492, with the surrender of Granada to the Castilians under King Ferdinand of Aragon. A few months later in that same year Christopher Columbus sailed westwards

Pizarro and Almagro sail for Peru

into the Atlantic and made a landfall in the Caribbean. The ensuing years were spent establishing a Spanish presence in the islands of the West Indies and exploring the northern coast of South America. Francisco Pizarro took part in many of these explorations, tough and unrewarding raids on the tribes of the American forests.

The European conception of the Americas – or Indies, as they were called – changed dramatically when in 1519 Hernán Cortés discovered and invaded the mighty Aztec empire in Mexico. Cortés led only some five hundred men and sixteen horses, but he skilfully won the alliance of rebellious subject tribes. By adroit diplomacy and the endurance and ruthless courage of his men Cortés conquered an empire of exotic brilliance. Spain, a country of under ten million inhabitants, had seized a land with a population and wealth as great as its own. Cortés's achievement fired the romantic Spanish imagination. Younger sons of feudal families and Spaniards of all classes sailed eagerly to seek adventure and riches across the Atlantic.

While Cortés was conquering Mexico, Spaniards were beginning to explore the Pacific coast of South America. In 1522 Pascual de Andagoya sailed some two hundred miles along the coast of Colombia and ascended the river San Juan. He was seeking a tribe called Virú or Birú; and the name of this tribe, altered to 'Perú', came to be applied to a country lying far to the south.

Three partners acquired Andagoya's ships and succeeded in raising money to finance another voyage. The three were Francisco Pizarro and Diego de Almagro, both citizens of Panama and holders of quotas of Indians there, and Hernando de Luque, a priest who was apparently acting as agent of the trio's financial backer, Judge Gaspar de Espinosa. Pizarro sailed in November 1524 with eighty men and four horses. This first expedition was not a success: it reached a place that the Spaniards called, for obvious reasons, Port of Hunger, and Almagro lost an eye in a skirmish with primitive natives at 'Burned Village'. No riches were found along the coast, and the adventurers had difficulty persuading Espinosa to finance a further attempt.

The three partners entered into a formal contract on 10 March 1526*

* References for all quotations will be found in the Notes and References on pages 547–624, indicated by the page-number of the text and the last words of the quotation. Notes to other passages are indicated by an asterisk in the text and are likewise printed at the end of the book. A dagger (†) indicates an explanatory note.

and Pizarro sailed eight months later. He took some 160 men and a few horses in two small ships commanded by the able pilot Bartolomé Ruiz. The expedition divided: Pizarro camped at the river San Juan, Almagro returned for reinforcements, and Ruiz sailed on southwards. Ruiz's ships crossed the equator for the first time in the Pacific, and then, suddenly, came the first contact with the Inca civilisation.

The Spanish ships encountered and captured an ocean-going balsa raft fitted with fine cotton sails. No one who saw that raft was in any doubt that it was the product of an advanced civilisation. The vessel was on a trading mission to barter Inca artefacts for crimson shells and corals. A breathless report of the raft was sent back to King Charles I, who was also Holy Roman Emperor Charles V. 'They were carrying many pieces of silver and gold as personal ornaments . . . including crowns and diadems, belts and bracelets, armour for the legs and breastplates; tweezers and rattles and strings and clusters of beads and rubies; mirrors decorated with silver, and cups and other drinking vessels. They were carrying many wool and cotton mantles and Moorish tunics . . . and other pieces of clothing coloured with cochineal, crimson, blue, yellow and all other colours, and worked with different types of ornate embroidery, in figures of birds, animals, fish and trees. They had some tiny weights to weigh gold . . . There were small stones in bead bags: emeralds and chalcedonies and other jewels and pieces of crystal and resin. They were taking all this to trade for fish shells from which they make counters, coral-coloured scarlet and white.'† Eleven of the twenty men on the raft leaped into the sea at the moment of capture, and the pilot Ruiz set six others free on shore. But he shrewdly kept three men to be taught Spanish and trained as interpreters for a conquest of this mysterious empire.

Ruiz rejoined Pizarro and ferried the expedition south to explore the coast of Ecuador. They returned to the uninhabited Isla del Gallo, Island of the Cock, in the Tumaco estuary. These coasts are humid, barren of food and often infested with noxious mangrove swamps. The Spaniards suffered terribly. Three or four a week were dying of hunger and disease. When the expedition had lost a large part of its men, a desperate appeal from the survivors reached the Governor of Panama. He opened a full-scale enquiry on 29 August 1527, and ordered that any men who wished to return should be evacuated. The expedition had been maintained largely by the fanatical determination of Francisco

Pizarro. He now drew a line across the sand of the Isla del Gallo and challenged his men to cross it and remain with him. Thirteen brave men did so. They stayed with Pizarro on the island and ensured the continuance of the expedition.

Pizarro's perseverance was rewarded the following year. He sailed south in a voyage of true exploration, with only a handful of soldiers and none of the baggage of an invasion. The expedition entered the Gulf of Guayaquil, and sighted its first Inca city at Tumbez. An Inca noble visited the ship and a Spaniard, Alonso de Molina, landed with a present of pigs and chickens. A tall and dashing Greek, Pedro de Candía, also landed to confirm Molina's description of Tumbez as a well-ordered town. Here at last was the advanced civilisation that the adventurers had been seeking so ardently. Candía astounded the inhabitants by firing an arquebus at a target, but this first contact between Spaniards and subjects of the Inca was very cordial.

Pizarro sailed on down the coast of Peru as far as the modern Santa river. Two further landings confirmed the magnitude of the discovery and the sophistication of this mysterious empire. The expedition returned with evidence: llamas, pottery and metal vessels, fine clothing, and more boys to be trained as interpreters. Pizarro's men had glimpsed the edges of a great civilisation, the product of centuries of development in complete isolation from the rest of mankind.

The explorers were excited by their discoveries and the potential for conquest, but they could not arouse the enthusiasm of the Governor of Panama. They decided to send Pizarro back to Spain to win royal approval, and to raise more men and money. Pizarro was well received by King Charles at Toledo in mid-1528. He was fortunate that his visit coincided with the return of Cortés, who charmed the court ladies with lavish presents of Mexican treasure, and was rewarded with a marquisate and other honours. Cortés encouraged Pizarro; but it was the brilliant inspiration of the conquest of Mexico that made it easy for Pizarro to recruit keen young adventurers in his native Trujillo de Extremadura. King Charles had to leave Toledo, but on 26 July 1529 the Queen signed a Capitulación authorising Pizarro to discover and conquer 'Peru'. Pizarro was named Governor and Captain-General of Peru, Almagro became commandant of Tumbez, and Luque was appointed Protector of the Indians, with a promise of becoming Bishop of Tumbez.*

Pizarro sailed from Seville in January 1530 with a flotilla full of would-be conquerors, and including his younger half-brothers Hernando, Juan and Gonzalo Pizarro and Francisco Martín de Alcántara. In Panama, Diego de Almagro was understandably disgusted with his meagre appointment in the Toledo agreement. He was persuaded to continue the enterprise only by being promised the title Adelantado (Marshal) and a governorship of territory beyond that of Pizarro.

Pizarro's third voyage sailed from Panama on 27 December 1530, but inexplicably chose to land on the Ecuadorean coast long before reaching Tumbez. Months of hardships followed: a wearisome march along the tropical coast, an epidemic of buboes, a stay on the dreary island of Puná in the Gulf of Guayaquil and many skirmishes with primitive natives. The most serious battle took place when the expedition attempted to cross on rafts from Puná to the mainland of Inca Peru. The conquistadores were finally beginning to invade the Inca empire, but they were in a remote corner, far from its fabulous cities and treasures. Tumbez, the site of the promised bishopric, was in ruins and there were no signs of a Spaniard who had chosen to remain there. The natives said that this destruction was the result of a civil war within the Inca empire.

The year 1531 and part of 1532 had elapsed since this third expedition left Panama, but Pizarro advanced cautiously. He left Tumbez in May 1532 and moved to the district of Poechos on the Chira river. He spent the ensuing months exploring the arid north-western corner of Peru. Weeks were spent ferrying some of the men from Tumbez to Tangarara, 120 miles to the south. Various reinforcements sailed down the coast to raise the spirits of Pizarro's men: the seasoned conqueror Sebastián de Benalcázar brought thirty men from Nicaragua, and the dashing Hernando de Soto came with another contingent. Pizarro killed a local chief called Amotape, apparently to intimidate the natives of this outlying province. He then selected a site for the first Spanish settlement in this strange new country: in mid-September a small ceremony marked the foundation of San Miguel de Piura near Tangarara. Some sixty Spaniards were left as the first citizens of San Miguel, and Pizarro struck out into the Inca empire with a tiny army: 62 horsemen and 106 foot-soldiers. The months of hesitation were over.

Pizarro's force marched out of San Miguel on 24 September 1532. It spent ten days at Piura, paused at Zarán (modern Serrán), Motux (Motupe), and reached Saña on 6 November. Up to now the Spaniards had remained on the coastal plain, a narrow strip of desert between the Pacific and the Andes mountains, but on 8 November they decided to turn inland and march up into the sierra. The Incas were mountain people, with lungs enlarged by evolution to breathe rarefied air. Although they had conquered the many civilisations of the hot coastal valleys, the true Inca empire lay along the ranges of the Andes and it was here that any conqueror must confront them.

With striking good fortune, Pizarro's Spaniards marched into Peru precisely at a moment of great passion in a war of dynastic succession. When Pizarro first sailed down the Pacific a few years earlier the Inca empire was ruled in tranquillity by one venerated supreme Inca, Huayna-Capac. His possessions stretched for almost three thousand miles along the Andes, from central Chile to the south of modern Colombia – a distance greater than that across the continental United States, or Europe from the Atlantic to the Caspian. With the Pacific Ocean to the west and the Amazonian forests to the east, the Incas were confident that they had absorbed almost all civilisation.

Huayna-Capac had for many years been leading the empire's professional army against tribes in the extreme north, Pasto and Popayán in Colombia. The fighting was stubborn and the campaigns dragged on. The Inca and his court had long been absent from the imperial capital Cuzco, and Huayna-Capac was considering establishing a second capital in the north at Quito or Tumibamba. It was during these campaigns that Huayna-Capac was first informed of the appearance of tall strangers from the sea. He was destined never to see any Europeans. His army and court were struck by a violent epidemic that killed Huayna-Capac in a delirious fever, at some time between 1525 and 1527. The disease may have been malaria, but it could have been smallpox. The Spaniards brought smallpox with them from Europe and it spread fiercely around the Caribbean among peoples who had no immunity. It could easily have swept from tribe to tribe across Colombia and struck the Inca armies long before the Spaniards themselves sailed down the coast. The epidemic 'consumed the greater part' of the Inca court including Huayna-Capac's probable heir, Ninan Cuyuchi. 'Countless thousands of common people also died.'†

The premature deaths of the great Inca Huayna-Capac and his heir left an ambiguous situation. The most likely successor was the Inca's son Huascar, and he succeeded as ruler of the capital city Cuzco. Another son, Atahualpa, was left in charge of the imperial army at Quito. He was probably acting as provincial governor of the area on behalf of his brother, although a number of chroniclers said that the dying Inca had decided to divide the vast empire into two sections, one ruled from Cuzco, the other from Quito. We shall never know the exact nature of the legacy. What we do know is that, after a few years of quiet, civil war broke out between the two brothers. Different chroniclers gave different versions of the origins of this conflict, depending on the sympathies of their native informants. Being Europeans, most chroniclers were at pains to discover which brother, Huascar or Atahualpa, had the best 'legitimate' claim to the throne. This was irrelevant, for the Incas did not stress primogeniture. They were concerned only that the new Inca should be of royal blood and fit to rule. If the eldest or favourite son designated as heir by his father proved weak or incompetent, he was soon deposed by a more aggressive brother in a civil war or palace revolution. Most of the eleven Incas who had ruled up to this time had succeeded only after some such struggle, and the result was a line of remarkable rulers.*

When the civil war broke out, Atahualpa had possession of the professional army, which was still fighting on the northern marches under its generals Chalcuchima, Quisquis and Rumiñavi. Huascar had the loyalty of most of the country. It took only a few years for relations between the two brothers to degenerate into open conflict. Huascar's militia army attempted to invade Quito, but after initial success was driven south through the Andes by the seasoned forces loyal to Atahualpa. A series of crushing victories by the Quitans culminated in the defeat and capture of Huascar in a pitched battle outside Cuzco. Many peoples of the empire regarded the victorious Quitans as hostile invaders, and the professionals responded with the brutality they had learned in the northern campaigns. Atahualpa ravaged the province of the Cañari tribe as punishment for its chief's intrigues. Quisquis, the general who conquered Cuzco, set out to exterminate all members of Huascar's family to dispose of any other pretenders. He sent the captive Inca northwards under strong escort. Chalcuchima, Atahualpa's supreme commander, held the area of the central Andes with another army at Jauja, while Rumiñavi was the general left in command of the Quitan homeland. Atahualpa himself marched triumphantly southwards in the wake of his generals.

Pizarro started his march down the Peruvian coast just as this fierce civil war was ending. His men saw ample evidence of the recent fighting. Tumbez was in ruins. When Hernando de Soto rode inland on a reconnaissance he reached a town called Cajas which was 'in considerable ruin from the fighting that Atahualpa had waged. In the hills were the bodies of many Indians hanging from trees because they had not agreed to surrender: for all these villages were originally under Cuzco [Huascar], whom they acknowledged as master and to whom they paid tribute.'

When Pizarro learned about the civil war he immediately grasped how useful it could be for him. Cortés had brilliantly manipulated rival factions during the conquest of Mexico twelve years before. Pizarro hoped to do likewise.

By another extraordinary coincidence, the victorious Atahualpa happened to be camped in the mountains at Cajamarca, not far from Pizarro's line of march. Reports reached Atahualpa as soon as the Spaniards landed on the mainland, and he was told that they were pillaging the countryside and abusing the natives. But Atahualpa was too engrossed in the civil war to be particularly concerned with the movements of the 150 strangers. He was fully occupied in leading his army, arranging the occupation of the newly won empire, planning his own journey to Cuzco, and awaiting reports from his commanders to the south. When Pizarro and his men marched out of San Miguel, Atahualpa did not yet know whether Quisquis had won or lost the battle for Cuzco. But he sent one of his close advisers to investigate the strangers. This Inca noble reached Cajas while Soto's reconnaissance was there, and at once impressed the Spaniards with his authority. They noted that the local chief 'became greatly frightened and stood up, for he did not dare remain seated in his presence'. And when the envoy reached Pizarro's camp, 'he entered as casually as if he had been brought up all his life among Spaniards. After having delivered his embassy . . . he enjoyed himself for two or three days among us.' Atahualpa's messenger brought presents of stuffed ducks and two pottery vessels representing castles. The more suspicious Spaniards assumed that the ducks, which had been skinned, were intended to represent the fate that awaited the intruders, while the model castles were to indicate that many more fortresses lay ahead.*

The envoy had also been ordered to report on Pizarro's force. During the two days he was in the Spaniards' midst, he went about examining every detail of their horses and armour and counting their numbers. He asked some Spaniards to show him their swords. 'He happened to go up to one Spaniard to do this, and put his hand on his beard. That Spaniard gave him many blows. When Don Francisco Pizarro heard of this, he issued a proclamation that no one should touch the Indian, whatever he did.' The envoy invited Pizarro to proceed to Cajamarca to meet Atahualpa. Pizarro accepted, and sent the Inca a present of a fine Holland shirt and two goblets of Venetian glass.*

The small force of invaders turned inland, away from the Pacific coast and up into the Andes. The Spaniards probably marched up an Inca trail ascending the Chancay stream past the town of Chongoyape. From the sands of the coastal desert they would have passed through plantations of sugar and cotton. As they climbed through the Andean foothills the valley narrowed into a canyon whose sides would have been covered in fields and terraces. At the source of the Chancay Pizarro's force probably swung south along the watershed of the Andes, crossing treeless savannah at some 13,500 feet. They were apprehensive, excited by the rapid change of altitude, and disquieted by the sight of Inca forts and watchtowers overlooking their route. But Atahualpa had decided to allow the strangers to penetrate the mountains, and his warriors did nothing to impede their advance.

The Spaniards were fortunate that Atahualpa had decided not to oppose their march into the mountains, for they were moving across difficult country, a region rarely penetrated to this day. Hernando Pizarro wrote: 'The road was so bad that they could very easily have taken us there or at another pass which we found between here and Cajamarca. For we could not use the horses on the roads, not even with skill, and off the roads we could take neither horses nor foot-soldiers.' This assessment was reasonable: less professional Inca armies destroyed a force as large as this in similar mountainous country four years later.

Finally, on Friday 15 November, the Spanish force emerged from the hills and looked down onto the valley of Cajamarca. This is a beautiful, fertile valley, only a few miles wide but remarkably flat – a very rare distinction in the vertical world of the Andes, where most rivers rush through precipitous canyons, and the only flat ground is on the high, infertile savannahs. The valley today has cows grazing beneath eucalyptus groves, and boasts a chocolate factory – all imported and unusual sights. The ground is strewn with millions of potsherds, painted with elaborate geometric designs of the pre-Inca Cajamarca civilisations, and on the desolate hills above the town are weird watercourses and incomprehensible designs cut into rock outcrops. Modern Cajamarca is a charming red-roofed Spanish town, with fine colonial monasteries and a lovely cathedral (plate 1).

Pizarro halted his men at the edge of the valley to await the rear-guard, and then rode down in three squadrons in careful marching order. Atahualpa had the tents of his army's camp spread out across a hillside beyond the town. 'The Indians' camp looked like a very beautiful city. . . . So many tents were visible that we were truly filled with great apprehension. We never thought that Indians could maintain such a proud estate nor have so many tents in such good order. Nothing like this had been seen in the Indies up to then. It filled all us Spaniards with fear and confusion. But it was not appropriate to show any fear, far less to turn back. For had they sensed any weakness in us, the very Indians we were bringing with us would have killed us. So, with a show of good spirits, and after having thoroughly observed the town and tents, we descended into the valley and entered the town of Cajamarca.'

Cajamarca itself proved to contain only four or five hundred of its normal two thousand inhabitants. There was a sun temple in an enclosure at its edge, and a series of buildings full of holy women. These chosen women formed part of the empire's official sun religion, but were also one of the privileges of the ruling Inca hierarchy. They were chosen as girls, for either their noble birth or outstanding beauty, and were then moved to cloistered colleges in the provincial capitals such as Cajas or Cajamarca. These chosen girls, acllas, spent four years weaving fine cloth or brewing chicha for the Inca and his priests and officials. Some then became mamaconas, remained in perpetual chastity and spent their lives in the service of the sun and shrines. Others were given as wives to Inca nobles or tribal chiefs, and the most beautiful became concubines of the Inca himself.

The Spaniards first saw a 'nunnery' of acllas and mamaconas when Hernando de Soto led a reconnaissance inland to Cajas. It is easy to imagine the effect of this building full of beautiful girls on men who had been without women for months. Diego de Trujillo recalled that 'the women were brought out on to the square and there were over five hundred of them. Captain [Soto] gave many of them to the

Spaniards. The Inca's envoy grew very indignant and said: "How dare you do this when Atahualpa is only twenty leagues from here. Not a man of you will remain alive!" '

Francisco Pizarro assembled his men in the square of Cajamarca, which was surrounded on three sides by long buildings each of which had a series of doors on to the open space. It began to hail, so the men took shelter in the empty buildings. The Spaniards were apprehensive but eager to behave correctly. Francisco Pizarro therefore sent Hernando de Soto to visit Atahualpa with fifteen horsemen and Martín, one of the interpreters acquired on the second voyage. They were to ask him how he wished the strangers to lodge. Shortly after Soto's departure, Hernando Pizarro grew alarmed. As he explained: 'I went to talk to the Governor, who had gone to inspect the town in case the Indians should attack us by night. I told him that in my view the sending of fifteen of the best horsemen was a mistake . . . If Atahualpa decided to do anything, [the fifteen] were not enough to defend themselves; and if some reverse befell *them* it would be a very serious loss to him. He therefore ordered me to go with a further twenty horsemen who were in a fit state to go, and once there to act as I saw fit.' Fortunately for us, the contingents sent to visit Atahualpa on that first evening included some of the chroniclers who left eyewitness accounts: Hernando Pizarro, Miguel de Estete, Juan Ruiz de Arce, Diego de Trujillo and possibly Cristóbal de Mena and Pedro Pizarro.

A paved road ran for the few miles between Cajamarca and the Inca's residence. Atahualpa was in a small building close to some baths, the natural hot springs of Kónoj that still hiss and bubble out of the ground to this day. The Spaniards advanced with trepidation through the silent ranks of the Inca army. They had to cross two streams, and left the bulk of the horsemen at the second stretch of water while the leaders rode in to find Atahualpa. 'The pleasure house . . . had two towers [rising] from four chambers, with a courtyard in the middle. In this court, a pool had been made and two pipes of water, one hot and one cold, entered it. The two pipes came from springs . . . beside one another. The Inca and his women used to bathe in the pool. There was a lawn at the door of this building and he was there with his women.' The moment had finally come when the first Spaniards were to confront the ruler of Peru. Here was 'that great lord Atahualpa . . . about whom we had been given such reports and told so many things'. 'He was seated on a small stool, very low on the ground, as the Turks and Moors are accustomed to sit,' 'with all the majesty in the world, surrounded by all his women and with many chiefs near him. Before arriving there, there had been another cluster of chiefs, and so forth with each according to his rank.'

Atahualpa was wearing the royal insignia. Every important Peruvian wore a llautu, a series of cords wound round the head. But the Inca alone had a royal tassel hanging from the front of this circlet. It consisted of 'very fine scarlet wool, cut very even, and cleverly held towards the middle by small golden bugles. The wool was corded, but below the bugles it was unwound and this was the part that fell on to the forehead ... This tassel fell to the eyebrows, an inch thick, and covered his entire forehead.' Because of the tassel, Atahualpa kept his eyes downcast and Soto could get no reaction from him. 'Hernando de Soto arrived above him with his horse, but he remained still, making no movement. [Soto] came so close that the horse's nostrils stirred the fringe that the Inca had placed on his forehead. But the Inca never moved. Captain Hernando de Soto took a ring from his finger and gave it to him as a token of peace and friendship on behalf of the Christians. He took it with very little sign of appreciation.' Soto delivered a prepared speech to the effect that he was a representative of the Governor, and that the Governor would be delighted if he would go to visit him. There was no reaction from Atahualpa. Instead, one of his chiefs answered for him and said that the Inca was on the last day of a ceremonial fast.

At this point Hernando Pizarro arrived and delivered a speech similar to Soto's. Atahualpa apparently gathered that the new arrival was the Governor's brother, for he now looked up and began to converse. He told of the first report about the Christians that he had received from Marcavilca, chief of Poechos on the Zuricari river (the modern Chira) between Tumbez and San Miguel. This chief 'sent to tell me that you treated the chiefs badly and threw them into chains, and he sent me an iron collar. He says that he killed three Christians and one horse.' Hernando Pizarro responded hotly to the latter claim. 'I told him that those men of San Miguel were like women, and that one horse was enough [to conquer] that entire land. When he saw us fight he would see what sort of men we were.' Hernando Pizarro warmed to his subject and became more expansive. He told Atahualpa that 'the Governor [Francisco Pizarro] loved him dearly. If he had any enemy he should tell [the Governor] and he would send to conquer that person. [Atahualpa] told me that four days' march away, there were some very savage Indians with whom he could do nothing: Christians should go there to help his men. I told him that the Governor would send ten horsemen, which was enough for the entire land. His Indians would be needed only to search for those who hid. He smiled, as someone who did not think much of us.'

Atahualpa invited the Spaniards to dismount and dine with him. They refused, and he offered them drink. After some hesitation for fear of poison, they accepted. Two women immediately appeared with golden jugs of the native maize beverage, chicha, and ceremonial drinks were exchanged with the Inca. The sun was now setting, and Hernando Pizarro asked permission to leave. The Inca wanted one Spaniard to remain there with him, but they claimed that this would be contrary to their orders. So they took their leave, with Atahualpa's instructions that they were to make their quarters in three houses on the square, leaving only the main fortress for his own residence. He also gave them the assurance they most wanted: he himself would go to Cajamarca the following day to visit Pizarro.

During the meeting, Atahualpa had been 'closely examining the horses, which undoubtedly seemed good to him. Appreciating this, Hernando de Soto brought up a little horse that had been trained to rear up, and asked [the Inca] whether he wanted him to ride it in the courtyard. He indicated that he did, and [Soto] manoeuvred it there for a while with good grace. The nag was spirited and made much foam at its mouth. He was amazed at this, and at seeing the agility with which it wheeled. But the common people showed even greater admiration and there was much whispering. One squadron of troops drew back when they saw the horse coming towards them. Those who did this paid for it that night with their lives, for Atahualpa ordered them to be killed because they had shown fear.'

The Spaniards now had time to ponder the seriousness of their situation. 'We took many views and opinions among ourselves about what should be done. All were full of fear, for we were so few and were so deep in the land where we could not be reinforced. . . . All assembled in the Governor's quarters to debate what should be done the following day. . . . Few slept, and we kept watch in the square, from which the camp fires of the Indian army could be seen. It was a fearful sight. Most of them were on a hillside and close to one another: it looked like a brilliantly star-studded sky.' Cristóbal de Mena recalled how the danger broke down class differences among the Spaniards. 'There was no distinction between great and small, or between foot-soldiers and horsemen. Everyone performed sentry rounds fully armed that night. So also did the good old Governor, who went about encouraging the men. On that day all were knights.'

The Spaniards now realised, for the first time, the sophistication of the empire they had penetrated. They found themselves isolated from the sea by days of marching over difficult mountains. They were in the midst of a victorious army in full battle order, which Soto and Hernando Pizarro estimated at forty thousand effectives – 'but they said this to encourage the men, for he had over eighty thousand.' Added to this was the fear of the unknown, 'for the Spaniards had no experience of how these Indians fought or what spirit they had'. From what they had seen of Atahualpa himself, his well-disciplined army, and the brutality of the recent civil war, they had no reason to hope for a friendly reception of any long duration.

The men Pizarro was leading were skilled and seasoned soldiers. Many had gained experience in the conquests in and around the Caribbean, Mexico and Central America. Pizarro himself had first arrived in the Indies in 1502 and was now, in his mid-fifties, one of the richest and most important citizens of Panama.† Although he could not read and was a poor horseman, his command of the expedition was never in question for a moment – any friction that occurred was between his captains, Diego de Almagro, Hernando de Soto, Hernando Pizarro and Sebastián de Benalcázar. Other members of the expedition had gained experience in the battles of northern Italy and north Africa that were making Spain the leading nation in Europe and the Spanish tercios its most dreaded soldiers. Even the youngest members – for most of the Spaniards were in their twenties – compensated for any lack of fighting experience by skill in military exercises and by courage and dash. In the feudal structure of Spanish society an ambitious man could rise only by marrying an heiress or by warfare. There was the spirit of a gold rush about this expedition, fortified by some of the conviction of a crusade.

Despite their experience, Pizarro's 150 men had marched into an impasse and were now thoroughly frightened and desperate. All that they could decide during that anxious night was to employ the various tactics and advantages that had proved successful in the Caribbean. They could use surprise, attacking first without provocation, and take advantage of the novelty of their appearance and fighting methods. Their weapons – horses, steel swords and armour – were far superior to anything they had encountered so far in the Indies, although they were not so sure about the Incas. They had in mind the tactic that had succeeded so well in the conquest of Mexico: the kidnapping of the head of state. They could also try to make capital of the internal dissensions within the Inca empire – Hernando Pizarro had already offered the services of Spaniards to help Atahualpa in his inter-tribal fighting. Possibly their greatest advantage lay in the self-assurance of belonging to a more advanced civilisation and the knowledge that their purpose was conquest: to the Indians, they were still an unknown quantity of uncertain origin and unsure intentions.

It was agreed that Governor Pizarro should decide on the spur of the moment the course of action to be adopted when Atahualpa was in Cajamarca the following day, Saturday 16 November. But careful plans were made for a surprise attack and capture of the Inca. 'The Governor had a dais on which Atahualpa was to sit. It was agreed that he should be enticed on to it by kind words and that he should then order his men to return to their quarters. For the Governor was afraid to come to grips when there were so many native warriors and we were so few.' The attack was to be made only if success appeared possible or if the natives made any threatening move. There were two more peaceful options. Atahualpa might be persuaded to make some act of political or spiritual submission. Or, if the natives seemed too powerful, the Spaniards could maintain the fiction of friendship and hope for a more favourable opportunity in the future.

The square of Cajamarca was ideally suited to the Spaniards' plan. Long low buildings occupied three sides of it, each some two hundred yards long. Pizarro stationed the cavalry in two of these, in three contingents of fifteen to twenty, under the command of his lieutenants Hernando de Soto, Hernando Pizarro and Sebastián de Benalcázar. The buildings each had some twenty openings on to the square, 'almost as if they had been built for that purpose'. 'All were to charge out of their lodgings, with the horsemen mounted on their horses.' Pizarro himself, being a poor horseman, remained in the third building with a few horse and some twenty foot. His contingent was 'charged with the capture of Atahualpa's person, should he come suspiciously as it appeared that he would'.

Roads ran down from the town and entered the square between these buildings. Groups of foot and horse were concealed in these alleys to close them. The lower side of the square was enclosed by a long wall of tapia, compacted clay, with a tower in the middle that was entered from the outside; beyond this lay the open plain. At the middle of the square, apparently on the upper side, was a stronger stone structure that the Spaniards regarded as a fort. Pizarro had the remainder of the infantry guard the gates of this, possibly to preserve it as a final refuge. Inside he stationed Captain Pedro de Candía with 'eight or nine musketeers and four small pieces of artillery'. The firing of these arquebuses was the pre-arranged signal for the Spaniards to charge into the square.

Atahualpa was in no hurry to make the short journey across the plain to Cajamarca. He had just finished a fast and there was drinking to be done to celebrate this and the victory of his forces at Cuzco. The morning went by with no sign of movement from the native encampment. The Spaniards became increasingly jittery. The familiar noble envoy arrived from Atahualpa saying that he intended to come with his men armed. 'The Governor replied: "Tell your lord to come . . . however he wishes. In whatever way he comes, I will receive him as a friend and brother." ' A later messenger said that the natives would be unarmed. The Spaniards 'heard mass and commended ourselves to God, begging him to keep us in his hand'. Atahualpa's army finally began to move at midday and 'in a short while the entire plain was full of men, rearranging themselves at every step, waiting for him to emerge from the camp'. The Spaniards were concealed in their buildings, under orders not to emerge until they heard the artillery signal. The young Pedro Pizarro recalled: 'I saw many Spaniards urinate without noticing it out of pure terror.'

Atahualpa had clearly decided to turn his visit to the extraordinary strangers into a ceremonial parade. 'All the Indians wore large gold and silver discs like crowns on their heads. They were apparently all coming in their ceremonial clothes.' 'In front was a squadron of Indians wearing a livery of chequered colours, like a chessboard. As these advanced they removed the straws from the ground and swept the roadway.' 'They pointed their arms towards the ground to clear anything that was on it – which was scarcely necessary, as the townspeople kept it well swept. . . . They were singing a song by no means lacking grace for us who heard it.'

As the tension mounted, Atahualpa paused on a meadow half a mile from the town. The road was still full of men, and more natives were still emerging from the camp. There was another exchange of messengers. Atahualpa started to pitch his tents, as it was by now late afternoon: he said that he intended to stay the night there. This was the last thing Pizarro wanted, for the Spaniards were particularly frightened of a night attack. In desperation, Pizarro sent one Hernando de Aldana 'to tell him to enter the square and come to visit him before night fell. When the messenger reached Atahualpa, he made a reverence and told him, by signs, that he should go to where the Governor was.' He assured the Inca 'that no harm or insult would befall him. He could therefore come without fear – not that the Inca showed any sign of fear.'

Atahualpa complied. With the sun sinking low, he continued his progress into the town. He left most of the armed men outside on the plain, 'but brought with him five or six thousand men, unarmed except that they carried small battle-axes, slings and pouches of stones underneath their tunics'. Behind the vanguard, 'in a very fine litter with the ends of its timbers covered in silver, came the figure of Atahualpa. Eighty lords carried him on their shoulders, all wearing a very rich blue livery. His own person was very richly dressed, with his crown on his head and a collar of large emeralds around his neck. He was seated on the litter, on a small stool with a rich saddle cushion. He stopped when he reached the middle of the square, with half his body exposed.' 'The litter was lined with parrot feathers of many colours and embellished with plates of gold and silver. . . . Behind it came two other litters and two hammocks in which other leading personages travelled. Then came many men in squadrons with headdresses of gold and silver. As soon as the first entered the square they parted to make way for the others. As Atahualpa reached the centre of the square he made them all halt, with the litter in which he was travelling and the other litters raised on high. Men continued to enter the square without interruption. A captain came out in front and went up to the fort on the square which contained the artillery' and 'in a sense took possession of it with

The first confrontation between Atahualpa and Hernando Pizarro

a banner placed on a lance'. This stiff banner was Atahualpa's royal standard, with his personal coat of arms.

Atahualpa was surprised to see no Spaniards. He later admitted that he thought they must have hidden from fear at the sight of his magnificent army. 'He called out "Where are they?" At this Friar Vicente de Valverde, of the Dominican order . . . emerged from the lodging of Governor Pizarro accompanied by the interpreter Martín', 'and went with a cross in one hand and his missal in the other. He advanced through the troops to where Atahualpa was.'

The various chroniclers who were present gave slightly different versions of the conversation that ensued between Valverde and Atahualpa. Most agreed that the priest began by inviting the Inca to advance into the building to talk and dine with the Governor. Ruiz de Arce explained that this invitation was made 'in order that he would emerge more from his men'. Atahualpa did not accept. He told Valverde that he would not move forward until the Spaniards had returned every object that they had stolen or consumed since their arrival in his kingdom. He may have been establishing a *casus belli* with this difficult demand.

Valverde began to explain his own function as a minister of the Christian religion, and launched into an exposition of 'the things of God'. He also mentioned that he, a friar, had been sent by the Emperor to reveal this religion to Atahualpa and his people. In effect, Valverde was delivering the famous Requirement, an extraordinary document that the royal council had ordered to be proclaimed in any conquest before resorting to bloodshed.* The priest said that the doctrine he was describing was contained in the breviary he was holding. 'Atahualpa told him to give him the book to examine. He gave it to him closed. Atahualpa did not succeed in opening it and the friar extended his arm to do so. But Atahualpa struck him on the arm with great disdain, not wishing that he should open it. He himself persisted in trying to open it and did so.' He was 'more impressed, in my opinion, by the writing itself than by what was written in it. He leafed through [the book] admiring its form and layout. But after examining it he threw it angrily down among his men, his face a deep crimson.' 'The boy who was interpreter and was there translating the conversation went running off to fetch the book and gave it to the priest.'

The critical moment had come. Xerez and Hernando Pizarro wrote that Atahualpa stood up on his litter, telling his men to make ready. The priest Vicente de Valverde returned to Pizarro, almost running, raising a call to battle. According to Mena he shouted: 'Come out! Come out, Christians! Come at these enemy dogs who reject the things of God. That chief has thrown my book of holy law to the ground!' According to Estete he cried to Pizarro: 'Did you not see what happened? Why remain polite and servile toward this overproud dog when the plains are full of Indians? March out against him, for I absolve you!' And for Trujillo it was: 'What is Your Honour going to do? Atahualpa has become a Lucifer!' For Murúa: 'Christians! The evangels of God are on the ground!' Juan Ruiz de Arce wrote, simply, that Valverde returned 'weeping and calling on God'.

Pizarro launched the ambush with the prearranged signal. He 'signalled the artilleryman [Pedro de Candía] to fire the cannons into their midst. He fired two of them but could not fire more.' The Spaniards in armour and chain mail charged their horses straight into the mass of unarmed natives crowding the square. Trumpets were sounded and the Spanish troops gave their battle cry 'Santiago!' 'They all placed rattles on their horses to terrify the Indians. . . . With the booming of the shots and the trumpets and the troop of horses with their rattles, the Indians were thrown into confusion and panicked. The Spaniards fell upon them and began to kill.' 'They were so filled with fear that they climbed on top of one another – to such an extent that they formed mounds and suffocated one another.' 'The horsemen rode out on top of them, wounding and killing and pressing home the attack.' 'And since the Indians were unarmed they were routed without danger to any Christian.'

'The Governor armed himself with a quilted cotton coat of armour, took his sword and dagger and entered the thick of the Indians with the Spaniards who were with him. With great bravery . . . he reached Atahualpa's litter. He fearlessly grabbed [the Inca's] left arm and shouted "Santiago" . . . but he could not pull him out of his litter, which was on high. All those who were carrying Atahualpa's litter appeared to be important men, and they all died, as did those who were travelling in the litters and hammocks.' 'Many Indians had their hands cut off but continued to support their ruler's litter with their shoulders. But their efforts were of little avail for they were all killed.' 'Although [the Spaniards] killed the Indians who were carrying [the litter], other replacements immediately went to support it. They continued in this way for a long while, overpowering and killing the Indians until, becoming exhausted, one Spaniard stabbed [at the Inca] with his dagger to kill him. But Francisco Pizarro parried the blow, and from this parry the Spaniard trying to strike Atahualpa wounded the Governor on the hand. . . . Seven or eight [mounted] Spaniards spurred up and grabbed the edge of the litter, heaved on it and turned it on to its side. Atahualpa was captured in this way and the Governor took him to his lodging.' 'Those who were carrying the litter and those who escorted [the Inca] never abandoned him: all died around him.'

Meanwhile the terrible carnage continued in the square and beyond. 'They were so terrified at seeing the Governor in their midst, at the unexpected firing of the artillery and the irruption of the horses in a troop – which was something they had never seen – that, panic-stricken, they were more concerned to flee and save their lives than to make war.' 'They could not flee in a body because the gate through which they had entered was small. They therefore could not escape in the confusion. When those at the rear saw how far they were from the sanctuary and safety of flight, two or three thousand of them flung themselves at a stretch of wall and knocked it to the ground. [This wall] gave on to the plain, for on that side there were no buildings.' 'They broke down a fifteen-foot stretch of wall six feet thick and the height of a man. Many horsemen fell on this.' 'The foot-soldiers set about those who remained in the square with such speed that in a short time most of them were put to the sword. . . . During all this no Indian raised a weapon against a Spaniard.'

The cavalry jumped the broken curtain wall and charged out into the plain. 'All were shouting, "After those with the liveries!" "Do not let any escape!" "Spear them!" All the other fighting men whom the [Inca] had brought were a quarter of a league [a mile] from Cajamarca and ready for battle, but not an Indian made a move.' 'When the squadrons of men who had remained in the plain outside the town saw the others fleeing and shouting, most of them broke and took to flight. It was an extraordinary sight, for the entire valley of four or five leagues was completely filled with men.' 'It was a level plain with some fields of crops. Many Indians were killed. . . . Night had already fallen and the horsemen were continuing to lance [natives] in the fields, when they sounded a trumpet for us to reassemble at the camp. On arrival we went to congratulate the Governor on the victory.'

'In the space of two hours – all that remained of daylight – all those troops were annihilated. . . . That day, six or seven thousand Indians lay dead on the plain and many more had their arms cut off and other wounds.' 'Atahualpa himself admitted that we had killed seven thousand of his Indians in that battle.' 'The man [killed] in one of the litters was his steward (the lord of Chincha), of whom he was very fond. The others were also lords over many people and were his counsellors. The cacique lord of Cajamarca died. Other commanders died, but there were so many of them that they go unrecorded. For all those who came in Atahualpa's bodyguard were great lords. . . . It was an extraordinary thing to see so great a ruler captured in so short a time, when he had come with such might.' Atahualpa's nephew wrote that the Spaniards killed Indians like a slaughterer felling cattle. The sheer rate of killing was appalling, even if one allows that many Indians died from trampling or suffocation, or that the estimates of dead were exaggerated. Each Spaniard massacred an average of fourteen or fifteen defenceless natives during those terrible two hours.*

Atahualpa had been hustled away from the slaughter of his subjects and placed under strong guard in the temple of the sun at the edge of Cajamarca. Some of the cavalry continued to patrol the town in case five or six thousand natives on the hill above might attempt a night attack. Meanwhile, with the bodies of thousands of natives lying in heaps on the square, the victors were paying extraordinary attention to their prisoner. 'The Governor went to his quarters with his prisoner Atahualpa. He disposed of his clothes, which the Spaniards had torn to pull him from the litter . . . ordered local clothing to be brought, and had him dressed. . . . They then went to dine and the Governor had Atahualpa sit at his table, treating him well and having him served as he was himself. He then ordered him to be given the women he wished from those who had been captured, to serve him, and ordered a bed to be made for him in the room where the Governor himself slept.'

All this solicitude was accompanied by speeches of amazing condescension. 'We entered where Atahualpa was, and found him full of fear, thinking that we were going to kill him.' 'The Governor . . . asked the Inca why he was sad, for he ought not to be sorrowful. . . . In every country to which we Christians had come there had been great rulers, and we had made them our friends and vassals of the Emperor by peaceful means or by war. He should not therefore be shocked at having been captured by us.' 'Atahualpa . . . asked whether the Spaniards were going to kill him. They told him no, for Christians killed with impetuosity but not afterwards.'

Atahualpa asked as a favour of the Governor to be allowed to speak to any of his people who might be there. 'The Governor immediately ordered them to bring two leading Indians who had been taken in the battle. The Inca asked them whether many men were dead. They told him that the entire plain was covered with them. He then sent to tell the troops who remained not to flee but to come to serve him, for he was not dead but in the power of the Christians. . . . The Governor asked the interpreter what he had said, and the interpreter told him everything.'

The Spaniards immediately asked the glaring question: Why had a ruler of Atahualpa's experience and power walked into such an obvious trap? The answer was quite clear. The Inca had totally misjudged and underestimated his opponents. Marcavilca chief of Poechos and the noble envoy who had spent two days with the invaders had both seen the Spaniards at their most disorganised. According to Atahualpa they 'had told him that the Christians were not fighting men and that the horses were unsaddled at night. If he [the noble] were given two hundred Indians he could tie them all up. [Atahualpa said] that this captain and the chief... had deceived him.'

The Inca admitted the fate he had planned for the strangers. 'He answered half smiling that... he had intended to capture the Governor but the reverse had happened, and for this reason he was so pensive.' 'He told of his great intentions: what was to have been done with the Spaniards and the horses.... He had decided to take and breed the horses and mares, which were the thing he admired most; and to sacrifice some of the Spaniards to the sun and castrate others for service in his household and in guarding his women.' There is no reason to doubt this explanation. Atahualpa, flushed with victory in the civil war, could afford to play cat-and-mouse with the extraordinary strangers that had marched from some other world into the midst of his army. He could not conceive that, with the odds so completely in his favour, the Spaniards would be the first to attack. Nor could he imagine that an attack would come without warning or provocation, before he had even held his meeting with Governor Pizarro.

The Spaniards themselves had acted in terror and desperation, and could scarcely believe the crushing success of their ambush. 'Truly, it was not accomplished by our own forces for there were so few of us. It was by the grace of God, which is great.'

2

The Inca Atahualpa a prisoner of the Spaniards

ATAHUALPA CAPTIVE

On the following morning the elated Spaniards followed up their success, securing its results with rapid efficiency. Hernando de Soto took thirty horsemen in battle formation to inspect Atahualpa's camp. The great native army was still there: 'the camp was as full of troops as if none had ever been missed'. But none of the stunned warriors offered any resistance. Instead, the captains of the various contingents made a sign of the cross to indicate their surrender: Pizarro had told Atahualpa to instruct them to do this.* Soto returned to Cajamarca before noon 'with a cavalcade of men, women, sheep [llamas], gold, silver and clothing... The Governor ordered that all the llamas should be released, for they were in great quantity and encumbered the camp: the Christians could still kill as many as they needed every day. As for the Indians who had been gathered up, ... the Governor ordered that they be brought into the square so that the Christians could take those they needed for their service.... Some were of the opinion that all the fighting men should be killed or have their hands cut off. The Governor would not consent. He said it was not good to do such a great act of cruelty.' 'All the troops were assembled and the Governor told them to return to their homes, as he had not come to harm them.... Atahualpa also ordered this.' 'Many of them did go, and it seemed to me that not more than twelve thousand Indians remained there.' 'Meanwhile the Spaniards in the camp made the Indian prisoners remove the dead from the square.'

The invasion of Peru was unique in many ways. The military conquest preceded peaceful penetration: no traders or explorers had ever visited the Inca court and there were no travellers' tales of its spendours. The Europeans' first glimpse of Inca majesty coincided with its overthrow. The Conquest began with the checkmate. From that first day onwards, the Peruvians were not only divided by their civil war but were also leaderless. To add to their confusion, their Inca continued to function in his captivity, issuing orders as absolute ruler.

Atahualpa was a clever man, and he immediately acted to try to extricate himself from an almost impossible situation. He noticed that the Spaniards appeared to be interested only in precious metals. Soto's men brought back all the gold and silver they could find in the Inca's camp. Its quality exceeded the wildest hopes of the conquistadores: their gold rush was already a dazzling success. From this military camp alone Soto brought 'eighty thousand pesos [of gold], seven thousand marks of silver and fourteen emeralds. The gold and silver was in monstrous effigies, large and small dishes, pitchers, jugs, basins and large drinking vessels, and various other pieces. Atahualpa said that this all came from his table service, and that his Indians who had fled had taken a great quantity more.' Atahualpa took note, and concluded that he could buy his liberty with more of these metals. He still could not conceive that these improbable 170 men were the spearhead of a full-scale invasion – and the Spaniards had no wish to disabuse him. 'He told the Governor that he knew perfectly well what they were seeking. The Governor told him that the fighting men were seeking nothing more than gold for themselves and their Emperor.'

The Inca now offered his famous ransom. 'The Governor asked him how much he would give and how soon. Atahualpa said that he would give a room full of gold. The room measured 22 feet long by 17 feet wide, and [was to be] filled to a white line half way up its height – [the line] he described must have been about 1½ estados [over eight feet] high. He said that up to this level he would fill the room with various objects of gold – jars, pots, tiles and other pieces. He would also give the entire hut filled twice over with silver. And he would complete this within two months.' The Spaniards were staggered by the unexpected proposal. 'Certainly an offer of vast proportions! When he had made it, Governor Francisco Pizarro, on his own and his captains' advice, had a secretary summoned to record the Indian's offer as a formal pledge.'

The room described by Pizarro's secretary Xerez had a capacity of some 3,000 cubic feet, 88 cubic metres. A visitor to modern Cajamarca is shown a longer and larger room of fine coursed Inca masonry in one of the narrow streets on the slope above the main square. This larger room was being shown to tourists by the seventeenth century. The local native chief took Antonio Vázquez de Espinosa there in 1615 and told him that 'the room remains and will remain untouched as a memorial to Atahualpa'. This larger room was probably part of the sun temple and may have been Atahualpa's prison chamber.†

Atahualpa made his offer initially to save his own life 'because he feared that the Spaniards would kill him'. Pizarro could have killed Atahualpa, but he obviously appreciated his value as a hostage and had been at pains to capture him alive. Pizarro was relieved to see that the native chiefs still obeyed Atahualpa in captivity, and was naturally delighted to learn that a fantastic ransom would also be brought to the Spaniards' camp. He accepted Atahualpa's offer with alacrity. 'The Governor promised to restore him to his former liberty, provided he did no treason' and 'gave him to understand that he would return to Quito, the land that his father had left him.'

Once again the Spaniards duped the Inca. By tempting him with the prospect of restoration to his Quitan kingdom, they made Atahualpa a willing hostage, even a collaborator. His life became their indemnity, and the orders he issued appeared to endorse their presence. Pizarro and his men needed time to send news of their incredible success to their compatriots in Panama, to attract reinforcements with which to push deeper into Peru. The longer Atahualpa took to collect his ransom, the better for Pizarro. Both sides settled down to wait: the Spaniards for the arrival of the reinforcements and the gold, and Atahualpa for the settlement of his ransom, the restoration of his liberty and the departure of the odious strangers.

During the months in Cajamarca, the Spaniards were able to observe their royal captive. 'Atahualpa was a man of thirty years of age, of good appearance and manner, although somewhat thick-set. He had a large face, handsome and fierce, his eyes reddened with blood. He spoke with much gravity, as a great ruler. He made very lively arguments: when the Spaniards understood them they realised that he was a wise man. He was a cheerful man, although unsubtle. When he spoke to his own people he was incisive and showed no pleasure.' Licenciate Gaspar de Espinosa wrote to the Emperor what he had heard about Atahualpa's intelligence: 'He is the most educated and capable [Indian] that has ever been seen, very fond of learning our customs, to such an extent that he plays chess very well. By having this man in our power the entire land is calm.'

To their great good fortune, the Spaniards had in their control an absolute ruler whose authority was unquestioned. The only curb on an Inca's powers was ancient custom and a tradition of benevolent rule. Atahualpa's father Huayna-Capac and his royal predecessors had made a genuine effort to ensure the well-being and happiness of their subjects. The Inca's prestige was heightened by a claim to be the descendant of the sun, with whom he was supposed to be in constant communication. This identification with the most powerful influence on men's lives has been common for rulers throughout history, most notably in Egypt and Japan, and the result for the Inca was that he was worshipped during his lifetime. By Atahualpa's time the Inca's divine status was enhanced by being surrounded by a protective screen of women, and by the use of only the finest possible objects for personal use. He 'was served by one sister for eight or ten days, and a great quantity of chiefs' daughters served these sisters... These women were constantly with him to serve him, for no Indian man entered where he was.... These lords and the sisters whom they considered as wives wore very fine, soft clothing, as did their relatives. [Atahualpa] placed his cloak over his head and fastened it under his chin, covering his ears. He did this to hide one ear that was broken, for when Huascar's men took him they damaged it.' Atahualpa's sister Inés Yupanqui reported years later that

'Atahualpa's wives were so greatly respected that no one even dared to look them in the face. If they did anything irregular they were immediately killed, and so was any Indian who committed any excess involving them.'

Pedro Pizarro observed the rituals surrounding the Inca's daily life with fascination. When Atahualpa ate, 'he was seated on a wooden duho little more than a span [nine inches] high. This duho was of very lovely reddish wood and was always kept covered with a delicate rug, even when he was seated on it. The ladies brought his meal and placed it before him on tender thin green rushes . . . They placed all the vessels of gold, silver and pottery on these rushes. He pointed to whatever he fancied and it was brought. One of the ladies took it and held it in her hand while he ate. He was eating in this manner one day when I was present. A slice of food was being lifted to his mouth when a drop fell on to the clothing he was wearing. Giving his hand to the Indian lady, he rose and went into his chamber to change his dress and returned wearing a dark brown tunic and cloak. I approached him and felt the cloak, which was softer than silk. I said to him, "Inca, of what is a robe as soft as this made?" . . . He explained that it was from the skins of [vampire] bats that fly by night in Puerto Viejo and Tumbez and that bite the natives.'

On another occasion the young Pedro Pizarro was taken to see the royal storehouses full of leather chests. 'I asked what the trunks contained and [an Indian] showed me some in which they kept everything that Atahualpa had touched with his hands, and the clothes he had discarded. Some contained the rushes that they placed in front of his feet when he ate; others the bones of meat or birds he had eaten . . . others the cores of the ears of corn he had held in his hands. In short, everything that he had touched. I asked why they kept all this there. They told me that it was in order to burn it. Anything that had been touched by the rulers, who were sons of the sun, had to be burned, reduced to ashes and thrown to the air, since no-one was allowed to touch it.'

'These lords slept on the ground on large cotton mattresses. They had large woollen blankets to cover them. In all Peru I saw no Indian to compare with Atahualpa for ferocity or authority.'

The adulation surrounding Atahualpa could go to extraordinary lengths. Juan Ruiz de Arce recalled that 'he did not spit on to the ground when he expectorated: a woman held out her hand and he spat into it. The women removed any hairs that fell on to his clothing and ate them. We enquired why he did that in spitting, [and learned that] he did it out of grandeur. But with the hairs he did it because he was very frightened of sorcery: he ordered them to eat the hairs to avoid being bewitched.'

This carefully cultivated aura of divinity helped to support the Inca's absolute rule, and Atahualpa had the self-confidence and assurance to make full use of his great powers. The chiefs of his faction continued to look to him for leadership, and he exercised it from his Spanish prison. 'When the chiefs of this province heard of the arrival of the Governor and the capture of Atahualpa, many of them came peacefully to see the Governor. Some of these caciques were lords of thirty thousand Indians, but all were subject to Atahualpa. When they arrived before him, they did him great reverence, kissing his feet and hands. He received them without looking at them. It is remarkable to record the dignity of Atahualpa and the great obedience they all accorded him.' 'He behaved towards them in a most princely manner, showing no less majesty while imprisoned and defeated than he had before that befell him.' 'I recall that the lord of Huaylas asked permission to go to visit his territory and Atahualpa gave it to him, but allowed a limited time in which he was to go and return. He delayed somewhat longer. On his return, while I was present, he arrived with a present of fruit from his province. But when he entered the Inca's presence he began to tremble to such an extent that he could not remain standing. Atahualpa raised his head a little and, smiling, made him a sign to leave.'

The deification and glorification of the Inca were essential props in the Incas' rule of their great empire. Only a century before the Spaniards' arrival, the Incas had been an insignificant mountain tribe occupying only the valley of Cuzco. In about 1440 they were attacked and almost overwhelmed by the neighbouring Chanca tribe, but defended themselves and won a major victory on the plain above Cuzco. This success launched the tribe on a course of headlong expansion. The ruling family produced a series of Incas who combined an insatiable appetite for conquest with military skill and a genius for government and administration.† The Incas instinctively adopted many of the most successful devices of colonial and totalitarian regimes. They avoided bloodshed wherever possible, preferring to absorb tribes into the empire by attraction. But their well-disciplined armies were devastatingly effective when necessary. They imposed an official religion of sun worship throughout the empire, and claimed solar descent for the Inca and the entire royal family. Members of the family itself were allowed to know that the connection with the sun was based on deception: their ancestor Manco Capac had used a suit of shining armour to reflect the sun's rays, and a similar golden disc caught that reflection in the sun temple at Cuzco.

Members of the royal family occupied all important administrative positions throughout the empire. Immediately below them was the caste of the Inca nobility, distinguished by golden plugs in their ear-lobes – from which the Spaniards knew them as orejones or 'big ears'. The orejones occupied lesser administrative positions. The Incas ruled conscientiously, but they also enjoyed every available form of luxury and privilege. They had the finest food and clothing, magnificent table services, ornaments and palaces, beautiful women, servants, a private language, permission for incest denied to ordinary people, the use of roads and special bridges, travel in litters, a different scale of punishments, the right to chew the mild narcotic coca, and so forth. The chiefs of conquered tribes were gradually admitted to this privileged class. Their sons were taken to Cuzco to be educated and to participate in court rituals. The families of subject rulers thus perpetuated the caste distinction, enjoying its privileges while surrendering much of their power. Members of the entire Inca tribe also enjoyed special prestige, and a feeling that they were part of an élite. Groups of them were transplanted into newly won areas to form a nucleus of unquestioned loyalty. As the empire expanded, other Quechua-speaking tribes were designated as honorary Incas. Inca society was thus sharply class-conscious, even if the distinctions were not based on a monetary system or on private property, and even if the ruling caste provided a benign welfare state.

The charisma of the privileged class depended on its uninterrupted success. This was damaged by the ravages of the Quitan epidemic, the internecine war of succession, and above all by the slaughter in the square of Cajamarca and capture of the Inca. The mass of Andean natives became disillusioned and increasingly indifferent to the fate of the former ruling class. They could not appreciate that Pizarro's Spaniards

Huascar Inca being led towards Cajamarca by Atahualpa's generals Quisquis and Chalcuchima

represented the vanguard of an invasion that could eventually subjugate them all. So they stood aside, and the Spaniards became aware that the empire's class distinctions could work in their favour just as much as the family passions of the civil war.

The policy that Atahualpa decided to enforce with all the authority at his command was the fulfilment of the ransom to save his own life. He apparently reasoned that the Spaniards, who had not killed him in the first flush of victory, would abide by their promise to release him when the ransom was collected. He would then be free to enjoy the empire that his generals had just won for him. He therefore issued orders to his commanders to remain at their posts in southern Peru, to expedite the sending of the ransom gold, and not to attempt to rescue him by force. Atahualpa himself was content to exist in his accustomed comfort at Cajamarca while the gold was accumulated.

Very soon after his capture, word came that his brother Huascar was being brought captive from Cuzco and was only a few days' march from Cajamarca. Pizarro told Atahualpa how much he was looking forward to seeing this rival Inca, and ordered Atahualpa to ensure his safe arrival. The Spaniards thought that they would soon have both claimants to the Inca throne in their possession. Atahualpa was still obsessed with the politics of the civil war and confident that the Spaniards represented no threat of external invasion. So instead of ordering Huascar's release to organise a national resistance, he thought only of the danger to himself of having his rival in Cajamarca. Huascar was therefore killed by his escort at Andamarca in the mountains above the Santa valley between Huamachuco and Huaylas and not far south of Cajamarca. Atahualpa protested to the Spaniards that Huascar's guards had acted on their own initiative, and Pizarro accepted this highly unlikely explanation. It was inconceivable that any Peruvians would have dared to kill Atahualpa's brother without his explicit orders, particularly when they were so close to the Inca's presence.*

The killing of Huascar gave Atahualpa an immediate personal advantage, and was the culmination of his extermination of the Cuzco branch of the royal family. 'It was something he was ordinarily accustomed to do to his brothers . . . for, as he himself said, he had killed many others of them if they followed his brother's [Huascar's] faction.' One of Atahualpa's favourite possessions was the head of Atoc, one of Huascar's generals who had imprisoned Atahualpa at Tumibamba and who had been defeated in the first battle of the civil war at Ambato, south of Quito. Cristóbal de Mena saw this 'head with its skin, dried flesh and hair. Its teeth were closed and held a silver spout. On top of the head a golden bowl was attached. Atahualpa used to drink from it when he was reminded of the wars waged against him by his brother. They poured the chicha into the bowl and it emerged from the mouth through the spout from which he drank.'

Atahualpa continued to enforce his triumph in the civil war in other ways while he was at Cajamarca. Pedro Pizarro said that two half-brothers called Huaman Titu and Mayta Yupanqui asked permission of Governor Pizarro to leave Cajamarca for their home in Cuzco. Although they were armed with swords by the Spaniards, Atahualpa sent after them and had them killed on the road.* Two other brothers reached Cajamarca in mid-1533 and one of these was Tupac Huallpa, the man with the best claim to succeed Huayna-Capac now that Huascar was dead. 'They came very secretly for fear of their brother. . . . They slept near the Governor because they did not dare to sleep elsewhere' and 'pretended to be ill throughout the time that Atahualpa was there, not leaving their room. [Tupac Huallpa] did this from fear that Atahualpa might send to have him killed as he had his other brothers.'

Civil wars breed deep passions and violent hatreds. Atahualpa's behaviour was understandable in terms of his own claim to rule as Inca, but was a tragedy in the face of the foreign menace. The country was deprived of leadership and unity at the moment when it had greatest need of both. Had the Spaniards arrived a year later they would have found the country solidly ruled by Atahualpa. And, as Pedro Pizarro wrote, 'Had Huayna-Capac been alive when we Spaniards entered this land it would have been impossible to win it, for he was greatly loved by his subjects. . . . Also, if the land had not been divided by the wars of Huascar and Atahualpa, we could not have entered or conquered it unless over a thousand Spaniards had come simultaneously. But it was impossible at that time to assemble even five hundred because so few were available and because of the bad reputation of this country.'

Atahualpa once tried to achieve a trial of strength with the Spaniards. He challenged Pizarro to have one of his men wrestle with a gigantic native called Tucuycuyuche. Pizarro accepted, and nominated the tough Alonso Díaz. The Indian warrior arrived naked and with his hair cut short and was at first victorious. But Díaz extricated himself, caught Tucuycuyuche in a murderous grip and strangled him. Indian awe of the Spaniards increased still further.*

It took some time for the gold to be amassed and transported across the empire towards Cuzco. The latter half of November and the month of December 1532 elapsed with little activity beyond the arrival of a load of gold 'in very remarkable and large pieces, with jars and pitchers of up to two arrobas' capacity. Some Spaniards whom the Governor had assigned for the task began to crush these objects and to break them so that [the chamber] would hold more. [Atahualpa] asked them: "Why do you do that? I will give you so much gold that you will be satiated with it!"' The impatient conquistadores began pestering the Inca to deliver the gold he had promised. He whetted their appetites by describing the treasures of the two great temples of the empire: the sun temple of Coricancha at Cuzco and the great shrine and oracle of Pachacamac on the coastal desert, south of modern Lima. He suggested that Pizarro should send Spaniards to oversee the ransacking of these holy places – Atahualpa himself had probably seen neither temple and could afford to be callous about them. He was more concerned with the worship of his own Inca ancestors, and issued strict orders that nothing connected with his father Huayna-Capac should be touched.

His immediate objective was the completion of the ransom which could be achieved only with the inclusion of gold from the temples. Atahualpa may well have feared that the priests of the temples would prefer to hide their treasures than sacrifice them to save a usurping Inca.

Pachacamac was a pre-Inca shrine, so greatly venerated along the coastal plain that the Incas had not dared to tamper with it when they conquered the coast in the late fifteenth century. They incorporated it instead into their own religious canon, building an enclosure for holy women alongside the great step pyramid of adobe bricks. The idol Pachacamac was also identified with the Inca's own creator god, who had no name but whose titles included Ilya-Ticsi Viracocha Pacayacacic, meaning Ancient-founded Creator Lord, Instructor of the World.

The high priest and the chief of Pachacamac appeared in Cajamarca at the end of 1532. They were given a chilly reception by Atahualpa, who asked Pizarro to throw the priest into chains and taunted him to have his god arrange his release. Atahualpa explained to Pizarro why he was so angry with Pachacamac and its priest. The oracle there had recently delivered three disastrously wrong predictions: it advised that Huayna-Capac would recover from his illness if he were taken into the sun, but he died; it told Huascar that he would defeat Atahualpa; and, most recently, it had advised Atahualpa to make war on the Christians, saying that he would kill them all. Atahualpa concluded that a shrine which was so fallible could contain no god, and Pizarro told him he was a wise man to reach this conclusion.*

The Spaniards waiting in Cajamarca became bored and restive. They were also increasingly nervous about their isolation and anxious lest attacks might be mounted against them. 'The Governor kept receiving reports every day that fighting men were coming against him.' 'The lord Governor and all of us . . . saw ourselves in great danger every day. That traitor Atahualpa continually made troops come upon us. They came, but did not dare approach.' There was a report that native troops were massing at Huamachuco, a few days' march to the south of Cajamarca. Hernando Pizarro was sent with twenty horse (including the authors Miguel de Estete and Diego de Trujillo) and some foot-soldiers to investigate.* The expedition left Cajamarca on 5 January 1533 but found no hostile troops at Huamachuco. Some Indian chiefs were tortured and said that Atahualpa's commander-in-chief Chalcuchima was not far to the south with a great army. Hernando Pizarro sent three men back to report to his brother and transport some gold, but 'on the road a disaster befell them. The companions who were bringing the gold quarrelled over certain pieces that were missing. One of them cut off another's arm – something that the Governor would not have wished for any amount of gold.' Francisco Pizarro now gave his brother permission to proceed to the temple of Pachacamac. Nothing more was seen of Chalcuchima's army, although the small force of Spaniards was apprehensive that he was 'close by with 55,000 men'.

Hernando Pizarro's contingent rode on, deep into the Inca empire. They moved up the modern Callejón de Huaylas, with the superb snow-covered peaks of Huascaran and the Cordillera Blanca to their left and the turbulent grey Santa river rushing through its warm canyon below. They were well received in the towns through which they passed, and were able to admire the tranquil efficiency of the country. Hernando Pizarro and Estete both wrote in praise of the suspension bridges, roads and storehouses they passed. They were clearly delighted by what they saw of this strange empire.

Inca Peru was the product of thousands of years of isolated development. It lay along the ranges of the Andes and the bone-dry belt of coastal desert between them and the Pacific. To the west lay the world's broadest ocean, to the east the overwhelming barrier of the Amazon forests, and to the south the bleak wilds of Araucania and Patagonia. The Peruvians developed a unique civilisation in this vacuum. Recent archaeology has dug back to the remote origins, to a time before the discovery of ceramics or of farming. In the thousands of years since then, Peruvian man had developed his skills in a steady progression, possibly stimulated by infusions of outside talent but most probably in complete isolation. Great civilisations had flourished and gone long before the florescence of the Inca tribe. We know these civilisations only by the names of their most famous archaeological sites, but can reconstruct their way of life from the mass of excavated material. At the time of the golden age of Greece, northern and central Peru were dominated by a culture called Chavín, named after a great stone temple in the hills above the valley being penetrated by Hernando Pizarro's contingent. Chavín, famous for its fierce, highly stylised representations of baleful pumas and condors, gave way to a series of different cultures in the valleys along the Pacific coast. In the north there was Mochica, a vibrant civilisation about which we know much, for its pottery and textiles have survived in their tens of thousands in dry coastal tombs. Many Mochica pots were effigies, showing with freshness and naturalism the facial types, everyday life, plants, warfare and sexual practices of its people. In southern Peru at this time the Nazca civilisation produced pottery and textiles of supreme beauty.

Mochica, Nazca and many other valley cultures were overcome, in about 1000 A.D., by a civilisation that probably originated at Tiahuanaco, beside Lake Titicaca on the Bolivian altiplano. The type site of Tiahuanaco, with its stone-faced platforms, monolithic statues, and the famous gateway of the sun, was an ancient ruin at the time of the Conquest. The Incas claimed that their tribe had its origins at Lake Titicaca, and they probably learned many of their building and masonry techniques from Tiahuanaco. For a time Tiahuanaco dominated Peru, but its unity gave way to a number of tribal or city states. In the sierra there was a series of powerful tribes – Cañari, Chachapoya, Conchuco, Yarivilca, Huanca around Jauja, Chanca at Abancay and Andahuaylas, Inca, the Colla and Lupaca at Lake Titicaca, and many more. But the most sophisticated state was Chimu on the northern coast of Peru. In art it continued, with less brilliance, from Mochica; but in its vast symmetrical cities, elaborate irrigation and defensive works, and political structure, Chimu developed much that was assimilated by the Incas.

What Hernando Pizarro's men saw as they marched into the Chimu part of the Inca empire was a well-ordered agricultural society. The ordinary Peruvians lived simple peasant lives. They farmed and lived collectively, with no private property, strongly bound to their families and clans, villages and fields. Because of Peru's isolation, its plants, animals and even diseases were unique, all unknown to the European invaders. The Peruvians had no draught animals to help with their farming. They ploughed with foot-ploughs, long poles hardened at the point and equipped with foot-rests and handles. The men stood in lines to plough, prising up the earth with the poles; their wives crouched opposite, breaking the sod, hoeing and planting. It was a cheerful occasion, with chanting and drunken celebrations. Each month of the year had its tasks and festivities in the agricultural calendar. In January, when Hernando Pizarro rode towards Pachacamac, maize was growing, protected by the farmers and their children from birds and predators. The maize was harvested in May, the most important month of the agricultural year, and in June potatoes and oca (sweet potatoes) were dug from the ground. Although the Spaniards could never have guessed it, potatoes were to be Peru's greatest legacy to the world. They originated in Peru and grow there in a profusion of varieties and colours. It has been calculated that the world's annual potato harvest is worth many times the value of all the treasures and precious metals taken from the Inca empire by its conquerors.

The Peruvian peasants lived in simple thatched huts, smoky and smelly, full of guinea-pigs, dogs and fleas. Apart from the guinea-pigs, and occasional dried llama meat or fish, their diet was vegetarian – mostly maize, potatoes or rice-like quinua. Peru is a hard country: most of its level ground is desolate puna, too high for normal cultivation, or the strip of coastal plain where miles of desert separate the river valleys. This desert has formed because the cold Humboldt current runs close to the Peruvian coast and the land is warmer than the sea: moisture is sucked away from the land instead of the reverse. The towering ranges of the Andes rise sheer above the narrow plain and rain-clouds from the Amazon are always caught in the barrier of the sierra. All that is left for normal cultivation and habitation are the river valleys – tight, crumbling canyons in the mountains, or shallow beds of vegetation on the Pacific side. Almost nowhere in Peru are there stretches of rich farming land as found in Europe or North America. To these topographical difficulties are added the relative meanness of natural endowment. Peru had few domestic animals, few crops, and few trees outside the Amazon forests.

The Incas applied their extraordinary organisational genius to overcome these natural deficiencies. The agricultural collectives were organised to build and maintain elaborate terraces, shoring up the hillsides in great flights of rough fieldstone terracing. The water resources of the dry coastal plain were husbanded, and the heavy rains of the sierra were tamed by fine irrigation canals and ditches. The imperial administration kept storehouses full of food and herds of llamas and alpacas, primarily for the use of its own administrators and armies but also as insurance against bad harvests. It moved the rural population to equalise the standard of living throughout the empire, and also to plant colonies of loyalists amid potentially restive tribes.

As a result of this administrative efficiency, and a steady regime of disciplined agricultural labour, the population of the Inca empire flourished. But it lived on a weak diet, badly deficient in proteins: the

Indians sadly lacked the milk, eggs and meat of European diet. Their quinua had some protein and so had potatoes (which were dehydrated by freezing and crushing into white meal called chuño); the various plants supplied a reasonably balanced range of vitamins. The natives ate twice a day, in the morning and evening, seated on the ground and eating from bowls. Most of their meals were stews or soups. Anyone now wishing to sample Inca food has only to stay in an Indian hut off the main roads. The wife throws her greens, potatoes and maize into a bubbling pot and ladles this out to her family and guests. Guinea-pigs scurry about and nibble up every scrap that drops on to the mud floor; occasionally they themselves find their way into the brew. Boiled potatoes are handed out with the mud still clinging to them, and boiled or roasted cobs of corn are also eaten in the fingers. Inca households brewed their own chicha, with old women chewing the maize so that their saliva would start the fermentation. It is a pleasant, murky drink like stale cider – not the 'sparkling chicha' of Prescott's imagination. The common people were forbidden stronger alcohol, such as sugar-based spirits, and the chewing of coca was a privilege reserved to the Inca nobility. So everyday life in the Inca empire was that of most peasant communities: a steady struggle for subsistence, punctuated by the religious festivals of the agricultural calendar. The farmers then were little different from the Andean Indians of today: resigned and passive, sturdy and fatalistic. They formed a perfect proletariat, docile, obedient and deeply conservative. Their descendants look impassive, even melancholy, but there is a quizzical almost mocking expression in the faces of the more intelligent. They are handsome people, with lovely copper skin and high Mongolian cheekbones. Their noses are proud and Roman, but their foreheads and chins recede. Andean children are delightful, with bright black eyes and perpetually rosy cheeks. They look sturdy, for the race has evolved enlarged lungs and rib-cages to breathe the thin mountain air.

Ordinary Indians in Inca times wore a standard uniform and were forbidden any variations. They were issued with their clothes from common stores and wore them day and night. When sleeping, the Indians, then as now, removed only their outer garments. Their clothes – one suit for everyday wear, one for festivals – were constantly darned but rarely washed. Men wore a breechclout: a piece of cloth passed between the legs and fastened to a belt in front and behind. Above they wore a white sleeveless tunic, a straight-sided sack with openings for the head and bare arms, hanging down almost to the knees: this gave them the appearance of Romans or medieval pages. Over the tunic they wore a large rectangular cloak of brown wool, knotted across the chest or on one shoulder. The women wore a long belted tunic, rather Grecian, hanging to the ground but slit to expose the legs when walking. This was actually a rectangular piece of cloth wound around the body across the breast, with the ends fastened by pins on each shoulder. It was held at the waist by a broad sash with decorative squares or patterns. Above it the women wore a grey mantle fastened at the breast by a large decorative pin, and hanging behind to the level of their calves. Both sexes went barefoot or wore simple leather sandals bound to their ankles.†

In their isolated development the Peruvians had evolved many of the attributes of other civilisations: textiles, pottery, dress, metals, architecture, roads, bridges and irrigation. But they had failed to achieve three discoveries that we would regard as fundamental: the wheel, the arch and writing. They used rollers to move vast building blocks, but never invented the wheel spinning on an axis. With no animals as strong as horses or oxen, the Incas had little potential for wheeled vehicles. Peru is such a mountainous country that roads are for ever climbing or descending: Inca transport was all done by human runners and porters, or by columns of llamas carrying light loads. Most of the Andes have been without roads for wheeled traffic until the present century. The mass of superb pre-Columbian pottery was shaped by coiling, in the absence of a potter's wheel. The lack of the arch and keystone was less serious: the Incas were magnificent masons and could cut fine rectangular lintels. They sloped the sides of doorways and niches inwards, to lessen the distance to be covered at the top. The resulting trapezoidal or sentrybox openings are a hallmark of Inca construction.

The closest that the Peruvians came to writing were mnemonic devices used to record numerical statistics or historical events. The Mochica apparently used bags of marked beans in this way. The Incas had the famous quipus, rows of strings in which the colour of the threads and the loops of the knots represented arithmetical units or recording categories.† The Incas also had a sophisticated system of public record, with a caste of professional historians who handed down verbal traditions, rather like Homeric bards or medieval troubadours. This lack of writing is a terrible handicap to historians of the Conquest: everything is recorded by the pens of Spaniards. Fortunately for us the Spaniards often interrogated Incas about their past, either through official enquiries or on the initiative of individual chroniclers. Atahualpa's nephew, Titu Cusi Yupanqui, dictated a long narrative which is the only historical record by a member of the Inca royal family. Some chroniclers had Inca mothers or wives, notably Garcilaso de la Vega, Felipe Guaman Poma de Ayala and Juan de Betanzos. Others became expert in the Peruvian language, Quechua, and learned much through friendship with members of the Inca family. A notable chronicler in this latter category was the priest Martín de Murúa, whose sympathies were with the Indians and who provided much detail from the Indian side of the decades after the Conquest.†

Such were the civilisation and people being discovered by Hernando Pizarro's small expedition in January 1533. It spent fifteen days riding through the mountains, and descended on to the coastal plain for a further week of travel towards Pachacamac. A severe disappointment awaited it at the great temple. As Atahualpa feared, the priests had hidden most of their treasure, if they ever had any. The sanctuary lay at the top of a vast adobe step pyramid, with enclosures at each level. Anyone climbing to the very top was supposed to have fasted for a year, and communication with the oracle was permitted only through the intervention of its priests. The Spaniards strode past the outraged guardians and forced their way straight to the topmost level. Miguel de Estete recalled the anticlimax to their journey. The sanctuary proved to be a small cubicle 'of cane wattle, with some posts decorated with gold and silver leaf and some woven fabrics placed on the roof as mats to protect it from the sun ... Its locked door was closely studded with a variety of objects – corals, turquoises, crystals and other things. This was eventually opened, and we were certain that the interior would be as curious as the door. But it was quite the contrary. It certainly seemed to be the devil's chamber, for he always lives in filthy places. ... It was very dark and did not smell very pleasant. Because of this they brought a candle. And so we entered with it into a very small cavern, which was rough and of no craftsmanship. In the middle a post was planted into the ground with the figure of a man at its head, badly carved and badly formed. ... Seeing the filth and mockery of the idol, we went out to ask why they thought highly of something so dirty and ugly.' As for the communications with the devil, Hernando Pizarro wrote, somewhat hesitantly: 'I do not believe that they speak to the devil there... for I took pains to investigate this. [There was] one ancient priest, one of those most intimate with their god, who... had said that the devil told him not to fear the horses for they caused terror but did no harm. I had him tortured, but he remained obdurate in his evil sect.... As far as one can see, the Indians worship the devil not from devotion but from fear.... I made all the chiefs of the region come to witness my entry, so that they would lose their fear. And having no preacher, I myself made them a sermon, explaining the deception in which they had been living.' The Spaniards remained at Pachacamac for most of the month of February, searching in vain for its treasure.† One cannot help being impressed by the audacity and effrontery of Hernando Pizarro and his handful of followers. They blithely overthrew a sanctuary so sacred and venerable that it had been left intact even by the conquering Incas (plate 11). They did so knowing that weeks of difficult marching separated them from their isolated compatriots at Cajamarca.

Atahualpa had originally promised that his ransom would be completed in two months—the time needed to send messengers to Cuzco and to transport the gold from its temples back to Cajamarca. After the initial weeks of delay there was a steady trickle of treasure into the city. 'On some days twenty thousand, on other days thirty thousand, fifty, or sixty thousand pesos de oro would arrive. [It consisted of] large pitchers or jars of from two to three arrobas [50 to 75 pounds] in size, and large silver jars and pitchers, and many other vessels. The Governor ordered it all to be placed in a building where Atahualpa had his guards. ... To keep it more safely, the Governor placed Christians to guard it by day and night, and it was all counted as it was placed in the house so that there should be no fraud.'† Atahualpa must have been impressed by all this, and convinced that the Spaniards were taking his offer seriously. Once the storehouse could be filled the Spaniards would presumably depart with their loot, Atahualpa would be released to rule as Inca, and his native army might even annihilate the strangers before they left Peru. He was therefore anxious to complete the ransom, and this depended on the inclusion of the slabs of gold that lined the temple of the sun of Cuzco. One of Atahualpa's close brothers arrived with a convoy of treasure and reported that much more was delayed at Jauja. He may also have said that nothing had yet been done about stripping the temple of Coricancha. Atahualpa therefore suggested that Pizarro should send some Spaniards to Cuzco to oversee the dispatch of its treasures. Pizarro was reluctant to commit more Spaniards on ventures into the unknown empire, but relented when a messenger arrived from his brother Hernando at Pachacamac. Atahualpa promised to send one of his relatives with any envoys and to order his generals, Chalcuchima at Jauja and Quisquis at Cuzco, to guarantee their safety. Three men finally volunteered: Martín Bueno and Pedro Martín, both from Moguer, and one of the various Zárates in Pizarro's army.* They left Cajamarca on 15 February 1533. 'The Governor dispatched them, commending them to God. They took natives who carried them in hammocks, and were very well served.'

The envoys were given a frosty reception by Atahualpa's general Quisquis, the recent conqueror of Cuzco. 'He liked the Christians very little, although he marvelled greatly at them.... This captain told them that if they refused to release the cacique [Inca] he himself would go to rescue him.' Atahualpa's orders were explicit: the gold was to be stripped from the temple, but nothing connected with Huayna-Capac's own mummy was to be disturbed. Quisquis therefore sent the envoys to the sun temple Coricancha. They found, as Atahualpa had suspected, that it was still intact. 'These buildings were sheathed with gold, in large plates, on the side where the sun rises, but on the side that was more shaded from the sun the gold in them was more debased. The Christians went to the buildings and with no aid from the Indians – who refused to help, saying that it was a building of the sun and they would die – the Christians decided to remove the ornament ... with some copper crowbars. And so they did, as they themselves related.' The Spaniards prised off seven hundred plates, which Xerez reported as averaging some 4½ pounds of gold each when melted down. 'The greater part of this consisted of plates like the boards of a chest, three or four palmos (2–2½ feet) in length. They had removed these from the walls of the buildings, and they had holes in them as if they had been nailed.' The envoys were not allowed to visit the entire city, but what they did see intoxicated them. 'They said there was so much gold in all the temples of the city that it was marvellous.... They would have brought much more of it if this would not have detained them longer, for they were alone and over 250 leagues from the other Christians.' They did, however, 'take possession of that city of Cuzco in the name of His Majesty'. They locked one building full of gold and silver vessels 'and placed a seal of His Majesty on it and another for Governor Pizarro, and also left a guard of Indians'. They reported seeing one golden sacrificial altar that weighed 19,000 pesos and was large enough to hold two men. Another great golden fountain was beautifully made of many pieces of gold: it weighed over 12,000 pesos and was dismantled for transportation to Cajamarca. The envoys even penetrated a sanctuary containing the mummies of two Incas. An old lady wearing a golden mask was responsible for fanning flies off the bodies. She insisted that the intruders remove their boots before entering. After meekly complying with this formality, 'they went in to see the mummies and stole many rich objects from them'.

Quisquis's attitude towards the Spanish envoys revealed the dilemma of the Inca's victorious generals. To save their sovereign's life they had to co-operate with his kidnappers. And having just completed their triumphant advance to Cuzco, they did not dare to leave their posts to attempt his rescue. Quisquis was occupying Cuzco with an army of 30,000* and the commander-in-chief Chalcuchima was at Jauja, midway between there and Cajamarca, with 35,000.* Other garrisons of a few thousand men were holding strategic centres such as Vilcashuaman and Bombón. To the north, between Cajamarca and Quito, the third commander Rumiñavi was in charge of Atahualpa's base. His forces had been swollen by the troops sent home from Cajamarca by Pizarro, and he could draw on levies from a friendly population. Rumiñavi was the only commander not occupying a hostile part of the empire: his was therefore the only army with relative mobility.

Hernando Pizarro at Pachacamac was not far from Chalcuchima across the mountains at Jauja. He used native runners to invite the Inca commander to come down to the coast to meet him. Chalcuchima replied that he would meet the Spaniards at the point where their route back to Cajamarca joined the mountain highway, at the southern end of the Huaylas valley. Hernando Pizarro accepted this rendezvous. His force left Pachacamac at the beginning of March but turned inland up the Chincha (modern Pativilca) valley to the large town of Cajatambo at its head. The natives claimed that Chalcuchima had passed through the town on his way to the rendezvous. 'But since it was believed that these Indians rarely tell the truth, the Captain determined to go out on to the royal road' which ran from Jauja to Huánuco and thence to Cajamarca by the Marañon route. This involved crossing the desolate Cordillera Huayhuash at an altitude of almost 5,000 metres. 'The road was mountainous and so covered in snow that we experienced great difficulty.' 'The men were very weakened and the horses tired and unshod.'

The Spaniards reached the royal highway at Bombón on 11 March and learned that Chalcuchima was indeed still at Jauja. They moved south-eastwards to reach that city on Sunday the 16th. For all his reckless bravery, Hernando Pizarro was frightened at approaching the headquarters of Atahualpa's most formidable commander. 'One of Atahualpa's chiefs whom I was taking with me, and whom I had treated well, warned me that I should make the Christians advance in formation because he believed that the captain [Chalcuchima] was at war. Climbing a small hill close to Jauja we saw a great black mass in the square, which we thought was something burned. But when we asked what it was they told us it was Indians.' 'We did not know whether they were warriors or townspeople.' 'All the men advanced thinking that we were going to fight the Indians. But on entering the square some chiefs came out to receive us in peace' and the ominous dark mass proved to be 'townspeople who had assembled for a festival'.

Hernando Pizarro had gone to Jauja to 'attempt through sweet talk to persuade Chalcuchima to accompany him to where Atahualpa was'. The tiny force hoped 'to bring the gold, disperse the army he had, and bring [Chalcuchima] in person for his own good; or if he were unwilling to attack him and seize it'. Without dismounting, Hernando Pizarro asked for Chalcuchima and learned that he had retired across the river from the city. Pizarro had with him one of Huayna-Capac's sons, possibly Atahualpa's brother Quilliscacha. This prince was sent to reason with the elusive Chalcuchima. The Spaniards meanwhile cleared the square of natives and camped there with the horses saddled and bridled throughout the night. Pizarro had told the local chiefs that the horses were angry and would destroy any native who blundered on to the square. But nothing can come between Andean Indians and their celebrations: the Spaniards found themselves surrounded by continuous dancing, singing and drinking revels throughout the five days of their stay at Jauja.*

Chalcuchima returned to Jauja with the prince the following morning, riding in his litter and accompanied by a fine retinue. He went to Hernando Pizarro's quarters and the two leaders had a day of fruitless negotiation. Pizarro tried to persuade Chalcuchima to accompany him to Cajamarca, claiming that Atahualpa wanted his commander by his side. Chalcuchima explained that Atahualpa had sent to order him to stay at Jauja, and he would not move until he received a definite counter-order. If he left Jauja the district would certainly rise behind him in favour of Huascar's faction. The two sides were still deadlocked when night fell. The Spaniards again spent the night ready for instant battle, while Chalcuchima pondered the arguments that had been put to him during the day. For some unknown reason he decided to give way. Next morning he returned to tell Hernando Pizarro that he would accompany him to Cajamarca, since he was so anxious for him to do so. They would set out in two days' time, taking a large consignment of gold and silver and leaving Jauja under the command of the prince who had accompanied Pizarro.

Chalcuchima's decision was a tragic mistake – one of the turning points in the collapse of resistance to the Spanish invaders. Here was the most formidable commander in the Inca empire handing himself over voluntarily into what proved to be captivity. Chalcuchima at the moment of his surrender was a victorious general in the midst of a devoted army. He was almost as big a catch for the Spaniards as the Inca himself had been, for Chalcuchima's military reputation was already established under Huayna-Capac and might have been great enough for him to lead a united resistance against Pizarro. He was perhaps the only man in Peru with sufficient stature to overcome the hatreds of the civil war – despite his own command of the victorious Quitan army, and despite his part in the execution of Huascar.

What made Chalcuchima change his mind after a day of stubborn debate with Hernando Pizarro? Some Spaniards, with typical self-assurance, thought that 'this commander was afraid of the Christians, particularly of those on horses'. Hernando Pizarro wrote that 'in the end, when he saw that I was determined to bring him, he came of his own free will'. But it seems improbable that the Goliath Chalcuchima could have been afraid of Pizarro's tiny isolated force. Estete admitted that he 'was so strong in men that he would have caused terrible damage had he launched a night attack on the Christians . . . Chalcuchima had majordomos responsible for provisioning his army; he had many carpenters doing woodwork; many other aspects of his personal guard and service were on a massive scale; he had three or four porters in his household. In short, he imitated his sovereign in his establishment and in all other respects. He was feared throughout the land, for he was a very valiant man and had conquered on the orders of his lord over six hundred leagues of land. In the course of this he had had many engagements in the field and in difficult passes, and had been victorious in all of them. There was nothing left for him to conquer anywhere in the country.' 'This captain had many fine men: in the presence of the Christians he had counted them on his knots [quipus] and had found 35,000 Indians.'

Chalcuchima must therefore have been deluded by Hernando Pizarro's 'sweet talk' to which the native prince added authority. He apparently believed the claim that Atahualpa wanted his general to accompany the Spaniards back to Cajamarca. Possibly he was curious to learn in a personal meeting what Atahualpa really wanted of him and his army, and what attitude he should adopt towards the Spaniards. He may have feared that if any Spaniards were killed in a clash with his army Atahualpa would suffer – although had Chalcuchima captured Hernando Pizarro he might have bargained him in exchange for the Inca. He woefully underestimated the consequences of his action. For by riding out of Jauja with this deceptively small band of strangers he delivered himself into captivity and eventual death.

Chalcuchima arranged for his men to make silver and copper horseshoes for the Spaniards. The journey from Jauja to Cajamarca was an amicable affair during which the Spaniards had the privilege of seeing the country with its greatest general as their guide. Lodgings and supplies for men and horses were on hand at each night's resting place. There were particularly splendid festivities during the two days that the party spent at Huánuco. The ruins of this city, now known as Huánuco Viejo, lie above La Unión, a remote village on the upper Marañon. They contain superb stonework, and are unique in being the only ruins of an important Inca city to remain untouched by later occupation (plate 10). The tumbled grey stones of the city's houses and platform temples lie disturbed only by the deterioration of time at the edge of a flat stretch of pale treeless savannah. From Huánuco the travellers rode north through the beautiful country between the eastern slopes of the Cordillera Blanca and the great gorge of the Marañon. This area is not yet penetrated by motor roads, and almost every precipitous hilltop is crowned by the curious ruined towers of the pre-Inca Yarivilca civilisation. It is a sunny place where a traveller rides past hilltop villages and gazes into stupendous valleys dropping towards the Marañon. Chalcuchima had fought his way down this road a few months previously. At one bridge over a difficult canyon he described to his travelling companions how Huascar's men had defended the position for three days, burning the bridge and forcing his Quitans to swim the river.*

Hernando Pizarro marched into Cajamarca on 25 April after an absence of three months.* 'The Spaniards came out to meet us with great happiness and rejoicing.' They had every reason to be pleased: Pizarro may have failed to find much treasure, but he had brought a prize captive. Chalcuchima's status now changed abruptly. On the journey he had been a companion and a host. He now became a virtual prisoner, with never less than twenty Spaniards guarding him throughout the remainder of his life. His first action was to have an audience with Atahualpa, and the Spanish onlookers were impressed by the protocol observed between the captive ruler and his commander-in-chief. 'When Chalcuchima entered the doors behind which his lord was imprisoned, he took a normal load from one of the Indians that he had brought with him, and placed this on his back, as did the many other chiefs who came with him. He entered where his lord was, carrying this load, and when he saw [Atahualpa] he raised his hands to the sun to give thanks for having been allowed to see him again. He went up to him with great reverence, weeping, and kissed him on the face, hands and feet, and the other chiefs who had come with him did the same. Atahualpa showed great majesty. He did not look him in the face or pay him any greater attention than he would have paid the humblest Indian who might have come before him – although there was no one in all his kingdom that he loved as dearly.' 'The cacique Atahualpa was deeply distressed by the arrival of his commander, but since he was very astute he pretended that it pleased him.'

Now that the Spaniards had Chalcuchima in their power, they began to abuse him. They were convinced that when he conquered Cuzco he must have seized the gold of Huayna-Capac and of Huascar – the envoys had not yet returned to confirm that it was all still in the city. When Governor Pizarro insisted in demanding the gold, Chalcuchima could do no more than protest 'that he had no gold and that they had brought it all'. No one believed him. 'Everything he said was a lie. Hernando de Soto took him aside and threatened to burn him unless he told the truth. He gave the same answer as before. They then erected a stake and tied him to it, and brought much firewood and straw, saying that they would set fire to him unless he told the truth. He asked them to call his lord. [Atahualpa] came with the Governor and spoke to his captain, who was tied up.' Chalcuchima explained his obvious danger but the Inca said it was a bluff 'for they would not dare to burn him. They then asked him for the gold once more and he would not tell about it. But as soon as they set a little fire to him, he asked that his lord should be taken away from in front of him, for he was signalling him with his eyes not to tell the truth. Atahualpa was therefore removed.

'[Chalcuchima] then said that on the cacique's orders he had come on three or four occasions with a large force against the Christians. But, as the Christians knew, his ruler Atahualpa had himself ordered him to withdraw for fear that the Christians would kill him.... They then took that Indian captain to the house of Hernando Pizarro and kept a close guard on him. Such a guard was necessary, for the greater part of the army obeyed the orders of this captain even more than they did those of their lord the cacique Atahualpa himself.... And although he was half burned, many Indians came to serve him because they were his servants.' Hernando Pizarro later testified that Chalcuchima was brought to him 'with his legs and arms burned and his tendons shrivelled; and I cured him in my lodging'.

A procession of Inca warriors, weeping in penitence

3

EXPEDIENCY

On the eve of Easter, 14 April 1533, Francisco Pizarro went out of Cajamarca to greet his partner Diego de Almagro, who had marched inland with a force of 150 fresh Spaniards. 'Both friends and old companions received one another with great demonstrations of affection. The Marshal [Almagro] immediately went to visit Atahualpa, kissed his hand, doing him great reverence, and had a friendly conversation with him.' Atahualpa had nothing to celebrate, for this new arrival completely changed the balance of power at Cajamarca.

Although Almagro had been ill at Panama, he had upheld his part of the partnership agreement by equipping a force of 153 Spaniards and fifty horses, building a ship 'with two topsails', obtaining Hernán Ponce's ships and the famous pilot Bartolomé Ruiz – all recently returned from Peru – and sailing down the Pacific coast.* Almagro, like Pizarro before him, landed on the Ecuadorean coast and exhausted his expedition searching along it; he moved on to Tumbez, but the natives were uncommunicative; and only when one of his ships sailed to San Miguel did he and his men learn of the undreamed-of success of their colleagues. Pizarro, for his part, had sent his secretary Pedro Sancho and others to watch for Almagro on the coast. He even thought to send gold for the new arrivals to pay off their ships – for there were ugly rumours that Almagro might attempt a conquest of his own.*

Hernando Pizarro's contingent returned to Cajamarca eleven days after Almagro's arrival. The Spanish forces in the city were thus almost doubled, and the small force of strangers began to look like the vanguard of an invasion. The Inca immediately began to suspect that he might never buy off the Spaniards. 'When Almagro and these men arrived, Atahualpa became anxious and feared that he was going to die.' He asked whether permanent settlement was contemplated, and 'how the Indians were to be divided among the Spaniards. The Governor told him that they were going to allot a cacique to every Spaniard. Atahualpa asked whether the Spaniards were going to settle with their caciques. The Governor said no, they would make towns where the Spaniards could stay together. Hearing this, Atahualpa said 'I shall die'. . . . The Governor reassured him, saying that he would give him the province of Quito for himself – the Christians would take the land from Cajamarca to Cuzco. But since Atahualpa was a clever Indian, he realised that he was being deceived and became very friendly with Hernando Pizarro who had promised that he would not consent to his being killed.'

Atahualpa clung to the hope that the ransom agreement was still valid, although he now realised that the Spaniards had been waiting as much for the arrival of Almagro's reinforcements as for the ransom gold. The treasure trains were arriving more frequently now, and on 3 May, Pizarro ordered the gold and silver that had been accumulated to date to be melted down. On 13 May the first of the three Spaniards who had gone on the reconnaissance to Cuzco returned with thrilling reports of the gold that was on its way from the great city. A month later, Hernando Pizarro left Cajamarca for Spain, taking with him 100,000 castellanos for the King and a report on the success of the expedition.

The Spaniards now had a great hoard of pieces of precious metal, accumulated from the time they first landed in Peru. The entire treasure was carefully guarded by Pizarro, and no individual Spaniard was allowed to retain any. At long last, on 17 June, the Governor issued decrees ordering the distribution of the silver, and the melting and assaying of the gold; the distribution of the gold was not made until 16 July.* Indian smiths carried out the melting on nine forges under the supervision of Pizarro's servant Pedro de Pineda. The melting of silver and gold continued from 16 March to 9 July, and on many days the smiths were melting 60,000 pesos – over 600 pounds – of gold. Over eleven tons of gold objects were fed into the furnaces at Cajamarca, to produce 13,420 pounds of 22½-carat 'good gold'; the silver objects yielded some 26,000 pounds of good silver. Much of this consisted of vases, figures, jewellery and furnishing ornaments, the masterpieces of Inca goldsmiths. Its destruction was an irreparable artistic loss. We can only judge the splendour of what was destroyed by the known quality of Inca ceramics and textiles and by a few surviving objects of precious metals.*

The gold and silver that emerged from the furnaces was officially marked with the royal mark, to show that it had been legally melted and that the royal fifth had been paid. The treasure was meticulously recorded by the army's notaries and by royal officials who had arrived with Almagro. The quota for a horseman was some 90 pounds of gold and 180 of silver, and foot-soldiers received half this amount. Francisco Pizarro himself took almost seven times the horseman's quota, and was given 'as a present' the throne on which Atahualpa travelled: it was 15-carat gold and weighed 183 pounds (83 kilos). Hernando Pizarro was given some three-and-a-half times the quota and Hernando de Soto double. The melter and marker received one per cent of the total, and the assayer was given a bonus. The crown received one fifth, of which Hernando Pizarro had already taken less than half to Spain. But the ecclesiastics received less than the share of a foot-soldier, and only token awards were made to the men who had just arrived with Almagro or to those who had remained on the coast at San Miguel.

'When Atahualpa heard of Hernando Pizarro's departure he wept, saying that they would kill him since Hernando Pizarro was leaving.' On the day after Hernando left Cajamarca, 13 June, the two Spaniards at last returned from Cuzco, escorting an extraordinary convoy of 225 llama-loads of the city's gold and silver; a further sixty loads of baser gold arrived a few days later. It is impossible to give the value of the ransom in modern terms. The purchasing power of gold and silver have altered since the sixteenth century, and so have the relative values and desirability of goods and services. The treasure was worth far less in Peru than it would have been in Europe. Nevertheless, it is interesting to know what Atahualpa's ransom would fetch on the present bullion market. The gold would be worth £2,570,500 ($6,169,200) and the silver £283,850 ($681,240), a total of £2,854,350 ($6,850,440).*

The sudden release of this vast treasure turned Cajamarca into a gold-rush town where European goods fetched dizzy prices. 'If a man owed another something he paid with a lump of gold, without weighing the gold and quite unconcerned whether it was worth double the amount of the debt or not. Those who owed money went from house to house followed by an Indian laden with gold, seeking out their creditors to pay what they owed them.'

As Atahualpa saw the ransom treasure melted down and himself still a prisoner, he grew desperate. He must now have become increasingly sure that the Spaniards had no intention of releasing him. His only hope was to be rescued by force. The only commander in a position to do this was Rumiñavi, the general left in charge of Quito when Chalcuchima and Quisquis marched south. Atahualpa may have ordered Rumiñavi to advance close to Cajamarca, to prepare to attack his captors there and any Spaniards who attempted to transport gold to the coast. The Spaniards were highly suspicious that some such rescue would be attempted. Their suspicions soon became convictions. 'It had often been thought almost certain that he had given orders for warriors to assemble to attack the Spaniards. This had in fact been ordered by him, and the men were all in readiness with their captains. But the cacique [Inca] only delayed the attack because he himself was not free and his general Chalcuchima was also a prisoner.' The rumours spread. The chief of Cajamarca came to Governor Pizarro and told him that Atahualpa had definitely sent orders to muster the men of his Quitan homeland. ' "All these men are coming under a great commander called Rumiñavi and are very close to here. They will come by night and will attack this camp, setting fire to it on all sides. The first person they will endeavour to kill will be you, and they will release their lord Atahualpa from his prison. Two hundred thousand natives of Quito are coming, with thirty thousand caribs who eat human flesh." . . . When the Governor heard this warning he thanked the cacique warmly and did him much honour. He ordered a secretary to record it all, and he made a report about it for him. This report was taken to an uncle of Atahualpa and to some leading nobles and Indian women. It was found that what the cacique lord of Cajamarca had said was all true.' Pizarro's secretary Pedro Sancho confirmed the Spaniards' investigations of this terrifying rumour. 'Long reports of all this were received from many caciques and from [Atahualpa's] own chiefs. They all revealed and confessed the plot voluntarily, without fear, torture or coercion.' The informants even gave details of the difficulties being experienced in feeding this army. They said that it had been divided into separate contingents but that it had still been found necessary to harvest green maize and dry it to provide reserves of food.

The Spaniards treated the rumours with the utmost seriousness. Pizarro ordered a strong guard on the camp. 'Fifty horsemen patrolled during each quarter and the entire hundred and fifty [stood to] at dawn. During all these nights the Governor and his captains did not sleep: they were inspecting the watches and arranging what was necessary. During the watches when it was the men's turn to sleep they did not remove their arms, and the horses remained saddled.'

Pizarro confronted Atahualpa with the damning reports. The Inca made a vigorous denial, arguing that he would never dare order an army to attempt his rescue from men as powerful – and as ruthless – as the conquistadores. And without his order no army would make a move. 'Atahualpa answered: "Are you joking? You are always telling me jokes. What chance would I or all my men have of disturbing men as brave as you? Stop mocking me like this." [He said] all this without showing any sign of embarrassment, but smiling to conceal his cunning. He made many other brilliant remarks of a quick-witted man during the period after his capture. Spaniards who heard them are amazed to see such wisdom in a barbarian.' A young member of Pizarro's force, Pedro Cataño, remembered hearing the Inca argue with powerful logic: 'It is true that if any warriors were to come they would be coming from Quito on my orders. But find out whether it is true. And if it *is* true, you have me in your power and can execute me!' Despite this spirited defence, Pizarro put a chain round his captive's neck to forestall any attempted escape – Atahualpa was known to have escaped from captivity at the outset of the civil war. Pizarro's secretary Xerez claimed that it was later learned that when Atahualpa was first put in chains, he sent to halt Rumiñavi's advance, but later countermanded this and 'sent instructions on how and where and at what time they were to attack the camp; for he was alive, but if they delayed they would find him dead'.

Pizarro summoned a meeting of the Spanish leaders: his own captains, Diego de Almagro, the royal officials including the treasurer Alonso Riquelme, the Dominican friar Vicente de Valverde, the scrivener Pedro Sancho, the inspector Miguel de Estete and others. The debate raged more about the expediency of keeping Atahualpa alive than about the existence of Rumiñavi's army. Now that the treasure had been melted everyone wanted to leave Cajamarca for the eldorado of Cuzco. 'We were planning how to take Atahualpa along the road and what guard to place on him. We discussed and debated whether we would be capable of defending him in the bad passes and rivers should his people try to rescue him.' Many felt that Atahualpa had become a liability, a Mary Queen of Scots to embarrass the would-be rulers of Peru.

Pizarro himself and most of the men who had lived in close proximity to Atahualpa during the previous eight months wanted to keep him alive. They knew the value of the captive Inca as an indemnity. Some even felt that since the ransom had been paid the Spaniards should honour their side of the bargain. The Inca could scarcely be accused of having harmed the Spaniards in any way. The only Spaniard who had been injured since Pizarro advanced into Peru was the man whose arm was cut off by one of his compatriots. Some Spaniards may even have grown fond of the captive with whom they had spent some cheerful evenings. The newer arrivals were less sentimental. They were eager to advance deeper into Peru to win fortunes for themselves and were frightened that as long as Atahualpa was alive the invaders would be in hourly danger. 'With his death it would all cease and the land would be calmed.'

The arguments reached an impasse and the debate returned to the immediate danger of Rumiñavi's army. 'Wishing to know the truth . . . five notables volunteered to go in person to investigate and to see whether those warriors were in fact coming to attack the Christians. In the end the Governor . . . agreed, and Captain Hernando de Soto Captain Rodrigo Orgóñez, Pedro Ortiz, Miguel Estete and Lope Vélez went to find those enemies who were said to be approaching. The Governor gave them a guide or spy who said that he knew where [the enemy] was, but after two days' march the guide fell to his death off a precipice. ... But the five horsemen continued until they reached the place where it was said that they would find the enemy army.'

The departure of this reconnaissance did little to calm the mounting hysteria at Cajamarca. The young soldier Pedro Cataño claimed that he had been highly indignant when he first heard the rumour that the Inca might be killed. He hurried off to protest to the Governor; but Pizarro put him in chains and in prison to punish his presumption and to cool his youthful ardour. Almagro visited him in prison, and Pizarro then flattered him with an invitation to the great honour of dining with Almagro and himself. During the dinner there were emotional speeches in which Pizarro thanked young Cataño for having helped dissuade him from harming the Inca. Cataño, overcome, declared that 'in the name of all the conquistadores, he kissed the hands of His Lordship for having acted in this way'. The dinner was followed by a game of cards. While they were playing, the Biscayan Pedro de Anades burst into the room dragging a Nicaraguan Indian. He explained that this Indian had been three leagues from Cajamarca and had seen a vast horde of native troops advancing towards the town. Pizarro interrogated the Indian who repeated his story with corroborative details. Almagro burst out: 'Is Your Lordship going to allow us all to die, out of love of Cataño?' Pizarro left the room in silence, followed shortly afterwards by Almagro. Xerez and Mena also recorded the arrival at sunset on a Saturday night of 'two Indians who were in the Spaniards' service'. These said they had met other natives fleeing before an approaching army. This army – which the natives themselves had not seen – 'had arrived three leagues from there and would come that night or the next to attack the Christians' camp for it was approaching at great speed'.

There was a further emergency meeting of the council. 'Captain Almagro insisted on [Atahualpa's] death, giving many reasons why he should die.' 'The royal officials demanded the death penalty, and a learned doctor who was with the army judged that the evidence was sufficient.' 'They decided, against the will of the Governor who was never happy about it, that Atahualpa must die since he had broken the peace and plotted treason in bringing men to kill the Christians.' 'They decided to kill that great cacique Atahualpa immediately' and so 'the Governor, with the agreement of the royal officials, the captains and men of experience, sentenced Atahualpa to death. He ordered as his sentence, because he had committed treason, that he should die by burning unless he were converted to Christianity.'

There was no trial, nothing but a panicky decision by Pizarro, who succumbed to Almagro's and the royal officials' opportunist demands for death. 'Those native lords had certainly read few laws and did not understand them, and yet [the Spaniards] delivered this sentence on an infidel who had not been instructed. Atahualpa wept and said that they should not kill him, for there was not one Indian in the country whom they could manage without his command. Since they had him prisoner what did they fear? If they were doing it for gold or silver, he would give them twice what he had already commanded. I saw the Governor weep from sorrow at being unable to grant him life, because of the consequences and risks in the country if he were released.'

Once the decision was made the Spaniards acted with chilling speed, as if they feared there might be second thoughts if they hesitated. Pizarro's secretary Pedro Sancho wrote a detailed account of the execution. It took place in the ill-fated square of Cajamarca, as night was falling on Saturday, 26 July 1533.† Atahualpa was 'brought out of his prison and led to the middle of the square, to the sound of trumpets intended to proclaim his treason and treachery, and was tied to a stake. The friar [Valverde] was, in the meantime, consoling and instructing him through an interpreter in the articles of our Christian faith. . . . The Inca was moved by these arguments and requested baptism, which that reverend father immediately administered to him [christening him Francisco after Governor Pizarro]. His exhortations did [the Inca] much good. For although he had been sentenced to be burned alive, he was in fact garrotted by a piece of rope that was tied around his neck.'

The conquistador Lucas Martínez Vegaso recalled an extraordinary scene that took place on that evening. While Atahualpa was tied to his chair, with the garrotte tied about his neck, 'he said that he commended his sons to the Governor don Francisco Pizarro. Fray [Vicente] de Valverde advised him to forget his wives and children and to die like a Christian: if he wished to be one he should receive the water of holy baptism. But he continued to persist in commending his sons, with great weeping, indicating their size with his hand, showing by the signs he made and by his words that they were small, and that he was leaving them in Quito. The Father again tried to induce him to become a Christian and forget his children, for the Governor would look after them and would keep them in place of any children of his own. [Atahualpa] said that, yes, he wanted to be a Christian; and he was baptised . . .' Atahualpa's own sister Inés Yupanqui confirmed that she saw Atahualpa leave his sons to Pizarro.*

'With these last words, and with the Spaniards who surrounded him saying a credo for his soul, he was quickly strangled. May God in his holy glory preserve him, for he died repenting his sins, in the true faith of a Christian. After he had been strangled in this way and the sentence executed, some fire was thrown on to him to burn part of his clothing and flesh. He died late in the afternoon, and his body was left in the square that night for everyone to learn of his death. On the following day the Governor ordered all the Spaniards to attend his funeral. He was carried to the church with a cross and the rest of the religious ornaments, and was buried with as much pomp as if he had been the most important Spaniard in our camp. All the lords and chiefs in his service were very pleased with this: they appreciated the great honour that had been done to him.'

Far from 'appreciating the great honour' of Christian burial, Atahualpa's immediate followers were stunned by his death. 'When he was taken out to be killed, all the native populace who were in the square, of which there were many, prostrated themselves on the ground, letting themselves fall to the earth like drunken men.' There were further moving scenes during the funeral. 'When we were in the church singing the funeral service for Atahualpa with his body present, certain ladies – his sisters, wives and other intimates – arrived with great clamour, to such an extent that they interrupted the divine office. They said that the tomb must be made much larger: for it was the custom when the chief lord died for all who loved him to be buried alive with him. They were told that Atahualpa had died a Christian and was being given that service as one. What they were asking could not be done, for it was very evil and against Christianity: they must go away and not interrupt, and should allow him to be buried.' Pedro Pizarro remembered that 'two sisters remained who went about making great lamentations, with drums and singing, recording the deeds of their husband. Atahualpa had told his sisters and wives that if he were not burned he would return to this world. They waited until the Governor had gone out of his room, came to where Atahualpa used to live, and asked me to let them enter. Once inside they began to call for Atahualpa, searching very softly for him in all the corners. But seeing that he did not answer, they went out making a great lamentation. . . . I disabused them and told them that dead men do not return.'

Only after Atahualpa was dead did Soto and his reconnaissance party return. 'He brought news that he had seen nothing and that there was nothing.' 'They found no fighting man, nor any with arms, but everyone was at peace. . . . Therefore, seeing that it was a trick, an infamous lie and palpable falsehood, they returned to Cajamarca. . . . When they reached the Governor, they found him showing much emotion, with a large felt hat on his head for mourning, and his eyes wet with tears.' When he heard that Soto had found nothing, Pizarro 'was very sad at having killed him; and Soto was even more sad, for he said – and he was right – that it would have been better to have sent him to Spain: he himself would have undertaken to put him to sea. This would have been the best thing to have done with that Indian, for it was not possible to leave him in that country.'

It is fascinating to notice the changing attitudes of Spanish writers towards Atahualpa's death. The first wave of eyewitnesses – the official secretaries Francisco de Xerez and Pedro Sancho, the overseer Miguel de Estete, Cristóbal de Mena, and the town council of Jauja – all omitted any mention of Soto's reconnaissance and its negative findings. Xerez alone spoke vaguely of two native scouts being sent to investigate. These writers already felt unsure of themselves. In the face of possible censure, they closed ranks to insist on the reality of the danger from Rumiñavi's army.

Francisco Pizarro himself wrote to the Emperor Charles V, on 29 July 1533, to explain his actions. He said that he had executed Atahualpa because he had been advised that the Inca 'ordered a mobilisation of fighting men to come against [me] and against the Christians who went there and were present at his capture'.† Pizarro wrote the same explanation to his brother Hernando who was by then in Panama. Hernando repeated it without comment: 'According to what the Governor [Francisco Pizarro] writes me, it was learned that Atahualpa was making an assembly to wage war on the Christians; and he says that they executed him.'

The immediate outside reaction to the execution was highly critical. Licenciate Gaspar de Espinosa, the experienced Governor of Panama, wrote the King a letter that was carried by the ship in which Hernando

Pizarro travelled to Spain. Espinosa's reaction was typical of that of the Spanish court and of educated opinion throughout Europe. He was dazzled by Pizarro's achievements: 'the riches and greatness of Peru increase daily to such an extent that they become almost impossible to believe . . . like something from a dream.' But he was totally unimpressed by the circumstances surrounding Atahualpa's execution. 'They killed the cacique [Inca] because they claim that he had made a great assembly of men to attack our Spaniards. Because of this the Governor was persuaded – almost forced – to do it, with great pleas and demands put to him by Your Majesty's officials . . . In my opinion [the Inca's] guilt should have been very clearly established and proved, and there should have been no possible alternative whatever, before it became necessary to kill a man who had fallen into their hands and who had done no harm to any Spaniard or other person.' Espinosa knew that 'the greed of Spaniards of all classes is so great as to be insatiable: the more the native chiefs give, the more the Spaniards try to persuade their own captains and governors to kill or torture them to give more . . . But they can never achieve any of this with *me*.' Espinosa suggested that Atahualpa should have been exiled to other Spanish-occupied territories. 'They could have sent him to this city [Panama] with his wives and servants, as his rank deserved. There was no lack of his own gold [to pay for this]. We here would all have honoured him and treated him with as much deference as we would a great noble of Castile.'

The Emperor was moved by views as strong as these. He wrote Pizarro a cold despatch: 'We noted what you said about the execution you carried out on the cacique Atahualpa whom you captured.' Charles acknowledged Pizarro's claim that the Inca had ordered a hostile mobilisation. 'Nevertheless we have been displeased by the death of Atahualpa, since he was a monarch, and particularly as it was done in the name of justice. . . . After being informed about the matter, we shall order what is necessary.' The King was mindful of the sanctity of the divine right of kings. This was seriously undermined when a successful upstart like Pizarro could execute with impunity a man who had been one of the world's most powerful emperors.

Writers during the 1540s reflected this official disapproval by attacking Pizarro. Gonzalo Fernández de Oviedo roundly condemned the killing of 'so great a prince' as an 'infamous disservice to God and the Emperor: an act of great ingratitude . . . and outstanding evil'. Pascual de

Andagoya openly accused Pizarro and his officials of deception: they 'made Indian sorcerers who bore ill-will against Atahualpa declare that he had an army ready to kill them. Atahualpa replied that it was a lie . . . they should send someone to the plain where it was said that the army was [massing] to verify the story. Captain Soto set out for this purpose with some companions. But Pizarro and his councillors, as was prearranged, killed Atahualpa before Soto's return.' Juan Ruiz de Arce, writing privately for his own family, also accused Pizarro of 'practising deception on the conquistadores' by sending Soto off on a wild-goose chase.

About 1550 there was another change of emphasis. The first conquerors were by now established as glamorous, almost legendary heroes. Chroniclers therefore sought scapegoats, and ceased accusing Pizarro and his men of blatant fraud. The proof of Atahualpa's supposed guilt lay in the testimonies of natives interrogated about Rumiñavi's army. These interrogations were conducted through Indians whom the Spaniards had trained as interpreters. The chief interpreter, an ingratiating youth known affectionately as Felipillo, was arrested by Almagro in 1536 and garrotted for treachery to the Spaniards. Before dying he confessed other instances of having incited natives against the Spaniards.* Agustín de Zárate and the jingoistic Francisco López de Gómara developed a story that Felipillo had intentionally distorted the native testimonies in 1533. It was said that he was caught making love to one of Atahualpa's wives; only his value as an interpreter saved him from instant death for this act of *lèse-majesté*; he then engineered Atahualpa's execution to save his skin. The story provided the necessary scapegoat and saved the conquistadores' reputations. No one considered that officials as astute as Xerez and Sancho were unlikely to be deluded by diametrical mistranslation of such crucially important testimonies. The story of the false interpreter was therefore eagerly accepted. It originated in the 1550s and is still being retailed to this day.*

The reputations of the conquistadores were also partly redeemed by a tradition that Atahualpa's execution resulted from an orderly trial – not simply a hasty decision by the leading Spaniards. Fernández de Oviedo inadvertently started this concept by writing that 'they began to make a process, badly composed and worse written, its chief authors a factious, uncontrolled, dishonest priest, an amoral and incompetent scrivener, and others of the same stamp'. López de Gómara, whose *Hispania Victrix* glorified Spain's conquests, amplified this by saying that Pizarro had given the Inca a correct trial – a notion with strong appeal to the legalistic sixteenth-century Spanish mind.* Garcilaso saw the germ of a good story. In his *Comentarios reales*, published in 1617, he claimed that Gómara had scarcely done justice to the 'trial'. He himself inflated it into a 'solemn and very lengthy' affair complete with two judges, prosecuting attorney, defence lawyers, procurators, many witnesses (all of whose testimonies were of course distorted in translation), a twelve-point investigation, protracted courtroom debate, appeals to the Emperor Charles V, and the appointment of an official protector for Atahualpa. The indictment of the Inca by now included charges concerning his killing of Huascar, of the Cañari and other tribes, alleged human sacrifice and incestuous polygamy. Garcilaso also produced the names of eleven Spaniards said to have stood forward bravely in defence of the Inca. In this way, the trial, originally evolved by Gómara to whitewash the Spaniards' behaviour, was used by the eloquent Garcilaso to heighten the drama, irony and horror of Atahualpa's execution.* Garcilaso's 'trial' made an intensely dramatic story and it dies hard. Prescott repeated it, but with a note of caution: mystified that the Incas' great apologist should have stressed this legal foundation for Atahualpa's execution, Prescott concluded that 'where there was no motive for falsehood, as in the present instance, [Garcilaso's] word may probably be taken'. Later historians have been far less hesitant in repeating it. Markham, Helps, Means and others all waxed indignant over the 'judicial murder'. A book written in 1963 even produced a modern comparison: 'As in Stalinist Russia in our own times, the trial was mounted with an ostentation of correct forms.' Markham made much of Garcilaso's eleven defenders, 'the few men of honour and respectability' whose names were 'worthy of remembrance'. Hyams and Ordish praised the 'great moral courage' of this 'ten per cent' of the Spanish community at Cajamarca for being 'ready to stand up, at some risk to themselves from the majority and from the Establishment, for justice and mercy'. Unfortunately for a good story, it has been shown that only two of the eleven valiant men named by Garcilaso were actually in Cajamarca at that time.* The eyewitness accounts all in fact gave a quite different impression. The early chroniclers implied that most Spaniards favoured leaving the Inca alive or were indifferent: it was a determined minority headed by Almagro and the treasurer Riquelme who bullied Pizarro into permitting the tragic execution.

The existence of Rumiñavi's army will always remain an open question. There are various reasons for believing in it: the testimonies of the Inca nobles, the conviction of many Spaniards that it was approaching, the likelihood that Atahualpa would have summoned it in his desperation. It was also recorded that an Inca force entered Cajamarca soon after the Spaniards' departure, to remove the Inca's body and destroy the city that had seen the humiliation of the empire. Rumiñavi certainly did command an army that fought the Spaniards with distinction the following year, and he probably had it ready for an attack in 1533.*

But there are powerful arguments to support the traditional view that the Spaniards were thrown into panic by a chimera. The native testimonies are inconclusive: the witnesses may have been playing civil war politics against Atahualpa, they may have thought that they were being asked whether Rumiñavi was in a position to threaten Cajamarca, or they may simply have enjoyed kindling the Spaniards' obvious jitters. No Spaniard and no Indian in the Spaniards' service ever saw any hostile army. Rumiñavi, coming from Quito, would almost certainly have advanced southwards along the main highway, and if Soto made his reconnaissance as far as Cajas, as Pedro Pizarro said he did, he should have met any native force.*

More important, perhaps, than the truth behind the rumours was the effect that they produced on the Spaniards at Cajamarca. The men there felt vulnerable now that the gold was collected and they had failed to release the Inca. None of them had ever actually fought an Inca army, and half the camp consisted of new arrivals who had not even participated in the slaughter on the square of Cajamarca. It is therefore possible that many Spaniards were in the grip of panic, heightened by a mixture of paranoia, guilt and uncertainty. This panic, reinforced by the sudden corroboration by the Nicaraguan Indian, may momentarily have distorted Pizarro's judgement.

In normal circumstances the Spaniards would have opposed a native threat by mounting their horses and charging out against it. These were not normal circumstances. Whether or not Pizarro believed in the approach of Rumiñavi's army, he was obviously influenced by the expediency of destroying the captive Inca. Atahualpa found himself at the point of impact between two alien worlds. He had fully expected to be killed immediately after his capture, and this conviction grew again when he saw that instead of departing with the ransom gold the Spaniards at Cajamarca were reinforced by their compatriots. Some Spaniards thought that the Inca had ceased to be of value as a hostage and a co-operative puppet administrator. Everyone at Cajamarca now wanted to march to Cuzco. Almagro's men were desperate to make their own fortunes. By killing Atahualpa they ensured that any further treasure was not considered as part of his ransom to which they had no claim.* Now that Atahualpa had been cruelly and cynically deceived about his restoration, he could no longer be relied upon to endorse Spanish rule. During the long, difficult march to Cuzco, the presence of the Inca could be a magnet for attacks on the Spanish column; and his release to rule at Quito was unthinkable. Expediency demanded that the Inca be eliminated, and expediency was the only possible justification for his murder.

Many believed and believe the accusation first made by Juan Ruiz de Arce: that Pizarro became convinced of the need to dispose of Atahualpa and deliberately sent Soto on a futile mission to be rid of the Inca's leading champion. But the weight of evidence from men who were present was that Pizarro, already impressed by the arguments and demands of Almagro and Riquelme, suddenly gave way in a moment of excitement as night was falling on that fateful Saturday. The execution of a ruler was a momentous decision that should never have been made so hastily. Pizarro knew that the two men who had always been his principal captains, Hernando Pizarro and Hernando de Soto, would have opposed it. Atahualpa could be accused of no hostile act of any sort, despite the outrageous provocation he had had from the invaders. And the squalid public garrotting and ceremonial hypocrisy of the funeral did nothing to enhance his death. If the Inca had to be a political victim it would have been better to have disposed of him in the seclusion of his prison, or to have sent him to die in exile.

4

TUPAC HUALLPA

UNJUST and callous though it was, Atahualpa's execution *did* achieve the immediate purpose of Almagro and those who had insisted on it. The combined forces of Pizarro and Almagro were now free to push their conquest into the heart of the Inca empire. The first reaction to Atahualpa's death was one of relief among much of the local population. It seemed to them to mark an end to oppression by the Quitan victors of the civil war. Pizarro lost no time in consolidating the goodwill of the Huascar faction, for he was about to march towards Cuzco and wished to appear as its liberator. He knew by now the extent to which the administration of the empire depended on the Inca himself. And he had the good fortune to have the eldest remaining legitimate son of Huayna-Capac with him in Cajamarca. This was Tupac Huallpa, the younger brother of Huascar, and the man who had taken much care to avoid assassination by Atahualpa's followers after his arrival in Cajamarca.* Pizarro made sure that this puppet Inca would be crowned

An act of homage by Inca orejones to the King of Spain

with every traditional attribute of an Inca investiture. As soon as Atahualpa was buried, Pizarro 'immediately ordered all the caciques and chiefs who were residing in the city in the court of the dead ruler – there were many of them, some from remote districts – to assemble in the main square so that he could give them another ruler who would govern in the name of His Majesty'. They said that Tupac Huallpa was next in line of succession and would be acceptable to them.

The coronation started on the following day. 'The appropriate ceremonies were performed, and each [chief] came up to offer him a white plume as a token of vassalage, for this had been the ancient custom among them ever since the country had been conquered by the Incas. After this they sang and danced and held a great feast, during which the new cacique king did not wear any precious clothing and did not carry the fringe on his forehead, as the dead ruler had done. The Governor asked him why he did this. He replied that it was the custom of his ancestors for the man taking possession of the empire to mourn the dead ruler. They traditionally spent three days fasting, shut up inside a house, and then emerged with great majesty and ceremonial to hold a great feast.... The Governor told him that if this was an ancient custom, he should observe it.' The natives rapidly built a large house in which the new Inca was to make his retreat. 'Once the fast was over he emerged magnificently dressed and accompanied by a great crowd of people, caciques and chiefs who were guarding him; and any place where he was to sit was decorated with very valuable cushions and with tapestries beneath his feet. Atahualpa's great general Chalcuchima... was seated alongside the Inca, with Captain Tiso beside him and various brothers of the Inca on the other side. And so on in succession on alternate sides sat the other caciques, military commanders, provincial governors and lords of large districts. In short, no one was seated there who was not a person of quality.' 'They then received him as their lord with much humility, and kissed his hand and cheek, and turning their faces to the sun they gave thanks to it, saying that it had given them a natural lord. They then placed a very fine fringe on him, which is the equivalent of a crown among them.' 'They all ate together on the ground, for they do not use tables, and after they had dined the cacique said that he wished to do homage in the name of His Majesty, just as his own chiefs had done to him. The Governor told him to do whatever he saw fit, and the Inca then offered him a white plume that his caciques had given him . . . The Governor embraced him with great affection and accepted it.'

On the following day it was the Spaniards' turn to conduct the ceremonies. 'The Governor presented himself to the assembly dressed as finely as he was able in silk clothing, accompanied by the royal officials and some hidalgos* from his army who attended in their best clothes to add to the solemnity of the ceremony.' Pizarro spoke to each chief in turn, and had his secretary record their names and provinces. He then delivered a declaration known as the Requirement, in which Spanish captains were supposed to inform native populations that the conquerors had been sent by the Emperor Charles in order to bring them the teaching of the true religion, and that all would be well if they submitted peacefully to the Emperor and his God. 'He read all this out to them and had it proclaimed word by word through an interpreter. He then asked them whether they had fully understood it and they answered that they had. . . . The Governor then took the royal standard in his hands, raised it above him three times, and told them that they must each do the same.' The caciques all dutifully raised the royal standard, to the sound of trumpets. 'They then went to embrace the Governor who received them with great delight at their prompt submission. . . . When it was all over the Inca and the chiefs held great festivities. There were daily celebrations and entertainments with games and parties which were generally held in the Governor's house.'

While the native leaders celebrated what appeared to many of them as a restoration of their legitimate royal house, the Spaniards made final preparations for the conquest of central Peru. A number of the less adventurous conquistadores asked permission to return to Spain with their share of the treasure. Pizarro felt confident enough to let them depart. He gave them llamas and Indians to help transport their gold across the mountains to San Miguel. Francisco de Xerez was among them, and he reported the sad news that some Spaniards lost treasure to a value of over 25,000 pesos when some llamas and Indians ran off with it. The returning conquistadores sailed to Panama and thence in four ships to Spain. The first ship sailed up the Guadalquivir to Seville on 5 December 1533, carrying Cristóbal de Mena and the first Peruvian gold to reach Europe. Hernando Pizarro arrived on the second ship on 9 January with the first treasure for the King. In addition to the gold and silver that had already been melted down into bars, Governor Pizarro had thought to send some works of art, to demonstrate the sophistication of this unknown civilisation. There were 'thirty-eight vessels of gold and forty-eight of silver, among which was one eagle of silver whose body had a capacity of two cántaras [eight gallons] of water; two huge urns, one gold and one silver, each of which could hold a dismembered cow; two golden pots each of which could hold two fanegas of wheat; a golden idol the size of a four-year-old boy; and two small drums'. The treasure was unloaded on to the jetty of Seville, and transported by ox-cart to the Chamber of Commerce. Hernando Pizarro eagerly wrote to King Charles, on 14 January 1534, that he was bringing these precious objects, 'pitchers, vases and other rare shapes that are worth seeing'. He assured the King that no prince had ever possessed such a beautiful and fascinating collection. Even the Council of the Indies was excited. It wrote to the King: 'Since the news is so great, we beg Your Majesty to see [Pizarro's letters], and to make provision... whether it pleases Your Majesty that he should come before your royal person with the pieces of silver and other jewels he is bringing.'

The King's first reaction was negative: he ordered the Chamber to melt all but a few light objects and to mint the gold into coins immediately. But he partially reversed this order a few days later, permitting the collection to be displayed to the general public and allowing Hernando to bring him a few more pieces. One young man who viewed the works of art was Pedro Cieza de León. His imagination was fired. He later became one of the most important and perceptive chroniclers of Peru, but he always remembered 'the magnificent specimens exhibited in Seville which had been brought from Cajamarca'. Hernando Pizarro went to Toledo at the end of February, taking a small selection of objects. These included one huge silver urn and the two heavy gold pots (all of which were later sent 'to make coins'), two small gold drums, the bust of an Indian plated in gold and silver, and a golden stalk of maize. The King recorded no pleasure over these few lovely objects. They were entrusted to the royal jeweller after their brief public display, and were probably melted down, as were the other objects left in Seville.*

The return of the first conquistadores caused intense excitement. Hernando Pizarro was given a magnificent reception at Court, where he negotiated concessions highly favourable to Pizarro, and he then went on a recruiting drive to fire the enthusiasm of the young men of his native Extremadura. Mena and Xerez each produced books that rapidly became best-sellers and were translated into other European languages. Post-Renaissance Europe was dazzled by the discovery and sudden conquest of an unimagined empire of such brilliance.

The men at Cajamarca were now ready to abandon the city they had occupied for the past eight months and to set out towards Cuzco. They were attempting one of the most staggering invasions in history. Without supplies, communications or reinforcements, this tiny contingent was going to try to force its way into the heart of an enormous hostile empire, to seize its capital city. The road from Cajamarca to Cuzco lies along the line of the central Andes. It crosses and recrosses the watershed between the Amazon basin and the Pacific, and traverses half a dozen subsidiary ranges of mountains and wild torrents. The distance as the crow flies between the two cities is some 750 miles, and the journey was comparable to travelling from Lake Geneva to the eastern Carpathians or from Pike's Peak to the Canadian border, in each case following the line of the mountains.

Pizarro, Almagro, Soto and their men marched out of Cajamarca on 11 August 1533.* Progress was uneventful in the early stages. Two days were spent at Cajabamba and four at Huamachuco. The army made its way through the rolling country between Cajamarca and the hills above the Huaylas valley, an area without spectacular views, but green and nowadays relatively wooded for Peru, with low, gnarled native Peruvian trees and plantations of tall imported eucalyptus. Two ruins survive at Huamachuco: close to the colonial town is a compact ruin whose walls intersect at right angles and whose rectangular enclosures possibly served as an Inca army camp;* and on a rocky ridge above the town is a wild series of mud and fieldstone walls towering out of a tangle of bushes and brambles. This is Marca Huamachuco, a citadel dating from the period before the Inca conquest when Peru was divided into city states.* From Huamachuco Pizarro's force marched on to Andamarca, the town where Huascar was killed by Atahualpa's men, and rested there for three days. They decided not to take the main highway through the Conchucos to the east of the Cordillera Blanca because of its many hills. They descended instead into the deep valley of Huaylas, at the point where the turbulent Santa river turns westwards to slice through dry gorges of pink rock towards the Pacific.

Huaylas was reached on the last day of the month, after its river had been crossed on one of the famous Inca suspension bridges. 'At a point where the rivers are narrowest and most terrifying, and their waters most compressed, they make a great stone foundation on either bank. Thick wooden beams are laid across this stonework, and they fasten across the river cables of a thick osier, made like anchor ropes except that these cables are each some three palmos [3½ feet] thick. When half a dozen of these have been joined and laid across the river, to the width of a cart, they are interwoven with strong hemp and reinforced with sticks . . . When this is done, they place edges on either side like the sideboards of an oxcart. And so it lies suspended in mid-air, far above the water. It seemed impossible to make the horses – animals that weigh so much and are so timorous and excitable – cross something suspended in the air. . . . Although they refused at first, once they were placed on it their fear apparently calmed them and they crossed one behind the other, and there was no mishap at this first bridge.' Pedro Sancho recalled how this first crossing terrified him: 'to someone unaccustomed to it, the crossing appears dangerous because the bridge sags with its long span . . . so that one is continually going down until the middle is reached and from there one climbs until the far bank; and when the bridge is being crossed it trembles very much; all of which goes to the head of someone unaccustomed to it'.

Pizarro's men rested for eight days at Huaylas before marching up its radiant valley. It was warm in the valley itself, with a profusion of wild flowers, good crops of maize and nowadays even palm trees. But the sides of the valley rise steeply and evenly to towering ranges of mountains on either side: to the west the bare crests of the Cordillera Negra, and to the east the permanent snows of the Cordillera Blanca crowned by Peru's highest mountain, Huascarán.* The valley walls are too steep: for the high mountain tarns occasionally burst their banks of glacial moraine and the hillsides collapse in murderous landslides. Pizarro's men were in no hurry to leave the valley, and spent twelve days of mid-September resting at Recuay. Beyond here one Inca road ran down the Fortaleza valley to reach the coast near the great adobe fortress-temple of Paramonga. But Pizarro followed the higher road that skirted the mountains to the south east, climbing across the heads of the Pativilca and Huaura rivers through Chiquián, Cajatambo and Oyón.*

The Spaniards were now almost half-way to Cuzco, and there had been few difficulties during the eight weeks since they left Cajamarca. The part of Peru through which they passed had been strongly loyal to Huascar's faction, and 'as far as Cajatambo the caciques and lords of the road gave the Governor and the Spaniards a good reception, providing all that they needed'. Despite this, the conquistadores had advanced with caution, 'always using great vigilance . . . and always maintaining a vanguard and rearguard'. Riding in litters in the midst of the Spanish column were the leading survivors of the two sides of the civil war: the young new Inca Tupac Huallpa and Atahualpa's great general Chalcuchima. The former thought that he was being borne towards a restoration of his family on the imperial throne at Cuzco, and was an eager collaborator with the conquistadores. But Chalcuchima had seen himself lured away from his command at Jauja, tortured on arrival at Cajamarca, and had witnessed the execution of his lord Atahualpa. It was hardly surprising that the Spaniards should have feared and suspected every move by this formidable figure. No sooner had they left Cajamarca than they learned that a friendly Inca prince, Huaritico, whom they had sent ahead to ensure the repair of bridges and preparation of the route, had been killed by Quitan troops for his collaboration. 'The cacique [Tupac Huallpa] showed great grief at his death and the Governor himself regretted it because he had liked him and because he was very useful to the Christians.' Chalcuchima was blamed for the killing.* The Incas maintained storehouses along the highways to feed passing imperial armies. When some of these were found to be empty Chalcuchima was again blamed; but he protested that this inefficiency resulted from having Tupac Huallpa in charge of the native part of the expedition.* The suspicions about Chalcuchima heightened as the invaders marched closer to his former headquarters at Jauja. Cajatambo and Oyón were found to be almost deserted: their inhabitants had fled to avoid Atahualpa's troops. An Indian now reached the column with news that Chalcuchima's former army at Jauja was under arms and preparing to resist under the command of one Yucra-Hualpa. Quitan patrols were trying to prevent word of these preparations from reaching Tupac Huallpa. Pizarro decided to ensure that Chalcuchima could not escape to lead the resistance, and had him put in chains. Part at least of the Peruvian nation was finally trying to oppose the invasion. As Pizarro's secretary naively explained: 'The reason why these Indians rebelled and were seeking war with the Christians was that they saw the land being conquered by the Spaniards, and they themselves wished to govern it.'

Pizarro's men now left the eerily empty towns on the Pacific side of the Andes, and marched up through the same desolate pass that Hernando Pizarro had crossed in March. There was still snow on the ground, and some men suffered from the terrible nausea of the altitude sickness, soroche. To the east of the pass the country continued to be bare, a plateau of wet treeless savannah and lichen-covered rocks. The anxieties of the invaders were heightened when yet another village was found deserted. There were further reports of Quitan troops massing ahead. 'It was considered certain that this force had been moved on the advice and at the orders of Chalcuchima – he intended to escape from the Christians to go to join it.' By Tuesday 7 October, the Spaniards rejoined the main imperial highway at Bombón on Lake Chinchaycocha (modern Junín).

In view of the increasing rumours, Pizarro decided that he must accelerate the advance. He therefore left the unwieldy part of his column – the infantry, artillery, precious metals, even the tents – to proceed more slowly under the command of the royal treasurer Alonso Riquelme. Pizarro himself pressed ahead with the best 75 horsemen and his able lieutenants Diego de Almagro, Hernando de Soto, his own brother Juan, and Pedro de Candía, with a special contingent of twenty foot-soldiers to guard the chained Chalcuchima.*

The Inca highway ran further to the east than does the modern dirt road: it climbed over the hills and down into the tight warm valley of Tarma. It was all ideal topography for an ambush. 'The pass proved difficult – it looked as though we could never climb it. There was a difficult stretch of rock to descend into the canyon where all the horsemen had to dismount. Afterwards we had to climb to the top of an ascent, most of which was precipitous, difficult mountainside.' Modern Tarma is a pretty town closely surrounded by hillsides of flower nurseries. But Pizarro feared that its cramped location would leave his horses no room for manoeuvre. He paused only long enough to feed the horses and pushed on to spend the night of 10 October on an exposed hillside. Sancho recalled it vividly. The men 'remained continuously on the alert, with the horses saddled and the men themselves unfed. They had no meal whatsoever, for they had no firewood and no water. They had not brought their tents with them and could not shelter themselves, so they were all dying of cold – for it rained heavily early in the night and then snowed. The armour and clothing they were wearing were all soaked.' Next day the weary men rode on through Yanamarca and saw the corpses of over four thousand natives killed in one of the battles of the civil war – a further reminder of the fighting qualities of the Quitan professional soldiers.* They moved through hills covered with the ruins of pre-Inca settlements of Huanca Indians, and at last looked down on to the surprisingly level Jauja valley, with the Inca city cradled between sharp hills at its northern end.*

They also saw below them dark masses of Quitan soldiers, the men once commanded by their captive Chalcuchima and now led by Yucra-Hualpa.* But as they advanced down into the valley there was a vivid illustration of the suicidal cleavage that was paralysing Peru. 'The natives all came out on to the road to look at the Christians and greatly celebrated their arrival, for they thought that it would mean their escape from the servitude in which they were held by that foreign army.' Meanwhile the 'foreign army' of Quitan Incas was bracing itself for an act of resistance – the first military action in the seventeen months since the Spaniards landed on the Peruvian mainland or in the two months since they left Cajamarca. The bulk of the native army was massed on the far bank of the Mantaro river. But six hundred soldiers had been sent into Jauja in a last-minute attempt to destroy the city's great storehouses. As the first two Spanish horsemen rode into Jauja they met armed natives running between the houses. The Spaniards reacted with the tactic whose effectiveness they had learned in their conquests in Mexico and Central America. They charged immediately. There was a skirmish in the narrow alleys of the town, and as more Spaniards galloped in the native troops were driven back to the river. They had just succeeded in setting fire to the thatched roof of one large storehouse and some other buildings. Juan Ruiz de Arce recalled that they entered Jauja just as the city was catching fire. Pedro Pizarro remembered that various golden vases were recovered from the embers of the burned storehouse, and Martín de Paredes and Toribio Montañés both wrote from San Miguel that Pizarro had taken '300,000 pesos of burned gold in Jauja'. Almagro continued the pursuit by driving his horses into the river, which was beginning to rise at the start of the rainy season. The Indians on the far bank were in two minds as to whether to stand and fight or flee to defensible positions. Some fled northwards into the hills, others attempted to fight and were cut down. The battle ended in a field of maize at the river's edge, with the slaughter of the frightened warriors who had tried to take refuge there: 'On investigation it was found that not fifty had escaped.'

The native army must have been demoralised by this first savage encounter. Its leaders determined to march south to try to join Quisquis's forces at Cuzco. But the Spaniards again acted quickly. After resting the exhausted men and horses for only one day, Pizarro sent eighty mounted men in pursuit. Riding hard, the invaders soon reached the Indians' camp where the fires were still smoking. The great column of the native army was moving down the broad Mantaro valley a few miles beyond, with its soldiers marching 'in squadrons of a hundred with the women and serving people between the squadrons'. The rearguard – 'a squadron of good men' – put up a resistance, but when they were ridden down the rest of the troops ran for cover in the rocky hills bordering the valley. Many men were too slow, and the Spaniards were merciless. 'The pursuit continued for four leagues [sixteen miles] and many Indians were speared. We took all the serving people and the women ... there was a good haul of both gold and silver.' Herrera noted that the captives included 'many beautiful women, and among them two daughters of Huayna-Capac'.

Francisco Pizarro stayed at Jauja for two weeks, from Sunday 12 October to Monday the 27th. A week after his arrival the slower-moving infantry under Riquelme reached the town with the army's baggage and treasure. There was much activity during this brief stay. Jauja was tentatively 'founded' as a Spanish municipality and designated as the first Christian capital of Peru. Eighty Spaniards, half of whom had horses, were to be left in the town as its first citizens, and buildings were earmarked for a church and a town hall. Now that the invaders had encountered organised resistance, Pizarro decided to pare down his army, leaving the less useful elements to guard the treasure at Jauja. The royal treasurer Riquelme was also left behind: Pizarro preferred to be unencumbered by his advice and free from his observation; Riquelme did not mind being left in the rear of the fighting. Since many of the conquistadores were leaving their hoards of gold, there was a flurry of will-making and other legal activity by those about to strike deeper into unconquered Peru.*

During the stay at Jauja, the young Inca Tupac Huallpa died of an illness that had been weakening him since Cajamarca. The Spaniards were deeply distressed at the loss of this compliant puppet. They looked for a scapegoat. The leading citizens of Jauja wrote to the King: 'It was commonly supposed that Captain Chalcuchima had given him herbs or some drink from which he died – although there was no proof or certainty about this.' The young Inca probably died from natural causes, although Chalcuchima had good reason to kill this collaborating member of the Cuzco branch of the royal family. The death was a real embarrassment for the Spaniards. Pizarro had chosen Tupac Huallpa as the man most acceptable to Atahualpa's chiefs, and he now had no idea whom to elevate. He was unaware that Peru was seething with plots to crown other claimants. In Quito, Atahualpa's military commanders were considering the coronation of the Inca's brother Quilliscacha while the general Rumiñavi was about to seize power for himself. In Cuzco Quisquis was rumoured to have offered the royal fringe to Atahualpa's brother Paullu, who had shown certain sympathies to the Quitan cause.* Pizarro hastily summoned a meeting of native chiefs, including the two generals Chalcuchima and Tiso. The meeting reached deadlock, with two possible candidates for Inca. The Huascar faction suggested a full brother of the dead Tupac Huallpa, presumably one called Manco who was in the Cuzco area; the Quitans favoured Atahualpa's young son. Pizarro tried to give secret encouragement to both sides. He told Chalcuchima that he would make him regent if he could lure Atahualpa's son to the Spanish camp for coronation. Chalcuchima said that he would send envoys to Quito to try to bring the boy. Both leaders were probably lying and the proposal came to nothing.*

This haggling over the succession revealed how low the majesty of the Inca had sunk since the outbreak of the civil war and humiliations of Atahualpa. As the Inca lost prestige, so did the entire ruling caste of Peru. Another disruptive tendency was the decline of Cuzco and the resurrection of regional centres and tribal capitals. The Inca empire was just as much the triumph of one tribe and city as of a ruling dynasty. Cuzco was therefore fostered as spiritual and administrative heart of the empire, just as Rome and Byzantium had been in their day. It contained the superb residence of each successive Inca, a pantheon of mummified rulers, an enormous central square for daily ceremonial, a court attended by representatives of every assimilated tribe and the administrative councils of the Inca's court. The Inca language, Quechua, was imposed throughout the empire and has proved the most durable legacy of the Incas: over half the present population of Peru speaks it as first language.

Cuzco was equally paramount as the religious heart of the empire. It contained the chief temples of the official creator god Viracocha and of the sun and moon worship that the Incas were trying to impose in place of local tribal deities. The animism that had existed before this sun worship was still preserved in innumerable shrines and holy places – rocks, springs, caves, trees – throughout the valley of Cuzco. These included the hills of Huanacauri and Qenco and the cave of Tambo-toco at Pacaritambo. All these holy places were now identified with the legend of the acquisition of Cuzco by the Incas' ancestor Manco Capac.* The Incas showed skill and tact in dealing with the deities of conquered tribes. Portable gods and holy objects were accorded the honour of being transferred to Cuzco with their attendant priests. Once there, they acted as hostages for the good behaviour of their tribes. Cuzco also formed the focal point of many religious ceremonies that studded the Inca calendar. At the start of the rains in December there was the ceremony of Capac Raymi, in which foreigners were required to leave Cuzco while adolescent Inca nobles underwent their coming-of-age initiations. May witnessed Aymoray celebrating the maize harvest, and in June there was Inti Raymi, the important festival in honour of the sun. The Situa or Coya Raymi in September was a ceremony of purification: gods of subject tribes participated in the main ceremony, and relays of runners radiated from Cuzco throughout the empire performing symbolic rites of exorcism. Ploughing could take place in outlying provinces only after the Inca had first broken the ground at Cuzco with a golden hand-plough.*

Cuzco's pre-eminence was temporarily eclipsed by Huayna-Capac's long residence in the north, but Atahualpa had been on his way to be crowned and to rule in the former capital. The civil war and the Spaniards' capture of Atahualpa damaged the prestige of Cuzco as well as of the Inca dynasty and tribe. The result was a resurgence of tribes that had been only half digested into the empire. This centrifugal tendency was only just beginning. The Spaniards saw the first signs of it in the hostility of the Jauja Huancas towards the Inca army of occupation. Regionalism and tribalism became increasingly important with the melting away of recently-imposed Inca systems of government. It was invaluable to the Spanish invaders – just as useful as the dynastic schism of the civil war, and the indifference of the native masses to the fate of the upper classes of Inca society.

Another factor commonly thought to have worked in the Spaniards' favour was the natives' identification of them with the returning creator god Viracocha. There is little evidence to support this idea. Atahualpa and his military commanders clearly regarded the Spaniards as ordinary mortals and had little hesitation in fighting them. None of the contemporary accounts of the Conquest showed that the native leaders hesitated for fear that the intruders might be divine. Atahualpa said after his capture that he had allowed the Spaniards to penetrate as far as Cajamarca only because of their small numbers. For the peasants the Spaniards were awesome strangers, but not divinities.

The legend of divinity grew when later chroniclers noticed similarities between Inca origin myths and their own biblical stories. Two of the most conscientious, Pedro de Cieza de León and Juan de Betanzos, both of whom wrote shortly after 1550, were struck by the fact that the natives called Spaniards 'Viracochas' like their god. 'When I asked the Indians what this Viracocha was like when the ancients saw him . . . they told me that he was a tall man who had a white robe reaching down to his feet; that he wore this robe with a belt; that his hair was short and there was a crown around his head in the manner of a priest; and that he carried in his hands something that now appears to them like the breviaries that priests carry in their hands.' Viracocha was 'a white man, large of stature, whose air and person aroused great respect and veneration. . . . The Indians further relate that he travelled until he came to the shore of the sea, where, spreading his cloak, he moved on it over the waves, and never appeared again nor did they see him. Because of the manner of his departure they gave him the name of Viracocha, which means "foam of the sea".' Cieza wrote that the name Viracocha was first applied to the Spaniards by Huascar's followers, to whom the conquistadores appeared as god-like deliverers from Atahualpa's Quitans. Atahualpa's own nephew agreed: 'They seemed like Viracochas, which was our ancient name for the universal creator.' And they had appeared mysteriously from the same sea into which the creator god had disappeared.

Pizarro and his Spaniards were only dimly aware of the forces that may have been acting in their favour, apart from the obvious dynastic struggle that they tried to exploit. To them Cuzco was still the all-important hub of the empire. Natives spoke of the city with veneration, and the three Spaniards who had been there gave spellbinding descriptions of its treasures. Cuzco became an irresistible magnet for every man in Pizarro's force. Its inaccessibility and its armies of native defenders were discounted in a passionate frenzy to conquer this supreme prize.

5

THE ROAD TO CUZCO

PIZARRO's small force now embarked on the most thrilling part of its conquest, the final march from Jauja to Cuzco. The total force involved, after a garrison of the more feeble had been left at Jauja, was 100 horse and 30 foot, together with some native auxiliaries. Pizarro had a reasonable idea of the country that lay ahead. The three Spaniards who went on the reconnaissance to Cuzco in April had carefully recorded the towns and physical features along the route. This central section of the Andes is wild, magnificent country, a vertical land of mountains deeply cut by fierce rivers plunging towards the Amazon. The topography changes with altitude, descending from snow-capped mountains to bare, misty puna high above the tree-line, down to pretty Andean valleys full of maize and flowers, and on down to suffocating heat and cactus in the

Atahualpa's great general Chalcuchima firing his sling

depths of the canyons. The road from Jauja runs for a time alongside the Mantaro river, constantly climbing into and out of the valleys of its tributaries. The Mantaro turns abruptly northwards towards the Amazon, and the road to Cuzco has to continue across a succession of great rivers separated by ranges of hills.

This mountainous region would have been almost impassable had it not been for the Incas' own superb roads. An efficient people, the Incas excelled in civil engineering and depended on roads to control the empire. The main royal highway ran along the line of the Andes, from Colombia through Quito, Cajamarca, Jauja and Vilcashuaman to Cuzco, and on through modern Bolivia to Chile. A parallel highway followed the Pacific coast, and the two were joined by many lateral connections, particularly from Cuzco to the coast through Vilcashuaman. Europe had seen no roads like these since Roman times. Hernando Pizarro wrote, 'the mountain road really is something worth seeing. Such magnificent roads could be seen nowhere in Christendom in country as rough as this. Almost all of them are paved.' With no draught animals or wheeled vehicles, the Incas built their roads only for walking men and trains of llamas. The roads climbed the slopes of the Andes with flights of steps or tunnels unsuitable for the horses. They were well graded and often supported by fine stone embankments to cross steep mountainsides or marshes. Pedro Sancho described the terrifying ascent of Parcos that Pizarro's force had to climb four days after leaving Jauja. After fording the river we 'had to climb another stupendous mountainside. Looking up at it from below, it seemed impossible for birds to scale it by flying through the air, let alone men on horseback climbing by land. But the road was made less exhausting by climbing in zigzags rather than in a straight line. Most of it consisted of large stone steps that greatly wearied the horses and wore down and hurt their hooves, even though they were being led by their bridles.' The Inca roads were narrow, averaging only some three feet in width in difficult mountain country, but the flagstone paving was good, and so were the long flights of steps that worried the Spanish horses.

The Inca government built posthouses at regular intervals along the roads for the use of its officials, porters and armies. These tambos consisted of sleeping shelters and rows of rectangular storehouses, and the local population had to keep them serviced and provisioned. Important messages and loads were carried along the highways by relays of chasquis, runners stationed in huts some four to five miles apart. The chain of chasquis, each running at full speed, could carry messages across the country extraordinarily quickly. But the messages themselves had to be oral or quipu string records because of the Incas' lack of writing.

The road from Jauja to Cuzco crossed the river barriers in a series of fibre-rope suspension bridges. Pizarro hoped that he could capture some of these bridges before the retreating Quitans had time to destroy them. The cavalry force that overtook the Inca column just outside Jauja was supposed to have pressed ahead to secure the first bridge, but it turned back because of the exhaustion of the horses and lack of fodder. Now that the army was rested and ready for the final expedition, Pizarro sent his seventy best horsemen ahead under Hernando de Soto to try to take the bridges. He himself and Almagro followed with the other thirty horse and thirty foot-soldiers guarding Chalcuchima. Soto left Jauja on Thursday 24 October, and Pizarro the following Monday. Of our various eyewitness informants, Pedro Pizarro, Diego de Trujillo and Juan Ruiz de Arce were with Soto, while the meticulous Pedro Sancho and Miguel de Estete were with their master Francisco Pizarro.

The Inca force that had been caught marching down the Jauja valley continued southwards to join forces with the main Quitan army that was occupying Cuzco. Its commanders were determined to prevent the Spaniards from reaching Cuzco, and equally keen to preserve the Atahualpan faction's hold on the imperial capital. This is why they moved deeper into the Inca empire instead of retreating northwards towards their base at Quito. It was a courageous decision, for they were well aware that the local population would rise against them as they departed, and they were leaving an ever-greater stretch of hostile country between themselves and their homeland. When they burned the bridges against Pizarro's advance they were also burning them against their own retreat.

The civil war was still the issue of the moment. By killing Atahualpa the Spaniards had cast themselves as champions of Huascar's cause. The native population welcomed them as such, and the Quitans probably fought them more as the champions of their defeated opponents than as the spearhead of a foreign invasion. Pizarro was of course well aware of this attitude which he exploited continually. His soldiers often enjoyed a reception as liberators. This was especially true at Jauja where the inhabitants were merciless in hunting down any Quitan survivors and handing those they found over to the Spaniards.* The Quitans in revenge adopted a scorched earth policy as they moved southwards. The burning of strategic suspension bridges was an obvious tactical move, but the burning of villages and food stores along the line of march was the parting blow of a retreating army of occupation. This destruction made the Spaniards' advance uncomfortable, but any inconvenience was far outweighed by their gaining the support of the local population.

Below Huancayo the Mantaro river drops into an ugly gorge and runs for some sixty miles between walls of crumbling yellow clay and outcrops of black rock. The Inca road crossed this canyon near its upper end, and the Quitans very properly burned the suspension bridge. But they did not discover that the bridge's guardians had hidden their supplies of repair materials. When Soto's men reached the place, the guards were able to build a rapid temporary bridge, and this structure was also able to carry Pizarro's contingent, even though the hooves of Soto's horses had left it full of holes. On the night after this crossing, Pizarro's men camped in a deserted village that had been burned and sacked by their retreating enemies. They were without water, for the Quitans had destroyed the aqueducts. The following night they reached a village called Panaray and were dismayed to find no inhabitants or food there either, despite the fact that its chief had marched with them from Cajamarca to Jauja and had gone ahead to prepare supplies in his village. It was only on the following day, at Parcos, that their hardships were eased by a chief who provided them with badly needed food, maize, wood and llamas.

The Quitan occupation of southern Peru was based on the cities. The army that retreated from Jauja therefore made its next stop at Vilcashuaman, the next administrative centre, 250 miles to the south-east. Soto's men covered this distance in only five days, meeting no resistance on the way. They camped five hours' march before Vilcas and rode into the city at dawn on 29 October. Once again the speed of their movements caught the Indians by surprise, and they eluded the sentries posted at the approaches to the town. The Quitan warriors were away hunting. 'They had left their tents, their women and a few Indian men in Vilcas, and we captured these, taking possession of everything that was there at the hour of dawn, which was when we entered Vilcas. We thought that there were no more troops than those who had been there then. But at the hour of vespers, when the Indians had been informed [of our arrival] they came from the steepest direction and attacked us, and we them. Because of the roughness of the terrain, they gained on us rather than we on them, although some Spaniards distinguished themselves – for instance Captain Soto, Rodrigo Orgóñez, Juan Pizarro de Orellana and Juan de Pancorvo, and some others who won a height from the Indians and defended it strongly. The Indians on that day killed a white horse belonging to Alonso Tabuyo. We were forced to retreat to the square of Vilcas, and all spent that night under arms. The Indians attacked next day with great spirit. They were carrying banners made from the mane and tail of the white horse they had killed. We were forced to release the booty of theirs that we were holding: the women and Indians who were in charge of all their flocks. They then withdrew.'

Soto in a dispatch to Francisco Pizarro described how he had been reluctant to fight in the difficult country surrounding Vilcas, but had eventually left ten men in the town and had led thirty others through a defile and down a difficult slope. The enemy's bold attack had killed one horse and wounded two others, as well as wounding a number of Spaniards. But, although the natives recaptured their baggage, the battles cost them over six hundred dead, including one of their commanders called Maila. Despite its bravery, the Quitan army had been defeated in its first two clashes with the Spaniards. The only consolation was that the dreaded horses were seen to be mortal, and enough was now known of Spanish tactics for the preparation of an ambush to annihilate them. With this in mind the Indian army marched eastwards to join its companions at Cuzco. 'Counting those who went, those who remained there, and the natives of the district, a vast quantity of Indians was assembled. We all agreed that there could have been 25,000 Indian warriors.'

Vilcashuaman is on a plateau, a promontory overlooking the deep cleft of the Vischongo river a few miles from its junction with the larger Pampas river. The country above Vilcas is rolling, treeless puna, the present-day home of the Morochucos, a race of remarkable horsemen said to be directly descended from the conquistadores themselves. A modern visitor sees them riding their tough ponies across their pale green steppes, and every year their spectacular horsemanship provides the main attraction at the fairs during Ayacucho's Holy Week. Around Vilcas itself the valleys are richer, full of wild flowers, maize-fields and clusters of tuna prickly pears. The village has no motor road: a visitor must walk the final few miles after a half day's drive from Ayacucho. No one has excavated Vilcashuaman, and it is full of half-buried Inca terraces and palace walls. One fine stretch of coursed masonry runs along the lower part of the façade of the village church. At one edge of the village is a stone step pyramid, the only surviving Inca structure of its kind. It was either a sun temple, or an usno, a mound on which the Inca presided over his court (plate 13).*

Below Vilcashuaman the canyons plunge toward the hot, airless bed of the Pampas river, 6,000 feet below. Pizarro's and Almagro's men spent most of 6 November making their way down this spectacular descent, with the stone steps of the Inca highway cutting their horses' hooves. They just succeeded in swimming the river. The Quitans had destroyed the bridge, but did not remain to contest the crossing.

Soto now decided to disobey his instructions by leaving Vilcashuaman before the arrival of his compatriots. In his letters to the Governor he explained that he wanted to race ahead to try to capture the Apurímac bridge, to prevent a junction between the Jauja army and Quisquis's force. Diego de Trujillo and Pedro Pizarro who were with him gave a different explanation: Soto, Orgóñez and various other hotheads decided that 'since we had endured the hardships, we should enjoy the entry into Cuzco without the reinforcements that were coming behind'. Because of this disobedience and greed, wrote Pedro Pizarro, 'we were all almost lost'.

Soto's advance went smoothly for some days. He crossed the Pampas, Andahuaylas and Abancay rivers unopposed. Quisquis had sent a force of 2,000 men to reinforce the troops of Chalcuchima's command, but these turned back when they met the army retreating from Vilcas. Pizarro followed a few days behind Soto. Two days after leaving Vilcas he decided to split his force yet again, sending Almagro ahead with thirty horse to overtake and reinforce Soto. He himself continued the march with only ten horse and twenty infantry guarding the wretched Chalcuchima. The following day his men were alarmed to find two dead horses, but Soto had left a message explaining that these died from exposure to the great extremes of heat and cold – the Spaniards were not aware that the army was also being affected by the high altitudes of the Andes.

The Spaniards now had to cross the greatest obstacle of all, the mighty canyon of the Apurímac, the river whose Quechua name means Great Speaker. The Inca highway crossed this gorge on a high suspension bridge, the bridge whose successor was the subject of Thornton Wilder's novel, *The Bridge of San Luis Rey*. The approaches to the ancient structure can still be seen half-way up the sides of the valley, with the narrow Inca road running through a tunnel before it turns, past massive stone buttressing, out into the present void. The bridge was burned by the time the Spaniards reached it, but they were able to ford the river despite its strong current and slippery stone bed. The water came up to the breasts of the horses. The conquistadores were extraordinarily fortunate to cross these mighty rivers with such ease. Herrera, Philip III's official historian of the Conquest, wrote: 'It was remarkable that they crossed the rivers with their horses even though the Indians had dismantled the bridges, and although the rivers are so powerful. It was a feat that has never been seen since, particularly not with the Apurímac.' The conquistadores' luck was partly due to the fact that they made their march at the driest time of the year, just before the start of the rains. A few months later the rivers they swam and forded would become swirling grey torrents rising high up the walls of their canyons.

No Indians opposed the crossing of the Apurímac. Soto and his men pressed eagerly onward towards Cuzco. On the eastern bank of the Apurímac the ground rises in stages, with a series of steep ascents interspersed by more gradual climbs up tributary valleys. On Saturday 8 November, at noon, Soto's men began to climb the last of the steep hillsides, the ascent of Vilcaconga. 'We were marching along with no thought of a line of battle', wrote Ruiz de Arce. 'We had been inflicting very long days' marches on the horses. Because of this we were leading them up the pass by their halters, marching in this way in groups of four.' Men and horses were tired by the midday heat, which was intense. They stopped to give the horses some maize that had been brought to them by the inhabitants of a nearby town. There was a small space with a little brook, half way up the slope. Just before the Spaniards reached this, Hernando de Soto, who was a crossbow-shot's length ahead of the others, saw the enemy appear along the top of the mountain. Three or four thousand Indians came charging down, completely covering the hillside. Soto shouted to his men to form line of battle, but it was too late. The Indians were hurling a barrage of stones before them, and the Spaniards' first reaction was to scatter to avoid these missiles. They ran to either side of the path, and those who had time to mount spurred their horses up the hillside, thinking that they would be safe if they could reach the level ground at the top. 'The horses were so exhausted that they could not catch breath sufficiently to be able to attack such a multitude of enemy with any dash. [The natives] never stopped harassing and worrying them with [the barrage] of javelins, stones and arrows they were firing. They exhausted them all to such a degree that the riders could hardly raise their horses to a trot, and some not even to a walk. As the Indians perceived the horses' exhaustion, they began to attack with greater fury.' Five of the Spaniards – an eighth of the total force – were caught in the full crush of Indians before they could reach the top. Two were killed on top of their horses; the others fought on foot, but were cut down before they were seen by their companions. One man had failed to draw his sword to defend himself, and because of this the Indians were able to grab the tail of the horse above and to prevent its rider from passing forward with the rest. For once the Inca soldiers had caught the Spaniards in hand-to-hand fighting – the only form of battle that they knew well. At such close quarters they were able to use their armoury of clubs, maces and battle-axes. The five or six Spaniards killed in the battle all had their heads split open by such weapons.

Pizarro was by now leading an army of rich men, for all had shared on one of the most successful looting operations in history. One of the men killed at Vilcaconga was Hernando de Toro; his estate was valued on Pizarro's orders and found to consist of thirteen slabs of 15-carat gold, to a total of 4,190 pesos. Toro was a dashing young hidalgo, one of the most popular men in the army, but he had encouraged Soto to press ahead to occupy Cuzco. Another victim, Miguel Ruiz, left 5,873 pesos, of which 3,905 were in gold and the balance in IOUs from two fellow-conquistadores. The other dead, Gaspar de Marquina, Francisco Martín Soitino, Juan Alonso and a man called Hernández, had all received shares of the distributions of treasure up to that time.

The day's fighting ended with an attempt by Soto to lure the natives down on to the level plain. He ordered his men to retreat step by step down the mountainside. Some Indians pursued, firing stones from their slings, and the Spaniards were able to turn on these and kill some twenty of them. But the bulk of the Inca troops retired up the hill, and the exhausted Spaniards followed up to camp for the night on a hillock, with, as Ruiz de Arce vividly recalled, 'very little victory and plenty of fear'. The Indians camped a couple of crossbow-shots' distance away, on a higher hill, shouting abuse at the frightened band of invaders. The Spaniards spent the night armed and few slept. Soto posted sentries and saw that the eleven wounded men and fourteen wounded horses received attention – although it is difficult to imagine what can have been done for the battered and bleeding men on that cold, exposed hillside. He also tried to raise the spirits of his men with speeches of encouragement.

None of Soto's words can have produced an effect comparable to that of a sound heard by the sleepless Spaniards at one o'clock in the morning. It was the call of a European trumpet. The thirty horse that Pizarro had sent ahead under Almagro joined ten others left behind by Soto to escort the loot captured at Vilcashuaman. This combined force had heard about the battle and was advancing through the night with its trumpeter Pedro de Alconchel sounding his horn like a ship's siren.

The Indians greeted the new day confident of victory, only to find that the battered force of the previous afternoon had miraculously doubled. The jubilant Spaniards formed line of battle and advanced up the hillside. The Indians retreated and any who remained on the slope were killed. The arrival of a thousand men from Cuzco did not save the situation, and the natives' only salvation was the descent of a ground mist. Such silvery mists often cling to the edges of the Apurímac canyon on cool mornings. One rides through them huddled against the cold, longing for the Andean sun to break through the cloud, to glisten on the damp grasses and shine with dazzling brilliance on the snows of Soray and Salcantay.*

The battle of Vilcaconga was described by its survivors as 'a fierce fight, a highly dangerous affair in which five Spaniards were killed and others wounded, as well as many horses'. Quisquis's men at last made use of steep terrain to come to grips with the enemy. They proved that Spaniards and their horses were vulnerable and mortal. They destroyed part of Soto's tiny vanguard. Had they gone on to destroy his entire squadron, they might have become sufficiently emboldened and experienced to have crushed Almagro's and Pizarro's smaller isolated contingents. Native troops annihilated far larger Spanish forces in similar difficult country in later years. But this is speculation. The fact was that Quisquis acted too late. He failed to exploit the many earlier river crossings, steep ascents and tight valleys where his men could have trapped Pizarro's impudent force of invaders.

Almagro and a chastened Soto paused in a fortress at the top of the ascent to await Pizarro. The Governor crossed the Apurímac on Wednesday 12 November and slept at Limatambo, just below the scene of the battle. He joined his lieutenants next day and the combined force advanced to Jaquijahuana, modern Anta, a village only twenty miles from Cuzco itself. The bitter fighting at Vilcaconga had shown that Chalcuchima was an ineffectual hostage. The Spaniards became convinced that he was somehow controlling the movements of their enemies. As soon as Pizarro heard news of the battle, he had Chalcuchima put in chains and delivered a chilling announcement to his prisoner. 'You have seen how, with the help of God, we have always defeated the Indians. It will be the same in the future. You can be certain that they will not escape and will not succeed in returning to Quito whence they came. You can also rest assured that you yourself will never see Cuzco again. For as soon as I arrive where Captain [de Soto] is waiting with my men, I shall have you burned alive.' Chalcuchima listened attentively to this harangue. He then replied, briefly, that he was not responsible for the Indians' attack. Pizarro, sure of Chalcuchima's complicity, left him without pursuing the conversation. The fate of the great Inca general was sealed when the two Spanish forces reunited, for both Almagro and Soto were convinced that Chalcuchima had been behind the Indian resistance. At Jaquijahuana on the evening of Thursday 13 November he was brought out to be burned alive in the middle of the square. Friar Valverde tried to persuade him to imitate Atahualpa in a deathbed conversion to Christianity. But the warrior would have none of it. He declared that he had no wish to become a Christian and found Christian law incomprehensible. So Chalcuchima was once again set alight with 'his chiefs and most familiar friends the quickest in setting him on fire'. As he died he called on the god Viracocha and on his fellow commander Quisquis to come to his aid.

Quisquis still presented a serious threat. His army lay between the Spaniards and the Inca capital. He might attempt another pitched battle on the grassy hills above the city, or a desperate defence of Cuzco itself. Friendly Indians reported that the Quitans intended to set fire to the thatched roofs of the city, as they had attempted to do at Jauja. Cuzco lies in the fold of a valley and is invisible to a traveller from the north-west until he is immediately above it. But as the Spanish column rode closer, a cloud of smoke became visible above the line of hills. This seemed to be the start of the burning of Cuzco. Forty horse spurred ahead to prevent a contingent of the Quitan army descending into the city to complete its destruction. They found Quisquis's main army drawn up to defend a passage on the road in a last attempt to prevent the invading army from reaching Cuzco. 'We found all the warriors waiting for us at the entrance to the city.' A fierce fight ensued, in which the natives, 'in the greatest numbers, came out against us with a great shout and much determination'. The Indians drove the Spaniards back from the pass leading towards Cuzco. Juan Ruiz de Arce wrote bitterly that 'they killed three of our horses, including my own, which had cost me 1,600 castellanos; and they wounded many Christians'. Some Spanish horsemen were driven back down the slope. 'The Indians had never before seen Christians retreat, and thought that they were doing it as a trick to lure them on to the plain.' They therefore remained in the security of the hills and waited while Pizarro came up with the remainder of the Spaniards. The two armies camped on hills close to one another, and the invaders spent the night with their horses bridled and saddled. Pizarro himself described the battle outside Cuzco as a 'full-scale encounter'.

The four battles on the road to Cuzco – Jauja, Vilcashuaman, Vilcaconga, and the pass above Cuzco – had demonstrated the immense superiority of mounted, armoured Spaniards over native warriors. The Inca empire did not, as is sometimes supposed, go under without a struggle. Whenever the native armies were led by a determined commander they fought with fatalistic bravery. In the course of the Conquest the Incas, who were themselves formidable conquerors against other Andean tribes, tried to adapt their fighting methods to meet the extraordinary challenges of invasion by a more advanced civilisation. The mounted knight had dominated European military history since Roman times. This formidable figure could be stopped only by other horsemen using similar equipment, by archers, pikemen or elaborate defences. His domination of the battlefield ended only with the evolution of rapid-firing firearms. Whenever American natives had time to assimilate European weapons they were able to mount an effective resistance – for instance the natives of southern Chile, who acquired pikes and horses, or those of North America who adopted horses and firearms. But the Incas did not have time to make these adaptations to their fighting techniques, and their bare mountainous country did not possess suitable wood for pikes or bows.

The Inca armies were now confronting the finest soldiers in the world. Spanish tercios were considered the best in Europe throughout the sixteenth century. They had behind them the successful expulsion of the Moors from Spain, and many who now fought in Peru had participated in the defeat of Francis I at Pavia or of the Aztecs in Mexico. The men who were attracted to the American conquests were the most adventurous – as tough, brave and ruthless as the members of any gold rush. In addition to greed they possessed the religious fervour and unshakeable self-confidence of a crusading people which had been fighting the infidel for centuries and was still on the advance. Whatever one may think of their motives, it is impossible not to admire their bravery. In skirmish after skirmish their first reaction – almost a reflex – was to charge straight into the thick of the enemy. Such aggressiveness was intended as a psychological shock-tactic, and its effect was heightened by the invaders' reputation for success, invincibility, almost divinity.

Atahualpa's nephew Titu Cusi tried to describe the awe felt by his people in the face of these strangers. 'They seemed like viracochas, which was our ancient name for the universal creator. [My people] gave this name to the men they had seen, partly because they were very different from us in clothing and appearance, and also because we saw that they rode on enormous animals that had feet of silver – we said 'silver' because of the shine of the horses' shoes. We also called them this because we had seen them expressing themselves on to white sheets, just as one person talks to another – this referred to their reading books and letters. We called them viracochas because of their magnificent appearance and physique; because of the great differences between them – some had black beards and others red ones; because we saw them eat off silver; and also because they possessed yllapas (our name for thunder) – we said this to describe the arquebuses which we thought to be thunder from heaven.'

During the actual fighting of the Conquest, the Spaniards owed everything to their horses. On the march their horses gave them a mobility that continually took the natives by surprise. Even when the Indians had posted pickets, the Spanish cavalry could ride past them faster than the sentries could run back to warn of danger. And in battle a mounted man has an overwhelming advantage over a man on foot, using his horse as a weapon to ride down the enemy, more manoeuvrable, less exhausted, inaccessible and continually striking downwards from his greater height.

At the time of the Conquest there was a revolution in the method of riding. The pike and arquebus had made the fully armoured knight too vulnerable. He was now replaced by the trooper, jinete, on a lighter, faster horse. Instead of riding 'a la brida' with legs stretched out to take the shock of jousting, the riders of the Conquest adopted a new style called 'a la jineta'. This method had the rider in 'the position of the Moors, with short stirrups and the legs bent backwards so as to give the appearance of almost kneeling on the horse's back . . . With the high Moorish saddle, the rider used the powerful Moorish bit, a single rein, and always rode with rather a high hand. The reason was that the horses were all bitted on the neck, that is to say they turned by pressure on the neck and not by pulling at the corners of the mouth . . . As the bit had a high port, and often a long branch, the raising of the hand pressed the port into the palate . . . and a horse turned far more rapidly and suffered less [than under] the modern system.'

Both Spaniards and Indians attached immense importance to horses, the tanks of the Conquest. To Spaniards the possession of a horse elevated a man, entitling him to a horseman's share of conquered treasure. During the months of waiting at Cajamarca, Spaniards had paid fantastic prices for the few available horses. Francisco de Xerez described these prices 'even though some people may find them unbelievably high. One horse was bought for 1,500 pesos de oro and others for 3,300. The average price for horses was 2,500, but there were none to be found at this price.' This was sixty times the price being paid for a sword at Cajamarca at the same time, and the inflated values of Peru of course represented small fortunes in contemporary Spain. Many deeds of sale that have survived from the period confirm them.*

For the Indians, their enemies' great horses assumed a terrible value. They thought little of a Spaniard on foot, cumbersome in armour and breathless from the altitude; but the horses filled them with dread. 'They thought more of killing one of these animals that persecuted them so than they did of killing ten men, and they always placed [the horses'] heads afterwards somewhere that the Christians could see them, decked in flowers and branches as a sign of victory.'

The Spanish conquistadores wore armour and steel helmets. Some of the infantry wore a simple steel cap called a salade, of which the barbute type was still common at the time of the Conquest. It looked like a steel Balaclava helmet, similar to a modern steel helmet, but lower over the forehead and nape of the neck. The cabasset was another simple helmet. Its high domed crown resembled a 1920s cloche, and it often had a small apical peak like a French revolutionary liberty cap. But the most famous helmet was the morion. This was a bowl-like chapel-de-fer to which an elongated brim had been added. This brim swept along the sides in an elegant upward curve, rising to a point at the front and rear. The crown was often protected by a steel crest running from front to rear like that on the helmet of a French poilu of the 1914–18 War.*

All Spanish soldiers wore armour, but this varied in elaboration. Many of the wealthy leaders wore full armour, which came in a wide variety of styles ranging from heavy gothic suits to the Maximilian suits of the 1530s and 1540s. The period of the Conquest was the high point of the art of making armour. Plates covering exposed areas of the body were brilliantly jointed with articulated lames and hinges to permit freedom of movement to every limb. Special protective plates covered the shoulders, elbows and knees; but the steel of the breastplates and leg and arm protections was as light as possible. A full suit of armour weighed only about sixty pounds, and this weight was quite tolerable, being evenly distributed over the entire body. In the latter half of the century some parts of the body were less thoroughly protected, in order to economise weight. Instead of head-to-foot armour, soldiers adopted a half-suit extending only to the jointed lames, called tassets, that formed a skirt below the breastplate, or a three-quarter suit extending to the knees. Suits of armour had their own helmets. A solid crown covered the head and extended over the neck where it joined a series of overlapping plates called a gorget. The cheeks and chin were defended by a piece called bevor, and a hinged visor covered the face. This helmet also became lighter, with the visor being replaced by a peak across the forehead and a series of protective bars across the face itself.*

Although most of the rich men in the Conquest owned full armour or acquired it when they received shares of treasure, they often used lighter substitutes when fighting Indians. Some wore chain mail shirts, which weighed between fourteen and thirty pounds. These varied according to the size of their links, but most could withstand a normal thrust. Some suits had thicker or flattened wire at vulnerable places to reduce the size of the holes. Other conquistadores abandoned even chain mail in favour of padded cloth armour called escaupil, which they adopted from the Aztecs. Escaupil normally consisted of canvas stuffed with cotton. Spanish soldiers also defended themselves with small shields, generally oval bucklers of wood or iron covered in leather.

The most effective Spanish weapon was a sword: either the double-edged cutting sword, or the rapier, which over the years gradually lost its cutting edge and became thinner and more rigid for thrusting. These were the weapons that slaughtered thinly protected Indians. Sword manufacture had reached perfection by the sixteenth century, and Toledo was one of the most famous centres for the craft. Strict regulations and apprenticeships ensured that high standards were maintained. A blade had to survive rigorous testing before being decorated and mounted in its hilt: it was bent in a semicircle and in an s-bend, and then struck with full force against a steel helmet before being passed. The sword was often decorated with a motto: 'never unsheathed in vain'; 'por mi dama y mi rey, es mi ley' ('for my lady and my King, this is my law'); or more blatant advertising such as 'Toledan quality, the soldier's dream'. The blade, some three feet long, light, flexible and extremely strong and sharp, was a deadly weapon in the hands of skilled swordsmen. And the Spanish conquistadores, acknowledged as the finest fighting men in Europe, made it their business to be supreme in this art. Throughout the century swords, like horses, were rigorously forbidden to Indians under any circumstances whatsoever.

In addition to his sword, and to supporting daggers and poniards, the cavalryman's favourite weapon was his lance. Along with the crouching, highly mobile jineta method of riding, came the 'lanza jineta'. This was ten to fourteen feet long, but light and thin, with a metal tip shaped like a diamond or olive leaf. The rider could charge with the shaft resting against his chest; he could hold it down level with his thigh, parallel to the galloping horse, with his thumb pointed forwards in the direction of the blow; or he could stab downwards with it. Each method was enough to penetrate Indian padded armour.

It has sometimes been said that the Spanish triumph was due to their firearms. This was not so. Arquebuses were sometimes fired during the Conquest, but there were very few of them, and they played no significant role beyond producing a great psychological effect when they did go off. It was not surprising that few arquebuses were used. The cavalry despised them as an ungentlemanly arm, and the Conquest was largely the work of horsemen. They were unwieldy, from three to five feet long, and often needing a support at the end of the barrel. They were difficult to load: a measured charge of powder had to be pushed down the muzzle, followed by the ball. And they were even more difficult to fire: fine powder led through a hole to the main charge, and this had to be lit by a wick. Arquebusiers carried the long rope-like wick coiled around themselves or around the weapon; they lit it by striking a flint and tinder; and they had to blow on the lighted end before applying it to the powder. Later innovations produced an s-curved piece of metal that slightly accelerated the process by pressing the wick on to the powder. But it was almost a century before the flintlock was introduced.

Crossbows were used in the Conquest, but again with limited effect. This weapon had been invented to shoot a missile with sufficient velocity to penetrate armour, but the thrust was gained at the expense of ease or speed. The steel bow had to be bent back mechanically, either by heaving on a system of pulleys or by winding back along a series of ratchets with the help of a wheel called a cranequin. All this involved a laborious process of upending the weapon, treading the head against the ground, and heaving up the bowstring. The metal bolt, once fired, killed any Indian it struck, but the natives were not impressed by this cumbersome device, which often misfired or suffered mechanical breakages.

What could Quisquis's men offer against this armoury? They were still fighting in the bronze age, and their use of metal was unimaginative. They simply copied shapes that had been developed in stone, and their bronze was sadly blunt when matched against Spanish steel. They used a variety of clubs and maces, massive, heavy clubs of some hard jungle palm, and smaller hand-axes or head-breakers called champis. These had stone or bronze heads, shaped either as simple circles or adorned with star-shaped spikes – such heads litter museums and collections of Inca artefacts. Some of the larger clubs had blades like butchers' choppers. Almost all the Spanish soldiers and horses were battered and wounded by these clubs. But it was all too rare for one of these biblical weapons actually to kill a mounted, armoured, slashing Spaniard.

The natives had more success with their missiles. A favourite among the highland tribes was the sling, a belt of wool or fibre some two to four feet long. This was doubled over the projectile, generally a stone the size of an apple, and twirled about the head before one end was released. The slingshot then spun off to its target with deadly force and accuracy. Coastal tribes used palm throwing-sticks to fire javelins with fire-hardened points. The most effective weapon against cavalry was the long bow, but this was rarely used in Inca armies. Forest Indians used bows and arrows, just as they do today – their forests produced the necessary springy woods for their manufacture, and the dense conditions made arrows ideal weapons for shooting forest game. Whenever Inca armies fought near the Amazonian forests they could enlist jungle tribes with deadly contingents of archers, but they failed to exploit this fine weapon in the highlands.

An Inca warrior was a splendid figure. He wore the normal male dress of a knee-length tunic and resembled a Roman or Greek soldier or a medieval page. His tunic was often adorned with a patterned border and a gold or bronze disc called canipu in the centre of the chest and back. He had bright woollen fringes around his legs, below the knee and at the ankle, and often a plumed crest across the top of his helmet. The helmets themselves were thick woollen caps or were made of plaited cane or wood. Many soldiers wore quilted armour similar to the escaupil of the Aztecs. Beyond this the only protection was a round shield of hard chonta-palm slats worn on the back, and a small shield carried on the arms. These shields added further colour to the Inca battle line, for their wooden bases were covered with cloth or feather-cloth and had a hanging apron, all of which was decorated with magical patterns and devices.

After their defeat in the fierce fight above Cuzco, Quisquis's men lost heart. While the Spaniards spent an anxious night on the hill above the city, the natives left their campfires burning and slipped away in the darkness. When dawn broke Quisquis's army had vanished. 'The

Governor drew up the infantry and cavalry at the first light of dawn the following morning, and marched off to enter Cuzco. They were in careful battle order, and on the alert, for they were certain that the enemy would launch an attack on them along the road. But no one appeared. In this way the Governor and his men entered the great city of Cuzco, with no further resistance or fighting, at the hour of high mass, on Saturday, 15 November 1533.'

6

CUZCO

On the very day that the Spaniards cruelly destroyed Chalcuchima, a new figure of great importance entered their orbit. On the hillside above Vilcaconga, the native prince Manco appeared with two or three orejones and advanced to the column of horsemen. He presented himself to Governor Pizarro, and the Spaniards learned to their delight that this Manco was 'a son of Huayna-Capac and the greatest and most important lord in the land . . . the man to whom all that province succeeded by right, and whom all the chiefs wanted as their lord'. Manco was almost twenty but looked boyish and 'was wearing a tunic and cloak of yellow cotton'. 'He had been a constant fugitive', 'fleeing from Atahualpa's men to prevent their killing him. He came all alone and abandoned, looking like a common Indian.'

Both leaders must have seen the meeting as heaven-sent. The arrival of the invincible strangers meant for Manco the end of flight from Quisquis's attempted extermination of the family of Huascar. Pizarro's

Master stone-masons at work

men were the only force that could rid Cuzco of its Quitan army of occupation and elevate Manco himself to his father's throne. For Pizarro, Manco's sudden appearance provided the pliable ruler that he had been seeking since Tupac Huallpa's premature death. It meant that the Spaniards could enter their goal, Cuzco, as liberators, bringing with them the prince that the local tribes fervently wanted as ruler. Manco's son wrote later that the prince embraced Pizarro, who had dismounted from his horse. 'And the two of them, my father and the Governor, made a confederation.' Pizarro's oratory warmed to the occasion. He assured Manco: 'You must understand that I have come from Jauja for no other reason than . . . to free you from slavery to the men of Quito. Knowing the injuries they were doing to you, I wanted to come and put a stop to them . . . and to liberate the people of Cuzco from this tyranny.' Pizarro's secretary Pedro Sancho explained the obvious: that 'the Governor made him all these promises solely in order to please him . . . and that cacique remained marvellously satisfied, as did those who had come with him.' It was two days after this first meeting that Manco rode with his Spanish allies into the city of Cuzco.

The Inca capital was a low city lying across the foothills at the upper end of a trough-like green valley. Few of its houses had more than one storey. At first it must have looked reasonably familiar to the Spaniards as they rode over the brow of Carmenca hill and gazed down on the city. Most of the houses had steeply-pitched thatched roofs, like some medieval northern European town, and there would have been wisps of smoke from the many cooking fires hanging over the dull grey of the ichu grass thatch. The houses at the edges of Cuzco were simple rectangles, with a base of stone and the upper walls of mud bricks or of mud packed down into tapia forms. The roofs rested on agave beams, and the thatch was fastened down by a trellis of reeds tied to projecting bosses of the roof beams. The houses had wide eaves to protect the mud walls from the Andean rains. Pedro Sancho's report to the King found nothing unusual to say about these ordinary houses of Cuzco. 'Most of the buildings are built of stone and the rest have half their façade of stone. There are also many adobe houses, very efficiently made, which are arranged along straight streets on a cruciform plan. The streets are all paved, and a stone-lined water channel runs down the middle of each street. Their only fault is to be narrow: only one mounted man can ride on either side of the channel.'

It was only as the invaders moved on into the centre of the city that its marvels were revealed to them. All Cuzco's monumental buildings were clustered on a tongue of high ground projecting into the valley between two small streams, the Huatanay and Tullumayo. These streams added to the clean, almost austere, character of the Inca city. Their swift mountain water rushed along gutters in the middle of the streets and provided excellent sanitation. All the earliest visitors were impressed by this, and by the fact that both streams flowed in artificial channels with stone flags lining the sides and bed. Only fifteen years after Pizarro's entry Pedro Cieza de León wrote, sadly, that 'at present there are great heaps of rubbish alongside the banks of this river [Huatanay, which is] full of dung and filth. This was not the case in the days of the Incas, when it was very clean, the water running over stones. At times the Incas went to bathe there with their women, and on various occasions Spaniards have found small gold ornaments or pins which they forgot or dropped while bathing.'

The Huatanay ran, in its stone culvert, across the great central square, dividing it into two sections. To the west lay Cusipata, the square of entertainment, where the people crowded to celebrate their festivals. To the east was the larger Aucaypata, surrounded on three sides by the granite walls of the Incas' palaces. The vast square was surfaced in fine gravel.* Beneath ran a series of sewers to evacuate libations poured into ceremonial fonts – and to rid thc square of any unwanted effluents during the often riotous festivities.

Pizarro's column of weary horsemen and foot-soldiers marched two abreast along the narrow streets towards the square. They were jubilant at this supreme moment, the final triumph of successful explorers and conquerors. They described their prize to King Charles with breathless pride. 'This city is the greatest and finest ever seen in this country or anywhere in the Indies. We can assure Your Majesty that it is so beautiful and has such fine buildings that it would be remarkable even in Spain.' They were almost humble in contemplating their achievement. 'The Spaniards who have taken part in this venture are amazed by what they have done. When they begin to reflect on it, they cannot imagine how they can still be alive or how they survived such hardships and long periods of hunger.' In the first hours and days they were wary, expecting the Quitan troops who had fought so hard to prevent their entry to launch a counter-attack. 'But we entered the city without meeting resistance, for the natives received us with goodwill.' For a month Pizarro made his men sleep in tents on the main square, with their horses ready day and night to repel any attack.

The central square of Aucaypata was flanked by the palaces and ceremonial buildings of the Incas. Each Inca ruler built himself a palace during his reign, and after his death the building was preserved as his spiritual resting-place. It was full of his furnishings and tended by servants from his own ayllu or lineage, presided over by the Inca's mummy and his effigy or huauque. These mummified bodies were regularly carried out to participate in ceremonies on the square, and were sustained by offerings of food and drink. The Incas were too confident of the security of their empire and the honesty of its citizens to hide their dead rulers' possessions. There is thus no hope of the discovery in Peru of a Tutankhamen's tomb. Instead, the palaces provided billets for the officers of Pizarro's army, with each of the leaders taking possession of one of the buildings on the square itself. This casual occupation on the day of the Spanish entry was later converted into title-deed in the Act of Foundation of Cuzco as a Spanish municipality.

Francisco Pizarro himself took the Casana, which was, appropriately, the palace of the great conquering Inca Pachacuti who led Inca expansion out of Cuzco in the mid-fifteenth century. The palace lay on the north-western side of the square, at the corner where the Huatanay flowed across it. The outstanding feature of this palace was an enormous baronial hall. Garcilaso de la Vega saw it when he was a boy in Cuzco in the middle of the sixteenth century. 'In many of the houses of the Incas there were vast halls, 200 yards long by 50 to 60 yards wide, in which the Indians celebrated their festivals and dances when rainy weather prevented them from holding these in the open air. I saw four of these halls that were still intact in Cuzco during my childhood.... The largest was that in Casana, which was capable of holding four thousand people.' The great hall of Casana was later destroyed to make way for colonial arcades and shops. Some of these collapsed in the earthquake of May 1950; behind lay the fresh, pale grey stones of some of the palace walls, which have been left on view.

Pizarro's younger brothers Juan and Gonzalo were quartered alongside him, in buildings that had been used by Huayna-Capac, and before him had belonged to earlier Incas.* Diego de Almagro, as Pizarro's partner and second-in-command of the expedition, was awarded the newest palace which had just been built for Huascar, on the hill immediately above the northern corner of the square and just beyond the quarters of the younger Pizarro brothers.*

Another great palace lay across the square opposite Francisco Pizarro's Casana. This was Huayna-Capac's main palace, Amaru Cancha. Pedro Sancho described it as the finest of the four palaces on the main square. 'It has a gateway of red, white and multicoloured marble, and has other flat-roofed structures that are very remarkable.' Miguel de Estete wrote that 'it has two towers of fine appearance, and a rich gateway faced with pieces of silver and other metals' and Garcilaso remembered one of these towers as having 'walls some four estados [30 feet] high, but the roof was far higher, built of the fine wood that was used in royal palaces. The roof and walls were round. Instead of a weather-vane at the top of the roof, there was a tall thick pole that added to the height and appearance of the building. The tower was over 60 feet in diameter.' Amaru Cancha possessed one of the great thatched halls in addition to its towers. It fell to the lot of Hernando de Soto and Hernando Pizarro, who was then sailing towards Spain. Hernando Pizarro eventually gained possession of the entire site and sold it, many years later, to the new Jesuit order. The Jesuits' lovely pink baroque church of the Compañía now occupies this side of the square.*

With Francisco Pizarro and his captains installed among the relics of the Incas in their empty palaces, the Governor granted property to the ecclesiastical and municipal authorities of the city. He designated a building on the terraces above the square to be the first town hall. The church received a more imposing site: the palace and hall of Suntur Huasi that dominated the eastern side of the square. Vicente de Valverde, Bishop of Tumbez and future Bishop of Cuzco, installed himself here with a chapel dedicated to Our Lady of the Conception. The property has never changed hands, although more than a century was to elapse before completion of the superb baroque cathedral that now glorifies this site.

The road to the southern quarter of the empire, the Collasuyo, left the square to the right of the Suntur Huasi. Other palace enclosures lay alongside it, and long stretches of their walls survive to this day. At the corner of the square was the massive enclosure called Hatun Cancha, the palace of the fifth Inca, Yupanqui. Beyond was the enclosure of Yupanqui's successor, Inca Roca which is known by the modern name Hatun

Rumiyoc or 'great stone'. The name commemorates an enormous boulder embedded into its northern enclosing wall. Every visitor to Cuzco is taken to see this rock, because its façade has no fewer than twelve corners, some convex and some concave, but all interlocking with uncanny precision into the adjacent stones of the wall (plate 24). Another great enclosure lay to the south of Hatun Cancha: Pucamarca, home of the great conqueror, the tenth Inca Tupac Yupanqui. These three royal corrals, Hatun Cancha, Hatun Rumiyoc and Pucamarca, formed an easily-defensible barracks for Pizarro's horsemen. They became the Spaniards' strong-point to control the centre of Cuzco, and many of the soldiers were awarded plots of land within them in the settlements of 1534.

The Quechua word 'cancha' means enclosure, and it helps us recapture the appearance of Inca Cuzco. The palaces of the Incas were elaborate corrals, surrounded by a masonry retaining wall and flanked by beautifully thatched chambers opening on to the central courtyard. This plan is common in any architecture derived from farming communities, but among the Incas such enclosures were a privilege of chiefs. 'Only the caciques' houses have large courtyards in which the people used to gather to drink during their festivals and celebrations.' Bernabé Cobo noted three characteristics of all Inca building. 'Firstly, each room or chamber was a separate entity: they did not interconnect with one another. Secondly, the Indians did not whitewash their houses as we do ours, although leading chiefs used to have walls painted in various colours and with simple decorations. Thirdly, neither nobles' nor commoners' houses had fixed doors mounted for opening and closing. The Indians simply used canes and wattle to shut the doorway when they closed it. . . . They used no locks, keys or protection, and were not concerned to make large, ornate gateways. All their doors were small and plain, and many were so low and narrow that they look like oven doors. When we go to give confession to the sick we have to crouch or even crawl on all fours to enter.'

It was a privilege of Inca royalty to have buildings with masonry walls, built by a guild of highly skilled masons. The simplicity of plan of the royal palaces was amply compensated by the glory of their stonework. The original Spanish conquerors and all subsequent visitors were deeply impressed. Cobo wrote: 'The only remarkable feature of these buildings was their walls. But these were so extraordinary that it would be difficult for anyone who has not actually seen them to appreciate their excellence.' Cieza de León echoed this wonder: 'In all Spain I have seen nothing that can compare with these walls and the laying of their stones.'

The Incas' skill as masons is their most impressive artistic legacy – in other spheres of the arts they were overshadowed by earlier Andean civilisations. They succeeded in cutting and polishing their stones with dazzling virtuosity. Adjoining blocks fit tightly together without visible mortar. Even when the blocks interlock in complicated polygonal patterns, their joints are so precise that the crevices look like thin scratches on the surface of the wall. And when earthquakes have brought flimsier later walls crashing down, the Inca ashlars have remained proudly unaffected, with each block still smoothly fixed between its neighbours.

The Incas used three types of stone in their public buildings in Cuzco. Most of the Incas' own palaces were made of rectangular ashlars of a black andesite that weathers to a deep reddish-brown. A greenish-grey diorite porphyry from Sacsahuaman hill provided most of the large polygonal blocks used in enclosure walls such as Hatun Rumiyoc. And Yucay limestone was the hard grey stone used extensively in the fortress of Sacsahuaman and for foundations and terraces throughout the city.

The surfaces of the stones were smoothly polished, but each individual block was bevelled inwards at the edges of its outer face. As a result the joins between blocks were indented and the wall as a whole was crisply rusticated. There was no structural purpose for this bevelling. It was purely decorative, to break an otherwise smooth wall with contrasts of shadow, to give full weight to each individual block, and to draw attention to the brilliant precision of the joining cracks. It proved a successful aesthetic device: the curved surfaces of the stones give a fluid grace to the walls of Cuzco.

Inca masons adopted two styles for their walls. In some the stones interlock in haphazard shapes, with no two stones identical and the joints between them undulating like an elaborate jigsaw puzzle. This style is known as polygonal. In the other style, the stones were cut into rectangles and laid in regular courses, generally with each successive course slightly smaller than the one below. This symmetrical style is known as coursed. The Incas themselves clearly preferred the tidiness of coursed masonry and used it in the walls of important buildings. But to the casual modern observer the polygonal walls are more baffling and impressive. It is almost alarming to see gigantic rocks fitting snugly together like great lumps of putty. Cobo gave the normal reaction to the sight: 'I assure you that although they appear rougher they seem to me to have been far more difficult to build than the walls of coursed ashlars. For they are not cut straight but are nevertheless tightly jointed to one another. One can imagine the amount of work involved in making them interlock in the way in which we see them. . . . If the top of one stone has a projecting corner there is a corresponding groove or cavity in the stone above to fit it exactly. . . . Such a work must have been endlessly tedious. For to make the stones interlock it must have been necessary to remove and replace them repeatedly to test them. And with stones as large as these, it becomes clear how many people and how much suffering must have been involved.'

It used to be thought that the polygonal walls were more ancient than the more familiar coursed walls, but recent archaeology has shown that both styles were common during the Inca empire in the late fifteenth century.† There is a plausible explanation for the two styles of stonework. Polygonal masonry was used only for terraces or retaining enclosure walls, where strength was required. The rocks were left in uneven shapes to waste as little as possible of large boulders, and in imitation of the rough fieldstone walls that are common for terracing throughout the Andes. Coursed masonry was used for free-standing house walls. The style may have been an imitation of the turf buildings found in the Cuzco area. The turfs were cut in rectangular pieces and laid with the grass downwards. As they dried the tops and bottoms contracted and the outer sides bulged. This would provide a precedent for the ornamental countersinking of the joints on Inca masonry.*

Inca architecture had one other unmistakable characteristic. Doors and niches were invariably built in trapezoidal shapes, with the sides tapering inwards towards the lintel at the top. This was a logical method for builders who had not discovered the principle of the arch. It reduced the length of the lintel stone and spread the thrust of the weight it supported. Rows of such trapezoidal niches broke the monotony of Inca walls. Sometimes the niches were the size of sentry-boxes, tall enough to accommodate a line of standing attendants, but more often they were smaller, sunk into the wall at chest height to form a row of convenient cupboard alcoves.

Francisco Pizarro faced many immediate problems after occupying Cuzco. He had to protect his new prize from a Quitan counter-attack. He had to settle the native government and provide for the administration of the native populace. And he had to reward his own victorious troops and persuade them to remain as settlers.

Now that Cuzco was his despite a spirited defence by Quitan troops, with Chalcuchima dead and Quisquis defiant, Pizarro made no further attempt to play off both sides in the civil war. He sided openly with the Cuzco faction of Huascar's branch of the royal family. He and his men delightedly donned the mantle of liberators. On the day after their entry, Pizarro made Manco ruler, 'since he was a prudent and spirited young man, was chief of the Indians who were there at the time, and was legitimate heir to the kingdom. He did this rapidly . . . so that the natives would not join the men of Quito, but would have a ruler of their own to revere and obey.'

Pizarro immediately encouraged the new ruler to organise an army to rid Cuzco of its Quitan invaders. Manco wanted nothing better than to avenge the persecution of his family. 'Within four days he had assembled five thousand Indians, all of them well equipped with all their weapons.' Fifty Spanish horse under Hernando de Soto accompanied this force in pursuit of Quisquis, who had retreated into the mountains of the Condesuyo, the western quarter of the empire, and was on the upper Apurímac some twenty-five miles south-west of Cuzco. The allied expedition lasted ten days but was a failure. Quisquis's advance guard defended a pass and redoubt, and warned the main Quitan army of the approach of Soto's cavalry. Quisquis retreated across the gorge of the Apurímac near a village called Capi, burned the suspension bridge and repulsed an attempted allied crossing with a hail of missiles. The Spaniards were appalled by the region, 'the wildest and most inaccessible they had ever seen', but Manco was pleased that his men emerged well from a savage battle with part of Quisquis's force.

Although the Quitan army eluded this punitive expedition, its morale was shattered by this third defeat. Quisquis could no longer force his men to remain near Cuzco, far less to launch a counter-attack on its foreign conquerors. They thought only of returning to their homes and began a long migration towards Quito.*

The expedition against Quisquis returned to Cuzco towards the end of December 1533. Soto's Spaniards were eager to participate in any looting, and Manco wanted to receive formal coronation as Inca. Manco retired to a mountain retreat for the requisite three-day fast. He then made his triumphal entry into the square, amid all the ritual that had attended the coronation of his half-brother Tupac Huallpa in Cajamarca four months earlier. The coronation was coupled with victory celebrations for the deliverance from the Quitan occupation. Days of riotous festivities ensued, and the conquistadores could observe the full panoply of Inca ceremonial. The mummified bodies of the Inca ancestors played a prominent part – the Christians did not yet feel strong enough to interfere with these heathen practices. Miguel de Estete left a vivid account of those days of jubilation. 'Such a vast number of people assembled every day that they could only crowd into the square with great difficulty. Manco had all the dead ancestors brought to the festivities. After he had gone with a great entourage to the temple to make an oration to the sun, throughout the morning he proceeded in rotation to the tombs where each [dead Inca] was embalmed. They were then removed with great veneration and reverence, and brought into the city seated on their thrones in order of precedence. There was a litter for each one, with men in its livery to carry it. The natives came down in this way, singing many ballads and giving thanks to the sun.... They reached the square accompanied by innumerable people and carrying the Inca at their head in his litter. His father Huayna-Capac was level with him, and the rest similarly in their litters, embalmed and with diadems on their heads. A pavilion had been erected for each of them, and the dead [kings] were placed in these in order, seated on their thrones and surrounded by pages and women with flywhisks in their hands, who ministered to them with as much respect as if they had been alive. Beside each was a reliquary or small altar with his insignia, on which were the fingernails, hair, teeth and other things that had been cut from his limbs after he had been a prince.... They remained there from eight in the morning until nightfall with no lull in the festivities.... There were so many people, and both men and women were such heavy drinkers, and they poured so much into their skins – for their entire activity was drinking, not eating – that it is a fact that two wide drains over half a vara [eighteen inches] in diameter which emptied into the river beneath the flagstones ... ran with urine throughout the day from those who urinated into them, as abundantly as a flowing spring. This was not remarkable when one considers the amount they were drinking and the numbers drinking it. But the sight was a marvel and something never seen before.... These festivities lasted for over thirty days in succession.' Pedro Sancho described, in particular, the mummy of Huayna-Capac as being 'wrapped in rich tapestries and almost intact – only the tip of the nose is missing'. Pedro Pizarro recalled that the daily ritual began with a procession carrying an effigy of the sun and headed by the chief priest Villac Umu. The ceremony also included a symbolic meal for each mummy. Food was burned on a brazier in front of the effigy, and chicha maize alcohol was poured into great pitchers of gold, silver or pottery. The ashes of the burned food and the chicha were then poured into a round stone font that emptied into the same hidden conduits that had carried off the urine.

The Spaniards again exploited the coronation to stage a demonstration of friendship and allegiance between natives and Europeans. 'After the friar [Valverde] had celebrated mass, he came out on to the square with many men from his army, and in the presence of the cacique [Inca], the chiefs of the land, the fighting men ... and his own Spaniards, with the Inca on a stool and his men on the ground around him, the Governor addressed them in the same way as he had done on similar occasions. I [Pedro Sancho], as his secretary and the army scrivener, read out the proclamation and the Requirement that His Majesty had ordered. The contents were translated by an interpreter and they all understood them and replied that they had.' Each chief then went through the ritual of raising the royal standard of Spain, twice, to the sound of trumpets, and Manco drank from a golden cup with the Governor and leading Spaniards. The natives 'sang many ballads and gave thanks to the sun for having allowed their enemies to be driven from the land, and for sending the Christians to rule them. This was the substance of their songs, although' – Estete added cautiously – 'I do not believe that it was their true intention. They only wished to make us think that they were content with the company of Spaniards....'

The document that was read out, translated and 'understood' by the native chiefs was an extraordinary statement called the Requirement. This proclamation resulted from a moral debate that had been raging in Spain and the Indies for over twenty years. The question at issue was whether the Spaniards had any right to conquer the native kingdoms of the Americas. Although Pope Alexander had divided the world with a line that awarded Africa and Brazil to Portugal and the remainder of the Americas to Spain, many argued that this donation was for religious proselytism only – not for conquest and invasion. 'The means to effect this end are not to rob, to scandalise, to capture or destroy them or to lay waste their lands: for this would cause the infidels to abominate our faith.'

As early as 1511, the Dominican friar Antonio de Montesinos had launched the debate with a startling, searching sermon to the settlers on the island of Hispaniola. 'You are in mortal sin,' he warned them. 'You live and die in it, because of the cruelty and tyranny you practise in dealing with these innocent people. Tell me, by what right or justice do you keep these Indians in such cruel and horrible servitude? On what authority have you waged a detestable war against these people who dwelt quietly and peacefully in their own land?'

The pro-Indian movement found a champion when Bartolomé de las Casas, who had been enjoying the life of a colonist for a dozen years, suddenly reversed his attitude in 1514. Las Casas championed the Indians throughout the remainder of his long life. Matías de Paz, professor of theology at the University of Salamanca, had written a treatise in 1512 that argued that the King had the right to propagate the faith but not to invade for wealth. But other authorities endorsed the monarchy's right to rule the Indies, since the childlike natives needed European paternalism. They quoted Joshua's defeat of Jericho as precedent for righteous extermination of infidels. They argued that heathen Americans should be treated in the same way as the Moors – even though the latter had violated and harassed Christian territory, while the Americans had lived in peaceful isolation.*

Spanish monarchs were profoundly troubled by the debate over their moral rights to conquer. In sixteenth-century Spain theologians were immensely influential, and all Spaniards, even humble soldiers, had an extraordinary respect for religion and for legal formalism. The King therefore appointed a commission containing members with both conflicting points of view. The results of their debates were the Laws of Burgos, 1512–13, which regulated many aspects of native life in the West Indies. The ordinances were reasonably humane when dealing with housing and clothing, and the protection of men, women and children from excessive working hours. But Indian men were compelled to work for nine months a year for Spaniards.

The debate on the moral rights of conquest continued. The King appointed a further commission to meet during 1513 in a monastery in Valladolid. Its brain-child was the Requirement, a proclamation to be read out to natives through interpreters, before Spanish troops opened hostilities against them. It represented a victory for the pro-Conquest thinking of its author, Juan López de Palacios Rubios. He claimed that the Requirement provided a means for Indians to avoid bloodshed by complete and immediate surrender.

The Requirement itself contained a brief history of the world, with descriptions of the Papacy and Spanish monarchy and of the donation of the Indies by Pope to King. The Indian audience was then required to accept two obligations: it must acknowledge the Church and Pope and accept the King of Spain as ruler on behalf of the Pope; it must also allow the Christian faith to be preached to it. If the natives failed to comply immediately, the Spaniards would launch their attack and 'would do all the harm and damage that we can', including the enslavement of wives and children, and robbery of possessions. 'And we protest that the deaths and losses which shall result from this are your fault . . .'

Spanish conquistadores sailed forth with the Requirement in their baggage, and it was read out in various extraordinary situations – to empty villages, to natives already captive, or, in the present instance, during a victory celebration on the square of a conquered capital city. Las Casas confessed that on reading the Requirement he did not know whether to laugh at its ludicrous impracticality or weep at its injustice. But Pizarro had his instructions as to its use, and it satisfied Pedro Sancho's sense of legal rectitude to perform the ritual of its proclamation.

Having given the natives a ruler and read them the Requirement, Pizarro could begin the hispanicisation of his prize, Cuzco. He had the pleasant task of supervising the plunder of its great treasures. His men had survived much to achieve this incredible success. As Sancho wrote to the King, 'the conquistadores endured great hardships, as the entire country is the most rugged and mountainous terrain that a horse is capable of crossing. . . . The Governor would never have dared to make this long and dangerous expedition had he not had the greatest confidence in all the Spaniards in his force.' Pizarro now had to see that the looting was carried out in an orderly manner, with strict supervision of the distribution of treasure among members of the expedition, with a

1 *The tiled roofs, eucalyptus trees and cathedral of modern Cajamarca, seen from a carved rock known as the Inca's seat. Pizarro approached across the hills in the distance*

2 *Cristóbal de Mena,* La conquista del Perú, *Seville 1534—the first book ever published about Peru*

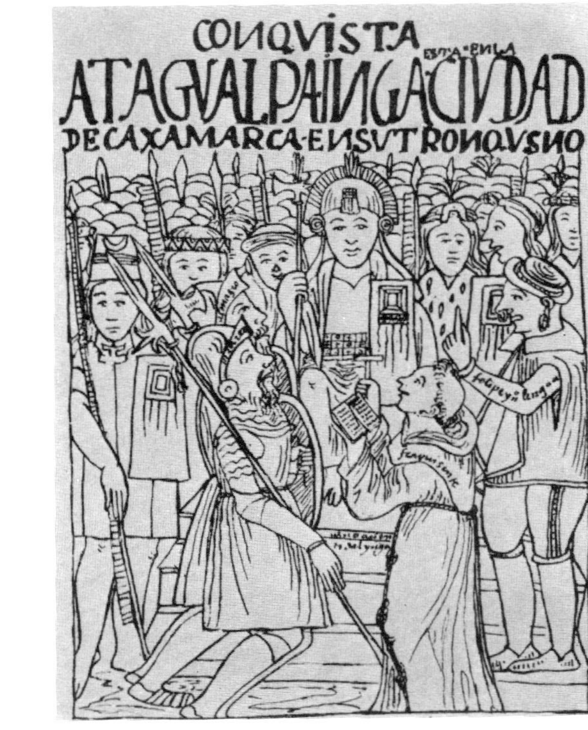

3 *Felipe Guaman Poma de Ayala,* Nueva corónica y buen gobierno, *drawn in Peru about 1600*

4 *Theodore de Bry's powerful anti-Spanish propaganda, an illustration for Benzoni's* Americae, *Frankfurt, 1596*

Three versions of the scenes on the square of Cajamarca. None is wholly accurate. The European artists imagined the Incas almost naked and their city surrounded by crenellated walls. The Peruvian Guaman Poma de Ayala, although accurate about dress, showed Atahualpa seated on a throne platform rather than a litter and surrounded by armed warriors

Some of the precious objects that escaped the conquistadores' furnaces

5 *A golden llama of the Inca period*

6 *A silver beaker of the earlier Chimu period*

7 *Golden idols show the pierced earlobes of noble Inca orejones*

8 *A Chimu ceremonial tumi knife, with the blade of gold and silver. The handle is surmounted by a jaguar of gold and turquoise*

9 *A Chimu god, of gold and turquoise, forms the handle of a ceremonial tumi knife*

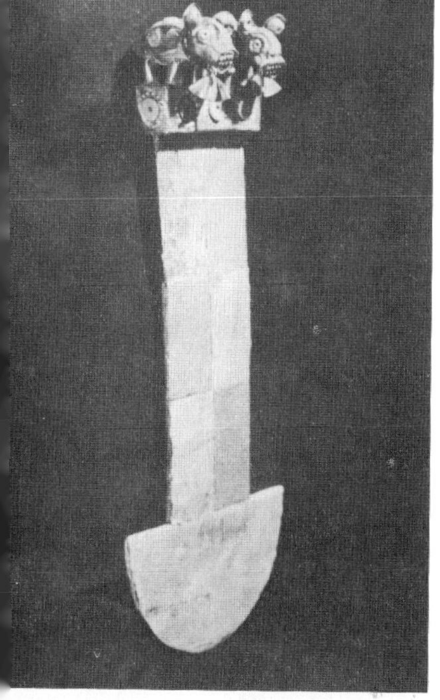

10 *Pumas guard a trapezoidal gateway in the remote ruin of Huánuco Viejo, once an important Inca city visited by Hernando Pizarro and Chalcuchima on the journey to Cajamarca in 1533*

11 *Hernando Pizarro strode past outraged priests to overthrow the idol of Pachacamac. The great temple stands beside the Pacific and consists of platforms of adobe bricks*

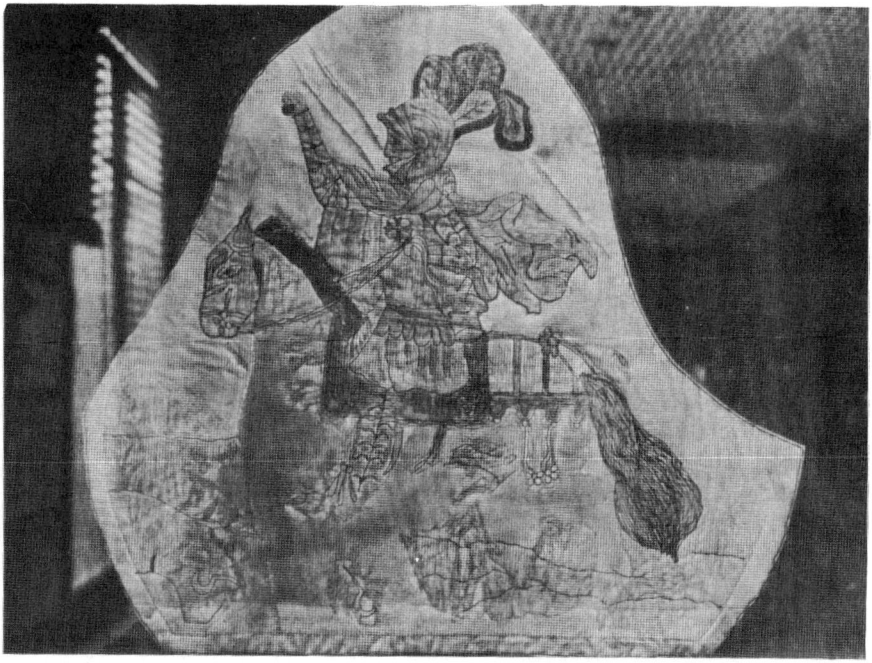

12 *The reverse of the royal standard carried by Francisco Pizarro throughout the conquest of Peru*

13 *The author in the gateway to the usno platform of Vilcashuaman, the only surviving structure of its kind in Peru. The Inca sat on the stone seat beside the top of the stairway*

14 *The unexcavated Inca palace at Vilcashuaman, where Hernando de Soto fought in November 1533*

15 *Polygonal Inca masonry in the terraces of Tarahuasi, Limatambo, below the hill of Vilcaconga*

fifth of everything reserved to the Crown, and with a minimum of provocation of the natives from whom it was being stolen. Pizarro and his officers had sufficient authority to enforce some degree of restraint on their men, and it was obvious to the men themselves that their situation in Cuzco was too precarious for them to risk undue excesses. Pizarro was also helped by the fact that the treasures of precious metal had to be laboriously melted down before they could be distributed or removed.

The melting and distribution of the Cuzco treasure was carried out with even more precautions than at Cajamarca. Rafael Loredo found ninety pages of documents relating to it, and these revealed twenty-two stages in the legal processes involved. The precious objects were collected into a large shed within Pizarro's lodging and each item was recorded by the acting treasurer Diego de Narvaez. Pizarro first ordered the melting to begin under the supervision of Jerónimo de Aliaga on 15 December 1533, and during the ensuing weeks there were many decrees swearing in the officers involved, weighing the precious metals and arranging for separate melting of the poor quality silver, finer silver and gold. The royal treasurer Alonso Riquelme was still at Jauja with the official marks. On 25 February 1534 Pizarro therefore had to authorise the fabrication of new marks bearing the royal arms. These should have been stored in a chest with two locks, but the conquistadores had to settle for one lock, since 'at present no chest with two keys could be found'. The crier Juan García was sent out on 2 March to summon all who still had any silver to bring it to the melting.

The silver was allocated at the discretion of Francisco Pizarro and of Vicente de Valverde according to each soldier's individual merits, with extra half-shares being awarded to the riders of certain outstanding horses. Pizarro himself allocated the gold between 16 and 19 March, generally in the same proportions as the silver. In addition to the men in Cuzco, shares were given to those who stayed at Jauja, or had ridden back to San Miguel with Sebastián Benalcázar, or had been killed at Vilcaconga. There was roughly half as much gold as at Cajamarca – much of Cuzco's gold had been transported to Cajamarca for Atahualpa's ransom – but over four times as much silver. The monetary value of the Cuzco melting was in fact slightly greater.† Francisco Pizarro received simply the share due 'for his person and two horses and the interpreters and for his page Pedro Pizarro'. He held in trust the share of his partner Diego de Almagro, who received more than anyone from the two distributions. The crown again received its fifth, including 'a woman of 18 carats who weighed 128 gold marks' (no less than 65 lb or 29·5 kilos) and a llama of 18 carat gold weighing over 58 lb (26·45 kilos) as well as other smaller figures. Juan Ruiz de Arce wrote that 'His Majesty got a further million pesos de oro of gold and silver.'

The sack of Cuzco was one of the very rare moments in world history when conquerors pillaged at will the capital of a great empire. It was an event to fire the imagination of every ambitious young man in Europe. Francisco López de Gómara wrote that, on entering Cuzco, 'some of them immediately began to dismantle the walls of the temple, which were of gold and silver; others to disinter the jewels and gold vases that were with the dead; others to take idols of the same materials. They sacked the houses and the fortress, which still contained much of Huayna-Capac's gold. In short, they took a greater quantity of gold and silver there and in the surrounding district than they had in Cajamarca with the capture of Atahualpa. But since there were many more of them now than there had been then, each man received less. Because of this, and because it was the second such occasion and did not involve the imprisonment of a king, it had less publicity.'

Pedro Pizarro recalled one of the most spectacular finds. 'In one cave they discovered twelve sentries of gold and silver, of the size and appearance of those of this country, extraordinarily realistic. There were pitchers half of pottery and half gold, with the gold so well set into the pottery that no drop of water escaped when they were filled, and beautifully made. A golden effigy was also discovered. This greatly distressed the Indians for they said that it was a figure of [Manco Capac] the first lord who conquered this land. They found shoes made of gold, of the type the women wore, like half-boots. They found golden crayfish such as live in the sea, and many vases, on which were sculpted in relief all the birds and snakes that they knew, even down to spiders, caterpillars and other insects. All this was found in a large cave that was between some outcrops of rock outside Cuzco. They had not been buried because they were such delicate objects.'

The greatest prize in Cuzco was the temple of the sun, the golden enclosure, Coricancha. It lay at the foot of the triangular promontory between the Huatanay and Tullumayo streams, a few hundred yards south of the main square. Although the golden cladding had been stripped from the temple for Atahualpa's ransom, it was still full of precious objects. Juan Ruiz de Arce recalled his entry into this treasure-house. 'Since Atahualpa had ordered that nothing of his father's should be touched [when the ransom was collected], we found many golden llamas, women, pitchers, jars and other objects in the chambers of the monastery. There was a band of gold eight inches wide running round the entire building at roof level.' Diego de Trujillo described his insolent entry into the temple. 'As we entered, Villac Umu, who was a priest in their canon, cried: "How dare you enter here! Anyone who enters here has to fast for a year beforehand, and must enter barefoot and bearing a load!" But we paid no attention to what he said, and went in.'

Coricancha is still a religious place, for the Dominicans soon acquired the site and built a monastery around the Inca building. The northern side of the temple is now occupied by the colonial church of Santo Domingo and by the reception rooms of the monastery. But the Inca wall running down the eastern side is virtually intact, a magnificent stretch of two hundred feet of coursed masonry, with each ashlar bulging slightly and interlocking snugly with its neighbours. Much of the wall stands to its original height of fifteen feet above the outer street and ten feet above the higher platform of the temple. It tapers towards the top and leans inwards, all to enhance the illusion of height and strength.* A series of rectangular chambers surrounds the central courtyard of the temple. Many of these have their Inca stonework intact, with rows of trapezoidal niches sunk into the wall at shoulder level. But the most dramatic architectural feature of Coricancha is a curved retaining wall at the north-western corner, beneath the western façade of the church of Santo Domingo. The dark grey stones are superbly fitted and finished, and rise to twenty feet with a slight entasis to correct any optical illusion. This smooth curve of wall has remained tightly intact through the various earthquakes in Cuzco's history, and some visitors try to draw moral conclusions from the fact that the mortarless Inca wall has stood firm while the Spanish church above has often crumbled (plate 27).

Apart from its masonry and golden cladding, the features of Coricancha that were described most often by the chroniclers were a garden of golden plants, a sacrificial font and a golden image of the sun. The artificial garden amazed the Spaniards with its delicate replicas of maize with silver stems and ears of gold. Cristóbal de Molina said that it was in the centre of the temple, before the room that housed the sun image.

Not surprisingly, none of its precious plants survived the melting of 1534.

The font was more substantial. Juan Ruiz de Arce witnessed ceremonies being performed at it during the first year of the Conquest. 'In the centre of the courtyard is a font, and beside this font is an altar, which was made of gold and weighed 18,000 castellanos. Beside it there was an idol. At noon the cover was removed from the altar and each nun [mamacona] brought a dish of maize, another of meat and a jar of wine and offered them to the idol. When they had finished offering their sacrifices, two Indian guardians came up carrying a large silver brazier. When this was lit they threw the maize and meat into it, and they threw the wine into the font. They sacrificed what had finished burning, raising their hands to the sun and giving thanks.' Reginaldo de Lizárraga, one of the Dominicans living in the monastery at the end of that century, confirmed that 'there remains in our convent a large stone font that is octagonal on the outer side. It is over a vara and a half [five feet] in diameter and over a vara and a quarter deep.'

There was greater confusion about the famous golden image of the sun. It was known as Punchao, which also meant daylight or dawn; the sun itself was called Inti. There were various sun images in Cuzco, and the temple of Coricancha also contained images of the moon, stars and thunder. The principal Punchao was 'an image of the sun of great size, made of gold, beautifully wrought and set with many precious stones'. This main effigy eluded the Spaniards. The boastful Biscayan Mancio Sierra de Leguizamo claimed that he had had it in his possession in Cajamarca, but had lost it in a night of gambling, and this produced the Spanish expression 'to gamble the sun before it rises'. Many chroniclers repeated this story, but no one believed Sierra. Pizarro did not allow any individual soldier to own precious objects from the ransom treasure before the official melting, least of all the most famous religious effigy of the empire. Spaniards continued to be tantalised by the missing Punchao, and Cristóbal de Molina of Santiago wrote in 1553 that 'the Indians hid this sun so well that it could never be found up to the present day'.

The looting of Cuzco was inevitable, since the capture of the city was the culmination of an invasion inspired by greed. But the artistic loss was tragic. The sensitive young priest Cristóbal de Molina condemned his compatriots. 'Their only concern was to collect gold and silver to make themselves all rich . . . without thinking that they were doing wrong and were wrecking and destroying. For what was being destroyed was more perfect than anything they enjoyed and possessed.'

The city of Cuzco also contained the immense storehouses of the Inca empire. Pizarro's men had often seen provincial storehouses as they marched along the royal highway. The Incas appreciated the importance of commissariat for their conquering armies, and they kept deposits of essential stores along their roads. The supplies were housed in rows of identical rectangular sheds, and some of these can be seen to this day. There is a perfect example at the remote village of Tantamayo above the right bank of the upper Marañon. The neat line of fieldstone sheds looks from a distance like a goods train making its way along the bare hillside.* But most of the provincial stores had been exhausted by the armies of the civil war or emptied by Quisquis.

Nothing had prepared the Spaniards for the gigantic stores they found still fully stocked at Cuzco itself. Pedro Sancho described 'storehouses full of cloaks, wool, weapons, metal, cloth and all the other goods that are grown or manufactured in this country. There are shields, leather bucklers, beams for roofing the houses, knives and other tools, sandals and breastplates to equip the soldiers. All was in such vast quantities that it is hard to imagine how the natives can ever have paid such immense tribute of so many items.' The young Pedro Pizarro was particularly struck by the collections of tiny feathers from which the Incas made vestments that still grace many museum collections. 'There were vast numbers of storehouses in Cuzco when we entered the city, filled with very delicate cloth and with other coarser cloths; and stores of stools, of foodstuffs or coca. There were deposits of iridescent feathers, some looking like fine gold and others of a shining golden-green colour. These were the feathers of small birds hardly bigger than cicadas, which are called "pájaros comines" [humming birds] because they are so tiny. These small birds grow the iridescent feathers only on their breasts, and each feather is little larger than a fingernail. Quantities of them were threaded together on fine thread and were skilfully attached to agave fibres to form pieces over a span in length. These were all stored in leather chests. Clothes were made of the feathers, and contained a staggering quantity of these iridescents. There were many other feathers of various colours intended for making clothing to be worn by the lords and ladies at the [Incas'] festivals. . . .

'There were also cloaks completely covered with gold and silver chaquiras (very delicate little counters), with no thread visible, like very dense chain mail, and there were storehouses of shoes with soles made of sisal and uppers of fine wool in many colours.'

The Spanish conquerors, dazzled by Cuzco's gold, paid no attention to these prodigious stores. They allowed them to be plundered by the yanaconas, the native auxiliaries who had attached themselves to the successful invaders.

A quipocamayo, official recorder of Inca history and statistics

7

JAUJA

THE year 1534 began with the two opposing forces – Spaniards and Quitan Incas – extended along the line of the Andes from Cuzco to Quito. The mass of the Peruvian population was indifferent, but the Cuzco branch of the Inca family was now firmly allied to the Europeans. The Spaniards were occupying three towns in Peru: Pizarro held Cuzco with the 150 best men; the royal treasurer Alonso Riquelme was at Jauja with eighty; and Sebastián de Benalcázar, who had escorted the gold from Cajamarca to the coast, was at San Miguel de Piura on the northern coast with a small force.

The Quitan troops were grouped into two armies. Rumiñavi was in control of Quito itself and the area covered by modern Ecuador. Quisquis was some thirteen hundred miles away in the mountains of the Condesuyo with the troops that had won the war against Huascar. His men had been defeated by the Spaniards at Vilcashuaman, Vilcaconga, outside Cuzco and at Capi, but none of these encounters had been conclusive.

Quisquis must therefore have had an army of some twenty thousand – many losses or desertions from his original force would have been replaced by contingents from Chalcuchima's command. But Quisquis had lost the initiative. It was now he and not Pizarro who was isolated far from his base. His men were clamouring to return to their Quitan homeland. They forced their reluctant general to abandon his conquest and undertake the long march homewards through the Andes. Quisquis's unwieldy army was accompanied by herds of llamas and droves of porters and women camp-followers – the same auxiliaries that kept being captured by the embarrassed Spaniards – and it had to traverse a hostile, mountainous countryside in the rainy season, with the bridges cut and the storehouses empty.

Manco's scouts soon reported that the Quitans had marched northwards down the left bank of the Apurímac to join the royal road. Such a move presented an obvious threat to Riquelme's men at Jauja. This was the weakest and most vulnerable Spanish force. A victory over these Spaniards could have far-reaching consequences. It would mean the elimination of a quarter of the foreigners then in Peru, and would leave Pizarro in Cuzco totally isolated from his compatriots in San Miguel. It would enable Quisquis to recover the part of Atahualpa's ransom not already on its way to Spain. Above all, it would show that Spaniards were not invincible and would restore the fighting spirit of the Inca army.

Pizarro appreciated the danger. 'It stung him greatly that he had left a great treasure in Jauja with a tiny garrison.' He decided to send his two leading lieutenants Diego de Almagro and Hernando de Soto northwards with fifty Spaniards. They were to be accompanied by a native force of some 20,000 under Manco Inca and one of his brothers.* But this relief force, which was supposed to leave Cuzco on the last day of 1533, did not in fact depart until late January. Its Spanish members were reluctant to leave the looting of the city, and Manco was enjoying the festivities of his coronation. It also seemed wise to wait until after the worst of the rainy season, 'for it rained a great deal every day'. When the force finally marched out its progress was slow. The rains had swollen the rivers and Quisquis had cut any remaining bridges. The Pampas river below Vilcashuaman presented an insurmountable obstacle. For twenty days Manco's men worked with antlike industry to reconstruct the suspension bridge. The master bridge-builders had terrible difficulties with the current which kept destroying their cables, but the Spanish onlookers were fascinated to watch the natives' skill. Manco himself returned to Cuzco taking a message that had been received from Riquelme in Jauja. Pizarro may have invited the Inca to return, feeling that it had been unwise to risk his loyalty in a battle against his Quitan cousins.* And so the relief contingent proceeded without Manco and did not cross the Pampas and reach Vilcashuaman until March, by which time the battle of Jauja had been decided.

Quisquis planned a pincer movement for his attack on Jauja. The city then lay alongside the Mantaro river, at a point where the flat and fertile valley is broken abruptly by grey rocky hills. A thousand natives were to advance through the hills, cross the bridge beside Jauja, and seize the heights behind the city. The remainder, some six thousand men, were to advance up the open valley. The plan did not proceed smoothly. Any element of surprise was soon lost, for 'such a great movement could not be kept secret' and native collaborators 'were diligent in reporting everything for their selfish interests'. Four Spanish cavalrymen sighted the Quitan army when it was crossing a bridge some fifty miles downstream of the city.* The timing of the attack also went wrong. Instead of attacking simultaneously from the two sides of Jauja, the thousand men coming through the hills arrived a day early and immediately tried to set fire to the city. The royal treasurer Riquelme's first concern was the royal gold. He placed it all in one house guarded by the least battleworthy of his eighty Spaniards. Ten horsemen and some crossbowmen repelled the Indians advancing across the bridge beside the city, and charged across to attack them on the far side.

The Spanish defenders spent that night and the following day armed and anxiously waiting. Only then did the main Quitan force appear and camp a mile from Jauja on the far side of a tributary of the Mantaro. Riquelme boldly marched out with half his men: eighteen horse, a dozen foot and two thousand friendly local natives. The Quitans had started to cross the swollen stream but retreated to its far bank. The Spaniards bravely advanced into the powerful current, to be met with a barrage of slingshots and arrows. Riquelme himself was hit on the head by a stone, knocked off his horse, swept downstream and saved with difficulty by some crossbowmen. Only one Spaniard was killed, but almost all were wounded. Three horses were killed and many Jauja natives died at the hands of the Quitans. But the Spanish cavalry and crossbowmen won the day: most of Quisquis's men ran for the security of the hills, and great numbers were cut down by the pursuing horsemen. The Spaniards were fortunate that a distinguished captain, Gabriel de Rojas, had ridden up from the coast a few days earlier. Alonso de Mesa also 'performed marvellously that day, for he was young and robust and had a good horse and fine arms'. The Spaniards even continued the pursuit into the hills, driving the Quitans from a hill fort they attempted to occupy. Quisquis rallied his men at Tarma but was dislodged from there. His men were eager to return to Quito, but Quisquis determined to try to occupy central Peru. He established himself in a mountainous stronghold near Bombón on Lake Junín.

So ended Quisquis's hopes for a dramatic victory. His soldiers were part of the empire's professional army and should have made a better showing. Instead, they quickly lost their aggressive spirit and became intent only on passing Jauja, an obstacle between them and their homeland. But the battle was really decided by the attitude of the local native tribes. Quisquis succeeded in winning some of them to his side, for the Spaniards found many among the dead after the battle. But the Indians in the town itself made no move against the Spaniards while the battle took place. They even provided two thousand auxiliaries for Riquelme's army. Their action was partly revenge for the Quitan occupation during the past year, but it was also a more fundamental revolt by the local Huanca tribe against the rule of the Incas from Cuzco. The hostile attitude of powerful tribes such as the Huanca was a decisive factor in the overthrow of Inca rule in Peru.

Almagro and Soto reached Jauja three weeks after the battle, which took place in mid-February. They duly moved off to attack Quisquis in his mountainous redoubt, taking forty Spaniards and a native contingent under one of Manco's brothers. They found the Quitans fortified in a pass called Maracayllo on the Bombón road. The defile was defended by walls against the horses, with narrow gates and only one means of entry on to its escarpment. Soto could do nothing against this citadel, and returned to Jauja.*

Francisco Pizarro and Manco Inca had by now completed their business in Cuzco. The treasures of the imperial city had all been melted down into bars bearing the royal mark of Spain. On 19 March the last act of distribution was signed, and on 23 March Pizarro performed the official ceremony of 'foundation' of Cuzco, the capital of the Incas, as a Spanish municipality.* Three days later Pizarro and Manco set out for Jauja, each leaving lieutenants in charge at Cuzco: Beltrán de Castro and Juan Pizarro for the Spaniards and Paullu Inca for the natives.*

The two leaders reached Jauja in mid-April and learned that Quisquis was still entrenched across the road to the north. The Inca sent to Cuzco for his finest troops: Pizarro made it clear that he did not want a vast native rabble. Four thousand fine warriors arrived, and Manco led these on another expedition against the Quitans. Hernando de Soto and Gonzalo Pizarro were in charge, leading fifty Spanish horse and thirty foot-soldiers. The local chiefs of Jauja also participated, just as they had done in the defence of the city against Quisquis. They kept careful count of their men: chief Guacra Paucar lent 417 warriors, and chief Apo Cusichaca of Hatun Jauja himself led 203 men.* The expedition marched north from Jauja in mid-May.* Quisquis had apparently abandoned his fortified pass and was continuing his march to the north. A series of sharp engagements took place, in which the curaca Guacra Paucar lost three-quarters of the men he had lent to the expedition. The Spaniards pursued Quisquis to the north of Huánuco, but disengaged when it was clear that he was abandoning central Peru and continuing his march towards Quito. Soto and his expedition returned to Jauja in early June. Although he had failed to destroy Quisquis, he had driven the last Quitan army from Huascar's part of Peru.*

With the Quitan army fleeing to the north and no other opposition in central or southern Peru, Pizarro could consider the conquest of Huascar's empire as complete. He and Manco spent six weeks at Jauja as triumphant allies. Together they had achieved undreamed-of success, and each congratulated himself on manipulating the other to win control of the country. It was only a matter of months since Pizarro had first led his tiny force out of Cajamarca and since Manco had been a princeling and fugitive hiding from the Quitan armies.

The Inca decided to celebrate the triumph by organising a great royal hunt for his friend and ally. An Inca hunt or chaco consisted of the encirclement of all the game in a vast area. Many thousands of beaters were sent out to surround the chosen site, and for days they moved inwards over hills and highland savannahs, driving the game before them towards the Inca. As the great circle tightened, the beaters formed concentric lines to prevent any animal from escaping. 'They encircled the thickets and fields, and with the noise of their shouting the animals came down from the hills on to the level ground. Here, little by little, the men closed ranks until they could join hands.' The game consisted of vicuñas and guanacos (both related to the domesticated llamas), roe-deer and mountain foxes, hares and even pumas. These animals were 'surrounded and trapped by a thick wall of men. A number of Indians entered into the enclosure and with sticks and other weapons killed or captured the quantity that the Inca wished – generally ten or fifteen thousand head – and released the remainder' after shearing the vicuñas of their precious wool. The entire occasion was one of the great festivities of the mountain people. Most of the population of the area participated, and from the chaco they supplied themselves with stocks of meat and wool, while also thinning the wild herds of old animals and surplus males.

Manco kept the preparations secret and gave a casual invitation to the Spaniards. 'One day the Inca asked the Governor whether he liked hunting, for he himself was so fond of it that he had ordered a hunt to be arranged some eight days previously. [He said that] he had told [Pizarro] nothing about it until he observed the circle [of beaters] near there; but it was now approaching, and if he wished to join in with some mounted men he should order them to get ready. And so, after eating, some fifty of us horsemen made ready – but in our fighting equipment for fear that the hunt might be on *us*.' But it was no trap, and the conquistadores witnessed the wild final scenes of one of the last great royal hunts, in which ten thousand natives took part, surrounding many miles of land and killing some eleven thousand head of game. It was a moment of great cordiality between Spaniards and Peruvians.*

Before leaving Cuzco in March, Francisco Pizarro had taken the first steps in organising his conquests into a Spanish province. He continued this task as soon as he reached Jauja. The Spaniards still occupied only two native cities, Cuzco and Jauja, and Pizarro transformed each of these into Spanish municipalities. At Cuzco, he performed an extraordinary ceremony to change the great Inca capital into a Spanish city. He described it in the official act of 'foundation':

'To mark the foundation I am making and possession I am taking, today, Monday 23 March 1534, on this gibbet which I ordered built a few days ago in the middle of this square, on its stone steps which are not yet finished, using the dagger which I wear in my belt, I, Francisco Pizarro, carve a piece from the steps and cut a knot from the wood of the gibbet. I also perform all the other acts of possession and foundation of this city . . . giving as name to this town I have founded: The most noble and great city of Cuzco.'

The Act of Foundation of Cuzco went on to describe the partition of the city among eighty-eight soldiers who chose to remain there as citizens. The Act also appointed a municipal government consisting of two alcaldes or mayors and eight regidores or aldermen, all of course officers of the invading army. This council met on 29 October and debated the length of frontage that should be given to each plot. A length of 200 feet was agreed, and the streets and palaces of central Cuzco were distributed among its conquerors in units of this width.* Pizarro performed a similar foundation at Jauja on 25 April, designating the city as Spanish capital of Peru, and distributing it among 53 Spaniards who elected to remain as citizens.*

The documents relating to the Spanish settlement of Cuzco and Jauja insisted that the native residents of those cities should be well treated. Pizarro's foundation of Cuzco advised the citizens to build a church and a city wall using materials from unoccupied palaces and storehouses, 'without taking their houses away from the native inhabitants'. In his preamble Pizarro reminded his men that 'the native inhabitants of this country . . . were, in God's words, created as our brothers and are descendants of our first ancestor'. The new town council met on 1 April and passed a series of resolutions protecting the houses, property and persons of the native chiefs. In particular the freedom of the Inca Manco must be preserved 'and his command over the natives must not be taken from him in any way whatsoever'.

When he left Cuzco, Pizarro ordered that no Spaniard was to search for gold and silver, to remove any from the natives or to leave Cuzco to hunt for it. Gold-rush fever naturally grew among the soldiers left in the city. They soon angrily demanded the repeal of this order.* Pizarro reproved them and explained that looting must cease, 'because if the natives are molested by being asked for gold and silver they might rebel. This must be avoided now and until there are more Spaniards.' Even then gold should be taken only from native rulers and not from ordinary people.* Pizarro now learned that a greedy conquistador, Gonzalo Maldonado, had imprisoned the high priest Villac

Umu to extort treasure. Pizarro was furious. He decreed that disobedience of his earlier order would carry the death sentence and total confiscation of property. Villac Umu was released, and a large quantity of extorted treasure was surrendered and taken to Jauja 'to help the King in his wars'. To enforce good behaviour among his restless men, Pizarro sent Hernando de Soto back to Cuzco as his lieutenant-governor at the end of July. His instructions to Soto stressed that he should prevent any Spaniard demanding gold from the natives or forcing them to mine it. 'You will take particular care that the Indians are well treated, not permitting them to suffer hardship of any sort by the Spaniards in charge of them.' When Soto presided over a meeting of the town council of Cuzco on 29 October, it ordered that 'no building or wall of a house belonging to the mamaconas [holy women] or Indian natives may be moved or torn down by anyone who finds them on his land-holding. Also that every one of these people should remain in the dwelling in which he has lived up till now, until the Governor may order to the contrary.'

These pious injunctions on behalf of the native population were half-hearted. Their authors admitted that restraint was necessary only until enough Spanish reinforcements had reached Peru to ensure its complete subjection. They also revealed that the conquerors were continuing to harass the natives now that the first haul of treasure had been melted down.

Spanish reinforcements were already arriving. Hernando Pizarro had appealed for more colonists when he passed through Panama and the Caribbean on his way back to Spain. His recruiting campaign brought dramatic results. The reports of treasure passing back to Spain electrified the Spanish settlements of the Caribbean. There was soon a stampeding gold rush. The officials of Puerto Rico lamented that 'the news from Peru is so extraordinary that it is making the old men move, and the youths even more so . . . there will not be a single citizen left unless they are tied down'. García de Lerma complained from Santa Marta that the 'greed of Peru' was gripping everyone, and the Audiencia of the island of Española deplored the exodus. Gonzalo de Guzman told the Emperor that everyone on his island of Fernandina wanted to go to Peru.* Francisco Manuel de Lando described his desperate measures to stop the rush from Puerto Rico: 'I am on guard day and night to see that no one leaves, but am not sure of containing them. Some two months ago I learned that some had rebelled with a boat to leave. [They were recaptured and] it was necessary to see three of them flogged and others maimed in my presence: some were whipped and others had their feet cut off.' The King decreed that no one could leave for Peru unless he were a substantial merchant or were married and taking his wife. But every available ship was soon sailing down the Pacific crammed with eager adventurers. Early in 1534, 250 reached San Miguel; but only thirty of these went on to join Pizarro in Jauja in late April.

By the time of Pizarro's conquest of Peru, Spain had had colonies in the Americas for a full generation. Spaniards were far ahead of other European peoples in their aggressive colonialism: they had an urge to create miniature Spains thousands of miles from the mother country. The recent reconquest of Spain itself from the Moors taught the importance of permanent settlement of newly won territory. The problem was to persuade Spanish explorers to stay in the Indies as permanent colonists.

In the first Spanish settlements, such as the island of Española (modern Haiti and the Dominican Republic), the expenses and initiative for the conquest had come from the Spanish Crown. The first settlers were induced to stay by being awarded allotments of natives to help them farm. The allotments, or repartimientos, were granted to the custody, or encomienda, of the recipient Spaniard, who was known as the encomendero. The Spanish encomendero was responsible for the religious instruction of the Indians in his encomienda, and in the first settlements there were generally only fifty or a hundred such Indians in each repartimiento. There was considerable debate about the morality of the encomienda system, which soon degenerated into a form of conscript or forced labour. Finally, in 1520, Charles V decreed the abolition of encomiendas: the intention was to leave the Indians as free men with rights equal to other Spaniards.

But before the system could be changed, Cortés made his spectacular conquest of Mexico. His followers went at their own expense and had to conquer a great hostile empire. In these different conditions, Cortés divided the Indians in encomienda among his followers as a reward for their extraordinary achievements. He wanted the few Spaniards in Mexico to remain together for defensive reasons, and therefore decreed that the Indians of each encomienda should pay tribute consisting of produce brought to the town house of their encomendero. This new system was endorsed by the Crown in ordinances for the government of Mexico published in November 1526. Indians were now obliged to pay tribute in kind, of goods they produced locally, but were not to give any personal service to their encomendero.

The Mexican system of encomienda was repeated for Peru, in the capitulación agreed with Pizarro on 29 July 1529. Because Pizarro was going to undertake the conquest at his own expense, he was empowered to grant encomiendas to his followers, provided that he observed the restrictions on forced labour and personal service contained in the 1526 ordinances. This expedient seemed to satisfy the immediate needs of Crown and conquerors alike. The King of Spain gained an empire with no risk or expense. He induced his subjects to stay and settle the new territory by granting them a life of luxury. And he kept his conscience clear by decreeing that the natives pay no more tribute to their encomenderos than they would have paid to the state in Spain. The grants of encomienda were rewards for the remarkable exploits of the conquerors, or, rather, were an attempt by the Spanish Crown to gain control of a conquest by endorsing the spoils of the conquerors.

Pizarro, as leader of the conquest, wanted to keep his followers grouped in European communities. The citizens of each municipality formed an effective militia if they remained together and did not scatter to their isolated estates. The conquistadores themselves were happy to settle in communities, to enjoy gregarious lives without having to exert themselves mentally or physically. They lived at a time when wealth was normally derived from landed income: trade and manual labour were socially unacceptable. The only honourable ways for a man to grow rich were by inheritance, marriage or the spoils of war and conquest. Many of the conquistadores were peasants or artisans to whom loads of Indian produce were the height of luxury.

There thus evolved in Mexico and Peru the extraordinary phenomenon known as encomienda. The natives living within a designated area, or subject to a certain chief, were awarded to the 'protection' of an encomendero. The natives themselves continued to own the land, and the Crown and its officials had legal jurisdiction over them. The encomendero's reward, and the bribe to induce him to settle in the Indies, was to be supported in luxury by the natives of his encomienda. They had to deliver to his town house vast quotas of local produce and precious metals. He resided in a Spanish municipality and was actually forbidden to live within the boundaries of his repartimiento. He was expected only to enjoy himself among his Spanish friends, provide for the religious instruction of his charges, and be ready to fight in the militia. But some fundamental questions were not yet settled: How much 'tribute' did the natives owe their encomendero? Who was to perform physical labour in the mines, roads and other public works? When did the original grants of encomienda expire?

Once the obvious treasures of Cajamarca, Jauja and Cuzco had been melted down and distributed, the conquerors wanted to receive encomiendas. Pizarro judged that the time had come when he could begin making such awards to his men. This was the transition from conquest to permanent occupation. 'He wished to encourage his men to remain and populate Cuzco. They undoubtedly stayed at great risk to their lives, since they were so few and the natives so many. He therefore gave very large repartimientos, which amounted to provinces and were whatever anyone requested.' 'One conquistador was given a repartimiento of 40,000 vassals, and all who remained as citizens received enormous repartimientos, none of less than 5,000 vassals.' The Crown itself was included in the bounty. 'Some twelve thousand married Indians in the Collao were given in repartimiento to serve His Majesty, near the mines, so that they could mine gold for His Majesty.' Pizarro made these lavish awards in March 1534, before leaving Cuzco, and continued after he reached Jauja.* On 20 April he invited Spaniards to settle in Jauja in return for grants of Indians, and fifty-three accepted the offer. Virtually any soldier who had been in Cajamarca could have an encomienda if he chose to remain in Peru. This applied to all combatants regardless of social origin. It was one of the few occasions when Spanish peasants or artisans could suddenly become rich men. Pizarro tended to give the largest and best encomiendas to his own relatives or servants, or to men from his native Extremadura.

On 23 May 1534, Pizarro's former secretary Rodrigo de Mazuelas reached Jauja. He had come directly from Spain and brought royal grants and instructions that had been issued a year before, at a time when the King had no idea of the extent of Peru or success of the Conquest. These instructions clearly told Pizarro to treat Indians as free persons with similar tribute obligations as owed by Spanish subjects to their King. Tribute was to go to encomenderos only in areas not directly placed under the Crown. Above all, encomenderos were not empowered to assess their own tribute. Before awarding encomiendas, Pizarro was to visit the areas in question, examine the population and conditions, and then assess tribute in moderate quantities and only in local produce. In practice Pizarro did not even attempt such inspections. He awarded his huge encomiendas on vague hearsay, contented himself with awarding them 'in deposit' pending a final distribution, and enjoined the recipients to obey the royal ordinances protecting the Indians.*

On 27 June the conquistadores on the new town council of Jauja appealed to Pizarro on behalf of their fellow-citizens to award more encomiendas. They argued that this was necessary to give the Indians the 'protection' of an encomendero against abuse by other Spaniards. Pizarro felt confident enough to accede. Many thousands of Peruvian Indians were entrusted to the protection of members of the victorious Spanish expedition. Although the natives did not yet know it, they had changed masters. Henceforth the produce they used to deliver to the civil and ecclesiastical storehouses of the Incas would be required at the house of a Spanish conquistador, and natives would have to spend much of every year working for their encomendero. Pizarro awarded the Indians of Bombón and Tarma to the treasurer Alonso Riquelme, the man who had defended Jauja against Quisquis. In the grant he wrote: 'I deposit these Indians on you ... so that you can use them on your estates and fields, mines and farms. I give you licence, power and authority for this ... on the understanding that you are obliged to convert and instruct them in the articles of our Holy Catholic Faith, and to treat them all well, in conformity with the ordinances issued for their benefit.'

Towards the end of August, Pizarro rode down from Jauja to the coast. He wanted to see the temple of Pachacamac and to make a further attempt to discover the treasures that had eluded his brother Hernando eighteen months earlier. A rumour of a native uprising brought him hurrying back into the mountains, but it proved a false alarm. On the journey back to Jauja, Pizarro observed lines of native porters struggling up the Lunahuaná valley to carry coastal foods and European supplies to the Spaniards in the mountains. It seemed illogical for the Spanish capital to be so remote from the sea. Pizarro decided to move his capital to the coast, to a climate and altitude more congenial to Spaniards, and to a place where he could never be separated from his maritime lines of communication. The Governor ventilated the idea at an assembly on 29 November. The Spanish citizens of Jauja elected three men to explore possible sites for a coastal city. Pizarro himself went down in late December and selected a site at the mouth of the Rimac river. The new city was founded on 5 January 1535 and named Ciudad de los Reyes in honour of the Magi because of its foundation at Epiphany. Streets and squares were marked out in a characteristic Spanish grid, and a Spanish city soon developed. Only the Spanish name did not survive. By the later sixteenth century the city reverted to a corruption of its native name: Lima.

A number of Spaniards did not accept Pizarro's inducements to settle in Peru. They preferred to take their shares of the treasure of Cajamarca, Cuzco and Jauja back to their homes in Spain. Pizarro felt secure enough to allow them to leave. The group left Jauja in mid-July and included the chronicler Juan Ruiz de Arce. The contingent was led by the royal paymaster Antonio Navarro, who took Pedro Sancho's brilliantly detailed narrative and a report from the town council of Jauja. Miguel de Estete and Diego de Trujillo ended their narratives at this point in time. Of all the eyewitnesses who left such a fine record of the first years of the Conquest, Pedro Pizarro alone continued his narrative beyond this date, and he wrote it almost forty years later. The history of the Conquest suffers seriously from the resulting hiatus. The first sensational narratives of the Conquest were immediate best-sellers throughout Europe. The letter from Gaspar de Espinosa of Panama, written 15 July 1533, was rapidly published in Nuremberg and Venice. The shrewd Sevillian printer Bartolomé Pérez published Cristóbal de Mena in April 1534 and Xerez in July. Venetian cartographers immediately reproduced maps of Peru, and by October Mena was circulating in Italian.

The returning conquerors received heroes' welcomes befitting immensely rich young men who had just conquered an unknown world. The Emperor Charles V was about to fight the Moors in Tunis, and borrowed no less than 80,000 ducats from these conquistadores. Some of them continued to Madrid to kiss the hands of the Empress. Juan Ruiz de Arce never forgot the reception they were given by the court ladies. 'There were twelve of us conquistadores in Madrid and we spent a great deal of money, for the King was absent and the Court was without knights. We had so many parties every day that some were left with no money. There were jousts and hooplas and juegos de cañas [a form of pig-sticking] all so lavish that it was a wonder to see them.' Ruiz de Arce was one of the prudent ones who saved enough to live a life of luxury, surrounded by horses and slaves, from his share of the treasures of the Incas.

8

THE QUITAN CAMPAIGN

FRANCISCO Pizarro and his ally Manco Inca controlled the central section of the Inca empire – the area roughly corresponding to modern Peru. The southern section, modern Chile and Bolivia, had not yet been occupied by Spaniards. But it was loyal to Manco and particularly to his brother Paullu, whose influence was strong in this southern quarter called Collasuyu or Collao. The only region still under arms was the province of Quito, modern Ecuador and southern Colombia. Quisquis was leading his men back towards this area, and some five thousand troops of Atahualpa's command at Cajamarca had returned to Quito when Pizarro dismissed them on the morning after the seizure of the Inca. This force was led back to Atahualpa's general Rumiñavi, who established himself as military ruler of the province.

The 'traitor' Rumiñavi flays Atahualpa's brother Quilliscacha (Illescas)

Shortly before his death Atahualpa sent his brother Quilliscacha to Quito to fetch his young sons, but Rumiñavi refused to release the boys. After Atahualpa was executed and given Christian burial, a native force descended on Cajamarca to destroy the city in a pathetic show of frustrated vengeance. It exhumed the body of the dead Inca and transported it for reburial at Quito. This force may have been Rumiñavi's army attempting to rescue Atahualpa, or it may have been a local contingent led by Zope-Zopahua, another leading general of the northern army.* When the funeral cortège reached its destination, Rumiñavi held a great drinking ceremony – heavy drinking is still a feature of any Andean wake – at which he made Quilliscacha and his followers insensate on potent sora and murdered them. Quilliscacha had probably been proposed as Inca or as regent for Atahualpa's sons. After nine months of power, Rumiñavi refused to submit to this higher authority. Or he may have killed Quilliscacha from disgust at his and Atahualpa's failure and collaboration with the Spanish invaders – such a man was not fit to lead the resistance. To flaunt his defiance of Atahualpa's government, Rumiñavi desecrated Quilliscacha's body. 'He extracted all the bones through a certain part leaving the skin intact, and made him into a drum. The shoulders formed one end of the drum and the abdomen the other, so that, with the head, feet and hands embalmed, he was preserved intact like an executed criminal – but transformed into a kettle-drum.' Rumiñavi thus remained as independent warlord of the northern section of the Inca empire.

The Spaniards were not long in turning their attention to this northern area. There were alluring rumours that Atahualpa had sent convoys of treasure to this province, and that its capitals of Tumibamba and Quito rivalled Cuzco itself in wealth. Martín de Paredes wrote in February 1534 that 'the riches of this Quito are said to be very great'. There was also uncertainty whether this area lay within the dominion granted to Pizarro in the royal concession of 26 July 1529. That document spoke of territories beyond – to the south of – 'Santiago', presumably Tumbez or Puna, the only Peruvian places known at that time; but Quito lay to the north.*

One of the mightiest conquistadores of Mexico and Central America, the Adelantado Pedro de Alvarado, decided to recruit a great expedition in his territory of Guatemala and to invade this northern province. He seized some ships on the Pacific coast of Nicaragua and sailed on 23

January 1534, at the time when Manco, Almagro and Soto were marching out of Cuzco. He landed on the Ecuadorean coast on 25 February. He probably landed in Ecuador in order to remain well to the north of Tumbez, or he may simply have made the same mistake that caused Pizarro and Almagro so many months of delay. The cold Humboldt current that makes a rainless desert of the Peruvian coast does not strike Ecuador, whose coastline consists of tropical forests and mangrove swamps. Alvarado himself wrote from Puerto Viejo on the Ecuadorean coast, on 10 March 1534: 'I left La Poseción on 23 January with twelve sails and five hundred fighting Spaniards, 119 of them with horses, a hundred crossbowmen and the rest foot-soldiers, and among them many hidalgos and persons of quality, all accustomed to warfare in these parts and expeditions into the interior.' His letter was addressed to Francisco de Barrionuevo, who had already written to the Emperor that Alvarado 'is taking four thousand Guatemalan Indians. Although there are so many I believe that they will die in a short time, both because they are from a hot country and are going to a cold one, and because Peru is sterile in food supplies. They say that Guatemala and Nicaragua have been left very depopulated.' After permitting his veterans to indulge in unnecessary cruelty to the coastal tribes, Alvarado plunged inland. He struck through dense forests towards the Andes, leading hundreds of chained natives as porters.

When Pizarro marched out of Cajamarca the previous August, he sent his Captain Sebastián de Benalcázar with nine horsemen to accompany some treasure to the port of San Miguel de Piura.* Eager Spaniards were by now sailing towards the Peruvian treasure hunt as fast as they could find passage, and Benalcázar soon found the number of Spaniards at San Miguel swollen to over two hundred, most of them clamouring to be led inland on the conquest of Quito.* Benalcázar was reluctant to move without orders from Pizarro or news of the fate of his companions' march on Cuzco. But the pilot Juan Fernández sailed on to San Miguel after depositing Alvarado's mighty army and brought the electrifying news of this rival invasion. Benalcázar hesitated no longer. He immediately set out for Quito with some two hundred men and sixty-two horses.*

There were now two unauthorised invasions of Quito in progress. News of Alvarado's threat was carried southwards by the newly-arrived Gabriel de Rojas. He bravely rode into the heart of the unknown empire, reached Jauja in time to help Riquelme defend it against Quisquis, and hurried on to tell the alarming news to Almagro at Vilcashuaman and to Pizarro at Cuzco. Almagro at once rushed north with a handful of men to try to rectify the situation. He reached Saña on the northern coast on 7 April and learned the further disturbing news of Benalcázar's expedition. He turned inland in hot pursuit, but was driven back to San Miguel by native opposition along the Quito road.*

Sebastián de Benalcázar marched out of San Miguel probably in mid-February, and was by now well advanced. There were no serious topographical barriers along the early part of his route. From San Miguel he marched across the dreary coastal plain of northern Peru. This is a barren, ghostly country of gnarled, stunted trees strewn across virtual desert. A dull haze hangs over this pale brown land, and its chief inhabitants today are herds of lean goats. Once across this flat wasteland, Benalcázar's army entered the hills and struck the main royal highway north of Cajamarca and Cajas.

The Quitan campaign was the finest hour of the Inca army. Now at last the issues were clear-cut. Although they did not have the inspiration of a crowned Inca at their head, Rumiñavi's men were fighting to avenge the murder of Atahualpa. The Spaniards no longer had a hostage to cause the natives to hesitate – Atahualpa and Chalcuchima were both dead. The myopic hatreds of the civil war no longer helped the Spaniards: Rumiñavi's men were not an army of occupation holding down freshly-conquered territory. Regionalism still aided the conquerors, for some of the Quitan tribes had only recently been absorbed into the Inca empire and thought that the arrival of the Spaniards was an opportunity to regain their autonomy. But for the most part the native armies were fighting in defence of their homeland. They now had no illusions about the divine status or peaceful intentions of the Christians – they recognised them for the ruthless invaders they were. The result was that the Quitan armies mounted a determined and heroic resistance.

The Quitan campaign has suffered from the lack of a chronicler. There was not a single first-hand account by a member of Benalcázar's expedition, none of the brilliantly detailed records by qualified men such as Xerez, Estete or Sancho, nor the soldiers' reminiscences of Mena, Ruiz de Arce, Hernando or Pedro Pizarro, Trujillo or Cataño. All that we have are chronicles written long after the event by men who took no part in the expedition. Only three gave any details of the events of this war. Best among these was Gonzalo Fernández de Oviedo, a distinguished historian and geographer who wrote a fine general history during the decades after the Conquest. He was an important official in the West Indies and interrogated travellers returning to Spain, so that much of his material came from eyewitnesses. Another good account was by Antonio de Herrera, the official seventeenth-century 'Cronista Mayor de Indias' (Chronicler-in-Chief of the Indies), who almost certainly copied his very detailed description of the campaign from a missing work by Pedro Cieza de León. The third detailed account was a heroic epic poem by Juan de Castellanos, written at the end of the sixteenth century. Beyond these we have only brief reports by Zárate and López de Gómara from the 1550s and one by the eloquent but often fanciful Garcilaso de la Vega at the beginning of the next century. With such inadequate records, the Quitan war has received little attention, and the gallant efforts of the native troops are not as well known as they deserve to be.

Rumiñavi and Zope-Zopahua made their first defensive move when the Spaniards were in the territory of the Palta tribe, around Saraguro in southern Ecuador. They sent an Inca commander called Chiaquitinta against the Spaniards when they were quartered at a place called Zoro-Palta – possibly the fortress of Paquishapa, which Cieza knew as Las Piedras because of the many fine Inca ashlars to be found there. The attack was a failure: Chiaquitinta ran into Benalcázar himself scouting ahead with thirty horse, and the native troops panicked and ran at their first sight of the dreaded animals. In the ensuing pursuit the Spaniards captured one of the wives of the dead Inca Huayna-Capac.*

Benalcázar marched on, now entering the territory of the Cañari tribe. He rested his men for a week at Tumibamba, modern Cuenca. Huayna-Capac's father, the great conqueror Inca Tupac Yupanqui, had started to build a sumptuous city, a second Cuzco, at Tumibamba. Cieza de León marched through the place in 1547, and was impressed by what he saw: 'These famous lodgings of Tumibamba were among the finest and richest to be found in all Peru. . . . The temple of the sun was of stones put together with the subtlest skill, some of them large, black and rough, and others that seemed of jasper. . . . The fronts of many of the buildings are beautiful and highly decorative, some of them set with precious stones and emeralds. Within, the walls of the temple of the sun and the palaces of the Lord Inca were covered with sheets of the finest gold and encrusted with many statues of this metal. . . . Whatever the Indians said about these residences fell short of reality, to judge by the remains. Today all is cast down and in ruins, but it can still be seen how great they were.' This town, which was a ruin in the mid-sixteenth century, has now vanished apart from one outlying building known as Inca-pirca.

Of greater interest to Benalcázar's invaders was an alliance offered to them at Tumibamba. The local Cañari Indians had been under Inca rule for little over half a century, and were still restless. The initial Inca conquest had been a violent affair. At one time the Cañari threw back Tupac Yupanqui's army to Saraguro, and when they were finally overcome the triumphant Inca was said to have killed thousands of these and other Ecuadorean Indians and thrown the bodies into Yaguar Cocha, the Lake of Blood.* More recently, the Cañari had sided with Huascar and his Cuzco faction against Atahualpa and his professional army. When Atahualpa routed the southerners at Ambato, in the first battle of the civil war, he took terrible vengeance on this tribe, ordering his commanders to kill most of its men and boys even though they had come to surrender and beg for mercy.* It is therefore understandable that three thousand Cañari now joined the Spanish force as eager volunteers. They performed throughout the Quitan campaign with savage glee. Of all the local tribes who turned against the Incas, the Cañari had the greatest justification for doing so.

Marching north from Tumibamba, Sebastián de Benalcázar's men climbed the 14,000-foot pass separating Cañar on the Pacific watershed from Riobamba, where the rivers drain towards the Amazon. Up here, in the desolate páramo, Rumiñavi had prepared his army for a pitched battle. He occupied the tambo of Teocajas, below the highest part of the pass but still above the tree-line, in chill, hail-swept uplands – a land of tough, slippery ichu grass, mountain tarns, mossy swamps and lichen-covered boulders soaked by mists and rain.

The battle of Teocajas began with a confrontation between ten Spanish horsemen scouting ahead under Rui Díaz and the brilliant mass of the Inca army. The natives began the shouting and shrill battle-cries that were always a feature of their warfare. One horseman managed to ride back to warn the main army while his companions held off the encircling hordes. Forty more Spanish cavalry rode up, quickly and silently, and then charged to a battle-cry of 'Santiago!' Although the Europeans must have suffered from the altitude, the battlefield was in their favour, with plenty of room for the cavalry to manoeuvre across the high savannah. The result was terrible. Slashing horsemen cut deep paths through the native ranks. 'The speed with which the Spaniards went to the rescue proved to have been most necessary.... They attacked fiercely, trampling the Indians under their horses and causing great bloodshed with their lances.... Terrible bravery and fury were shown by either side. The Indians rallied to a cry that this was the moment to preserve or lose their liberty. The Spaniards were saying that their very lives were at stake. The Indians' tenacity was exceptional. Although they saw the battlefield soaked in blood and covered with the bodies of their dead and wounded, and although they realised their perdition, they persisted in fighting with marvellous vigour, lacking neither force nor spirit.'

As the Spaniards retired after their devastating charge, a force of many thousands of Indians came towards them. The fleeing natives had been stopped and rallied by one captain, a magnificent figure. 'He wore a golden emblem on his chest and another on his head. He carried four batons in his left hand and his fighting mace in his right; and the batons were entwined from top to bottom with bands of beaten gold.' The Spaniards thought that he and his men must be coming to surrender, after the terrible slaughter of their companions. But the natives advanced fearlessly on to the level ground and attacked. The Spaniards turned their panting horses for another murderous charge, during which they captured that gallant captain. They began to retire to their camp with some of their horses wounded. But, to their dismay, yet another Indian commander attacked with 15,000 troops – 'such fine men that the Spaniards found themselves in considerable difficulties from them. They killed four Christians and as many horses. The Spaniards were, however, able to resist and to retreat, exhausted, to their camp with many horses wounded.' And even after the Spaniards had retreated to their camp and dismounted, more Indians appeared from the mountains and marched down upon them, advancing almost into the camp. A few horsemen rode out on the best remaining horses, and were able to hold out until nightfall, when the two forces camped within sound of one anothers' voices. The Spanish soldiers spent an uneasy night, but their horses slept and were able, next morning, to ride out and drive the natives off the level ground to the sanctuary of the mountains.

The battle of Teocajas took place on about 3 May 1534.* It was the greatest pitched battle of the Conquest, with, at the estimate of Gonzalo Fernández de Oviedo, 50,000 natives deployed.* It was an indecisive battle, because the Quitans did not stop Benalcázar's invasion and the Spaniards did not destroy the defenders' army. But it demonstrated that no amount of heroism or discipline by an Inca army could match the military superiority of the Spaniards. Once again, the dreaded horses had proved invincible. Rumiñavi's men had killed four horses at Teocajas. Unable to drag away the carcases, they cut off the heads and hooves and sent these on a victory demonstration throughout the countryside.*

In their desperation the natives invented ingenious traps to bring down the horses. At Teocajas and on other occasions during the Quitan campaign, they tried the stratagem that the English foot-soldiers had used to such effect at Crécy two centuries earlier. They prepared pits full of sharp stakes, camouflaged with branches and earth. They also tried mining the roads with small cuplike holes the size of a horse's hoof. But however temptingly Rumiñavi might expose his men on level ground beyond a line of pits, the Spaniards never fell into one of his traps.

At Teocajas they were saved by an Indian from Cajamarca who offered to guide them by a circuitous route to the west of the Incas' prepared position. The Spaniards seized on this means of escape from the great army surrounding them. They left under cover of darkness, and the epic poet Juan de Castellanos wrote that 'as the horizons grew sad and the light vanished behind them, a thousand fires shone in the Spaniards' camp as a show of preparing their food'. Five hundred native troops who were guarding the western exit of the valley were overpowered in the dark, and the Spaniards marched down from the bleak pass where the heavy fighting had taken place. The natives thought they had fled; but they descended towards the Pacific in a long detour before climbing back to the royal road at Lake Colta and Riobamba.* Again there was sharp fighting, and once again the natives tried to lure the invaders on to a defensive position protected by pits. But the holes were betrayed by one of Rumiñavi's eunuchs and the Spaniards left the road to ride along the crests of the hills.* Herrera ascribed this second lucky escape to the direct intervention of the Virgin Mary. Even so, five Spaniards were killed in fighting outside Riobamba and were buried there, while the invaders rested for a week.

From Riobamba onwards the Spaniards were fighting almost continuously until they reached Quito. Bands of natives shouting battle-cries would attack their flanks repeatedly during the march. There were more pits and traps and more battles as heroically fought as Teocajas. Five thousand Indians defended the crossing of the Ambato river and a half-hour battle ensued. There was another pitched battle at Latacunga, and another at Pancallo, where the natives defended a ravine against the invaders.

But the Spanish advance was inexorable. Benalcázar's men eventually reached their destination, Quito, on about 22 June. The distance as the crow flies from Piura to Quito is about four hundred miles, but the journey involves climbing in and out of a chain of Andean valleys, crossing and recrossing the watershed between the Pacific and the Atlantic drainages. It had taken the Spaniards four months. The conquerors were disappointed with their prize, for Quito had been systematically evacuated and burned. Rumiñavi had somehow succeeded in outmarching the mounted Spaniards and had left the city five days before their arrival. He took with him the treasure in the city, eleven of Atahualpa's relatives and a reputed four thousand women, and retired to the forested province of Yumbo. As he left, he set fire to the imperial palaces and storehouses. Gómara told a story about his departure that was intended to discredit Rumiñavi. According to him, Rumiñavi told the cloistered virgins of the sun that they must leave Quito to avoid being killed or dishonoured by the ruthless invaders that were about to enter the city. Some of the women either said that they would remain and suffer whatever fate was in store for them, or smiled inanely at what they were being told. Rumiñavi misconstrued their reaction, accused them of relishing the prospect of rape by these superhuman bearded men, and ordered three hundred of them to be executed.*

The Spaniards were never men to lose the initiative. They immediately dispatched Rui Díaz with sixty mounted men in pursuit of the Inca army. But Rumiñavi managed to elude them, and rallied his forces for a counter-attack. Together with Tucomango chief of Latacunga, Quingalumba of the Chillos, and Zope-Zopahua, Inca governor of the

Rumiñavi abandons native women to the Spaniards

region north of Ambato, he assembled 15,000 warriors and marched against the Spaniards remaining in Quito. It was a night attack – a form of fighting that the natives used too rarely. Surprise was lost because the Spaniards' Cañari auxiliaries had learned about the threatened attack. Benalcázar posted sentries along the moat that the Incas had built to defend the city. He arranged that his cavalry and infantry should silently take up posistions around the main square without being summoned by drums or trumpets. Although Rumiñavi's men saw that they were expected, they pressed home the attack, setting fire to the thatched roofs of the city. The Spaniards were forced to fight hand-to-hand and on foot to defend their quarters, while the Cañari advanced to counter-attack by the light of the burning houses. It was a night of heroic battle, but the coming of dawn restored the usual imbalance of warfare. The ferocious fighting of the darkness gave way to unopposed charges by the Spanish horse and a bloody pursuit. Rumiñavi was forced to flee again, abandoning his camp to the Spanish cavalry. These were pleased to find that it contained a haul of gold and silver vessels and many beautiful women. On the following day seven local chiefs came to surrender to the conquerors.

Benalcázar was still frantically searching for the treasure reputed to be in this northern province. He moved north of Quito in July, to Cayambe and Otavalo. Cieza de León recorded the impressions of these natives. 'After their first fright and wonder and amazement over what they heard about the horses and the speed with which they travelled – believing that the riders and the horses were one and the same thing – the fame of the Spaniards aroused great excitement in these people.' The anticipation of some of them was to be badly shattered. When Benalcázar reached a village called Quinche near Puritaco he found that all the men were away fighting with the national army. To make an example of these people – and to vent his frustration at finding so little treasure – he ordered all the women and children left in the place to be slaughtered. This was supposed to terrify other natives into returning to their homes. 'A feeble excuse to justify cruelty unworthy of a Castilian', was the verdict of Herrera, the official chronicler of the Conquest.

Two further leaders now made their appearance on the Spanish side in Ecuador. One was Diego de Almagro, who had left San Miguel de Piura with all available men in early May and followed Benalcázar to Quito.* He reprimanded Benalcázar for having left San Miguel without orders, but was profoundly relieved to find that he was still loyal to Pizarro and Almagro and had conquered Quito in their name.

The other Spanish leader was Pedro de Alvarado. His splendid army had landed confidently on the Ecuadorean coast and marched along it towards modern Guayaquil, pressing the local tribes into service as porters. But instead of moving directly inland, Alvarado apparently moved northwards into the jungles beyond Daule. His men hacked towards the river Macul, short of food, plagued by insects and disease, and with their weapons and armour rusting in the humid heat. They reached Tomabela famished and debilitated, having even passed through a rain of volcanic ash from an eruption of Mount Cotopaxi. Far above them, Benalcázar's men saw the same eruption while fighting their way along the final stretch of the road to Quito. Once again Alvarado apparently chose the wrong route to penetrate the Andes, and climbed one of the highest passes, between Chimborazo and Carihuairazo. He struck heavy snow and the hardships of great altitude. There were terrible scenes, with men, women and horses falling behind the march and freezing to death huddled together in the cold Andean nights.* Eighty-five Spaniards died, and the expedition lost most of its horses. But the greatest sufferings and deaths were among the wretched coastal Indians snatched from their tropical lowlands into this terrible bitter mountain world. All their agonies ended in anti-climax. When the shattered survivors of Alvarado's army finally stumbled on to the Inca highland road, their spirits sank to see the tracks of Benalcázar's and Almagro's horses' shoes. They knew that other Spaniards had reached Quito before them.

Because of the animosity between the intruder Alvarado and the incumbents Pizarro and Almagro, we have a record of the abuses perpetrated on the natives by Alvarado's conquistadores. Almagro held a judicial enquiry, interrogating some of the men who had been on Alvarado's expedition about the cruelties they had witnessed. Their answers were sent to Spain to discredit Alvarado in the eyes of King Charles and of humane opinion at his court. In the coastal towns of Sarapoto, Manta and Puerto Viejo 'the Indians welcomed the Spaniards into their houses and came out to them with food and maize for the horses.... Despite this, the Spaniards put those towns to the sack and the witness [Hernando Varela] saw men, women and children thrown into chains and brought into the camp bound with chains and ropes.' Diego de Vara said that he had seen many of these forced labourers die on the march to Quito: 'Some were killed by sword thrusts, others by stabbing and others from the loads they were carrying.' Many more died of exposure in the snowbound passes, and, as Barrionuevo had predicted, 'The Indians whom Alvarado took with him almost all perished, even though there had been many of them.' Pedro de Alvarado was more personally guilty of the hanging of the most powerful coastal chief, the curaca of Manta and Puerto Viejo, on flimsy suspicion of inciting natives to flee. Pedro Brabo remembered that 'when they were taking him along to be hanged he had shouted very loudly, calling on the Captain [Alvarado]. But I do not know why they hanged him, except that it was rumoured that he had incited other nearby Indians to revolt. I heard this being told to the Adelantado Alvarado, but I do not know whether the chief had in fact done so, because there was no interpreter through whom he could be properly understood.... This witness also said that he saw the chief of a town called Chonanan put to the dogs, and he saw another Indian burned alive. He said that they repeatedly burned and tortured Indians to be told the route.' Such cruelties were commonplace when a force of ruthless adventurers was invading such difficult country. But many of the men involved were shocked by what they had seen, and the incidents were considered sufficiently disgraceful to be reported to the King.

The native defence of the province of Quito had not ceased, despite the failure of the counter-attack on Quito itself. Although Almagro himself succeeded in reaching Quito, three other Spaniards who tried to follow him were killed. When he and Benalcázar left Quito to march back towards the coast, they encountered continued native opposition. There were skirmishes in the Chillo valley and on the right bank of the Pinta river. When the Spaniards reached the Liribamba or Chambo river, they found the passage defended, with warriors shouting defiance from the far bank of the torrent. The Cañari auxiliaries were eager to show their zeal against their Inca enemies. They plunged into the river, and over eighty of them drowned. The weaker Spanish horses also found the current too strong and had to turn back. But a dozen horses reached the far side and dispersed the defenders. 'They killed innumerable Indians, and the serving Indians whom the Christians had with them were the ones who made a great slaughter on the enemy.' The prisoners taken in this engagement told Almagro of the approach of Alvarado's rival Spanish army.

Although they continued to fight, the native forces lacked the cohesion that might have come from a royal leader, and were splintering into isolated pockets of resistance. It was the chief of Riobamba who led his own tribesmen in the attempt to defend the Liribamba crossing. Zope-Zopahua had retreated with his own forces to a fortified hilltop near Sicchos. Rumiñavi was still commanding some contingents of the regular Inca army in the Quito area, but was looking for a suitable redoubt from which to continue the defence. He eventually established himself in an almost inaccessible crag near Píllaro. Meanwhile, to the south, Quisquis's army – which had been driven from the Bombón–Huánuco area by Soto in June – was marching northwards through Cajamarca and on towards its Quitan base.

After he reached the mountains, Pedro de Alvarado moved northwards along the royal highway, dejectedly following the hoofprints of the Spanish forces that had already reached Quito. He learned that Zope-Zopahua was fortified in the Sicchos on his left flank, and prepared to march against him with a contingent that included many crossbowmen and arquebusiers – two arms in which his force was particularly strong, now that it had lost many of its horses. The move was delayed when his men captured eight scouts sent ahead by Almagro. The two Spanish expeditions now confronted one another. Alvarado had more men, and having survived gruelling hardships on the march inland they were desperate for plunder and rewards. Almagro, on the other hand, represented the Christian forces already in possession of Quito. He consolidated his position by officially founding a Spanish city called Santiago de Quito, even though he was at Riobamba when he did so.* The situation was ugly. Both sides mounted armed guards on their camps, and prepared for a battle that would have been far more evenly matched than any against the native defenders. Had it taken place, the Spanish survivors might have been sufficiently weakened to fall to a native counter-attack. But bloodshed was avoided in a series of complicated negotiations. Almagro agreed to buy Alvarado's ships and equipment for 100,000 pesos de oro. Alvarado was to return to Guatemala but his men were allowed to remain in Peru under Pizarro. The agreement was signed on 26 August, and two days later Almagro and Alvarado rode back to Quito and founded a Spanish city there called San Francisco de Quito. The two leaders then began to march back towards Peru, where Alvarado was to receive his payment, and Benalcázar was left in the new city with four or five hundred men from the two armies.

Quisquis was by now entering southern Ecuador. He and his men had accomplished an extraordinary retreat, covering a distance on the map of well over a thousand miles since they left the Condesuyo. The army still had between 12,000 and 20,000 effectives* and vast quantities of camp-followers, conscripted porters and animals on the hoof. It was rounding up llamas, guinea-pigs and other food from villages along its path, and was also burning and destroying the country it was abandoning. This was partly to prevent pursuit by Manco and Soto and to diminish the value of the land being abandoned to the Spaniards. It was also a parting blow of the civil war, bitter punishment for the collaboration of Huascar's supporters.

The first report of the approach of Quisquis's army came from the Cañari, whose chief warned Almagro and Alvarado as they were marching south through Tumibamba. Almagro did not believe the news but continued his march. It was only by chance that a contingent of Alvarado's men surprised Quisquis's vanguard under Sotaurco while it was occupying a pass in Chaparra province along their route. Sotaurco himself was captured and forced to reveal the position of the rest of the Quitan army. The Spaniards realised that Quisquis, after weeks of unmolested marching, was not expecting to meet any Spaniards here. They therefore acted with speed. The two commanders set off on a night march with all the cavalry who were in a fit state to follow. They had to pause for part of the night 'because as they were descending a ravine their horses lost their shoes from the great outcrops of rock in it, and they had to stop to shoe them with fires. But they continued their march at full speed and did not stop until late the following day when they came in sight of Quisquis's camp.'

Quisquis did not hesitate when the dreaded Spaniards suddenly appeared. He immediately divided his forces, sending the fighting men up a steep hillside under one of Atahualpa's brothers called Huaypalcon. He himself led the women and serving people in the opposite direction. The Spaniards pursued the warriors, rapidly surrounding the hill they were occupying. But Huaypalcon's men fortified their rugged position and held the Spaniards at bay until nightfall, causing casualties with boulders and missiles rolled down the hillsides. Quisquis had by now escaped, and Huaypalcon's men later left their position and succeeded in rejoining him.

The Spaniards 'continued their march and encountered Quisquis's rearguard. The Indians fortified themselves at a river passage and prevented the Spaniards from crossing throughout the day. Instead it was the Indians who crossed above the Spaniards' position and occupied a steep escarpment. The Spaniards suffered heavy casualties when they went to fight them there. Although they now wished to retreat they were prevented by the difficult terrain. Many were wounded as a result, particularly Alonso de Alvarado [of Burgos] whose thigh was pierced, and another Knight Commander of San Juan.' Three horses were killed and twenty wounded. The following day the natives fortified themselves on another steep hill and Almagro abandoned the engagement. 'It was later learned that the three thousand Indians who had been moving on Quisquis's left flank had cut off and beheaded fourteen Spaniards.'

Quisquis's army thus performed well despite the fact that it had been surprised while on the march. It inflicted considerable damage on the Spaniards, avoided the slaughter of a cavalry charge, and remained intact to continue its march to Quito. It had been forced to burn large quantities of clothing and other stores being transported, and allowed the Spaniards to seize a vast herd of over fifteen thousand llamas, together with over four thousand conscripted male and female porters.*

Quisquis did not know that the province of Quito had already been occupied by the foreigners. It was a terrible blow to the morale of his weary army to find that the Spaniards had been pouring into their homeland and that they could not rest from the hopelessly uneven struggle. They were defeated in their first skirmish with Benalcázar's men. Their will to fight now collapsed completely, even among officers of Inca origin. They had been away for two years and could think of nothing but dispersing to their homes. 'His commanders told Quisquis to ask the Spaniards for peace since they were invincible.' Quisquis refused, upbraided them for cowardice, and ordered them to follow him into a remote retreat from which they could continue the stubborn, desperate defence of their country. The officers now rebelled, saying that they would sooner die fighting than of starvation in the wilds. 'Quisquis reviled them for this and swore to punish the mutineers. Huaypalcon then struck him on the chest with his lance. Many others immediately ran up with clubs and battle-axes and killed him. Thus ended Quisquis and his battles, he who had been so celebrated a commander among the orejones.' It was a tragic end for one of the empire's finest generals, a man who passionately resented the menace and humiliation of the Conquest. It was equally tragic for the native cause that its finest veterans, the men who had gained most experience in fighting Spaniards, should have mutinied. Without the support of his officers or men Quisquis's position was untenable. They were probably right to abandon the hopeless attempt to attack armoured horsemen with primitive hand-held weapons, but one's sympathies are with the stubborn general who refused to capitulate. The only consolation in his death was that, as Pedro Pizarro wrote, 'the Spaniards never had him in their grasp'.

Quisquis never succeeded in joining forces with Rumiñavi or Zope-Zopahua, and both these commanders soon found themselves having similar difficulties. The Spaniards pursued Rumiñavi to his fortified position near Píllaro. A long, hard battle took place. But the defenders exhausted their supplies of missiles and ammunition and most fled by night towards Quijos. The remainder, being 'without arrows, lances or battle-axes' surrendered. Expelled from here, Rumiñavi 'attempted to assemble forces to continue the war, but all were very exhausted and wished only to live in peace. In the end someone told Sebastián de Benalcázar where he was to be found. [Benalcázar] sent some horsemen who discovered him with little more than thirty men and many women in charge of his baggage. [The Spaniards] suddenly attacked them, and those who could fled. Rumiñavi himself hid, very forlornly, in a small hut.' He attempted to cross a snow-covered mountain between Panzaleo and Umbicho in a bid to join Zope-Zopahua on his hilltop in the Sicchos. A spy recognised him and advised Alonso del Valle who sent a group of Spaniards in pursuit. Miguel de la Chica rode ahead and, as he told it, 'emerged through a short cut that led to a lake. When I reached the lake, the lord Rumiñavi was beside a small hillock, leaning against a tree. I recognised him by the insignia he was wearing. I closed with him and after struggling for a very long time, I captured him.'

There now remained only Zope-Zopahua, fortified in a crag probably to the north of Muliambato, with a good force of local troops and with Quingalumba, chief of the Chillos. The Spaniards attacked the position for two days. They finally made an entry using scaling ladders, 'by night, being guided in the ascent by certain stars, for we could do nothing by day on account of the multitude of Indians who were in the crag'. With this capture the resistance of the Inca forces in the province of Quito came to an end.

Although they had surrendered, the Quitan commanders suffered cruelly at the hands of the Spaniards. According to Marcos de Niza, who had been a chaplain with Alvarado's army, Benalcázar 'summoned Luyes, a great lord of those then in Quito, burned his feet and inflicted many other tortures on him, to make him reveal the whereabouts of Atahualpa's buried treasures – about which he knew nothing. He burned [the chief of] Chambo, another very important lord, who was innocent. He also burned Zope-Zopahua, who had been governor of the province of Quito . . . because he did not give as much gold as [Benalcázar] demanded, and knew nothing about the buried treasure.' Herrera said that the captive generals accepted their tortures with stoicism. 'They behaved with great composure and left him with nothing but his greed. He had them inhumanly killed, because he could not rid his mind of his first impression' that there must be treasure to be found. After the futile tortures, Rumiñavi was led out to execution on the city square of Quito.* He was the last of Atahualpa's great generals, the leader of the most determined resistance to Spanish invasion.

Benalcázar moved back to Quito in early December and divided the city among his followers. In February 1535 he sent Diego de Tapia to pacify the Quillacinga Indians on the Angasmayo river. In June he himself moved down to found Guayaquil on the coast, and to occupy, almost without bloodshed, the Huancavilca province. He later followed his lieutenants Pedro de Añasco and Juan de Ampudia northwards into Pasto and Popayán, and beyond the northern frontier of the Inca empire. They had to fight hard against the tribes of southern Colombia. But the Spaniards had now surpassed the Incas by advancing beyond the farthest point of Huayna-Capac's conquests.

9

PROVOCATION

MANCO Inca returned from Jauja to Cuzco with his thirty-four-year-old companion Hernando de Soto at the end of July 1534. His relations with the Spaniards were excellent. He was grateful to them for having placed him on the throne and driven the Quitans out of Huascar's Peru. The Spaniards were still on good behaviour towards the natives – a temporary restraint imposed by Pizarro's draconian edicts. They were delighted by their protégé Manco. Sancho wrote in June that his elevation as Inca 'has proved highly successful, for all the caciques and chiefs come to serve him and pay homage to the Emperor because of him'. Almagro wrote in May that 'a peace treaty has been made with him in the name of your Catholic and Imperial Majesty' and the town council of Jauja wrote in July acknowledging that Quisquis had been expelled thanks to 'the assistance and advice and friendship of this cacique'.

Once in Cuzco, Manco began to rule as Inca. He had to restore the country's blind faith in the Inca, to gather up the reins of power, and to

The reality of conquest: a Spaniard seizes an Indian girl from her parents assert himself as supreme ruler. He had also to restore the prestige of the Inca capital Cuzco, of the official religion, and of the imperial administration. All these props of Inca rule had been shattered by the upheavals of the civil war and the Spanish invasion. Centrifugal forces were tearing at all parts of the empire, particularly in the areas beyond Jauja and Titicaca which had been absorbed by the Incas within living memory.

The Incas had been experimenting in the administration of their rapidly-growing empire during the decades before the arrival of the Spaniards. They tried to supplant the chiefs of conquered tribes with Inca officials and a decimal structure of administration. The decimal system was based on a census of tribute-paying adult males. These were formed into pyramids of 10,000 with officials at every decimal level from ten to ten thousand. At the lowest levels were foremen representing ten and fifty tribute-payers; above these came the curaca class, with hereditary officials representing units of 100, 500, 1,000, 5,000 (picqua-huaranca curaca), and 10,000 (hunu curaca). The system sounded utopian, and it is difficult to tell how effectively it had been imposed at the time of the Conquest. It was operational in the Inca heartland of the central Andes, less so in the newer conquests to the north and south. The theoretical efficiency of the system appealed to the chroniclers, most of whom expounded it with admiration.* Above the officials in charge of decimal divisions, the Inca appointed senior governors from among his own blood-relatives. Each of the four suyos or quarters of the empire was administered by an Apo, a title also conferred on army commanders. Within the four suyos were a number of provinces or huamani, roughly corresponding to pre-Inca tribal areas, and each of these was governed by a centrally-appointed toqricoc. Although the Incas appointed only members of the Inca nobility to the top posts, Pachacuti had extended the title Inca-by-privilege to include all the Quechua-speaking tribes of the central Andes, and many of the decimal curacas were drawn from these.

The hereditary chiefs of conquered tribes continued to function alongside the centrally-appointed toqricocs. They were given honours and luxuries and their sons were specially schooled at the court in Cuzco. But their powers were shorn. Most functioned only in overseeing the collection of tribute and in selecting candidates for civil or military service. Each province was to witness a struggle for power between the tribal chief and the Inca toqricoc during the troubled years after the Conquest. The governors who had participated in Manco's coronation had by now returned to their provinces and were reasserting the central authority. In some instances they were establishing themselves as autonomous local rulers. But in many parts of Peru the thin crust of Inca rule had melted away for good. Until it could be imposed again by Inca armies, the local tribes were reverting to the rule of their traditional chiefs and were resurrecting tribal deities.

Just above the lowest level of the Inca social structure was a class of people known as yanaconas. These did not farm the land or pay tribute in recognised communities, but acted as personal servants to the Inca nobility. Some were specialist craftsmen; others were semi-skilled workers who travelled about in the service of their masters. The yanaconas quickly recognised the new rulers of Peru. They immediately attached themselves to the Spaniards as personal retainers. They formed an invaluable source of information for the invaders. In return for their services, the Spaniards continued to exempt them from tribute and gave the yanaconas opportunities for plunder and self-aggrandisement at the expense of their fellows. With such incentives, the numbers of tribute-free yanaconas naturally swelled.

The Incas had sewn their empire together and balanced its population by a system of transmigration. Pockets of colonists from the Inca homeland were settled in outlying provinces to establish a loyal nucleus. These were known as mitma-kona, which the Spaniards rendered as mitimaes. In parts of the empire the mitimaes formed as much as a third of the population, some originating from the Inca homeland, others transplanted from different parts of the empire. With the collapse of central authority, the villages of mitimaes throughout Peru turned in upon themselves to form isolated communities. They were now Manco's chief vehicle for re-establishing the authority of his governors.

Less useful to him were the communities of subject tribes that resided in Cuzco. The most important of these was the Cañari, the Ecuadorean tribe that had been cruelly decimated by Huayna-Capac and Atahualpa. The chief of the Cuzco Cañari welcomed Pizarro's column as it approached Cuzco, and the Cañari who had helped Benalcázar with such enthusiasm went on to serve the Spaniards as loyal auxiliaries throughout Peru. Manco therefore found himself surrounded in Cuzco itself by natives of dubious loyalty and a mass of collaborating yanaconas attached to the Spanish citizens.

The Spaniards were hardly aware of the Inca administration and its stresses. They dealt with the natives through a handful of interpreters. But they tacitly supported Manco's attempts to reconstruct imperial rule, since they trusted him and preferred to deal with one puppet Inca. Manco, in return, endorsed their rule and advised his officials to co-operate in the collection of tribute for Spanish encomenderos.

Manco started building himself a palace in Cuzco, as was the custom for each new ruler. The site allotted to him was on the slope above the main square, between the back of Francisco Pizarro's Casana and Huascar's palace of Colcampata, which lay at the highest point of the city below the cliff crowned by Sacsahuaman.*

Manco was also allowed to practise the ceremonies of the Incas' religious calendar. In April 1535 he held the great annual feast of Inti Raymi to celebrate the harvesting of the maize crop. Cristóbal de Molina, a young priest who had just arrived in Peru, was able to witness this splended occasion. His account of it is worth repeating at length. 'The Inca opened the sacrifices and they lasted for eight days. Thanks were given to the sun for the past harvest and prayers were made for the crops to come.... They brought all the effigies of the shrines of Cuzco on to a plain at the edge of the city in the direction of the sun's rise at daybreak. The most important effigies were placed under very fine, beautifully-worked feather awnings. These awnings were arranged in an avenue with one canopy a good quoit's throw from the next. The space [between] formed an avenue over thirty paces wide, and all the lords and chiefs of Cuzco stood in it.... These were all magnificently-robed orejones wearing rich silver cloaks and tunics, with brightly-shining circlets and medallions of fine gold on their heads. They formed up in pairs ... in a sort of procession ... and waited in deep silence for the sun to rise. As soon as the sunrise began they started to chant in splendid harmony and unison. While chanting each of them shook his foot ... and as the sun continued to rise they chanted higher.

'The Inca had a canopy in an enclosure, with a very rich stool for a seat, a short distance from the route of these men. When the time came for the chanting he rose with great dignity, placed himself at their head, and was the first to open the chant. They all followed his lead. After he had been there for some time he returned to his seat and dealt with those who came up to him. From time to time he would go to his choir, remain there for a while and then return. They all stayed there, chanting, from the time the sun rose until it had completely set. As the sun was rising towards noon they continued to raise their voices, and from noon onwards they lowered them, keeping careful track of the sun's course.

'Throughout this time, great offerings were being made. On a platform on which there was a tree, there were Indians doing nothing but throwing meats into a great fire and burning them up in it. At another place the Inca ordered ewes [llamas] to be thrown for the poorer common Indians to grab, and this caused great sport.

'At eight o'clock over two hundred girls came out of Cuzco, each with a large new pot of one and a half arrobas [six gallons] of chicha, plastered and with a cover. The girls came in groups of five, full of precision and order, and pausing at intervals. They also offered to the sun many bales of a herb that the Indians chew and call coca, whose leaf is like myrtle.

'There were many other ceremonies and sacrifices. It is sufficient to say that when the sun was about to set in the evening the Indians showed great sadness at its departure, in their chants and expressions. They allowed their voices to die away on purpose. And as the sun was sinking completely and disappearing from sight they made a great act of reverence, raising their hands and worshipping it in the deepest humility. All the apparatus of the festival was immediately dismantled and the canopies were removed. Everyone returned to their homes and the effigies and terrible relics were returned to their houses and shrines.

'These effigies that they had under the awnings were those of former Incas who had ruled Cuzco. Each had a great retinue of men who stayed there all day fanning away flies with fans like hand mirrors, made of swans' feathers. Each also had its mamaconas, who are like nuns: there were some twelve to fifteen in each awning.

'They came out in this same way for eight or nine days in succession. When all the festivals were over, they brought out on the last day many hand ploughs – these had formerly been made of gold. After the religious service the Inca took a plough and began to break the earth, and the rest of the lords did the same. Following their lead the entire kingdom did likewise. No Indian would have dared to break the earth until the Inca had done so, and none believed that the earth could produce unless the Inca broke it first.'

The ritual breaking of the earth by the Inca was one of the ways in which the personal mystique and authority of the Inca were asserted throughout the empire. But Manco Inca was having some difficulty in re-establishing that authority. He had been elevated by foreign troops during a period of turmoil. Some members of the native aristocracy were not yet sure that the testing period was over or that Manco had proved himself the most worthy of possible claimants. The surface calm in Cuzco during late 1534 and 1535 therefore concealed ruptures within the native community. Still deeper dissensions were also growing between the Spanish commanders and, above all, in relations between Indians and Spaniards.

When Manco marched out of Cuzco with Soto and later with Francisco Pizarro, he left his half-brother Paullu as his deputy in the city. Paullu was only a few months younger than Manco, both aged about twenty. Paullu's standing as an Inca prince was slightly inferior, for although he was a son of Inca Huayna-Capac, his mother had been Añas Collque, daughter of the chief of Huaylas and not a princess of royal Inca blood.* Paullu somehow survived Quisquis's attempt to extinguish the Cuzco branch of the royal family. He probably sought sanctuary south of Cuzco in the Collao, for his influence was always strong in the south. 'He was recognised as ruler and as the son of Huayna-Capac by all the land as far as Chile.' Paullu was disappointed that the Spaniards had selected Manco as Inca, but although he made a flamboyant return to Cuzco in early 1534, he did nothing to dispute Manco's title. This brother in fact gave strong support to Manco in suppressing any threats to his authority.

Manco was more suspicious of his other relatives, but his quarrels with them were closely linked to a rupture that was growing between the Spanish commanders Francisco Pizarro and Diego de Almagro. In December 1534 Almagro sailed down to Pachacamac to ratify his agreement disposing of Alvarado. Pizarro was delighted with the outcome of the Alvarado episode and sent Almagro inland to replace Soto as Governor of Cuzco. He himself continued with the foundation of his new capital at Lima. At this juncture, early in 1535, news reached Peru of a settlement by Charles V that awarded the northern part of the Inca empire to Pizarro and the southern part to Almagro. The exact details were not yet known – Hernando Pizarro was to bring them from Spain at the end of 1535 – but it seemed possible that Cuzco itself lay inside

Almagro's jurisdiction. At any rate, one Diego de Aguero, on first hearing the rumour, hurried after Almagro and overtook him at Abancay with the news that the King had awarded him Cuzco. This ambiguous situation naturally led the citizens of Cuzco to take sides between Almagro and the two younger Pizarro brothers, Juan and Gonzalo, who were in the city. Almagro had brought many men who had transferred from Alvarado's army, and these resented the wealth of the established citizens of Cuzco. The friction increased rapidly until March 1535 when the Pizarro supporters almost provoked open violence. They armed themselves, fortified an Inca palace with artillery, and 'scandalously emerged on to the square, on the point of starting a great altercation'. Juan Pizarro was just prevented from striking Hernando de Soto, who appeared too sympathetic to Almagro. A royal official, Antonio Tellez de Guzman, mediated, threatening both sides with severe punishment. 'For', he wrote to the King, 'had the Christians fought one another the Indians would have attacked those who survived.' The Governor Francisco Pizarro hurried south to try to pacify the explosive situation.

Pizarro reached Cuzco in late May 1535 and immediately tried to find solutions to its many problems. He negotiated an agreement with his old partner Diego de Almagro. The Marshal Almagro was to lead a great expedition to explore Chile, which undoubtedly lay within his southern jurisdiction. Pizarro gave considerable financial assistance to the venture. The obvious hope was that Chile would prove rich enough to satisfy Almagro and his followers. Soto offered a huge sum to be second-in-command, but Almagro preferred Rodrigo Orgóñez.* Preparations for this great conquest of the remaining third of the Inca empire calmed Cuzco by absorbing the energies and stimulating the imaginations of its restless soldiery.

Pizarro also reopened the official furnaces to melt treasure acquired in Cuzco since the initial melting fifteen months earlier. Francisco Pizarro himself brought the largest haul of loot to be melted, and his greedy younger brother Juan brought the second largest. Hernando de Soto, Gonzalo Pizarro and Diego de Almagro had also succeeded in amassing large quantities, and the Crown took its fifth of everything.* The sight of all this treasure also did much to cool tempers in the city.

Manco Inca had now been ruling in Cuzco for almost a year. It was difficult for the young man to assert his sovereignty when his Spanish allies flaunted their complete control of the city. Some of Manco's royal relatives therefore remained unconvinced of his suitability. They felt that Manco had not yet weathered the probationary period at the start of his reign. It might still be possible to elevate a rival claimant, or to reduce Manco's autocracy. The native factions gravitated towards the rival Spanish groups. Manco made it clear that his sympathies were with Almagro. He had no reason to dislike the Governor Francisco Pizarro, who had elevated him to the throne and with whom he had been very friendly in Cuzco and Jauja in the first half of 1534. He may have been antagonised by provocation from the younger Pizarro brothers Juan and Gonzalo, both in their early twenties. Or he may have followed the lead of Hernando de Soto, the Spanish commander with whom he had spent almost all the past eighteen months. Soto supported Almagro's claim to Cuzco, and the Marshal was by all accounts a charming man, who probably made a special effort to court the Inca.

Pizarro tried to smooth the differences between the native leaders. He and Almagro summoned Manco and the opposition party led by his cousin Pascac. They were invited to air their grievances at a debate. Manco and Paullu tried to maintain that any form of argument was an insult to the Inca's divine authority. Paullu rebuked Pascac and his supporters: 'How dare you speak so freely to your lord the Inca, saying anything you please with the consent of the Christians? Get down on your knees before him and beg his mercy for your effrontery. Behave as befits your rank.' When Paullu's outburst was translated to Pizarro, the Governor struck the Inca's brother for having tried to stop the debate in this way. 'This greatly annoyed the Inca. In the end it proved impossible to make peace between the Inca and his relatives.' The rift grew more passionate. Manco's son Titu Cusi claimed that Pascac plotted to assassinate the Inca with a concealed dagger while doing obeisance to him.*

Manco now took action to smash the opposition. He enlisted the help of Almagro, who sent Martín Cote and other Spaniards by night to murder his powerful brother Atoc-Sopa in his bed.* This killing secured Manco's position, although tension continued among the native aristocracy. It was inflamed by the interpreters attached to Pizarro and Almagro, who confused the Indians by claiming supremacy for their respective masters. Pizarro's interpreter went so far as to threaten Manco for his partiality to Almagro. Manco was alarmed by the threats and by the possibility of revenge in the native vendetta. He left his house and Spanish bodyguard one night and hid with Almagro in his bedroom. Pizarro's Spanish partisans learned of his departure and 'a great mob of them went to rob and loot his house, causing much damage – and not even giving much of the loot to the Governor'. Almagro told Pizarro that the Inca had hidden under his bed. He insisted that the intimidation must stop and that the looters must be punished, but Pizarro took no action.* The outrage rankled with Manco. It marked a turning-point in his relations with the Spaniards. The citizens of Cuzco saw that they could plunder the Inca with impunity, and many dropped any pretence of deference to the native ruler. Manco for his part grew in self-assurance with the destruction of his opponents. He was maturing rapidly and becoming more aggressive – potentially dangerous and increasingly sensitive to Spanish insult.

Almagro left Cuzco for Chile on 3 July 1535 with 570 cavalry and foot-soldiers, excellently equipped and supported by great trains of native porters. Francisco Pizarro left soon afterwards, returning to the coast to continue the foundation of another city, Trujillo, between Piura and Lima. Soto also left. He returned to Spain, fabulously rich, and in 1538 obtained from King Charles a licence for his expedition to Florida, on which he met his death near the Mississippi. Other expeditions left for explorations of the unknown fringes of the empire: Alonso de Alvarado to the Chachapoyas, Juan Porcel to the Bracamoros and Captain Garcilaso de la Vega to the Cauca valley on the coast of Colombia.

While preparing the Chilean expedition, Almagro asked his protégé Manco to supply a native contingent. The Inca responded generously, sending 12,000 men under the command of the two most powerful figures at his court, his brother Paullu and the high priest Villac Umu. This native force was to prove invaluable to Almagro. The presence of Paullu ensured the friendship and co-operation of almost all the peoples of the southern part of the empire. The chief priest Villac Umu was a relative of Huayna-Capac and an eminence of great authority. Cieza de León wrote that this priest 'dwelt in the temple. He retained his dignity for life and was married, and was so esteemed that he was equal in debate to the Inca. He had power over all oracles and temples, and appointed and removed priests. These [chief priests] were of high birth and powerful parentage.' Spanish chroniclers invariably compared Villac Umu to their own Pope.

Various motives have been read into Manco's sending of this force.

One theory was that he was ridding Cuzco of two potential rivals – but Paullu and Villac Umu had been his most ardent supporters up to the time of departure. Another theory was that Paullu engineered his own selection in order to ingratiate himself with the Spaniards and gain influence in southern Peru. But Paullu had no reason to suppose that his presence on the expedition would make Almagro transfer his affection from Manco. A third theory was that Manco had determined to rebel and sent Paullu with instructions to annihilate Almagro's army at an appropriate moment. The true explanation was probably far simpler. Manco still hoped to rule the Inca empire in collaboration with the Spaniards. He was therefore glad to send a strong force under his most trusted supporters on this great expedition, just as he had done on the various campaigns during 1534. This seemed an excellent opportunity to demonstrate his personal hold on the southern part of the empire.

With the departure in rapid succession of the Spanish leaders and of many of the Spaniards in Cuzco, Manco was left in a city administered by the young Juan and Gonzalo Pizarro. With this irresponsible pair in charge, Manco became subjected during the last months of 1535 to increasing harassment and insult. This treatment of the Inca himself reflected a hardening attitude towards the natives throughout Peru. The restraint that Pizarro had imposed with difficulty at the outset of the Conquest was now gone. With the arrival of Alvarado's army from Quito and increasing numbers of Spanish adventurers from the Caribbean, the conquerors felt securely in control of the country. The newest arrivals, who had missed the great riches of the first conquistadores, were often the most brutal towards the natives.

This was particularly true of Alvarado's men who had shown such cruelty in the march towards Quito, and who were now on Almagro's Chilean expedition. The priest Cristóbal de Molina was on that expedition and recorded his disgust. 'Any natives who would not accompany the Spaniards voluntarily were taken along bound in ropes and chains. The Spaniards imprisoned them in very rough prisons every night, and led them by day heavily loaded and dying of hunger.' Many natives fled their villages to escape Almagro's press-gangs, and the chiefs resented his demands for gold transmitted through Paullu. Bands of Spanish horsemen hunted down the missing villagers. 'When they found them they brought them back in chains. They carried off their wives and children as well, taking any attractive women for their personal service –

and for more besides ... When the mares of some Spaniards produced foals, they had these carried in litters by the Indian women. Others had themselves carried in litters as a pastime, leading their horses by the halters so that they would become good and fat.' The native porters 'worked without rest all day long, ate only a little roast maize and water, and were barbarously imprisoned at night. One Spaniard on this expedition locked twelve Indians in a chain and boasted that all twelve died in it. When one Indian died they cut off his head, to terrify the others and to avoid undoing their shackles or opening the padlock on the chain.'

The men on this expedition perfected a system of raiding known as rancheando, which Pedro Pizarro described 'as meaning, in plain language, to rob'. 'The Spaniards encouraged their Indian and Negro servants to become great rancheadores and robbers. In the camp, a Spaniard who was a good raider and cruel and killed many Indians was considered a good man with a big reputation. Anyone inclined to treat the natives well or to stand up for them was despised.' As a result of their cruelties, isolated groups of Spaniards were ambushed and killed by some of the natives of the Bolivian altiplano and by the tribes on the south-eastern edge of the Inca empire. There was no organised resistance such as Benalcázar had met from the Inca armies on the road to Quito. But the expedition suffered great hardships and considerable losses in the high passes into Chile. The high priest Villac Umu escaped at Tupiza in late October and made his way back towards Cuzco. In the province of Copiapó 'all the Indians whom they had brought from Cuzco fled, and the Spaniards were left with no one to fetch them even a pot of water'.

The brutality of the Chilean expedition was being repeated throughout Peru. The population 'was becoming incensed wherever the Spaniards passed by. This was because the Spaniards were not content with the service of the natives but tried to rob them in every town. In many places the Indians would not tolerate this, but began to rebel and to organise themselves for defence. The Spaniards undoubtedly went too far in their abuse of them.' The strangers who had once been welcomed as providential allies against Atahualpa's Quitans were now taking possession of repartimientos awarded to them as conquerors. The conquistadores demanded great quantities of local produce – llamas, vegetables, cloth, wood, precious metals – as well as personal service from hundreds of native men and women. Wherever there were mines the natives were put to work mining, particularly in the gold mines of the Collao that had been designated as a royal repartimiento.

The Spaniards had come without European women. Although high altitude decreases sexual desire in lowland men, the Spaniards naturally wanted and took native women. In many cases they settled with one native manceba as their permanent concubine. The Indians showed no great objection to this, and the women themselves were often attracted by the dashing foreigners. 'The Indian woman who proved most attractive to the Spaniards prided herself on the fact.' But the ravages of the invaders disrupted the native social structure. 'From this time onwards they adopted the custom of having public prostitutes, and they abandoned their former practice of marrying, for no woman who was good-looking was safe to her husband – it would be a miracle if she escaped the Spaniards and their yanacona servants.'

The Spanish leaders selected for themselves Inca princesses, ladies who would normally have slept only with the Inca himself or with princes of royal blood. Francisco Pizarro himself lived with Huayna-Capac's daughter Quispe Cusi, who was usually known to the Spaniards as Inés Huayllas Ñusta: Huayllas or Yupanqui were the surnames given by Spaniards to the Inca royal family; ñusta meant royal princess, as opposed to coya, which referred to the sister-queen of the Inca, or palla, applied to noble ladies not of royal blood. Pizarro was a fifty-six-year old bachelor who had never married a European wife. He was overjoyed when the fifteen-year old Inés bore him a daughter at Jauja in December 1534. The girl was solemnly baptised Francisca in the tiny church of Jauja, and three Spanish wives of conquistadores were even found to be godmothers. There were tournaments and celebrations among Spaniards and natives, all of whom were delighted at this product of the union of Spanish leadership and Inca royalty. Pizarro often played with his daughter Francisca, and arranged to have her legitimised by royal decree of 27 May 1536.* In 1535 Inés bore Pizarro a son whom he called Gonzalo. The Governor later arranged for her to marry his follower Francisco de Ampuero; their modern descendants are the powerful Peruvian family of Ampuero.

Diego de Almagro had been moving too rapidly on the conquests of Cuzco and Quito to acquire a native concubine before 1535. But he made a most profitable liaison when he reached Cuzco as lieutenant-governor. Manco had 'a sister who was the most important lady in the country.

She was called Marca-Chimbo, was the daughter of Huayna-Capac and of his full sister, and would have inherited the Inca empire had she been a man. She gave Almagro a pit in which there was a quantity of gold and silver tableware, which yielded eight bars or 27,000 silver marks when melted down. She also gave another captain 12,000 castellanos from the contents of that pit. But the poor creature was not shown any greater respect or favour by the Spaniards because of this. Instead, she was repeatedly dishonoured – for she was very pretty and of a gentle nature – and she caught the pox. Finally, however, she married a Spanish citizen [Juan Balsa] at a later date, . . . died a Christian, and was a very good wife.'

Hernando de Soto found himself a lady who formed part of a legend. The Jesuit chronicler Miguel Cabello de Balboa heard her story from Atahualpa's cousin Don Mateo Yupanqui in Quito.* The romantic legend began with a union between the Inca Huascar and a great beauty known as Curicuillor, 'the golden star'. Their daughter, also called Curicuillor, had a passionate love affair with Atahualpa's ambassador and general, Quilaco, but this Inca Romeo and Juliet soon found themselves tragically separated in opposite camps of the civil war. They returned to one another after many adventures during the war, and were living in Jauja when Soto arrived there. He sponsored them at their baptism, when Quilaco took the name Hernando Yupanqui and his wife became Leonor Curicuillor, after which they were married by the laws of the church. Quilaco died soon after, and Soto attached himself to the lovely young widow Leonor. When Soto occupied the Amaru Cancha on the south side of Cuzco's main square, he installed his mistress nearby.* They had a daughter, called Leonor de Soto, who married a royal notary, García Carrillo, and lived in Cuzco.*

One of Atahualpa's sisters called Azarpay accompanied the new Inca Tupac Huallpa and the Spanish army as far as Jauja. When Tupac Huallpa died, the royal paymaster Navarro asked Pizarro to give her to him – thinking that she could reveal treasures for him. Pizarro agreed, but Azarpay fled back to Cajamarca. She was discovered there by some Spaniards and brought to Lima at the end of 1535. The Governor himself installed her in his chambers, thereby arousing the jealousy of Inés.*

Although most Spaniards had pretty native girls as mistresses, they were most reluctant to marry them. They preferred to await Spanish wives; and Spanish women soon arrived in Peru in considerable quantities.

Garcilaso de la Vega, for instance, had a royal Inca lady as a concubine and produced the famous historian Garcilaso by her. But he eventually married a Spanish woman. Alonso de Toro kept and openly favoured his Indian mistress after marrying a Spanish wife. This so outraged the wife's family that her father eventually killed Toro. Alonso de Mesa acquired a veritable harem of Indian women. When he started writing his will in 1544 he claimed five children by five women, all living in his house. But later in the will he recorded six children by six women, and then remembered that there was a seventh who was pregnant.*

The younger Pizarro brothers also arranged native ladies for themselves. Juan Pizarro wrote in 1536: 'I have received services from an Indian woman who has given birth to a girl whom I do not recognise as my daughter'. Gonzalo Pizarro decided that he must also have an Inca princess. He conceived a passion for Cura Ocllo, the full sister and coya or wife of the Inca Manco. His demand for this woman scandalised the native nobility. Manco's son described how the high priest Villac Umu and the general Tiso rebuked Gonzalo 'with severe and angry expressions on their faces'. Gonzalo's reply to Villac Umu was typically thuggish. 'Who told you to talk like that to the King's corregidor? Don't you know what sort of men we Spaniards are? By the King's life, if you don't shut up I'll catch you and play games with you and your friends that you'll remember all your lives. I swear if you don't keep quiet I'll slit you open, alive, and make little pieces of you!' Manco duly gathered together a quantity of treasure and handed it over, but Gonzalo wanted the other part of his demand. 'Well, señor Manco Inca, let's have the lady coya. All this silver is fine, but she is what we really want.' Manco, in desperation, persuaded one of his sister's companions called Inguill to dress up as the coya. 'When the Spaniards saw her come out so well dressed and looking so beautiful they were delighted and shouted: "Yes, that one, that one! Weigh her in – she's the coya."' Gonzalo became insistent. '"Señor Manco Inca, if she is for me, give her to me right away because I cannot stand it any longer." My father [Manco] had her well prepared. He said "Yes, congratulations. Do whatever you wish with her." So [Gonzalo] went up in front of everyone, oblivious to all else, and kissed and embraced her as if she were his legitimate wife. My father and the rest were amazed and laughed greatly at this, but Inguill was horrified and frightened at being embraced by someone she did not know. She screamed like a mad woman and said she would run away rather than face people like these. When my father saw her behaving so wildly and refusing to go with the Spaniards, he realised that his own freedom depended on her going. He angrily ordered her to go with them. Seeing my father so angry, she did what he said and went, more from fear than for any other reason.' Manco's deception did not succeed, for he himself wrote later: 'Gonzalo Pizarro took my wife and still has her.'

Manco was an obvious target for the greedy Spaniards in Cuzco. They all knew that Atahualpa had provided the legendary treasures of Cajamarca, and many therefore assumed that his brother could conjure up similar riches. The Spanish leaders were always pestering the Inca, and for a time Manco appeased them by revealing caches of precious objects. Even lesser Spaniards joined in the persecution, until the harassment became intolerable.* Villac Umu returned to Cuzco and reported the cruelty of Almagro's Chilean expedition. The general Tiso, the empire's greatest surviving military leader, brought stories from all over Peru of the crumbling of the Inca administration and of excesses by Spanish invaders. The Incas of Huascar's faction were now aware that they had been duped: far from being liberated from the Quitan occupation, their entire race was sinking into the control of a foreign invader. Manco had clung to his title of Inca in the hope that with the elimination of family rivals he could restore the prestige of the monarchy. His hand was forced by the personal insults of the Spaniards in Cuzco, and his determination was strengthened by the advice of the elders Villac Umu, Tiso and Anta-Aclla, leaders of the native church and army. They exhorted the young Inca with passionate patriotism. 'We cannot spend our entire lives in such great misery and subjection. Let us rebel once and for all. Let us die for our liberty, and for the wives and children whom they continually take from us and abuse.' The arguments of these elders were successful. In the autumn of 1535 Manco Inca reached the momentous decision to oppose the Spaniards, to try to lead his people in driving the conquerors out of Peru. The decision meant the reversal of the policy of collaboration practised by Manco himself and his predecessors Tupac Huallpa and Atahualpa. The young Inca would now be leading a resistance formerly waged by his bitter enemies Quisquis and Rumiñavi.

Manco's first move was to summon a secret meeting of native chiefs, particularly those of the southern Collao. He expounded the provocation suffered by him and his people, and noted that with the departure of

Almagro's army Cuzco was relatively free of Spaniards. He announced his determination to rebel.

That night, under cover of darkness, the Inca slipped out of Cuzco in his litter, accompanied by some of his wives, servants and orejones. Some yanacona spies who had been at the meeting informed Juan Pizarro, who went to investigate Manco's house and found it empty. He roused his brother Gonzalo and a number of other Spanish horsemen and immediately galloped off in the darkness along the Collao road to the south-east of the city. They soon overtook some of the Inca's party, who claimed that their lord had left in the opposite direction. Gonzalo Pizarro seized an orejón of the Inca's retinue and 'pressed him to declare where the Inca was going, and, when he refused repeatedly, they tied a rope around his genitals and tortured him astutely until he gave a great shout, saying that the Inca was not going that way'. But four horsemen had continued along the road, continually asking for the Inca. At the far end of Cuzco valley, near Lake Muyna, Manco heard the horses approaching and left his litter to hide in some reeds. The Spaniards rode back and forth searching systematically for him. When he saw that he must be found, Manco surrendered. He gave an improbable explanation for his nocturnal adventure: he had been on his way to join Almagro at the Marshal's request.

Manco was brought back under guard and imprisoned on his return to Cuzco. His son complained that 'Gonzalo Pizarro ordered his men to bring irons and a chain, with which they shackled my father as they pleased. They threw a chain around his neck and irons on his feet.' The natives were highly incensed at this treatment of their Inca. They fasted, sacrificed and offered special prayers to their gods for the Inca's release and for an opportunity to be rid of the Spaniards. The latter considered themselves lucky to have caught the Inca in time. Pedro Pizarro commented that 'had this Indian not been captured at this juncture all of us Spaniards in Cuzco would have perished, for the greater part of the Christians had gone out to inspect the Indians on their encomiendas.'

The abuse of the Inca increased during his captivity. Manco was later reported to have said that 'he was urinated upon by Alonso de Toro, [Gregorio] Setiel, Alonso de Mesa, Pedro Pizarro and [Francisco de] Solares, all citizens of this city. He also said that they burned his eyelashes with a lighted candle.' Manco said on another occasion: 'I rebelled more on account of the abuses done to me than because of the gold they took from me, for they called me a dog and they struck me, and took my wives and the lands that I farmed. I gave Juan Pizarro 1,300 gold bricks and two thousand golden objects: bracelets, cups and other pieces. I also gave seven gold and silver pitchers. They said to me: "Dog, give us gold. If not, you will be burned." ' Cristóbal de Molina reported that 'they stole everything he had, leaving him nothing. They kept him imprisoned for many days on this occasion, guarded day and night. They treated him very, very disgracefully, urinating on him and sleeping with his wives, and he was deeply distressed.' Almagro's son repeated these accusations and added that Manco's persecutors 'urinated and spat in his face, struck and beat him, called him a dog, kept him with a chain round his neck in a public place where people passed.' These squalid reports were all repeated by men who hated the Pizarros, but they had a basis of truth. A royal emissary, Bishop Berlanga, reported to the King: 'Any claim that the Inca should serve no one has been a fraud. For the Governor has exploited him, and so have any others who wished.'

Manco's confinement took place in early November 1535. Tiso and the Collao chiefs had apparently succeeded in escaping from Cuzco at the time when their Inca was recaptured. Tiso rapidly made his way to the highlands north of Jauja, and whipped up revolt at Tarma and Bombón, on the encomienda granted to the royal treasurer Alonso Riquelme. The Governor Pizarro was in his new city of Lima and immediately ordered his local lieutenant Cervantes to suppress this insurrection. Cervantes marched out of Jauja, but Tiso eluded him, slipping away into the jungles to the east.* Meanwhile, the Collao chiefs returned to their tribes and ordered them to oppose Spaniards riding out to inspect their repartimientos. The first encomendero to be killed was one Pedro Martín de Moguer, followed shortly by Martín Domínguez. News arrived of the killing of Juan Becerril in the Condesuyo. Simón Suárez was told that the Indians of his encomienda would pay him tribute if he went to collect it; he was murdered on arrival. Similar ruses during these months achieved the deaths of some thirty Spaniards, on isolated encomiendas or in the royal mines.*

The Spaniards acted with characteristic vigour. Gonzalo Pizarro immediately rode out to take reprisals on the murderers of Pedro Martín de Moguer. He found the natives fortified on a rocky outcrop called Ancocagua in bad, humid country. He called up his brother Juan, who arrived with reinforcements. The Spaniards attacked under cover of a siege blanket but this was pierced by the defenders' stones. Manco had been asked to send an orejón on this expedition, and the Spaniards now arranged for this man to call for the surrender of the besieged. The orejón, instead, bravely called on the defenders to resist to the death, and when the Spaniards learned this through an informer they burned the orejón alive. A second orejón was sent from Cuzco, and this man betrayed his compatriots. He advised four Spaniards to shave their beards, dye their faces and disguise themselves as Indians. One night, the orejón persuaded the defenders to admit him and his four companions within the outer circuit of their walls. While the four Europeans waited in terror of a betrayal, the orejón talked the defenders into opening their second gate. The Spaniards rushed in and the natives were overwhelmed. Many leaped to their deaths from the cliffs, and 'a cruel slaughter began at the hands of the yanaconas, who cut off legs and arms in a welter of bloodshed, while the Spaniards showed no greater mercy.' The Pizarro brothers rode westwards from Ancocagua, to exact further reprisals on the Indians of the Condesuyo for the murder of Becerril.

Hernando Pizarro, who had left Peru over two years previously to take the first consignment of Atahualpa's ransom to King Charles, now reappeared. He brought with him a large contingent of adventurers from the mother-country, the most notable of whom were Alonso Enríquez de Guzman and Pedro Hinojosa. His brother Francisco was delighted to see Hernando again, and soon sent him to be Lieutenant-Governor of Cuzco, in command of the younger Juan, who remained as the city's corregidor.

Hernando Pizarro reached Cuzco in January 1536 to find that his younger brothers were still away on their punitive expedition. Manco Inca had just been released from imprisonment on instructions from Juan Pizarro.* Hernando met the young Inca for the first time and immediately undertook to win his friendship. He showed every possible favour to the native ruler, to the distress of the Spanish extremists, and Manco apparently responded warmly to this kindness. Hernando was acting partly on instructions from King Charles, who had been told of the Inca's great co-operation with the Spaniards in 1534, and had ordered that he be given the respect due to a hereditary monarch. Hernando had himself witnessed the wave of sympathy for Atahualpa at the Spanish court. Cuzco was now less vulnerable, with the arrival of Hernando and his companions and with the hurried return of many encomenderos from visits to their repartimientos. It therefore seemed safe to release Manco and to attempt by kindness to atone for the insults that might have inspired the incidents in the Collao. Whatever Hernando's motives for courting Manco, the rumour inevitably spread that he was ingratiating himself in order to obtain treasures that his brothers had failed to extract by bullying. Manco did in fact give Hernando Pizarro some precious objects, although the quantity was grossly exaggerated in later years.

The first months of 1536 passed in apparent tranquillity. The abortive revolts in the Collao, Condesuyo and at Tarma seemed to be extinguished. But the appeasement of the Inca, which might have succeeded six months earlier, was too late. Manco had suffered too much personal insult and was by now too well aware of the enormity of the Spanish invasion. If his resolve needed strengthening, this was done by the high priest Villac Umu, who insisted that the Spaniards in Cuzco could be defeated. For once it was the Inca who deceived the Spaniards. Manco was simply waiting for the end of the rains before assembling a great popular army against the invaders. At length the time came and Villac Umu and Manco despatched messengers to order a national mobilisation. Ever since his release from imprisonment, 'Manco was plotting to kill the Spaniards and to make himself king as his father had been. He had quantities of weapons manufactured in secret, and arranged large plantings of crops to have enough food in the wars and sieges that he hoped to wage.'

As the rainy season ended, great contingents of native warriors moved towards Cuzco, totally undetected by the Spanish invaders and their collaborators. In the sonorous words of Manco's son Titu Cusi, 'Villac Umu sent Coyllas, Usca Curiatao and Taipi to bring the men of the Chinchaysuyo; Llicllic and many other commanders went to the Collao to recruit the men of that quarter; Surandaman, Quilcana and Curi-Hualpa to the Condesuyo; and Ronpa Yupanqui' to the forest tribes of the Antisuyo. The mobilisation and the native headquarters were to be at Calca in the Yucay valley. Here they would be protected by the Yucay river from the Spanish cavalry, but would be only fifteen miles due north of Cuzco.

When the time was ripe, Manco had to leave Cuzco to preside at a meeting of his chiefs and to launch the revolt. Remembering the swift recapture when he had tried to escape from Juan Pizarro, Manco now adopted a more subtle stratagem. He simply asked permission of Hernando Pizarro to go with Villac Umu to perform some ceremonies in the Yucay Valley. He promised to bring back a life-sized golden statue of his father Huayna-Capac. The hint of gold worked like magic. Hernando gave permission, and the Inca and High Priest left Cuzco, accompanied by two Spaniards and by Pizarro's personal interpreter Antonico.*

Manco left on 18 April, Wednesday of Holy Week, after the close of Mass. There was an immediate outcry in Cuzco. The Indians in the city who had collaborated most closely with the Spaniards – and these included some of Manco's own kindred – insisted that Manco would return with an army to slaughter them all. They transmitted their apprehension to the Spanish citizens, who formed a delegation to protest to Hernando Pizarro. He tried to allay their fears by insisting on his complete confidence in the Inca. He continued to take this stand two days later when one Alonso García Zamarilla arrived with the news that he had met the Inca's party in the wild hills leading towards Lares, some distance beyond Calca. Manco had told this Spaniard that he was going up into the hills to fetch some hidden gold, and Hernando said that this sounded plausible.*

But Lares was the site chosen for the Indians' final pre-rebellion meeting. Manco sat in state over an assembly of Inca chiefs and military men. Two great golden jars of chicha were produced and each leader drank an oath staking his life on the extermination or expulsion of every Christian in the empire. The occasion was marked by appropriate speeches doubtless delivered in the ponderous, rambling but immensely impressive manner that characterises any gathering of Andean Indians to this day.

Finally, on Saturday the eve of Easter 1536, Hernando Pizarro was informed for certain that the Inca had rebelled and that his intentions were extremely dangerous. The Lieutenant-Governor promptly made an announcement to the people, telling them the terrible news and acknowledging his own gullibility. He then held consultations with the leading Spaniards as to the best course of action.*

Climax of the siege of Cuzco: Manco Inca and his men ignite the roofs of the city

10

THE GREAT REBELLION

As always, the Spaniards' first reaction to a disturbance with the Indians was to try to seize the initiative. Hernando sent his brother Juan with seventy cavalrymen – virtually every horse then in Cuzco – to disperse the Indians in the Yucay valley. While riding across the plateau of rolling grassy hills that separates the valley of Cuzco from that of Yucay, they met the two Spaniards who had been with Manco. These had been beguiled by him into leaving when he continued towards Lares, and they were now returning in all innocence to Cuzco, unaware of any native rebellion. The first sight of the magnitude of the opposition came when Pizarro's men appeared at the brow of the plateau and looked down at the beautiful valley beneath them. This is one of the loveliest views in the Andes; the river below winds across the broad flat floor of the valley, whose rocky sides rise as abruptly as the fantastic scenery in the background of a sixteenth-century painting. The slopes are tightly contoured with neat lines of Inca terraces, and above them, in the distance, the snowy peaks of the Calca and Paucartambo hills shine brilliantly in the thin air. But now the valley was filled with native troops, Manco's own levies from the area around Cuzco. The Spaniards had to fight their way across the river, swimming their horses. The Indians retreated on to the slopes and allowed the cavalry to occupy Calca, which they found full of a great treasure of gold, silver, native women and baggage. They occupied the town for three or four days, with the natives harassing the sentries at night but making no other attempt to drive them out. The reason for this was appreciated only when a horseman from Hernando Pizarro galloped in to recall the cavalry with all possible speed; for irresistible hordes of native troops were massing on all the hills around Cuzco itself. The cavalry force was harassed continuously on the return journey, but succeeded in entering the city, to the relief of the remaining citizens.

'As we returned we found many squadrons of warriors continuously arriving and camping in the steepest places around Cuzco to await the assembly of all [their men]. After they had all arrived, they camped on the plain as well as on the hills. So many troops came there that they covered the fields. By day they looked like a black carpet covering everything for half a league around the city of Cuzco, and by night there were so many fires that it resembled nothing less than a very clear sky filled with stars.' This was one of the great moments of the Inca empire. With their genius for organisation, Manco's commanders had succeeded in assembling the country's fighting men and in arming, feeding and marching them to the investiture of the capital. All this had been done despite the fact that the empire's communications and supply depots were disrupted, and without giving any warning to the astute and suspicious foreigners occupying the land. All the Spaniards were taken by surprise by the mobilisation at their gates, and were staggered by its size. Their estimates of the numbers opposing them ranged from 50,000 to 400,000, but the accepted figure by the majority of chroniclers and eyewitnesses was between 100,000 and 200,000.*

The great colourful steam-roller of native levies closed in from every horizon around Cuzco. Titu Cusi wrote with pride that 'Curiatao, Coyllas, Taipi and many other commanders entered the city from the

Carmenca side ... and sealed the gate with their men. Huaman-Quilcana and Curi-Hualpa entered on the Condesuyo side from the direction of Cachicachi and closed a great gap of over half a league. All were excellently equipped and in battle array. Llicllic and many other commanders entered on the Collasuyo side with an immense contingent, the largest group that took part in the siege. Anta-Aclla, Ronpa Yupanqui and many others entered on the Antisuyo side to complete the encirclement of the Spaniards.'

The native build-up around Cuzco continued for some weeks after the return of Juan Pizarro's cavalry. The warriors had learned to respect Spanish cavalry on level ground, and they kept to the slopes. The royal general Inquill was in charge of the encircling forces, assisted by the high priest Villac Umu and a young commander Paucar Huaman. Manco maintained his headquarters at Calca.

Villac Umu pressed for an immediate attack, but Manco told him to wait until every last contingent had arrived and the attacking forces had become irresistible. He explained that it would do the Spaniards no harm to suffer confinement just as he had done: he himself would come to finish them off in due course. Villac Umu was distressed by the delay, and even Manco's son criticised his father for it. But Manco was applying Napoleon's dictum that the art of generalship is to come to battle with a force vastly superior to the enemy's. He thought that his warriors' only hope against the Spanish cavalry lay in overwhelming numbers. Villac Umu had to content himself with occupying Cuzco's citadel, Sacsahuaman, and with destroying the irrigation canals to flood the fields around the city.

The Spaniards inside Cuzco were suffering just as much anxiety as Manco had hoped. There were only 190 Spaniards in the city, and of these only eighty were mounted. The entire burden of the fighting fell on the cavalry, for the 'greater part of the infantry were thin and debilitated men'. Both sides agreed that a Spanish infantryman was inferior to his native counterpart, who was far more nimble at this high altitude.* Hernando Pizarro divided the horsemen into three contingents commanded by Gabriel de Rojas, Hernán Ponce de León and his brother Gonzalo. He himself was Lieutenant-Governor, his brother Juan was corregidor, and Alonso Riquelme, the royal treasurer, represented the Crown.

At the outset, while the native forces were still massing, the Spaniards tried their tactic of charging out into the thick of the enemy. This met with far less success than usual. Many Indians were killed, but the crush of fighting men stopped the onrush of the horses, and once the Indians saw that the cavalry was thoroughly embroiled they turned on it with savage determination. A group of eight horsemen fighting around Hernando Pizarro saw that it was being surrounded and decided to retreat to the city. One man, Francisco Mejia, who was then alcalde or mayor of the city, was too slow. The Indians 'blocked his horse and grabbed at him and the horse. They dragged them about a stone's throw away from the other Spaniards, and cut the heads off [Mejia] and off his horse, which was a very handsome white horse. The Indians thus emerged from this first engagement with a distinct gain.'

This success against cavalry on level ground greatly emboldened the attackers. They moved closer to the city until they were camped right up against the houses. In the tradition of intertribal warfare, they tried to demoralise the enemy by jeering and shouting abuse and by 'raising their bare legs at them to show how they despised them'. Such skirmishes took place every day, with great courage shown on either side but no appreciable gains.

Finally on Saturday, 6 May, the feast of St John-ante-Portam-Latinam, Manco's men launched their main attack. They moved down the slope from the fortress and advanced along the steep, narrow lanes between Colcampata and the main square. Many of these alleys still end in long flights of steps between whitewashed houses and form one of the most picturesque corners of modern Cuzco. 'The Indians were supporting one another most effectively, thinking that it was all over. They charged through the streets with the greatest determination and fought hand-to-hand with the Spaniards.' They even succeeded in capturing the ancient enclosure of Cora Cora which overlooked the northern corner of the square.* Hernando Pizarro appreciated its importance and had fortified it with a palisade the day before the Indian onslaught. But his infantry garrison was driven out by a dawn attack.

If the horse was the Spaniards' most effective weapon, the sling was undoubtedly the Indians'. Its normal missile was a smooth stone about the size of a hen's egg,* but Enríquez de Guzman claimed that 'they can hurl a huge stone with enough force to kill a horse. Its effect is almost as great as [a shot from] an arquebus. I have seen a stone shot from a sling break a sword in two when it was held in a man's hand thirty yards away.' In the attack on Cuzco the natives devised a deadly new use for their slingshots. They made the stones red-hot in their camp fires, wrapped them in cotton and then shot them at the thatched roofs of the city. The straw caught fire and was burning fiercely before the Spaniards could even understand how it was being done. 'There was a strong wind that day, and as the roofs of the houses were thatch it seemed at one moment as if the city were one great sheet of flame. The Indians were shouting loudly and there was such a dense cloud of smoke that the men could neither hear nor see one another. . . . They were being pressed so hard by the Indians that they could scarcely defend themselves or come to grips with the enemy.' 'They set fire to the whole of Cuzco simultaneously and it all burned in one day, for the roofs were thatch. The smoke was so dense that the Spaniards almost suffocated: it caused them great suffering. They would never have survived had not one side of the square contained no houses and no roofs. Had the smoke and heat come at them from all sides they would have been in extreme difficulty, for both were very intense.' Thus ended the Inca capital: stripped for Atahualpa's ransom, ransacked by Spanish looters, and now burned by its own people.

From the captured bastion of Cora Cora the Indian slingers kept up a withering fire across the square. No Spaniard dared venture on to it. The besieged were now cornered in two buildings facing each other at the eastern end of the square. One was the great galpón or hall of Suntur Huasi, on the site of the present cathedral, and the other was Hatun Cancha, 'the large enclosure', where many of the conquistadores had their plots. Hernando Pizarro was in charge of one of these structures and Hernán Ponce de León of the other. No one dared to move out of them. 'The barrage of slingshot stones coming in through the gateways was so great that it seemed like dense hail, at a time when the heavens are hailing furiously.' 'The city continued to burn on that and the following day. The Indian warriors became confident at the thought that the Spaniards were no longer in a position to defend themselves.'

By extraordinary chance, the thatched roof of Suntur Huasi itself did not catch fire. An incendiary projectile landed on the roof. Pedro Pizarro said that he and many others saw this happen: the roof started to burn and then went out. Titu Cusi claimed that the Spaniards had Negroes stationed on the roof to extinguish the flames. But to other Spaniards it seemed a miracle, and by the end of the century it became established as such. The seventeenth-century writer Fernando Montesinos said that the Virgin Mary appeared in a blue cloak to extinguish the flames with white blankets, while St Michael was by her side fighting off devils. This miraculous scene became a favourite subject for religious paintings and alabaster groups, and a church called the Triunfo was built to commemorate this extraordinary escape.*

The Spaniards were becoming desperate. Even Manco's son Titu Cusi felt a touch of pity for these conquerors: 'They secretly feared that those were to be the last days of their lives. They could see no hope of relief from any direction, and did not know what to do.' 'The Spaniards were extremely frightened, because there were so many Indians and so few of them.' 'After six days of this strenuous work and danger the enemy had captured almost all the city. The Spaniards now held only the main square and a few houses around it. Many ordinary people were showing signs of exhaustion. They advised Hernando Pizarro to abandon the city and look for some way to save their lives.' There were frequent consultations among the weary defenders. There was desperate talk of trying to break the encirclement and reach the coast via Arequipa, to the south. Others thought that they should try to survive inside Hatun Cancha, which had only one entrance. But the leaders decided that the only thing to do was to fight back, and if necessary die fighting.

In the confused street fighting the natives were resourceful and ingenious. They evolved a series of tactics to contain and harass their terrible adversaries; but they could not produce a weapon that could kill a mounted, armoured Spanish horseman. Teams of Indians dug channels to divert Cuzco's rivers into the fields around the city, so that the horses would slip and sink into the resulting mire. Other natives dug pits and small holes to trip the horses when they ventured on to the agricultural terraces. The besiegers consolidated their advance into the city by erecting barricades in the streets: wicker screens with small openings through which the nimble warriors could advance to attack. Hernando Pizarro decided that these must be destroyed. Pedro del Barco, Diego Méndez and Francisco de Villacastín led a detachment of Spanish infantry and fifty Cañari auxiliaries in a night attack on the barricades. Horsemen covered their flanks while they worked, but the natives maintained a steady barrage from the adjoining roofs.

The flat walls of Cuzco's houses were exposed when the thatch was burned off in the first great conflagration. The natives found that they could run along the tops of the walls, out of reach of the horsemen charging below. Pedro Pizarro recalled an episode when Alonso de Toro was leading a group of horsemen up one of the streets towards the fortress. The natives opened fire with a bombardment of stones and adobe bricks. Some Spaniards were thrown from their horses and half buried in the rubble of a wall overturned by the natives. The Spaniards were only dragged out by some Indian auxiliaries.*

With inventiveness born of desperation, the natives evolved another weapon against the Christians' horses. This was the ayllu, or bolas: three stones tied to the ends of connected lengths of llama tendons. The twirling missile tangled itself around the horses' legs with deadly effect. The natives brought down 'most of the horses with this device, leaving almost no one to fight. They also entangled the riders with these cords.' Spanish infantry had to run up to disengage the helpless cavalrymen, hacking the tough cords with great difficulty.

The besieged Spaniards survived the burning roofs, sling-shots, bolas and missiles of the Inca armies. They tried to counter each new native device. As well as destroying the street barricades, Spanish working parties smashed the flumes along which the natives were diverting the streams. Others tried to dismantle agricultural terraces so that the horses could ride up them, and they filled in the pits and traps dug by the attackers. They even began to recapture parts of the city. A force of Spanish infantry recaptured the redoubt of Cora Cora after a hard battle. In another engagement some cavalry fought its way under a hail of missiles to a square at the edge of the city, where another sharp fight took place.

The brunt of the Indian attacks came down the steep hillside below Sacsahuaman and on to the spur that forms the central part of Cuzco. Villac Umu and the other besieging generals had established their headquarters within the mighty fortress. Indians attacking from it could penetrate the heart of Cuzco without having to cross the dangerous level ground on other sides of the city. Hernando Pizarro and the besieged Spaniards deeply regretted their failure to garrison this fortress. They realised that as long as it remained in enemy hands their position in the roofless buildings of the city was untenable. They decided that Sacsahuaman must be recaptured at any cost.

Sacsahuaman – local guides have learned that they can earn a larger tip by calling it 'saxy woman' – lies immediately above Cuzco. But the cliff above Carmenca is so steep that the fortress needed only one curtain wall on the city side. Its main defences face away from Cuzco, beyond the brow of the cliff, where the ground slopes away to a small grassy plateau. On that side the top of the cliff is defended by three massive terrace walls. They rise above one another in forbidding grey steps, casing the hillside like the flanks of an armoured dreadnought. The three terraces are built in zigzags like the teeth of great saws, four hundred yards long, with no fewer than twenty-two salient and re-entrant angles on each level. Anyone trying to scale them would have to expose a flank to the defenders. The regular diagonal shadows thrown by these indentations add to the beauty of the terraces. But the feature that makes them so amazing is the quality of the masonry and the size of some of the blocks of stone. As with most Inca terrace walls, this is polygonal masonry: the great stones interlock in a complex and intriguing pattern. The three walls now rise for almost fifty feet, and excavations by the archaeologist Luis Valcárcel showed that ten feet more were once exposed. The largest boulders are on the lowest terrace. One great stone has a height of twenty-eight feet and is calculated to weigh 361 metric tons, which makes it one of the largest blocks ever incorporated into any structure. All this leaves an impression of masterful strength and serene invincibility. In their awe, the sixteenth-century chroniclers soon exhausted the mighty buildings of Spain with which to compare Sacsahuaman (plates 30, 32).

The ninth Inca, Pachacuti, started the fortress and his successors continued the work, recruiting the many thousands of men needed to manhandle the great stones into place. Sacsahuaman was intended to be more than a simple military fortress. Virtually the entire population of the unwalled city of Cuzco could have retreated within it during a crisis. At the time of Manco's siege the crest of the hill behind the terrace walls was covered in buildings. Valcárcel's excavations – made to mark the four-hundredth anniversary of the Conquest – revealed the foundations of the chief structures within Sacsahuaman. These were dominated by three great towers. The first tower, called Muyu Marca, was described by Garcilaso as having been round and containing a water cistern fed by underground channels. The excavations confirmed this description: its foundations consisted of three concentric circles of wall of which the outer was seventy-five feet in diameter. The main tower, Salla Marca, stood on a rectangular base, sixty-five feet long.* Pedro Sancho inspected this tower in 1534 and described it as consisting of five storeys stepped inwards. Such height would have made it the Incas' tallest hollow structure, comparable to the so-called skyscrapers of the pre-Inca Yarivilca culture along the upper Marañon. It was built of coursed rectangular ashlars, and contained a warren of small chambers, the quarters of the garrison. Even the conscientious Sancho admitted that 'the fortress has too many rooms and towers for one person to visit them all'. He estimated that it could comfortably house a garrison of five thousand Spaniards. Garcilaso de la Vega remembered playing in the labyrinth of its corbelled subterranean galleries during his boyhood in Cuzco. He felt that the fortress of Sacsahuaman could rank among the wonders of the world – and suspected that the devil must have had a hand in its extraordinary construction.*

The beleaguered Spaniards now decided that their immediate survival depended on the recapture of the fortress on the cliff above them. According to Murúa, Manco's relative and rival Pascac, who had sided with the Spaniards, gave advice about the plan of attack. It was decided that Juan Pizarro would lead fifty horsemen – the greater part of the Spaniards' cavalry – in a desperate attempt to break through the besiegers and attack their fortress. Observers from the Indian side remembered the scene as follows: 'They spent the whole of that night on their knees and with their hands clasped [in prayer] at their mouths – for many Indians saw them. Even those on guard in the square did the same, as did many Indians who were on their side and had accompanied them from Cajamarca. On the following morning, very early, they all emerged from the church [Suntur Huasi] and mounted their horses as if they were going to fight. They started to look from side to side. While they were looking about in this way, they suddenly put spurs to their horses and at full gallop, despite the enemy, broke through the opening which had been sealed like a wall, and charged off up the hillside at breakneck speed.' They broke through the northern Chinchaysuyo contingent under the generals Curiatao and Pusca. Juan Pizarro's horsemen then galloped up the Jauja road, climbing the hill through Carmenca. They somehow broke and fought their way through the native barricades. Pedro Pizarro was in that contingent and recalled the dangerous ride, zigzagging up the hillside. The Indians had mined the road with pits, and the Spaniards' native auxiliaries had to fill these in with adobes while the horsemen waited under fire from the hillside. But the Spaniards eventually struggled up on to the plateau and rode off to the north-west. The natives thought that they were making a dash for freedom, and sent runners across country to order the destruction of the Apurímac suspension bridge. But at the village of Jicatica the horsemen left the road and wheeled to the right, fought through the gullies behind the hills of Queancalla and Zenca, and reached the level plain below the terraces of Sacsahuaman.* Only by this broad flanking movement were the Spaniards able to avoid the mass of obstacles that the Indians had erected on the direct routes between the city and its fortress.

The Indians had also used the few weeks since the start of the siege to defend the level 'parade ground' beyond Sacsahuaman with an earth barrier that the Spaniards described as a barbican. Gonzalo Pizarro and Hernán Ponce de León led one troop in repeated attacks on these outer enclosures. Some of the horses were wounded, and two Spaniards were thrown from their mounts and almost captured in the maze of rocky outcrops. 'It was a moment when much was at stake.' Juan Pizarro therefore attacked with all his men in support of his brother. Together they succeeded in forcing the barricades and riding into the space before the massive terrace walls. Whenever the Spaniards approached these they were greeted by a withering fire of slingshots and javelins. One of Juan Pizarro's pages was killed by a heavy stone. It was late afternoon, and the attackers were exhausted by the day's fierce fighting. But Juan Pizarro attempted one last charge, a frontal attack on the main gate into the fortress. This gate was defended by side walls projecting on either side, and the natives had dug a defensive pit between them. The passage leading to the gate was crowded with Indians defending the entrance or attempting to retreat from the barbican into the main fortress.

Juan Pizarro had been struck on the jaw during the previous day's fighting in Cuzco and was unable to wear his steel helmet. As he charged towards the gate in the setting sun, he was struck on the head by a stone hurled from the salient walls. It was a mortal blow. The Governor's younger brother, corregidor of Cuzco and tormentor of Inca Manco, was carried down to Cuzco that night in great secrecy, to prevent the natives learning of their success. He lived long enough to dictate a will, on 16 May 1536, 'being sick in body but sound of mind'. He made his younger brother Gonzalo heir to his vast fortune, in the hope that he would found an entail, and left bequests to religious foundations and to the poor in Panama and his birthplace Trujillo. He made no mention of the native siege, and left nothing to the Indian woman from whom 'I have received services' and 'who has given birth to a girl whom I do not recognise as my daughter.' Francisco de Pancorvo recalled that 'they buried him by night so that the Indians should not know he was dead, for he was a very brave man and the Indians were very frightened of him. But although the death of Juan Pizarro was [supposed to be] a secret, the Indians used to say "Now that Juan Pizarro is dead" just as one would say "Now that the brave are dead". And he was indeed dead.' Alonso Enríquez de Guzman gave a more materialistic epitaph: 'They killed our Captain Juan Pizarro, a brother of the Governor and a young man of twenty-five who possessed a fortune of 200,000 ducats.'

On the following day the natives counter-attacked repeatedly. Large numbers of warriors tried to dislodge Gonzalo Pizarro from the hillock opposite the terraces of Sacsahuaman. 'There was terrible confusion. Everyone was shouting and they were all entangled together, fighting for the hilltop the Spaniards had won. It looked as though the whole world was up there grappling in close combat.' Hernando Pizarro sent twelve of his remaining horsemen up to join the critical battle – to the dismay of the few Spaniards left in Cuzco. Manco Inca sent five thousand reinforcements, and 'the Spaniards were in a very tight situation with their arrival, for the Indians were fresh and attacked with determination.' Below 'in the city, the Indians mounted such a fierce attack that the Spaniards thought themselves lost a thousand times'.

But the Spaniards were about to apply European methods of siege warfare: throughout the day they had been making scaling ladders. As night fell, Hernando Pizarro himself led an infantry force to the top of the hill. Using the scaling ladders in a night assault, the Spaniards succeeded in taking the mighty terrace walls of the fortress. The natives retreated into the complex of buildings and the three great towers.

There were two individual acts of great bravery during this final stage of the assault. On the Spanish side Hernán Sánchez of Badajoz, one of the twelve brought up by Hernando Pizarro as additional reinforcements, performed feats of prodigious panache worthy of a silent-screen hero. He climbed one of the scaling ladders under a hail of stones which he parried with his buckler, and squeezed into a window of one of the buildings. He hurled himself at the Indians inside and sent them retreating up some stairs towards the roof. He now found himself at the foot of the highest tower. Fighting round its base he came upon a thick rope that had been left dangling from the top. Commending himself to God, he sheathed his sword and started clambering up, heaving up the rope with his hands and stepping off from the smooth Inca ashlars with his feet. Half way up the Indians threw a stone 'as big as a wine jar' down on him, but it simply glanced off the buckler he was wearing on his back. He threw himself into one of the higher levels of the tower, suddenly appearing in the midst of its startled defenders, showed himself to the other Spaniards and encouraged them to assault the other tower.*

The battle for the terraces and buildings of Sacsahuaman was hard fought. 'When dawn came, we spent the whole of that day and the next fighting the Indians who had retreated into the two tall towers. These could only be taken through thirst, when their water supply became exhausted.' 'They fought hard that day and throughout the night. When the following day dawned, the Indians on the inside began to weaken, for they had exhausted their entire store of stones and arrows.' The native commanders, Paucar Huaman and the high priest Villac Umu, felt that there were too many defenders inside the citadel, whose supplies of food and water were rapidly being exhausted. 'After dinner one evening, almost at the hour of vespers, they emerged from the fortress with great élan, attacked their enemies and broke through them. They rushed with their men down the slope towards Zapi and climbed to Carmenca.' Escaping through the ravine of the Tullumayo, they hurried to Manco's camp at Calca to plead for reinforcements. If the remaining two thousand defenders could hold Sacsahuaman, a native counter-attack might trap the Spaniards against its mighty walls.

Villac Umu left the defence of Sacsahuaman to an Inca noble, an orejón who had sworn to fight to the death against the Spaniards. This officer now rallied the defenders almost single-handed, performing feats of bravery 'worthy of any Roman'. 'The orejón strode about like a lion from side to side of the tower on its topmost level. He repulsed any Spaniards who tried to mount with scaling ladders. And he killed any Indians who tried to surrender. He smashed their heads with the battle-axe he was carrying and hurled them from the top of the tower.'

Alone of the defenders, he possessed European steel weapons that made him the match of the attackers in hand-to-hand fighting. 'He carried a buckler on his arm, a sword in one hand and a battle-axe in the shield hand, and wore a Spanish morrión helmet on his head.' 'Whenever his men told him that a Spaniard was climbing up somewhere, he rushed upon him like a lion with the sword in his hand and the shield on his arm.' 'He received two arrow wounds but ignored them as if he had not been touched.' Hernando Pizarro arranged for the towers to be attacked simultaneously by three or four scaling ladders. But he ordered that the brave orejón should be captured alive. The Spaniards pressed home their attack, assisted by large contingents of native auxiliaries. As Manco's son wrote, 'the battle was a bloody affair for both sides, because of the many natives who were supporting the Spaniards. Among these were two of my father's brothers called Inquill and Huaspar with many of their followers, and many Chachapoyas and Cañari Indians.' As the native resistance crumbled, the orejón hurled his weapons down on to the attackers in a frenzy of despair. He grabbed handfuls of earth, stuffed them into his mouth and scoured his face in anguish, then covered his head with his cloak and leaped to his death from the top of the fortress, in fulfilment of his pledge to the Inca.

'With his death the remainder of the Indians gave way, so that Hernando Pizarro and all his men were able to enter. They put all those inside the fortress to the sword – there were 1,500 of them.' Many others flung themselves from the walls. 'Since these were high the men who fell first died. But some of those who fell later survived because they landed on top of a great heap of dead men.' The mass of corpses lay unburied, a prey for vultures and giant condors. The coat of arms of the city of Cuzco, granted in 1540, had 'an orle of eight condors, which are great birds like vultures that exist in the province of Peru, in memory of the fact that when the castle was taken these birds descended to eat the natives who had died in it'.

Hernando Pizarro immediately garrisoned Sacsahuaman with a force of fifty foot-soldiers supported by Cañari auxiliaries. Pots of water and food were hurried up from the city. The high priest Villac Umu returned with reinforcements, just too late to save the citadel. He counter-attacked vigorously, and the battle for Sacsahuaman continued fiercely for three more days, but the Spaniards were not dislodged, and the battle was won by the end of May.

Both sides appreciated that the recapture of Sacsahuaman could be a turning point in the siege. The natives now had no secure base from which to invest the city, and they abandoned some of the outlying districts they had occupied. When the counter-attack on Sacsahuaman failed, the Spaniards advanced out of the citadel and pursued the demoralised natives as far as Calca. Manco and his military commanders could not understand why their vast levies had failed to capture Cuzco. His son Titu Cusi imagined a dialogue between the Inca and his commanders. Manco: 'You have disappointed me. There were so many of you and so few of them, and yet they have eluded your grasp.' To which the generals replied, 'We are so ashamed that we dare not look you in the face.... We do not know the reason, except that it was our mistake not to have attacked in time and yours for not giving us permission to do so.'

The generals might possibly have been right. Manco's insistence on waiting for the entire army to assemble meant that the Indians lost the element of surprise they had preserved so brilliantly during the early mobilisation. It also meant that the professional commanders could not attack while the Spaniards had sent much of their best cavalry to investigate the Yucay valley. The hordes of native militia did not necessarily add much to the effectiveness of the native army. But Manco had clearly felt that as long as his men suffered a terrible handicap in weapons, armour and mobility, their only hope of defeating the Spaniards was by weight of numbers. The heavy, determined fighting of the first month of the siege showed that the Spaniards had no monopoly of personal bravery. Once again, it was their crushing superiority in hand-to-hand fighting and the mobility of their horses that won the day. The only arms in which the natives had parity were projectiles – slingshots, arrows, javelins and bolas – and prepared defences such as breastworks, terraces, flooding and pits. But projectiles and defences rarely succeeded in killing an armoured Spaniard, and the siege of Cuzco was a fight to the death.

Manco could also be criticised for not directing the attack on Cuzco in person. He apparently remained at his headquarters at Calca throughout the critical first month of the siege. He was using his authority and energies to effect the almost impossible feat of a simultaneous uprising throughout Peru, together with the feeding and supply of an enormous army. But the Inca's presence was needed at Cuzco. Although there were plenty of imposing fighting men in the various contingents, the army lacked the inspiration of a leader of the stature of Chalcuchima, Quisquis or Rumiñavi.

The fall of Sacsahuaman at the end of May was by no means the end of the siege. Manco's great army remained in close investiture of the city for a further three months. The Spaniards soon learned that the native attacks ceased for religious celebrations at every new moon. They took full advantage of each lull to destroy roofless houses, fill in enemy pits, and repair their own defences. There was fighting throughout this period, with great bravery displayed on either side.

One episode will illustrate the typical daily skirmishes. Pedro Pizarro was on guard duty with two other horsemen on one of the large agricultural terraces at the edge of Cuzco. At midday his commander, Hernán Ponce de León, came out with food and asked Pedro Pizarro to undertake another tour of duty as he had no one else to send. Pizarro duly grabbed some mouthfuls of food and rode out to another terrace to join Diego Maldonado, Juan Clemente and Francisco de la Puente on guard.

While they were chatting together, some Indian warriors approached. Maldonado rode off after them. But he had failed to see some pits the natives had prepared, and his horse fell into one. Pedro Pizarro dashed off against the Indians, avoiding the pits, and gave Maldonado and his horse, both badly injured, a chance to return to Cuzco. The Indians reappeared to taunt the three remaining horsemen. Pizarro suggested 'Come on, let's drive these Indians away and try to catch some of them. Their pits are now behind us.' The three charged off. His two companions turned half way along the terrace, and returned to their post, but Pizarro galloped on 'impetuously lancing Indians'. At the end of the terrace the natives had prepared small holes to catch the horses' hooves. When he tried to wheel, Pizarro's horse caught its leg and threw him. One Indian rushed up and started to lead off the horse, but Pizarro got to his feet, went after the man and killed him with a thrust through the chest. The horse bolted, running off to join the other Spaniards. Pizarro now defended himself with his shield and sword, holding off any Indians who drew near. His companions saw his riderless horse and hurried to help him. They charged through the Indians, 'caught me between their horses, told me to grab the stirrups, and took off at full speed for some distance. But there were so many Indians crowding around that it was useless. Wearied from all my armour and from fighting, I could not go on running. I shouted to my companions to stop as I was being throttled. I preferred to die fighting than be choked to death. So I stopped and turned to fight the Indians, and the two on their horses did the same. We could not drive off the Indians, who had become very bold at the thought that they had taken me prisoner. They all gave a great shout from every side, which was their normal practice when they captured a Spaniard or a horse. Gabriel de Rojas, who was returning to his quarters with ten horsemen, heard this shout and looked in the direction of the disturbance and the fighting. He hurried there with his men, and I was saved by his arrival, although badly wounded by the stone and spear blows inflicted by the Indians. I and my horse were saved in this way, with the help of our Lord God who gave me strength to fight and to endure the strain.'

Gabriel de Rojas received an arrow wound in one of these skirmishes: it went through his nose as far as his palate. García Martín had his eye knocked out by a stone. One Cisneros dismounted, and the Indians caught him and cut off his hands and feet. 'I can bear witness', wrote Alonso Enríquez de Guzman, 'that this was the most dreadful and cruel war in the world. For between Christians and Moors there is some fellow-feeling, and it is in the interests of both sides to spare those they take alive because of their ransoms. But in this Indian war there is no such feeling on either side. They give each other the cruellest deaths they can imagine.' Cieza de León echoed this. The war was 'fierce and horrible. Some Spaniards tell that a great many Indians were burned and impaled. . . . But God save us from the fury of the Indians, which is something to be feared when they can give vent to it!' The natives had no monopoly of cruelty. Hernando Pizarro ordered his men to kill any women they caught during the fighting. The idea was to deprive the fighting men of the women who did so much to serve and carry for them. 'This was done from then onwards, and the stratagem worked admirably and caused much terror. The Indians feared to lose their wives, and the latter feared to die.' This war on the women was thought to have been one of the chief reasons for the slackening of the siege in August 1536. On one sortie Gonzalo Pizarro encountered a contingent from the Chinchaysuyo and captured two hundred of them. 'The right hands were cut off all these men in the middle of the square. They were then released so that they would go off. This acted as a dreadful warning to the rest.'

Such tactics added to the demoralisation of Manco's army. The vast majority of the horde that massed on the hills around Cuzco were ordinary Indian farmers with their wives and camp followers – with few exceptions a thoroughly militia army, most of whose men had received only the rudimentary arms drill that was part of the upbringing of every Inca subject. Only part of this rabble was militarily effective, although the entire mass had to be fed. By August the farmers began drifting away to sow their crops. Their departure added to the attrition of heavy losses in every battle against the Spaniards. Weight of numbers was Manco's only effective strategy, so the reduction of his great army meant that further operations against Cuzco might have to wait until the following year. But Cuzco was only one theatre of the national uprising. In other areas the natives were far more successful.

While he attacked Cuzco, Manco entrusted Quizo Yupanqui and his captains Illa Tupac and Puyu Vilca with the conquest of the central highlands.* Another general, Tiso, had already been fomenting rebellion among the natives of the Jauja area with some success. At the first hint of trouble, the Spaniards had despatched a punitive expedition of some sixty men, mostly foot-soldiers, under one Diego Pizarro. These operated in the Jauja area while Tiso vanished into the eastern jungles and reported back to Manco at Ollantaytambo.

Governor Francisco Pizarro first heard about the attack on Cuzco on 4 May, in his new capital Los Reyes, or Lima. He at once feared for his brothers and the other Spaniards isolated in Cuzco, and began organising relief expeditions. He sent thirty men over the mountains to Jauja, under Captain Francisco Morgovejo de Quiñones, one of the two alcaldes of Lima. This force marched inland in mid-May, with orders to proceed along the royal road and garrison the strategic crossroads of Vilcashuaman.* It travelled through peaceful country beyond Jauja and as far as Parcos, an important tambo above the gorge of the Mantaro. Morgovejo de Quiñones learned here that the natives had killed five Spaniards travelling towards Cuzco. His reprisals were swift and vicious. He gathered twenty-four chiefs and elders of Parcos into a thatched building and burned them all alive. He hoped in this way to intimidate the natives so that he could proceed unmolested to a rendezvous with Diego Pizarro at Huamanga.

Pizarro also despatched another force of seventy horse under his relative Gonzalo de Tapia. This contingent took the middle route, descending the coast for some 120 miles and then climbing inland past Huaitará, which is still graced by Inca ruins, to cross the Andes at 15,000 feet and strike the royal road north of Huamanga. Tapia's force crossed the desolate puna of Huaitará but was caught by the Indians in a defile on the upper Pampas river – 'one of the worst passes in the land'. It had run into Quizo Yupanqui's new army marching north from Cuzco.

These various parties of marching Spaniards were tempting prey to the rebellious natives. The superiority of Spanish horses and weapons was nullified by the Andean topography: for these central Andes are one of the most vertical places on earth, an endless succession of crumbling precipices, savage mountain torrents, landslides and giddy descents. Here, at last, the natives had a really effective natural ally. 'Their strategy', wrote Agustín de Zárate, 'was to allow the Spaniards to enter a deep, narrow gorge, seize the entrance and exit with a great mass of Indians, and then hurl down such a quantity of rocks and boulders from the hillsides that they killed them all, almost without coming to grips with them.' Using this tactic, the natives now succeeded in annihilating Tapia's seventy horse – almost as many mounted men as were defending Cuzco. The few survivors were sent as prisoners to Manco Inca.

Quizo continued northwards and soon met Diego Pizarro's sixty men, who were marching down the Mantaro towards Huamanga. The Indians repeated their successful use of topography. Quizo trapped and exterminated Diego Pizarro's entire force near the same Parcos where Morgovejo had burned the chiefs a few weeks before. News of these great victories was sent back to Manco, together with some Spanish post, weapons and clothing, the heads of many dead invaders, 'and two live Spaniards, one Negro and four horses'. The messengers reached the Inca soon after he had heard the news of the loss of Sacsahuaman at the end of May. His son Titu Cusi remembered the great rejoicings over the victory. To show his appreciation, Manco sent his victorious general 'a coya wife of his own lineage, who was most beautiful, and some litters in which he could travel with more authority'.

When Pizarro learned of the rebellion of Manco Inca, the prince he had crowned and in whom he had such confidence, he resorted to the familiar tactic of finding a rival puppet Inca. He chose a royal prince. probably Cusi-Rimac, who was with him in Lima. This man was given a hasty coronation and dispatched towards Jauja, protected by thirty horse under Captain Alonso de Gaete. Shortly afterwards pro-Spanish native yanaconas started bringing alarming rumours of the fate of the other expeditions. Pizarro decided that Gaete's thirty men were too vulnerable. He therefore sent a further thirty foot-soldiers under Francisco de Godoy, the other mayor of Lima for the year 1536, and they left in mid-July.

The victorious Quizo Yupanqui was also marching on Jauja. Most of the original citizens of Jauja had descended to the coast to settle in Pizarro's new capital Lima, but there were still a number of Christians living in the former Inca city. According to Martín de Murúa, these Spaniards were too arrogant to post sentries or make preparations to defend themselves. 'Quizo Yupanqui arrived one morning at daybreak. He came upon the Spaniards so suddenly that the first they knew was that they were surrounded on all sides. They did not even have time to dress, for they were still in bed. In this tumult they positioned themselves in an usno [temple platform] they had there as a fortress with any weapons they found most readily to hand. Anyone can imagine the confusion: for they never thought that the Indians would have the courage to attack them. . . . The fighting lasted from the morning when the Indians arrived until the hour of vespers . . . and the Indians killed them all, and their horses and Negro servants.'

As the thirty foot-soldiers under Godoy approached Jauja they met the pathetic figure of Gaete's half-brother Cervantes de Maculas fleeing with a broken leg on a pack-mule. This man and one other Spaniard were the only survivors of Gaete's force, which was reported killed by its own Indian auxiliaries. The puppet Inca apparently succeeded in crossing to the native side in the heat of the battle: a brother called Cusi-Rimache was later prominent in Manco's camp. Francisco de Godoy decided not to risk sharing the fate of Gaete's, Diego Pizarro's and Tapia's men. He turned around and rode back into Lima in early August 'with his tail between his legs, to give Pizarro the bad news'.

Quizo had now succeeded in annihilating almost all the Spaniards between Cuzco and the sea, including the inhabitants of Jauja and many travellers and encomenderos along the Jauja–Cuzco road. He had also defeated three well-armed relief forces of Spanish cavalry – over 160 men. The only Spanish force still at large in the central Andes was the thirty men under Morgovejo de Quiñones. These continued down the Mantaro after massacring the elders of Parcos. At one crossing they were trapped by hordes of native warriors, who occupied both banks of the deep canyon. Night fell, and Morgovejo's men camped by the river bank. They left camp-fires burning and were able to slip away in the darkness. There was another skirmish in a defile before the expedition managed to reach the tambo of Huamanga. The men could not rest for long. Throughout the following day native forces massed on the slopes around the tambo, and Morgovejo's men saw a cluster of handsome litters that evidently contained the general Quizo Yupanqui and his staff. But the Spaniards again escaped under cover of darkness. They climbed the hills behind the tambo, and even took their weeping native women and yanaconas, who feared the reprisals of their compatriots if their European masters were annihilated.*

The exhausted expedition now attempted to ride back across the Andes to the coast. There were many days of marching and fighting around the ravines of the upper Pampas river. The Spaniards finally reached the last defile that separated them from the sanctuary of the coastal plain. But the local Indians had prepared another ambush. As the column entered the pass the air was filled with the echoing shouts of the warriors. Most of the Spaniards were trapped. The path was too narrow for them to fight, and they were caught in a barrage of boulders. Captain Morgovejo leaped from his exhausted horse and mounted the croup of another. But this animal was struck by a rock: both riders were thrown and Morgovejo's thigh was shattered. The Spanish captain fought on for hours before he and his native attendant were killed. Four more of his men were killed at the passage of a river, but the shattered remnants reached the coast road and returned to Lima. They were almost the only survivors of some two hundred men sent into the sierra to relieve Cuzco.*

It was now some months since Francisco Pizarro had heard from his brothers in Cuzco, and he feared that they must be dead. 'The Governor was deeply worried by the course of events. Four of his commanders had now been killed, with almost two hundred men and as many horses.' When the rebellion broke out, Pizarro at once tried to consolidate his forces in Peru. Alonso de Alvarado was recalled from the conquest of the Chachapoyas, beyond Cajamarca in north-eastern Peru;

he eventually reached Lima with thirty horse and fifty foot. Gonzalo de Olmos brought some men and seventy horse from Puerto Viejo on the Ecuadorean coast; and Garcilaso de la Vega, the father of the historian, abandoned an attempt to colonise the Bay of San Mateo and took eighty men back to Lima. Pizarro's half-brother Francisco Martín de Alcántara was sent along the coastal plain to warn Spanish settlers of the danger and to gather them back to Lima.

But Pizarro could look for help beyond Peru. His invasion of the country was only one tentacle of Spanish colonisation of the Americas. Pizarro hoped at first to crush the rebellion from his own resources: he sent Juan de Panes to Panama in July to buy arms and horses with 11,000 marks of his personal fortune that were lodged there. But he soon realised that the rebellion was too serious and decided to appeal for help to all the Spanish governors in the Americas. In his desperation he even wrote on 9 July to Pedro de Alvarado, Governor of Guatemala, the man who had attempted to invade Quito and been humiliated and bought off by Pizarro only eighteen months previously. Pizarro now wrote him an eloquent plea: 'The Inca has the city of Cuzco besieged, and for five months I have heard nothing about the Spaniards in it. The country is so badly damaged that no native chiefs now serve us, and they have won many victories against us. It causes me such great sorrow that it is consuming my entire life.' Diego de Ayala took this letter to Guatemala and sailed on to Nicaragua, where he eventually recruited many good men. At the same time Licenciate Castañeda passed through Panama on the way to Spain, and Pascual de Andagoya wrote to the Emperor from Panama at the end of July that 'the Lord of Cuzco and of the entire country has rebelled'. 'The rebellion has spread from province to province and they are all coming out in rebellion simultaneously. Rebellious chiefs have already arrived forty leagues from Lima. The Governor [Pizarro] is asking for help and will be given everything possible from here.'

News of the Peruvian revolt gradually reached Spain itself. As early as February 1536 the aged Bishop Berlanga of Tierra Firme, who had just returned from a royal mission to Peru, reported to King Charles that Governor Pizarro was allowing his conquistadores to violate the ordinances for good treatment of the natives. He also reported ominously that 'the Governors have exploited the Inca, Lord of Cuzco, and so have any other Spaniards who wished to'. In April, Licenciate Gaspar de Espinosa wrote from Panama telling the King of the first killings of isolated Spaniards in the Cuzco area.* But news of the full siege of Cuzco did not arrive across the Atlantic until late August. The first reports of the rising said that 'Hernando Pizarro caused this rebellion, for it is said that he tortured the chief cacique so that he would give him gold and silver.'

Manco Inca, delighted by Quizo's victories, ordered him to descend on Lima 'and destroy it, leaving no single house upright, and killing any Spaniards he found' except for Pizarro himself. But Quizo wanted to ensure that his rear would be secure. He therefore spent July recruiting Jaujas, Huancas and Yauyos, trying to persuade these anti-Inca tribes to join his rebellion.* Quizo was understandably reluctant to engage Spaniards on the level ground and unfamiliar surroundings of the coastal plain unless he had overwhelming strength. He succeeded in recruiting a large force from the tribes along the western slopes of the Andes, and eventually brought his great colourful army down to the lowest range of Andean foothills, within sight of the hazy waters of the Pacific. The first news of them was brought by Diego de Aguero. He 'arrived in flight at Los Reyes, reported that the Indians were all under arms and had tried to set fire to him in their villages. A great army of them was approaching. The news deeply terrified the city, all the more so because of the small number of Spaniards who were in it.' Governor Pizarro sent seventy horse under Pedro de Lerma to try to prevent the Indians moving down on to the plain. A sharp engagement took place, in which many Indians lost their lives, 'one Spanish horseman was killed and many others wounded, and Pedro de Lerma had his teeth broken'. The native army continued its advance on the European city.

'When the Governor saw this multitude of enemy he had no doubt whatsoever that our side was completely finished.' Quizo's warriors advanced across the plain and some even entered the outlying houses of the city. But 'the cavalry were hidden in ambush, and at the appropriate moment they charged out, killing and lancing a great number of them until they retreated and climbed into some hills. A strong guard was mounted by night, with the cavalry patrolling around the city'. Quizo now moved his army on to the Cerro de San Cristóbal, a steep sugarloaf hill just across the Rimac river from the heart of Lima. This hill is encrusted today with a dusty slum, a tawny wart of poverty rearing above the modern city. Another hill was occupied by troops from the valleys of the Atavillos to the north-east of Lima – men better accustomed to the low, sea-level altitude and heavy atmosphere of Lima than their companions from the highland tribes. Other native warriors occupied hills between Lima and its port, a few miles away at Callao. The Spanish settlement was thus surrounded and almost severed from its communications with the sea.

The Spaniards were desperately anxious to dislodge Quizo's men from San Cristóbal. Cavalry were useless against its steep flanks. A bold night attack, normally successful 'since Indians are very cowardly at night', seemed suicidal against so steep a hill occupied by so many enemy. Nevertheless, five days were spent preparing for such an attack. 'It was agreed to build a shield of planks as protection against rocks. But when this was made it proved impossible to carry it.'

There was the usual confusion of loyalties during the siege. 'Many Indian servants of the Spaniards went out to eat and live with the enemy and even to fight against their masters, but returned at night to sleep in the city.' Francisco Pizarro suspected an Indian concubine of treachery. Atahualpa's sister Azarpay, whom 'he had in his lodging', was accused by Pizarro's chief love Doña Inés of encouraging the besiegers. 'So, without further consideration, he ordered that she be garrotted, and killed her when he could have embarked her on a ship and sent her into exile.' There were, on the other hand, 'some friendly Indians in the city who fought very well. Because of them the horses were spared from over-exertion, which they could not otherwise have endured.'

On the sixth day of the investiture of Lima, the native commander decided to end the stalemate with a full-blooded attack on the city. This was the critical moment of the rebellion, the attempt to drive the Spanish invaders back into the Pacific. Quizo Yupanqui 'determined to force an entry into the city and capture it or perish in the attempt. He addressed his forces as follows: "I intend to enter the town today and kill all the Spaniards in it. . . . Any who accompany me must go on the understanding that if I die all will die, and if I flee all will flee." The Indian commanders and leaders swore to do as he said.' The native leader also promised his men enjoyment of the handful of Spanish women then in Lima – probably not more than fourteen women including Francisca Pizarro's three godmothers. 'We will take their wives, marry them ourselves and produce a race of warriors.' The Spaniards had been busily creating the same race of mestizos ever since they set foot in Peru – but their unions with native women were not for eugenic purposes.

Quizo Yupanqui planned a simultaneous attack on Lima from three sides. He himself advanced from the hills to the east; the tribes of the Atavillos and north-central sierra marched in along the coastal road from Pachacamac.* The most determined attack was that of the Incas under Quizo himself. It possessed splendour and reckless bravery, the magnificent futility of the French at Agincourt, the British at Balaclava or the Confederates at Gettysburg. 'The entire native army began to move with a vast array of banners, from which the Spaniards recognised their determination. Governor Pizarro ordered all the cavalry to form into two squadrons. He placed one squadron under his command in ambush in one street, and the other squadron in another. The enemy were by now advancing across the open plain by the river. They were magnificent men, for all had been hand-picked. The general was at their head wielding a lance. He crossed both branches of the river in his litter.

'As the enemy were starting to enter the streets of the city and some of their men were moving along the tops of the walls, the cavalry charged out and attacked with great determination. Since the ground was flat they routed them instantly. The general was left there, dead, and so were forty commanders and other chiefs alongside him. It was almost as if our men had specially selected them. But they were killed because they were marching at the head of their men and therefore withstood the first shock of the attack. The Spaniards continued to kill and wound Indians as far as the foot of the hill of San Cristóbal, at which point they encountered a very strong resistance from the Indian redoubt.'† Manco Inca had lost his most successful general, the gallant Quizo Yupanqui.

With the slaughter of so many of their leaders, the fight went out of the Indians. The more agile Spaniards were planning a night attack on San Cristóbal, but the native army melted away into the mountains before the venture was attempted. The highland Indians were uncomfortable in the hot, close atmosphere of the coast: their lungs are specially evolved to live at high altitude. They despaired of driving Pizarro from his coastal city: the superiority of the Spanish horsemen was too crushing on a flat, open plain at sea level. The coastal tribes conspicuously failed to join in the highland revolt against the invaders, and the Spaniards had a number of coastal curacas in protective custody in Lima.*

The Spaniards besieged far away in Cuzco at this same time had two urgent objectives. One was to attempt to advise their compatriots on the coast that they were still alive. The other was to strike boldly at the Inca's headquarters in a bid to destroy the men who were directing the siege. A group of citizens persuaded Hernando Pizarro to send fifteen of his finest horsemen towards the coast, riding an unexpected route, southwards to the altiplano and then west through Arequipa. The fifteen included 'the flower of the men', dashing young horsemen such as Pedro Pizarro, Alonso de Mesa, Hernando de Aldana, Alonso de Toro and Tomás Vásquez. The selected men regarded the mission as certain suicide. Alonso Enríquez de Guzman thought that he was included because Hernando Pizarro had a personal grudge against him and wanted him killed. In the end, a delegation headed by the royal treasurer Riquelme persuaded Pizarro that the departure of these fine men would seriously weaken the city's defences. Hernando wisely reversed the order and the fifteen were spared from annihilation by native ambush.*

Hernando Pizarro now attempted to strike at the Inca himself. He had learned that Manco had moved from Calca to Ollantaytambo, a remoter stronghold some thirty miles downstream on the Vilcanota–Yucay–Urubamba river. Pizarro assembled all his best men: seventy horse, thirty foot and a large contingent of native auxiliaries. Gabriel de Rojas was left in Cuzco with the remaining weaker Spaniards. Hernando Pizarro marched his force down the Yucay with great difficulty, for the meandering river often ran against the steep rocky hills that enclose its valley. 'It had to be crossed five or six times, and each ford was defended.' The Spaniards finally reached Ollantaytambo after continuous fighting, but they were appalled when they came in sight of its massive pale grey walls. 'When we reached Tambo we found it so well fortified that it was a horrifying sight.'

The great ruin still stands to this day with its superb Inca masonry almost completely intact. Below the citadel is the town of Ollantaytambo in the bed of a small tributary valley. It is one of the few surviving examples of Inca town planning, with the wall foundations and grid of streets intact. Each town block contained two plots with entrances on to the longitudinal streets, and the original Inca houses are still occupied. Even the Inca names of the blocks survive, as do the houses that once contained the acllas or chosen holy women. The town consisted of five terraced enclosures, all contained in a symmetrical trapezoidal outline, the same tapering quadrangular shape so beloved of Inca architects. The Patacancha stream runs towards the Vilcanota beside the town, and beyond it a great cliff juts out towards the main river. The prow of this hillside contained the fortress-temple of Ollantaytambo. Undulating granite terrace-walls encase the steep slope at the end of the spur, while the hillside overlooking the town is lined with a great flight of seventeen broad terraces. At the top are walled fortifications – a rarity in Inca architecture – and within this sanctuary a platform faced with seven vast monoliths of pale porphyry, each some eleven feet wide. From below, the entire hillside seems to be embellished with the regular rows of polygonal Inca masonry (plates 33, 34).*

Hernando Pizarro's men occupied the flat stretch of plain between the town and the Yucay river. Because Ollantaytambo lies close to the forested country of the upper Amazon basin, Manco had recruited archers from jungle tribes into his army. One of Pizarro's troops described the bravery of these terrifying savages: 'They do not know what is meant by flight – for they continue to fight with their arrows even when they are dying.'

The town was full of these archers, firing from every terrace, and so was the citadel. Across the stream were Inca slingers. 'The Indians were thus fighting them from three sides: some from the hillside, others from the far bank of the river, and the rest from the town. . . . The Inca was in the fortress itself with many well-armed warriors.' 'They amassed such a quantity of men against us that they could not crowd on to the hillsides and plains.' A single flight of steps led up to the citadel. The gate at its foot had been sealed with a fieldstone wall through which an Indian could pass only on all fours. Two of the older conquistadores bravely rode their horses up against the walls of the town, but 'it was amazing to see the arrows that rained down on them as they returned, and to hear the shouting'. Another group of horsemen tried to attack the terraces below the citadel. But the defenders 'hurled down so many boulders and fired so many slingshots that, even had we been many more Spaniards than we were, we would all have been killed'.

A missile broke the haunch of the leading horse, which rolled over, kicking, rearing and falling, and dispersed the horses trying to follow. Hernando Pizarro tried sending a party of foot-soldiers to seize the heights above the fortress, but the Europeans were driven back by a hail of rocks. As the Spaniards wavered, the natives attacked. They charged out on to the plain 'with such a tremendous shout that it seemed as if the mountain was crashing down. So many men suddenly appeared on every side that every visible stretch of wall was covered in Indians. The enemy locked in a fierce struggle with [Pizarro's men] – more savage than had ever been seen by either side.' The natives had acquired many Spanish weapons and were learning to employ them effectively. 'It was impressive to see some of them emerge ferociously with Castilian swords, bucklers and morrión helmets. There was one Indian who, armed in this manner, dared to attack a horse, priding himself on death from a lance to win fame as a hero. The Inca himself appeared among his men on horseback with a lance in his hand, keeping his army under control.' The natives even attempted to use captured culverins and arquebuses for which powder had been prepared by Spanish prisoners.* Manco now released his other secret weapon. Unobserved by the Spaniards, native engineers diverted the Patacancha river along prepared channels to flood the plain. The Spanish horsemen soon found themselves trying to manoeuvre in rising water that eventually reached the horses' girths.* 'The ground became so sodden that the horses could not skirmish.' 'Hernando Pizarro realised that it was impossible to take that town and ordered a retreat.'

Night fell, and the Spaniards tried to slide away under cover of darkness, leaving their tents pitched beneath Ollantaytambo. But the column of defeated horsemen was observed 'and the Indians came down upon them with a great cry . . . grabbing the horses' tails'. 'They attacked us with great fury at a river crossing, carrying burning torches. . . . There is one thing about these Indians: when they are victorious they are demons in pressing it home, but when they are fleeing they are like wet hens. Since they were now following up a victory, seeing us retire, they pursued with great spirit.' The Indians had littered the road back with thorny agave spines which crippled the horses.* But the Spaniards succeeded in riding out of the Yucay valley that night, and they fought their way back into Cuzco the following day. Titu Cusi said that the natives laughed heartily at the Spanish failure, and the Spaniards knew that native morale had been raised by the defeat of this powerful expedition.* 'The Inca was extremely sad that Hernando Pizarro had gone, for he was sure that had he delayed another day no single Spaniard would have escaped. In truth, anyone who saw the appearance of the fortress could have believed nothing else. . . . For . . . on such occasions, where horses cannot fight, the Indians are the most active people in the world.'

The Indians now provided an unexpected boost to the morale of the men besieged in Cuzco. The relief expeditions sent by Francisco Pizarro had been carrying a quantity of dispatches and letters. These were brought to Manco, who was going to burn them. But a cunning Spaniard who was a prisoner in the Inca's camp suggested that the letters could be used more effectively: Manco should have them torn up and conveyed to the besieged to show the fate that had befallen their compatriots.* A group of Indians therefore appeared on top of Carmenca hill on the morning of 8 September. Hernando Pizarro and other horsemen duly rode out to chase them into the hills. On their return they found that the Indians had left two sacks containing the dried heads of six Spaniards and the torn remnants of a thousand letters. Manco had failed to appreciate the importance of written communication. The arrival of these letters – even in this macabre manner – enormously heartened the besieged. They learned that the Spaniards still held Lima and were trying to relieve them. They also learned from one letter 'almost intact, from our Lady the Empress' that Charles V had won a victory against the infidel in Tunis. Alonso Enríquez de Guzman received a personal letter from Francisco Pizarro, dated 4 May, in which the old Governor admitted that the Inca's rebellion 'has caused me much concern, both on account of the detriment . . . to the service of the Emperor, of the dangers in which you are placed, and of the trouble it will cause me in my old age'. The receipt of these letters put an end to any attempt to communicate with the coast. The Spaniards in Cuzco could only hope that their compatriots would survive the rebellion and would eventually relieve them.

Manco Inca, encouraged by the defeat of the Ollantaytambo mission, attempted to reassemble the army that had almost captured Cuzco four months earlier. The farmers who had returned to their villages were recalled for a fresh attempt on the city, before the start of the rainy season. The Spaniards were now living more normally in the ravaged city. They had mended the roofs of some houses, replacing the original thatch with less inflammable flat roofs of peat and wooden beams. At the height of the siege they had had their horses constantly saddled and bridled, and each man stood watch for one quarter every night. Now his guard duty was only on alternate nights, and the besieged knew that they need fear night attacks only during a full moon – the natives were occupied with religious ceremonies at the advent of each new moon.*

The most acute problem for the besieged was food. 'The greatest hardship endured by the Christians was incredible hunger, from which some died. For the Indians had, with great foresight, set fire to any buildings that contained supplies or stores.' Hernando Pizarro was able to harvest some maize planted by the natives near Sacsahuana, a few miles north-west of Cuzco. Detachments of cavalry escorted the columns of native auxiliaries carrying the maize towards the city, and fought off Manco's warriors who tried to intercept the operation.* But this maize was not enough. It could be months before rescue arrived from the coast, and the Spaniards, unable to undertake intensive farming, needed immediate supplies to avoid starvation during this period. The Inca quipocamayos were interrogated by the Spanish authorities after the siege. They reported that some important native commanders passed to the Spanish side with large contingents of Indians. These included Pascac, Manco's cousin and enemy, whom Hernando Pizarro praised as "captain-general of the Indians who were with me in the defence of Cuzco". These traitors revealed that 'Manco Inca's men had brought over a thousand head of cattle, maize and other provisions', and that this was delayed not far from Cuzco. Hernando Pizarro immediately sent Gabriel de Rojas with seventy horse to seize these llamas and to raid in the Canchas country along the Collao road. Pedro Pizarro wrote of this raid: 'We went and remained there for some twenty-five or thirty days, rounded up almost two thousand head of cattle [llamas], and returned with them to Cuzco without having had any serious engagement.' The capture of this food emboldened the besieged. Hernando at once sent another six-day raiding and punitive mission into the Condesuyo, south-west of Cuzco. It was intended to avenge the murders there of Simón Suárez and other encomenderos at the start of the rebellion. 'But we could catch no one on whom to inflict punishment. So we collected some food and returned.'

These audacious raids saved the besieged from starvation. Although highly successful, they were a calculated risk and were very nearly disastrous. As soon as he learned that many of the best horsemen had left the city, Manco accelerated his mobilisation. The Spaniards left in Cuzco were anxious to learn about the natives' movements. Gonzalo Pizarro was sent on a night reconnaissance to capture some prisoners who could be tortured for information. He led eighteen horse on to the plateau to the north of the city, and in the darkness passed, without knowing it, between two large contingents of Manco's army. The Spaniards spent the night on their mission, divided into two small contingents under Gonzalo Pizarro and Alonso de Toro. When dawn came, they found themselves confronted by the enemy, Toro facing men from the northern part of the empire, and Pizarro a force of Manco's own levies, the finest men in the Inca army. In a running battle, the Spanish cavalry soon found themselves trying to escape to Cuzco, but the natives for once had the advantage. They had Pizarro's men exhausted. Valiant Indians succeeded in grabbing the tails of the horses while the riders tried to hack them off. The Spaniards were advancing only step by step through the crush of Indians, and were on the point of collapse. The situation was saved by native auxiliaries, some of whom had run back to the city to warn Hernando Pizarro of his brother's desperate plight. Hernando rang the bells to summon the citizens, and rode out with every remaining horse in Cuzco: eight in all. These trotted and galloped for three or four miles before coming to the scene of the engagement. They dispersed the native troops in a thundering charge. Toro's men rode up at the same time, and the Spaniards were able to make their way back to the city, exhausted and battered.

All agreed that this was the city's darkest hour. In another day the natives would renew their attack and would find the defenders with little food, their horses wounded, and many of their effectives away in the Condesuyo. The Spaniards' answer to this crisis was characteristic: they decided to take every horse capable of fighting and to attack the assembling native forces that very night. The attack was launched on Manco's own contingent, the best of the native troops. It achieved complete surprise. Gonzalo Pizarro caught a mass of Indians crossing a plain between two mountains and massacred them in 'one of the most beautiful skirmishes that was ever seen'. His charge ended with the horsemen riding out into the lake of Chincheros spearing swimming natives like fish. Hernando Pizarro encountered the Inca's guard of jungle archers and decimated them, despite arrow wounds to his own and other horses. The Spaniards had regained the initiative, and by demoralising Manco's own troops they emasculated the native attack. To heighten their psychological victory, they again cut the hands off hundreds of prisoners in the square of Cuzco.

The siege had now reached stalemate. The defenders had enough food to survive the rainy season but were too weak to break out of their encirclement. Manco's men had become convinced that they could not capture Cuzco by direct assault. They apparently hoped to trap the defenders during some sortie and 'they were waiting for the spring [of 1537] to assemble a more powerful army and complete the expulsion of the Spaniards'. But they had failed in the main purpose of the rebellion: the annihilation of the invaders in the Inca capital.*

While events in the mountains had reached a temporary stalemate, the balance of power in Peru was being altered by the arrival of seaborne Spanish reinforcements. Pizarro's first emissary had delayed for three months in assembling men and ships at Panama. The desperate Governor therefore sent Juan de Berrio in late September carrying letters of credit from himself and further appeals for assistance. The crusty old Francisco de Barrionuevo wrote to Spain from Panama that 'Berrio says that there are plenty of useless men in the city of Lima: some wounded, others fevered, others delicate and effeminate – but no men to go out against the Indians. . . . What's needed are men who will suffer hardships and hunger! There are plenty of effeminates down there!' More virile men began to reach Pizarro towards the end of 1536. Pizarro's own men returned from the Ecuadorean coast, and Alonso de Alvarado rode in with eighty men of his Chachapoyas expedition. The other Spanish governors also began to respond. 'Pedro de los Rios, brother of the Governor of Nicaragua, came in a large galleon with men, arms and horses.' The great Hernán Cortés in Mexico sent 'many weapons, shot, harnesses, trappings, silk cloth and a coat of marten fur in one of his ships under Rodrigo de Grijalva'. Licenciate Gaspar de Espinosa, Governor of Panama, sent men from Panama, Nombre de Dios and the Isthmus. In September the President of the Audiencia of Santo Domingo on the island of Española sent his brother Alonso de Fuenmayor with four ships containing a hundred cavalry and two hundred foot-soldiers.

When Pizarro's second appeal reached the Caribbean in November, two more ships went from Española, and its President wrote that 'the help from here now totals almost 400 [Spanish] men, 200 Spanish-speaking Negroes who are very good at fighting, and 300 horses'. But this proud force wasted three months trying to obtain ships on the Pacific, and did not reach Peru until the middle of the following year.* Juan de Berrio finally recruited four shiploads of men, but did not reach Peru until February 1537. Even the Spanish Crown responded to Pizarro's appeals: in November the Queen sent Captain Peranzures with fifty arquebusiers and fifty crossbowmen.* Manco Inca was trying to expel an invader supported by the resources of a huge empire: he could not hope to succeed against its united determination.

11

THE RECONQUEST

MANCO Inca hoped to reassemble his great native levies for a fresh attack on Cuzco after the rains in 1537. But he was no longer operating in a vacuum: two powerful Spanish armies were marching to the relief of the city.

In the north, Francisco Pizarro's appeals for help had attracted enough reinforcements for an overwhelmingly strong expedition into the mountains. The Governor was pessimistic about the chances of his brothers having survived in Cuzco, but he had to reconquer Peru. The captain-general of this relief expedition was Alonso de Alvarado, who had been recalled from colonising the Chachapoyas in north-east Peru. He left Lima on 8 November 1536 with 350 men, excellently equipped and including over a hundred horsemen and forty crossbowmen. The natives under Illa Tupac bravely tried to resist the advance of this powerful force. As it entered the mountains, twenty-five miles east of Lima, the Spaniards had 'a very fierce skirmish with the Indians'. A

Climax of the siege of Lima: the death of Manco's general Quizo Yupanqui soldier called Juan de Turuegano wrote to a friend in Seville: 'The Christians captured alive a hundred natives and killed thirty. They cut the arms off some of those that they captured, and the noses from others, and the breasts from the women. They then sent them back to the enemy, so that any who wanted to continue in rebellion could see that they too might have to submit to the knife.' Alonso de Alvarado was using the same tactic of mutilation that Hernando Pizarro was using in Cuzco: such terrorism was the Spaniards' ultimate psychological weapon. The Indians fought again on 15 November in the pass of the Olleros, but Alvarado's column continued to Jauja. The cautious commander paused there for a month to await reinforcements. He spent the time in the squalid business of exacting reprisals, torturing Indians for news of the Spaniards in Cuzco, and rounding up porters to replace coastal Indians who had died of exposure during the crossing of the Andes.

The Huanca tribe of Jauja, which had welcomed Pizarro in 1533 and refused to help Quisquis in 1534, had only reluctantly joined the rebellion when Manco's generals drove the Spaniards out of the mountains. The chiefs of Jauja were now firmly allied to the reconquering Spaniards. Turuegano said that two friendly Jauja chiefs accompanied Alvarado's column, 'and those caciques themselves burned any Indians they captured who were orejones, who are called Incas, and who were responsible for raising the country in rebellion'. After the battle on 15 November, 'they killed a thousand Indian orejones and took much booty from them'. Many Peruvian tribes still hated the Incas more than they did the Spaniards.

Alvarado was eventually reinforced by a further two hundred Spaniards under Gómez de Tordoya, all fresh men who had just arrived from Panama and Spain. He delayed a further month and then started to march slowly towards Cuzco. If the Indians had failed to defeat Hernando Pizarro's 190 men in Cuzco, they had little hope against Alvarado's 500. They did, however, try. Vast numbers of natives launched another determined attack on the Spanish column as it crossed Rumichaca, the bridge of stone over the Pampas river below Vilcashuaman. And when they were beaten off here, after killing some Spaniards and many native auxiliaries, they continued to harass the Spaniards as they marched eastwards towards Abancay.

The other Spanish army converging on Cuzco was Diego de Almagro's expedition returning from Chile. When Almagro marched out of Cuzco some twenty months previously, no one knew whether Cuzco fell within his governorship or that of Francisco Pizarro. Various royal decrees said that Pizarro was to govern the land for 270 leagues beyond the Spaniards' first point of entry at Puna. It did not specify whether this distance was to be measured due south in latitude or south-eastwards along the line of the coast; nor did it hint how this distance was to be calculated with the primitive surveying methods of the age, across the length of the Andes, and over country that was in rebellion. This bad drafting was to lead to years of civil war and the violent deaths of both governors, the former partners who had conceived and organised the original discovery and Conquest.†

When Almagro set out for Chile it was hoped that he would discover enough wealth within his own jurisdiction to make him and his followers forget Cuzco. They had shared in the distribution of loot from the first sack of the city, and that might have contented them. But the Chilean expedition proved a terrible disappointment. No one then suspected that the Bolivian altiplano concealed the fabulous silver mines of Potosí. Instead, Almagro's men marched back across the gruelling Atacama desert bitter and disillusioned after months of difficult exploration. Almagro's territory, New Toledo, proved to be a country devoid of spectacular cities or easy plunder.

As Almagro's men re-entered Peru the Indians kept telling them rumours of the rebellion and siege of Cuzco. Gonzalo Fernández de Oviedo, who wrote a long description of the Chilean expedition, said that Almagro hurried forward to obtain more certain news.* At Arequipa he began to correspond with the Inca Manco. Anxious to establish a close relationship, but hesitant about the difference in their ages, Almagro compromised by addressing the young Inca as 'My well-loved son and brother'. The letter that followed was a model of tact. Almagro had been told that Manco held Hernando Pizarro and the other Spaniards prisoner, and begged Manco to refrain from punishing them until his own arrival. He sympathised with the Inca for 'the abuse the Christians have done to your person, the robbery of your property and house, and the seizure of your beloved wives'. The guilty parties would certainly be punished in the name of the King, and Manco would probably be forgiven for his rebellion provided he desisted from it now. Almagro tried to present a picture of his own great strength, with the gross exaggeration that he had 'a thousand Christian men and seven hundred horses' in his command and was 'daily expecting two thousand more men'. But this image of power was weakened by his admission that 'I am come from a distant country and everything is used up, so that I have nothing to send you as a present. I know perfectly well that you are rich in Castilian cloth and wine and that you lack nothing. Nevertheless I am bringing a fur coat for you against the cold: I am keeping it for when we see one another. The King sent it to me, but I will give it to you.' Believing that Manco had the upper hand – and that he was in control of the coveted Cuzco – Almagro assured him that 'I would not think of doing anything without your approval and advice, and would never refuse you the friendship and goodwill I have always felt towards you.' The letter had a great effect among the natives.* Prisoners captured by the Spaniards in Cuzco boasted that their friend Almagro was on his way and would help them kill the survivors of the siege.

Almagro hurried inland from Arequipa on to the altiplano, over the pass of Vilcanota and down the river towards Cuzco. He paused at Urcos, twenty-five miles south-east of the city. Beyond here the road divides: to the right it continues down the Vilcanota–Yucay towards Calca and Ollantaytambo, to the left it passes into the valley of Cuzco through the Angostura, a defile protected by the Inca wall and gateway of Rumicolca. The three forces of Manco Inca, Hernando Pizarro and Diego de Almagro thus formed a triangle on the ground, each roughly equidistant from the others, each eager to seize some advantage from the impending clashes.

Hernando Pizarro's men were naturally overjoyed at the prospect of relief. They would have preferred to be rescued by forces sent inland by Francisco Pizarro rather than by the returning veterans of Almagro's expedition. They hoped to join forces with Almagro to defeat Manco Inca. But Hernando Pizarro and his brothers had no intention of surrendering the city they had been defending for the past year.

Almagro and his followers wanted Cuzco. They regarded Manco as a powerful potential ally in its seizure from the Pizarros. They also hoped by diplomacy to win the young Inca back to peaceful allegiance to Spain. Hernando, Juan and Gonzalo Pizarro could then be blamed for having provoked the Inca into rebellion, while Almagro would appear as the saviour of Peru and rightful governor of Cuzco.

Manco Inca was in a quandary. He still had a powerful army in the Yucay valley and had been mobilising the peasant levies for a fresh attack on Cuzco. But the approach of Almagro's and Alvarado's armies changed the balance of power at Cuzco in favour of the Spaniards. Manco was realistic enough to see that his rebellion was doomed. He could try to continue the struggle as a rebel chieftain in the wild country north of Cuzco. Or he could return to the comfort of ruling as puppet Inca under Almagro's protection in Cuzco. He was convinced that the Pizarro family would never forgive the killing of Juan Pizarro. Manco's return to the Spanish fold therefore depended on the total overthrow of the Pizarros, and on trust in Almagro's expressions of friendship. Manco probably preferred the idea of returning to Cuzco, but this meant risking his life on the security of Spanish promises.

Almagro opened proceedings by sending two of his men, Pedro de Oñate and Juan Gómez Malaver, on an embassy to the Inca at Ollantaytambo. They were to repeat to him that Almagro deplored the ill-treatment he had received from the citizens of Cuzco. If Manco would surrender, Almagro would ensure his pardon and the punishment of the guilty Spaniards by the royal authorities. This was exactly what Manco wanted to hear. He therefore gave a cordial reception to the two Spaniards, saw that they were well entertained during their stay, and even gave them some of the jewels and other objects taken from the men killed on Pizarro's relief expeditions.* He then proceeded to pour out his grievances. According to the account of the emissaries themselves, and of various other Almagrist sympathisers, the Inca complained of the outrages that had provoked him to revolt. He said that Gonzalo Pizarro had taken his wife, that Diego Maldonado had pestered him for gold, that he had been imprisoned with a chain round his neck. He named five citizens – including Pedro Pizarro and Alonso de Toro – whom he accused of having spat and urinated on him and of burning his eyelashes with a candle.* Almagro's men were impressed by all this; and also by the excellence of Manco's army and the defences of Ollantaytambo.

It was some time before Hernando Pizarro and his men believed the taunts of captured natives: that Almagro had returned from Chile and was marching against the city. They appreciated that something strange was afoot when the hordes of natives that had been massing for a fresh attack on the city vanished overnight. Hernando Pizarro sent a detachment of horsemen to reconnoitre along the Collao road and they confirmed Almagro's return. He also sent a young Indian with a letter to Manco himself. The Indian arrived while Almagro's envoys were with the Inca at Ollantaytambo. Pizarro's letter proved to be a warning to the Inca not to trust the word of Almagro, who was a treacherous liar and in any case was subordinate to Governor Pizarro. Such a warning from the man he had been besieging for the past year would seem to have been the most blatant of Trojan horses. But it had some effect. It reinforced Manco's reluctance to trust his life to any Spaniard. After much consultation among the native elders, Oñate was asked to give tangible proof of his anti-Pizarrist sentiment. He was to cut the hand off Pizarro's native envoy, a test that he performed without undue cavil, although he compromised by cutting off only the fingers at their bases.* Oñate was then sent to ask Almagro to come in person to negotiate with Manco at Yucay. Oñate reported all the native expressions of friendship, but warned Almagro to proceed with caution. Almagro therefore left part of his army under Juan de Saavedra at Urcos, and advanced with two hundred good horsemen down the Yucay valley towards the Inca. He paused at Manco's former headquarters, Calca, some twenty-five miles from Ollantaytambo.

One of Almagro's principal lieutenants now attempted a curious piece of personal diplomacy. Eager for the kudos of winning over the Inca, with whom he had formerly been friendly, one Rui Díaz took a a young Indian interpreter called Paco and made his way to the Inca camp. He went in defiance of Almagro, who wanted to keep the delicate negotiations in his own hands. Rui Díaz was well received at first, but Manco's attitude suddenly hardened. He demanded further proof of Almagro's hostility towards the Pizarrists in Cuzco. Almagro had captured four of Hernando Pizarro's scouts, and Manco now insisted that these must be executed as a demonstration of Almagro's good faith.

Different writers gave different reasons for Manco's changed attitude. One claimed that the Inca questioned the interpreter Paco and extracted from him an admission that Almagro planned to imprison the Inca and send him to Spain.* Cristóbal de Molina said that the rupture arose from Almagro's refusal to execute Pizarro's scouts. He could hardly agree to kill some of his compatriots, the gallant defenders of Cuzco. But Manco insisted on such an uncompromising display before he could dare to trust his person to Almagro's army. The doubts sown by

Pizarro's letter, the caution of his advisers, and his own suspicions made him realise the impossibility of marching into Cuzco alongside Almagro. Any last hopes of an alliance were dispelled when, according to Cieza, Almagro allowed his troops entering Calca to trample some natives, and when Almagro himself gave an insolent reply to a proud speech by Paucar, the young Indian commander of the Calca area.* Other more cynical Spaniards thought that Manco had never intended an alliance. He was simply waiting until Almagro had crossed to the eastern bank of the swollen Yucay river.*

Whatever the reason, five or six thousand of Paucar's men suddenly appeared on the hills around Calca and swept down to attack the Chileans shouting 'Almagro is a liar!' The Spaniards fought back with the customary counter-attacks, but had difficulty in recrossing the river on rafts in the face of this native aggression.* The Inca's anger also turned on the unfortunate Rui Díaz, the would-be negotiator of his surrender. Manco allowed his men to entertain themselves at his expense. They stripped Díaz naked and 'anointed him with their mixtures, and were amused to see his contorted features. They made him drink a great quantity of their wine or chicha and, tying him to a post, they shot at him from slings a fruit which we call guavas, distressing him greatly. In addition to this they made him shave off his beard and cut his hair. They wished to change him from a Spaniard and the good captain he was into an Indian with bare limbs.'

The attack on Almagro and his men ended Manco's chance of reconciliation with the Spaniards. He had not dared to surrender himself to Spanish good faith – a singularly debased commodity. Manco had staked everything on his valiant attempt to crush the scattered Spanish settlements in Peru. Now that his rebellion had failed he could not really hope to return as puppet Inca in Cuzco. There was no alternative for him but to retreat, to resume the life of a fugitive outlaw – an existence from which the Spaniards' arrival had rescued him three years earlier.

The Indians were given a respite during which to decide their future course of action. The two factions of Spaniards were embarked on a struggle for possession of Cuzco and, ultimately, of Peru. When Almagro's men escaped from Paucar's attack at Calca, they moved directly on Cuzco. On 18 April 1537 Almagro seized the city that Manco had been besieging unsuccessfully for the past year. The only resistance came from the Pizarro brothers, Hernando and Gonzalo, and a handful of their supporters. These were smoked out of a burning native palace and imprisoned. Almagro's next concern was to protect his prize from Alonso de Alvarado's relief expedition, which was still moving ponderously along the royal road from Jauja. The two armies of Spaniards, the two forces that had been hurrying to relieve Cuzco from the rebellious natives, came to battle at the crossing of the Abancay river on 12 July. Almagro's lieutenant Rodrigo Orgóñez won an almost bloodless victory over the Pizarrist troops, many of whom had just come to Peru in response to the Governor's panicky pleas for help.

At the moment when the Spaniards were falling into civil war, a rift occurred among the natives. Manco's half-brother Paullu now began to occupy the void left by the Inca's departure. The two men were of similar age, in their early twenties, but Paullu was of slightly inferior birth, although both were sons of the Inca Huayna-Capac. When Manco went to Jauja with Francisco Pizarro in 1534 he left Paullu as his lieutenant in Cuzco, and Paullu was always a staunch supporter of his brother's authority as Inca. Accordingly, when Manco was asked to send a native contingent with Almagro's Chilean expedition in 1535, he nominated Paullu and the high priest Villac Umu. For some reason Paullu did not join Villac Umu when he escaped from the expedition to incite Manco to rebel. Nor did he make any attempt to turn against Almagro's Spaniards when they were in Chile or struggling home through the Atacama desert. On the contrary, Paullu gave invaluable assistance to Almagro, and by his presence lent the foreign reconnaissance the respectability of a royal visit. The local population everywhere welcomed the Spaniards and supplied them with food and treasure. On the gruelling journey across the coastal desert, Paullu's men guided the expedition and cleaned the wells before its arrival. Even with this help, the eighteen-month exploration exhausted its members. Without it, Almagro and his men might never have returned. Two of them, Gómez de Alvarado and Martín de Gueldo, testified in 1540 that Paullu could easily have done them serious damage, had he wished, 'for he knows much about warfare and because so many men obey him'.

As Almagro's expedition returned, it was Paullu who informed it of the siege of Cuzco and gave accurate information about the progress of the rebellion. He was put under close guard for his pains. Despite this, Paullu continued to support Almagro loyally after his occupation of Cuzco. Paullu's men patrolled the royal road, reported on the progress of Alvarado's advance, and prevented any communication between Alvarado and the Spaniards in Cuzco.* At the battle of Abancay itself, Paullu's ten thousand auxiliaries helped in every way short of actual combat. They dug trenches, built two hundred rafts to help Orgóñez cross the river, and their shouting in the darkness caused Alvarado's men to hurry to the wrong part of the field.* But the chief advantage in having the support of the natives was that they provided an excellent fifth column. Orgóñez, for instance, was able to send Indians into Alvarado's camp with letters to individuals to persuade them to change sides.

Before marching on Lima, Almagro decided that he must dispose of Manco, who was still established only a day's march from Cuzco. It would not enhance his cause if, by marching against Francisco Pizarro, he left Cuzco so thinly defended that it fell to the rebellious natives. There was one further attempt at diplomacy. But Manco would not be lured into a peaceful surrender, even though the Pizarro brothers in Cuzco were now captive and his friend Almagro was clearly in the ascendant. It was said that Paullu sabotaged this negotiation by sending contradictory reports of Almagro's intentions: Paullu was enjoying power and had an interest in keeping his brother out of Cuzco. With the failure of this overture Manco decided that his position at Ollantaytambo was too exposed and vulnerable. The Spaniards knew his exact location and could reach him at any time by a few hours' riding. The Inca therefore wisely decided to retreat to somewhere more inaccessible.

Ollantaytambo lies at a crucial place in Peruvian geography. It is almost visibly at the junction between the Andes and the Amazon basin. Upstream lies the Yucay valley and the homeland of the mountainous Inca tribe – open country of rolling, grassy hills studded with rocky outcrops, or steep valleys rich in terraces of corn and potatoes. But downstream everything changes. As the ground falls away, the foothills of the Andes become matted with a tight fur of dense, low jungle. The tranquil Yucay changes its name and character to become the turbulent Urubamba. The climate becomes tropical, with heavy rains, electric storms and clammy mists shrouding the steep green hillsides. There are swarms of biting borrachudo flies and coral snakes in the forests. But, above all, there are trees, which begin to cover the ground here and do not relax their grip until the rivers reach the Atlantic. Below Ollantaytambo, the Yucay–Urubamba thunders through a mighty granite gorge. This used to be impassible, before its walls were blasted to make way for the modern narrow-gauge railway that takes tourists to Machu-Picchu. The ancient road avoided this gorge by climbing far behind the hills on the right bank of the river. It left the valley a few miles below Ollantaytambo, climbed to the pass of Panticalla, and then descended the Lucumayo stream to rejoin the Urubamba some thirty miles downstream and three thousand feet below Ollantaytambo. Another river joins the Urubamba from the west almost opposite the Lucumayo's entry from the east. This other tributary is now called Vilcabamba; its name has been applied to the entire wild region, and to the range of hills that lies between the Urubamba and the Apurímac to the north-west of Cuzco. Three places are important to us in this Vilcabamba–Lucumayo area. To the east, as the ancient road descended the Lucumayo towards the Urubamba, lay the town of Amaibamba; in the centre, between the mouths of the Lucumayo and Vilcabamba, the Urubamba was crossed by the strategic suspension bridge of Chuquichaca – the only easy route into the Vilcabamba area; and to the west, high up the Vilcabamba valley, lay the town of Vitcos, which has been identified with a group of ruins near the modern village of Puquiura. Vitcos was at well over nine thousand feet, a comfortable altitude for Incas, less than two thousand feet lower than Cuzco itself.

When Manco decided to abandon Ollantaytambo he was in effect relinquishing all the highland part of the Inca empire. He acknowledged that Spanish cavalry was invincible in this open country, and left the mass of his people under a foreign rule from which they have not escaped to this day. The departure from the great citadel was commemorated by poignant ceremonies and sacrifices. The diligent Pedro Cieza de León found two prominent native eyewitnesses to tell him about the ceremony and retreat. Villac Umu led the prayers, sacrifices and lamentations in the plain below the citadel while the idols were prepared for transportation. Manco's son Titu Cusi recalled that the Inca tried to gloss over the tragedy of the departure with brave talk. He claimed that the jungle provinces had long been importuning him to visit them, and 'I shall therefore give them this satisfaction for a few days.' There were in fact some shreds of hope for the natives. They had large armies still operating in the Collao and around Vilcashuaman and Huánuco, and the great levies that had attacked Cuzco the previous spring could theoretically be reassembled at a few weeks' notice. But their greatest hope came from daily reports that Pizarro was massing an army at Nazca for an attack on Almagro's highlands. The terrible invaders might destroy one another.

What Manco needed was an invulnerable, inaccessible refuge. In a meeting at Amaibamba, on the pass behind Ollantaytambo, he decided to try to reach a fortress called Urocoto that had been built by the Inca Tupac Yupanqui in the forests of the Chuis, east of Lake Titicaca. But after a month of difficult travel in the forested country beyond the Lares, Manco decided to return and seek refuge in the Vilcabamba valley. He descended the Lucumayo valley below Amaibamba and his men rebuilt the suspension bridge across the Urubamba at Chuquichaca. His followers crossed the river and established themselves in the town of Vitcos in the uplands at the head of the Vilcabamba valley.*

When Manco withdrew from Ollantaytambo, his men carefully demolished the road leading over the Panticalla pass to Amaibamba. But he underestimated his opponents. Almagro was anxious to win the kudos of subduing the Inca. As soon as Rodrigo Orgóñez returned from his bloodless victory at Abancay, Almagro sent him in immediate pursuit of Manco Inca. Orgóñez was one of the most energetic and dashing young conquistadores, and he was leading three hundred Spanish cavalry and foot-soldiers.† It would have taken a greater obstacle than broken roads to restrain such an adversary. Orgóñez's men soon reached Amaibamba, even though they were forced to travel much of the way on foot and had great difficulty clambering around places where the mountain path had been smashed or trees felled across it. At Amaibamba they routed the native troops who came out to defend the town. Manco escaped in his litter across the Urubamba towards Vitcos, leaving orders for the demolition of the bridge of Chuquichaca. This was only partly done: Orgóñez and his men were in hot pursuit, and many natives drowned in their haste to cross the river. In the confusion Rui Díaz, beardless and bedraggled, and other Christian captives of the Inca succeeded in hiding in some buildings and were able to rejoin their compatriots.

The few Spaniards who had reached Chuquichaca were too exhausted to cross the river that night: they were infantrymen who had done most of the fighting without benefit of horses. But next day Orgóñez himself arrived at the bridge and supervised its repair. The following dawn he crossed, entered the Vilcabamba valley, and soon penetrated as far as Vitcos itself. His men found the town full of spoil. It also boasted a sun temple with a crowd of frightened mamaconas and a sun image which Orgóñez took back to Paullu in Cuzco.

The attractions of this loot saved Manco's life, for while the Spanish troops ransacked Vitcos he escaped, as night was falling, into the mountains beyond the town. He fled like Prince Charles Edward after the failure of the '45, accompanied only by a handful of his most faithful followers. He had with him one devoted wife and possibly the priest Villac Umu. In the emergency he abandoned the traditional litter. Instead twenty fast runners of the Lucana tribe carried him in their arms. Orgóñez was soon in pursuit. He sent his four best horsemen up the pass behind the Inca, and, shortly after midnight, himself followed with twenty further horse. Although the Spaniards rode all night Manco eluded them – possibly because he had escaped by a different pass. When Orgóñez returned to rest in Vitcos he received orders from Almagro to return immediately to Cuzco, which he did towards the end of July 1537.*

He took back important booty. Manco's son Titu Cusi explained that the natives had removed the mummies of some of their Inca ancestors and the stone idol from the holy hill of Huanacauri from Ollantaytambo to the greater security of Vitcos.* Orgóñez's men took these sacred relics and also a golden image of the sun. They rescued Captain Rui Díaz and the Indians' other European prisoners, and recaptured a great quantity of Spanish clothing taken from Pizarro's ill-fated relief expeditions.* Such clothing and equipment were at a premium in Peru, and Almagro promptly distributed it among his veterans 'who had returned half naked from the Chilean expedition'. Rodrigo Orgóñez also brought over fifty thousand head of llamas and alpacas which Titu Cusi described as the cream of the royal herds. But most serious for Manco, the Spaniards led back 'over twenty thousand souls' so that the Spaniards could say complacently 'we are no longer concerned about any war that the Inca could wage'. These thousands of Manco's followers were released, to return gratefully to their villages.

Titu Cusi described the sad exodus. 'My father escaped as best he could with some followers, and the Spaniards returned to Cuzco highly satisfied with the loot they had taken, and with me myself and many coyas.' The boy Titu Cusi was entrusted to Pedro de Oñate, who was presumably a friend of Manco, for Almagro had selected him to deliver the delicate embassy after the return from Chile, and Manco had received him very hospitably. Titu Cusi later wrote that 'Oñate received me into his house in considerable comfort and treated me well. When my father heard this he sent to Oñate, thanked him greatly, and again entrusted me and some of his sisters to him, asking him to look after us, for which he would repay him.'

Manco had fallen pathetically low. It was only five months since he had been sending out messengers to recruit a great army for the final assault on Cuzco, and had been considering partnership with Almagro in attacking the city. Now he was a forlorn fugitive, driven from what had seemed a sufficiently remote corner of the empire. He was saved by the Spaniards' own dissensions.

When Rodrigo Orgóñez returned to Cuzco he found Almagro negotiating with a team of envoys from Francisco Pizarro. In mid-September the men of Chile marched down to the coast for an arbitration of the dispute between the two leaders. Gonzalo Pizarro escaped from captivity and succeeded in rejoining his brother in Lima. Hernando Pizarro was released during the course of the negotiations. With the three brothers reunited, the Pizarrist attitude hardened and the two factions were soon at war again, with Almagro controlling the sierra and Pizarro the coast. Hernando Pizarro led an invasion of the Almagrist territory and succeeded in reaching Cuzco. His campaign culminated in a complete victory at the edge of the city: the battle of Las Salinas, 26 April 1538. Rodrigo Orgóñez fought fiercely throughout the battle but was unhorsed and wounded by an arquebus shot. He was captured, beheaded, and his head was exposed in Cuzco after the battle. Diego de Almagro was also captured and, ten weeks later, Hernando Pizarro instigated a judicial condemnation of the 63-year-old Marshal. With Almagro's chief supporters under house arrest and a heavy guard around his prison, Hernando Pizarro ignored pleas for mercy from the astonished leader and had him garrotted. Spaniards in Peru were profoundly shocked. Indignant reports were soon on their way to Spain and contemporary writers became violently partisan in support or condemnation of the Pizarros. Cieza de León was one of the few who were impartial – he reached Peru a decade later when the furore had abated. He described Almagro as 'a man of short stature, with ugly features, but of great courage and endurance. He was liberal, but given to boasting, letting his tongue run on, sometimes without bridling it. He was well informed and above all much in awe of the King. A great part of the discovery of these kingdoms was due to him.' Almagro was a foundling, 'of such humble parentage that it could be said of him that his lineage began and ended with himself'.

During the year in which Almagro controlled Cuzco, Paullu served him as a loyal ally. In July 1537, Almagro arranged a formal ceremony in which he stripped Manco, *in absentia*, of his title of Inca 'and gave the fringe, which is the insignia and sceptre of the kingdom, to his brother Paullu Inca Yupanqui, a well-disposed and brave man. Once the fringe had been bestowed on the new Inca he was obeyed and revered by the Indians, particularly by those who obeyed the Adelantado [Almagro] or were on good terms with him.' Although Almagro had no possible right to transfer the title of Inca, Paullu was pleased to have it. Manco could not comprehend Paullu's desertion to the Spanish side, and sent him repeated entreaties to join him at Vitcos. 'Paullu replied that he must always retain his friendship for the Christians, who were so valiant that they could never fail to be victorious.' He added taunts about the failure of Manco's great army to subdue Hernando's two hundred Spaniards. Manco was bitterly disappointed and furiously angry. Paullu's motives were straightforward. He was convinced that the Spaniards were there to stay, and preferred a life of comfort under them to joyless retreat in Vilcabamba. He was an evident opportunist who saw in Manco's revolt a chance to supplant him as puppet Inca.

Paullu helped Almagro considerably during the campaign against Hernando Pizarro. He accompanied him to the coastal negotiations and his Indians kept Almagro informed of the Pizarros' moves. On one occasion they enabled Almagro to swoop down and capture a group of horsemen travelling between Lima and Ica.* His men helped defend the pass of Huaitará against Hernando Pizarro and to break up the road. He even made a fiery proposal to Almagro that they should attempt to catch Hernando Pizarro in some defile, in the way that Manco had destroyed the relief expeditions of 1536. 'In the passes I will defeat Hernando Pizarro and kill the greater part of his men. If your Christians do not wish to go, let me go alone with my Indians and I will do as I say.' Almagro refused the offer. When a citizen of Cuzco, Sancho de Villegas, approached Paullu with a proposition that they should both desert to Hernando's army, Paullu dutifully betrayed him to

Almagro, and Villegas was arrested and executed.* Paullu also disclosed the embassies that reached him from Manco. He told Almagro that Manco had said he would emerge from his retreat if Almagro were victorious, but that he dare not do so if Hernando Pizarro won, since his men had killed Juan Pizarro.

Paullu's loyalty to Almagro did not survive the Marshal's defeat. Having thrown in his lot with the Christians, Paullu could hardly be blamed for not wishing to be caught on the losing side in their absurd dispute. His natives fought on the Almagrist side in the battle of Las Salinas, and for a time engaged the native auxiliaries helping Pizarro. But when Paullu saw how the tide of battle was running he ordered his men to desist. He later explained that Almagro had told him not to fight Spaniards, and Hernando Pizarro had reinforced his native auxiliaries with five Spanish horsemen. Shortly after the battle the victorious Hernando Pizarro sent for Paullu, who readily agreed to support the new masters as fervently as he had their predecessors. Pizarro was pleased to have him as an ally. Paullu was not, of course, tainted by any association with the native rebellion, since he had been with the Spaniards in Chile throughout its duration. But this smooth switch of allegiance ended Manco's last hopes of reconciliation with his brother. Paullu had shown that his loyalties were to Spaniards, even Pizarros, and that he had no intention of surrendering his usurped title. Manco never forgave this collaboration and betrayal.

12

THE SECOND REBELLION

Manco Inca was shaken and unsure of himself during the year after his escape from Orgóñez. He was alarmed at the ease with which the Spaniards had penetrated to Vitcos, which Oviedo described as 'the place many claim to be the strongest possible or known anywhere in the world'. Manco now wanted to reorganise his forces in an even remoter stronghold. His son Titu Cusi reported that 'some chiefs of the Chachapoyas told him that they would take him to their town, Rabantu, where there was a fine fortress in which he could defend himself against all his enemies. He followed their advice.' The high priest Villac Umu did not wish to leave the Cuzco area, but Manco 'considered that [in the north] he would be safe from his enemies the Christians. He would not hear the neighing and stamping of their horses,

A battle between Indians and mounted Spaniards

and their sharp swords would no longer slash the flesh of his people.' 'No sooner had Rodrigo Orgóñez left the river [Vilcabamba] than the Inca assembled those of his people who were going with him and told them that they must retreat into the deepest recesses of the Andes, since the gods had allowed their enemies to conquer the empire of their ancestors the Incas Yupanqui. There they could live in security, without fear of destruction or of falling into the hands of the Christians. The Indians and principal orejones listened enthusiastically to the Inca and agreed to go into this voluntary exile with him.'

The proposal of the chiefs of Chachapoyas was a fascinating one. They were probably referring to the great hilltop fortress of Cuelape that rises high above the left bank of the Utcubamba river, not far to the south of Chachapoyas.* The choice would have been admirable. Of all the myriad ruins in Peru, Cuelape is the most spectacularly defended, the strongest by European standards of military fortification. The area itself is remote, for Chachapoyas lies east of the great canyon of the Marañon and of the road to Quito. It is on the final open stretch of the Andean foothills: a few miles further east the suffocating Amazonian forest takes a firm grip on the land. The Incas added the area to their empire only under Huayna-Capac; and the first road for wheeled traffic joining Chachapoyas with the Pacific seaboard was not completed until 1961. The fortress of Cuelape occupies a long mountain-top ridge, reached after a morning's stiff walk up from the Utcubamba river far below. Curiously, although the hillsides beneath are cultivated fields, and although the site is at an altitude of over 3,000 metres, the 800-yard-long fortress enclosure itself is now filled with dense forest. Cuelape's superb outer walls, rising in places to over fifty feet, are faced with forty courses of long, rectangular granite blocks. A steep ramp overshadowed by tall, inclined re-entrant walls leads into the mysterious gloom of the fortress enclosure. There, towering out of the tangle of trees and undergrowth, are the remains of the walls of inner enclosures, of watch towers, bastions and of some three hundred round houses. It has been calculated that the great walls of Cuelape contain 40 million cubic feet of building material – three times the volume of the Great Pyramid.

Manco apparently started north-westwards from Vitcos 'with all his men to travel by daily marches directly towards the town of Rabantu, which is towards Quito'. He may have travelled in the wild country bordering the lower Apurímac, for his son wrote that he visited the Pilcosuni Indians, who lived in the area between the river and the Pajonal plateau. Much of the area remains unexplored to this day. He would then have emerged on to the upper Huayllaga and 'fixed his quarters in the place where now stands the city of Huánuco' – well to the east of Inca Huánuco and the main highway. But for some reason Manco abandoned his northern exodus. He may have decided that Chachapoyas was too remote from the heart of the Inca empire, or that the forested crags of Vilcabamba still offered a safer refuge than the mighty walls of Cuelape. He may have been suspicious – rightly so – of possible treachery by the Chachapoyas. It is also probable that the quarrels between the Pizarro and Almagro factions restored his spirit. He knew that isolated groups of Spaniards could be destroyed if they were caught in difficult country. His travels also showed him that there were still great reserves of native manpower that had not been defeated after the relief of Cuzco.

Manco therefore determined to organise nothing less than a new rising against the invaders. This was a serious campaign, the last major attempt to dislodge the Spaniards from Peru. It was a remarkable demonstration of Manco's resilience, courage and determination. He and his commanders attempted to start a series of local uprisings along the length of the southern Andes, from the Marañon to south-eastern Bolivia and Chile.

It was easy for Manco to start the northernmost part of his revolt. The Inca general Illa Tupac, who commanded Manco's northern army, had never disbanded his men after Quizo Yupanqui was killed during the attack on Lima. He attacked Alonso de Alvarado's column as it marched towards Abancay in 1537, and then moved north to establish himself in the Huánuco area. Illa Tupac ruled this region as a local warlord and responded willingly when his relative Manco proclaimed a second uprising. North of Huánuco, the tribes of the Conchucos rebelled in the hills bordering the upper Marañon. Thousands of them descended towards the coastal city of Trujillo, killing any Spaniards or collaborating natives that they could catch. Their victims were sacrificed to the tribal deity Catequil.*

From Huánuco Manco moved southwards to Jauja and tried to inspire the local Huanca tribe to join his rebellion. He was unsuccessful. The Jauja Indians had been restless under Inca rule and had welcomed the first Spaniards. Many had helped Riquelme's Spaniards against Quisquis. They joined the 1536 revolt only when it was clear that the natives were in the ascendant, but later helped Alvarado in gleeful reprisals against the Incas.

Rebuffed at Jauja, Manco moved southwards and installed his men in a fertile valley not far north of modern Ayacucho. They may even have occupied for a time the great fieldstone enclosures of the Tiahuanacan ruin Huari.* From this base Manco's warriors launched a series of successful attacks on travellers moving along the main highway. Spanish merchants became terrified by rumours that captives were taken back to Vitcos, where the natives 'tortured them in the presence of their women, avenging themselves for the injuries they had suffered, by impaling them with sharp stakes forced into their victims' lower parts until they emerged from their mouths. Reports of this caused such terror that many persons with private business and others on public service did not dare to go towards Cuzco without a well-armed escort.'

As soon as Francisco Pizarro reached Cuzco, late in 1538, he sent the royal factor Illán Suárez de Carvajal with a large force to deal with this menace, to try to capture the elusive Manco and to relieve the fears of the Jauja tribes. Illán Suárez marched westwards to Vilcashuaman and then started north-westwards along the Jauja road.

Manco was by now in a hilltop village called Oncoy, to the north-west of Andahuaylas, and the inhabitants celebrated his arrival with a feast. His scouts reported that 'over two hundred fully-armed Spanish horse were out looking for him'. Illán Suárez soon learned of the Inca's presence. He turned eastwards off the royal road and descended to the turbulent Vilcas (Pampas) river. As Governor Pizarro wrote to the King, 'We all wish to capture the Inca, and [Illán Suárez], with this objective, attempted to take him from two sides. He himself remained on one side with the horsemen to occupy the upper escape route.' He sent an eager young captain called Villadiego with thirty arquebusiers and crossbowmen to reconstruct the bridge over the Vilcas and close the trap on Manco. Villadiego was to remain at the bridge until the attack was ready. Instead, he 'took the guardians of the bridge and put them to a rope torture' until they revealed that Manco was in the village of Oncoy, immediately above, with only eighty followers. 'When Villadiego heard this he was delighted, thinking that it would be easy to capture or kill the Inca, and that he would gain much honour and great rewards.' Villadiego immediately led his men up the steep path from the river canyon towards Oncoy. He had five arquebusiers and seven crossbowmen among his thirty men.*

According to Titu Cusi, it was Manco's sister-queen Cura Ocllo who came running to tell him of the approach of the enemy. He himself went to look and came hurrying back to organise his men. He had four captured horses and these were quickly saddled for himself and three chiefs of royal blood. The women in the Inca's camp were arranged at the top of the hill with spears in their hands, to look like more warriors. Manco had learned to ride with agility, wielding a lance in one hand, and he now had the satisfaction of riding into battle against Spaniards on foot. Villadiego and his men were utterly exhausted and winded from the steep, hot, waterless climb – Manco had cleverly destroyed an irrigation canal to deny them its water. The Spanish leader struck a flint as he heard the Indians approach and his arquebusiers quickly lit their wicks. But they were too slow in loading. Although one arquebus shot killed a native and some crossbows were fired, this was not nearly enough. A two-hour battle ended in a complete victory for the Inca. Villadiego, his arm broken by a battle-axe, was covered in wounds and died fighting. Twenty-four Spaniards were killed or forced over the edge of the mountainside; the remaining six fled down the slope. Manco's men stripped the bodies of their enemies and carried the spoils to Oncoy. 'They and my father', wrote Titu Cusi, 'were overjoyed by the victory they had won, and held celebrations and dances for some days in its honour.'

Manco was encouraged by the victory over Villadiego, and felt strong enough to punish the Huanca for their Spanish sympathies. In a series of skirmishes along the road to Jauja he 'killed and destroyed many of them, telling them most effectively, "Now get your friends to help you!" ' The Huancas did just that: they sent further pleas for help to the Spaniards. But the Incas were in the ascendant. Manco's generals Paucar Huaman and Yuncallo attacked and defeated a large force of Spaniards and Indian allies at Yuramayo, in the forests to the east of Jauja. Although Yuncallo died in the battle, many Spaniards were killed and their booty was taken back to Vilcabamba.*

Manco followed this second victory by exacting vengeance on the Huancas' principal shrine, Wari Willca. He destroyed its precinct, executed its priests, and dragged the stone idol across country before dumping it into a deep river. The Incas had always removed such tribal deities to Cuzco as hostages for the good behaviour of subject tribes.* The desecration of Wari Willca was Manco's most dramatic means of punishing the collaboration and treachery of the Huanca and Jauja Indians.

When Francisco Pizarro heard the news of Villadiego's defeat, he feared that 'the Inca might grow bold, assemble more men, defeat the Factor [Illán Suárez] and cause further damage'. He therefore raised seventy horse in Cuzco and himself marched out on 22 December 1538. Although he never made contact with the Inca, Pizarro went on to protect the vulnerable stretch of highway from Manco's raids. On 9 January 1539 he founded a Spanish city, originally called San Juan de la Frontera, at Huamanga, leaving twenty-four citizens and forty guards under the command of Captain Francisco de Cárdenas. This man wrote, later, 'at the time that I went there the town was in extreme danger, for there were very few Spaniards in it, and the Inca was close by with many warriors in rebellion . . . and they had killed some Spaniards'.

While Manco was active in the central Andes, his commanders were raising rebellions farther south: in the Condesuyo, beside Lake Titicaca, and in the Charcas, the easternmost tip of the Inca empire. The militant ecclesiastic Villac Umu operated in the mountains to the south and south-west of Cuzco, the area that had been the western quarter of the Inca empire, the Cunti-suyu or Condesuyo. Bishop Vicente de Valverde, who succeeded Villac Umu as spiritual head of Cuzco, described him as the man 'who ordered the sacrifices and temples of the sun and who was like a bishop or pope in this land' and noted that he had a powerful influence over the young Inca. Francisco Pizarro reported that Villac Umu was 'causing much damage and exciting the natives' in the area, which had been the scene of the outbreaks of resistance at the start of Manco's first rebellion.

The natives were even more successful farther south. Beyond the pass of Vilcanota lies a great plateau, the high, largely treeless altiplano. After the vertiginous slopes and claustrophobic valleys of the Andes, a traveller is relieved to see level ground and something approaching a horizon. The altiplano is at too great an elevation for maize or imported European cereals, but it produces abundant crops of potatoes, quinua and other indigenous plants, and its rolling expanses of pale grass support great herds of llamas and alpacas. The area is densely populated by Aymara- and Quechua-speaking Indians, but they live scattered in villages and small settlements, so that there is still a feeling of loneliness in the thin, clear air. At the northern edge of the altiplano lies the great Lake Titicaca. On clear days the lake can look Mediterranean, with steep, dry terraces on the island of Titicaca or the Copacabana peninsula. The air is still, Indian shepherd boys play baleful tunes on their reed pipes, there are sheep, dung-beetles, birds circling overhead, and imported eucalyptus trees. The icy water of the lake is an intense blue, reflecting the clear sky. But the illusion is quickly broken when a sudden squall whips across the lake, the water turns a turbulent grey and a cold wind dominates the high plateau.

Titicaca was more a cradle of Andean civilisation than Cuzco. Its Tiahuanaco culture had swept across Peru centuries before the development or expansion of the Inca tribe, and the Incas regarded the area with veneration. Their legends of the creation and the flood took place on the mysterious lake. During the period of Inca expansion, Pachacuti had first conquered south-eastwards as far as Lake Titicaca. His son Tupac Yupanqui Inca was confronted by a serious revolt of the tribes that lay along the fertile northern and western shores of the lake, notably the Lupaca and Colla. Tupac Inca crushed the revolt in battles at the butte of Pucará and at the Desaguadero river, the only outlet for the waters of the lake. That fighting had occurred only fifty years before the Spanish Conquest. Manco's revolt of 1538 began as a clash between these same Colla and Lupaca. Probably at the instigation of Manco's uncle Tiso, Cariapaxa, chief of the Lupaca, attacked the Colla and began devastating their lands as punishment for their collaboration with the Spaniards. The Colla appealed for help to the Spaniards in Cuzco.

Hernando Pizarro had just executed Almagro. He welcomed the prospect of a lucrative expedition into the Collao (as the Spaniards called the Collasuyu) for the area had never been thoroughly occupied or ransacked since Almagro marched through on his way to Chile. Fortunately for Hernando Pizarro, Cuzco now contained relatively few Almagrist sympathisers. Immediately after the battle of Las Salinas, Pedro de Candía – the gallant Greek who had been one of Pizarro's first supporters and who had fired the artillery on the square of Cajamarca – convinced himself that an eldorado of fabulous wealth lay in the jungles to the east of the Cuzco–Titicaca road. Candía was rich enough to organise a magnificent expedition of three hundred of the most dashing men in Cuzco. It plunged into appallingly difficult country and was soon devastated by famine, forest Indians, the ravages of the rain forest, and eventual mutiny. Only half its members returned, months later. The departure of these men left Hernando strong enough to kill Almagro. He now felt sufficiently secure to leave Cuzco for the Collao, taking with him his young brother Gonzalo and the puppet Inca Paullu, whose influence had always been particularly strong in the southern part of the empire.

The Lupaca determined to make a stand at the Desaguadero river, a weedy, unimpressive stream that drains the water of Lake Titicaca to the lake and marshes of Poopó and forms the modern boundary between Peru and Bolivia. Although apparently sluggish, the Desaguadero is deep and has a strong current. Things began badly for the Spaniards. A straggler from an advance party under Gonzalo Pizarro was captured and sacrificed by the natives at a shrine beyond the river – probably the ruins of Tiahuanaco itself. When Hernando Pizarro and the main force reached the Desaguadero, they found that the usual bridge of banana-shaped reed pontoons had been dismantled and the far bank was bristling with defenders.

Pizarro discovered a supply of balsa-wood logs at Zepita; Huayna-Capac had had this wood transported there on the backs of human porters along hundreds of miles of mountain paths. A raft was constructed from this wood and Hernando Pizarro pushed out on it with some twenty men, all heavily resplendent in sixteenth-century armour. A hail of stones and arrows greeted them from the far bank. This was too much for the native rowers. They stopped work and the raft was soon whirling helplessly downstream with its cargo of metal-clad veterans. Other Spanish horsemen rode out into the apparently placid waters to help their chief. But the water reeds and mud were too deep. The horses sank under the weight of their riders' armour and eight were drowned without trace. Hernando Pizarro eventually regained the northern bank amid the jeers of the Lupaca. He owed his rescue partly to Paullu and his men, who were eager to demonstrate their loyalty. As Paullu later testified, 'Hernando Pizarro was farther downstream, but when I, the Inca, saw him, I assisted him and did what I was and am still duty-bound to do for any servant of His Majesty – although I was capable of destroying them all utterly.'

That night two large rafts were made with more of the Incas' stores of balsa wood. These were launched on to the lake next day and propelled around the opening of the Desaguadero river. Hernando Pizarro was huddled on the first raft with forty armoured Spaniards. Behind came the second raft with Gonzalo Pizarro and a load of horses, which were to be kept out of range of the bank until a bridgehead had been established. The rafts must have looked like the Bayeux Tapestry's vivid illustrations of William's army sailing towards Hastings. Hernando jumped into chest-high water and waded ashore, fighting his way through the reeds. Even on foot the Spaniards were given immunity by their steel armour and helmets, 'for they could not be speared, and moved about from place to place suffering little damage'. The native auxiliaries now arrived on a flotilla of reed boats, the horses were landed from their raft, and the Spaniards were able to reassemble the pontoon bridge, which the natives had simply drawn across to the southern bank. Once the Spaniards were mounted the battle was won. The day ended with the usual terrible pursuit across the flat savannah towards the ruins of Tiahuanaco. The native chief Quintiraura was captured and his village burned.

The Pizarros progressed across the altiplano accepting the surrender of its natives, whom they treated with leniency. Hernando decided to return to Cuzco to welcome his brother Francisco and explain the execution of his partner Almagro. He left the campaign to Gonzalo Pizarro, not suspecting that yet another Indian revolt lay ahead of him.

Manco had sent his uncle Tiso to the Collao. This was the most formidable survivor of Huayna-Capac's generals, the man who had sat immediately below Chalcuchima at the coronation of Atahualpa's successor Tupac Huallpa at Cajamarca. Tiso began by executing Challco Yupanqui Inca, who had been Huayna-Capac's governor of the Collasuyo. Challco had been far too helpful towards Almagro in 1536, 'he led him to Chile, showing the roads . . . and making the Indians along the route obey him and do him no harm whatsoever'. As Challco's grandsons later testified, 'Manco Inca learned that our grandfather was favouring the Spaniards and sent one of his captains called Tiso with a great force of men to kill him. . . . Tiso caught him in Pocona, in the province of the Charcas and Chichas, and killed him there one night when I [his grandson] was with him.'

Inspired by the forceful Tiso and by a sense of self-preservation, the peoples of Consora and Pocona on the eastern foothills of the altiplano formed a federation with the warlike Chicha tribe. Under the leadership of the Chicha chief Torinaseo, they determined to resist the Spaniards' advance.

Gonzalo Pizarro now brought the Conquest to this area, which Governor Pizarro described to the King as 'a very wild country where Captain Tiso, the Inca's uncle, had fortified himself'. For five days Gonzalo Pizarro fought his way into the valley of Cochabamba. Once inside, he and his force of seventy Europeans and five thousand natives found themselves surrounded by warriors, who occupied all the passes out of the valley. The Spaniards spent an anxious night, with the fires of a native army flickering in the hills around the tambo they were occupying. Their leader encouraged them with speeches of bravado, but they kept their horses saddled and their sentries alert. And once again they opened the battle the following morning by charging into the thick of the hordes of warriors. Gonzalo Pizarro himself led one contingent. Another was under Captain Garcilaso de la Vega, who had just begotten, on an Inca lady Chimpu Occlo baptised as Palla Isabel Yupanqui, the son who was to be the most famous chronicler of the Incas: Garcilaso de la Vega Inca.† The third contingent was led by Oñate, the man who had cared for Manco's son Titu Cusi since his capture the previous year, and this contingent contained the zealous Paullu with his five thousand native auxiliaries.

The battle was furious, for the natives 'had succeeded in surrounding the camp with an infinite number of thick pickets so that the Spaniards could not use their horses'. The native auxiliaries had to dismantle these obstacles. The Indians gave blood-curdling war-cries, and the two sides fought 'entangled with one another, with the Castilians causing a notable slaughter with their lances and swords, and trampling the natives with their horses'. Gonzalo Pizarro and Oñate engaged the chiefs of Consora and Pocona, who had eight or nine thousand warriors.

The Chichas, meanwhile, attacked the Spanish infantry and native auxiliaries, who were in the tambo protected by a screen of horsemen under Gabriel de Rojas. The decisive factor in this latter engagement was the attitude of Paullu's men. Paullu clearly had to exercise his full authority to keep his men from joining their compatriots. He testified the following year that there was a moment in the battle when the natives had 'killed two Christians and wounded fifteen, all of whom fled. But I was bringing up the rear, reinforced them, and made them turn against the enemy. And because my own men took to flight I killed some [of my] Indians.' Martín de Gueldo confirmed Paullu's claim: 'This witness saw Paullu going about with a drawn sword in his hand and heard him say that he had wounded some of his own Indians because they took to flight. The witness also saw him protesting with his caciques because they were not going to fight, and saw him make them turn back against the enemy.' Alonso de Toro went even further in praise of Paullu. 'If Paullu had not been there the Spaniards would have suffered heavily. And if he had chosen to be treacherous they would have suffered far more – few or none of them would have escaped.' It was a tragedy for the native cause that Paullu chose to support the Spaniards with such determination, for he was clearly an excellent soldier. He knew that he was fighting his brother's supporters and yet, as Gueldo said, he did not turn his five thousand men against the seventy Spaniards even when they were utterly exhausted at the end of a day of fighting at this very high altitude. His conviction that the Spaniards would inevitably triumph, and his determination to cling to the Inca title outweighed any sense of patriotism.

The battle of Cochabamba lasted all day and night, 'with the greatest tenacity seen in any battle with the Indians in this country'. Paullu's resolution was decisive, for the Chichas, who had never before had to face Spanish forces, were the first to break and flee, with their chief crying 'we are all lost' as he ran to the hills. The Spanish horse then broke the spirit of the masses of Charcas Indians, and the afternoon was spent in a devastating pursuit that left eight hundred native dead on the battlefield.

Despite this, the natives still controlled the situation. They had fought bravely, killed and wounded many of the Spaniards and their horses, and killed 'a large part of the Indian auxiliaries'. They now sent word to Tiso, 'who was the Inca's captain-general in that province, and was one of the Christians' greatest enemies'. Tiso rapidly assembled 40,000 further native levies to press home the siege of Cochabamba. Gonzalo Pizarro, for his part, succeeded in sending a horseman galloping through the native lines to carry a plea for help to his brothers. Hernando was already marching south and Francisco had just returned to Cuzco from his expedition to found Huamanga in January 1539. The Governor sent forty-five citizens of Cuzco to reinforce his brothers.*

At this moment, the start of 1539, Manco's second rebellion was progressing satisfactorily. Groups of natives were under arms everywhere for a thousand miles from the Conchucos to the Charcas. But the successes were short-lived. The Spaniards were aroused to the danger and now moved out to crush each of the rebellious areas.

At Cochabamba the appeal to Tiso was too late. The tribes of the Charcas were demoralised by their battle with Gonzalo Pizarro. Garcilaso de la Vega inflicted further punishment on the Indians of Pocona, killing four hundred. The rebellion crumbled rapidly when strong reinforcements under Hernando Pizarro and Martín de Guzman succeeded in bypassing the main road and entered Cochabamba through the hills. The chief of Pocona was the first to surrender. He was well received by the Spaniards and welcomed as an ally. The other chiefs took note. One by one they surrendered and paid homage to King Charles: chiefs Anquimarca, Torinaseo of the Chichas, Taraque of the Moyos. And then, to the Spaniards' surprise, Tiso himself followed suit. 'Tiso said that he would surrender to Hernando Pizarro if he would pardon him and give his word to do him no personal injury. Hernando Pizarro gave his word, and with this Tiso came.' 'All were amazed by this, for Tiso had been the best subject that the Inca possessed.' Governor Pizarro described his capture as 'a stroke of the greatest good fortune'. On 19 March 1539 Gonzalo Pizarro entered Cuzco with his important captives and Bishop Valverde was able to describe Tiso and Paullu in the same breath as 'at peace and very good friends to us'.

The doughty Villac Umu was still in control of the Condesuyo. Pizarro sent Pedro de los Rios against him, and for eight months there was fierce fighting in this remote district. Much of the fighting had to be on foot, as the area was 'very wild and mountainous and could not be conquered on horses'. It was not until October 1539 that the high priest surrendered to the mercy of the Spaniards.

The revolt in the north continued even longer. Illa Tupac was in control of the entire area north of Jauja, including Bombón, Tarma and the Atavillos, the heads of the valleys on the Pacific side of the watershed. Alonso de Alvarado, the commander defeated by Almagro at the battle of Abancay, had been engaged on a conquest of the Chachapoyas in 1536, from which he was recalled by Francisco Pizarro at the outbreak of Manco's first rebellion. Immediately after the victory of Las Salinas, Alvarado decided to resume his conquest. Hernando Pizarro allowed him to leave Cuzco, and his men marched northwards from Jauja in mid-1538. Illa Tupac learned of their approach and rallied his Indians to resist. They surprised Alvarado's column on a snow-covered puna north of Lake Junín, and severely wounded at least one of his horsemen before being driven off. Only then was Alvarado able to continue his march to Chachapoyas.

Manco had sent another of his leading orejones, his cousin Cayo Topa, to win over the Chachapoyas tribe, presumably before he himself travelled there to seek refuge. But Cayo Topa failed: chief Guaman of the Chachapoyas refused to join the revolt – contrary to the invitation that some of his followers had delivered to Manco. Alvarado was therefore well received when he reached Chachapoyas, and his good treatment of the natives ensured the security of a new settlement that he founded at Rabantu. Cayo Topa himself was betrayed and captured in a dawn attack by a force of Spaniards and natives marching inland from Trujillo. Yet another powerful tribe had sided with the Spaniards against the Incas.*

The natives remained unsubdued at Huánuco. A strong force under Captain Alonso Mercadillo, instead of marching against them, remained at Tarma for seven months terrorising its peaceful inhabitants. These complained through the Treasurer Alonso Riquelme that Mercadillo's men had stayed there 'consuming their maize and cattle, robbing them of all the gold and silver they possess, taking their wives, keeping many Indians chained and making slaves of them . . . abusing, extorting and torturing the Indian chiefs so that they would declare their gold and silver'. Another force under Alonso de Orihuela defeated and captured the Inca Captain Titu Yupanqui, and reopened the road through the Chinchaysuyo.*

Gonzalo Pizarro marched north in mid-1539 to become Governor of Quito, but was delayed when 'he had to fight the Indians of the province of Huánuco. These came out in battle against him, and placed him in such danger that Marquis Pizarro had to send Francisco de Chaves to his rescue.' On arrival at Quito, Gonzalo Pizarro marched into the eastern jungles in search of the Golden Man, El Dorado. He descended the Napo river; and some of his men sailed on downstream under

Francisco Orellana to make the first astounding descent of the Amazon.*

At this time news reached Lima of a revolt in the Callejón de Huaylas where the Indians had killed two Spanish encomenderos. It looked like the start or renewal of another native rebellion. Francisco de Chaves was sent by the town council of Lima in July 1539 to suppress the revolt and punish its leaders.* Chaves rode out with savage fury on his mission of reprisal. For three months he swept through the valleys of Huaura and Huaylas and on to the Conchucos around Huánuco. The expedition was a bloodbath. Chaves sacked houses, destroyed fields, hanged men, women and children indiscriminately. 'The war was so cruel that the Indians feared they would all be killed, and prayed for peace.' Chaves was said to have slaughtered six hundred children under three years of age, and burned and impaled many adults. This was admitted in a royal charter of 1551, in which King Charles decreed that schools be founded and a hundred children be supported out of the estate of Chaves, who had himself been killed a decade earlier.*

Illa Tupac himself survived. Cristóbal Vaca de Castro told the Emperor in November 1542 that he was sending 'Captain Pedro de Puelles to the province of Huánuco, which was evacuated [by Spaniards in 1541] and is not yet at peace. I sent him so that he could repopulate and pacify it, and could conquer Illa Tupac, who is another Indian who is at large in revolt like the Inca, and is his relative.' Cieza de León wrote that Illa Tupac caused 'much mischief' but was finally captured 'after much difficulty' by Juan de Vargas in 1542. But Cieza may have been wrong, for Zárate reported that Illa Tupac was still at war in 1544.* Illa Tupac is one of the unsung heroes of the Inca resistance. He opposed the Spanish reconquest of the sierra in 1536 and then survived for at least eight years, maintaining Inca rule in the strategically important province of Huánuco.* There were more reprisals further south. The citizens of the new town Huamanga marched out and 'inflicted severe punishment on some villages that were in rebellion, killing and burning no small number of Indians'.

Manco himself had prudently retreated across the Apurímac when he heard that strong forces under Francisco Pizarro and Illán Suárez de Carvajal were hunting him. The Spaniards were convinced that he was on his last legs. Pizarro wrote to the King in February 1539 that he had retired 'with few men, totally disorganised and almost finished' but admitted in the same despatch that Manco was again 'sending messengers to the entire country so that it would revolt once more'. Bishop Valverde wrote: 'The Inca who is in rebellion is nearing the end, for he leads very few men. The Indian natives are so weary of war that they will not follow him, but remain in their villages.' The bishop hoped that Manco could be enticed out of his retreat, and recommended that he be treated generously if he came. Pizarro was less gentle: 'When summer comes he will not be able to defend himself against me, for all the land is now against him. I will have him in my hands, dead or a prisoner.' Both Valverde and Pizarro admitted that the spirit of rebellion was being kept alive only by the two courageous leaders, Manco Inca and Villac Umu. 'If they were captured the entire country would immediately serve as it should . . . But as long as there is a native leader [at large] the Indians' evil fancies can always take flight.'

Having abandoned the idea of finding sanctuary in Chachapoyas, Manco decided to build a new refuge in the Vitcos area, but one less vulnerable to Spanish cavalry than Vitcos had been. He chose to develop a settlement called Vilcabamba as his new capital. His son wrote: 'He returned to the town of Vitcos and from there went to Vilcabamba where he remained for some days resting. He built houses and palaces to make it his principal residence, for it has a warm climate.' The new capital was lower than Vitcos, deeper into the Amazonian forest. Manco probably chose it because he had fled there when Orgóñez penetrated to Vitcos.

Vilcabamba is a name full of mystery. It is the lost city of the Incas, the legendary last capital of Manco Inca and his successors. For over two hundred years, ever since the renewal of interest in pre-Columbian archaeology, there has been speculation about the location of Vilcabamba. Successive generations have accepted different sites as the most probable location. To understand the search for Vilcabamba it is necessary to know the history of its occupation during the sixteenth century. The fascinating story of that search will come later.*

Manco was granted little respite to build Vilcabamba at the beginning of 1539. The three Pizarro brothers were all active. Hernando left Cuzco for Spain on 3 April in order to justify his execution of Almagro to King Charles. He took some treasure as a sweetener, and also an anonymous *Relación* of the siege of Cuzco, which went so far in praising him that he often appeared to have resisted Manco single-handed.* The Governor

Francisco Pizarro left Cuzco to visit the Collao, which had just been pacified by his brothers. He had been created a marquis by King Charles but had not yet decided which territorial title to take for his marquisate – he was waiting to choose somewhere lucrative, preferably containing silver mines, but he never reached a decision.* With the award of the marquisate went a new coat of arms. It showed seven native chiefs chained with golden chains and forming a circle. In the middle was Atahualpa with a metal collar on his neck and his hands reaching down into two treasure chests – a heraldic celebration of the Conquest's subjection of the natives and seizure of their treasure.*

The youngest Pizarro brother, Gonzalo, set off in April to deliver the *coup de grâce* to Manco in Vilcabamba. The Spaniards had high hopes for Gonzalo's expedition. 'It is believed that, since [the Inca] is hemmed in, he cannot fail to be killed or taken prisoner. When this is done, order will be restored in the country; but until then everything is in a state of suspense.' We have eyewitness accounts of the campaign from both sides, Spanish and native. The two versions were written at the same time, thirty years later, by Gonzalo's cousin Pedro Pizarro, and by Manco's son Titu Cusi, who in 1539 had just been spirited from Oñate's house in Cuzco to join his father in Vilcabamba.

Gonzalo Pizarro was leading a massive expedition. 'Three hundred of the most distinguished captains and fighting men' according to Pedro Pizarro, while to the young Titu Cusi they appeared to be 'innumerable Spaniards – how many I cannot say, for the jungles were dense and it was impossible to count them'. He also had a native contingent led by his friend Paullu Inca.*

The Spaniards advanced into the Inca's territory as far as possible on their horses. They then had to leave the animals with a guard and proceeded on foot to the place where the Inca was said to be fortified. They had presumably crossed the Urubamba by the bridge of Chuquichaca and ridden up the ravine of the Vilcabamba river to the pass beyond Vitcos. From here onwards the road to Manco's new refuge of Vilcabamba must have run through jungle too dense for the horses: no other type of country would have forced the Spaniards to abandon their beloved mounts.

The invaders had many skirmishes with the natives as they advanced into the valley containing Vilcabamba. 'One morning they were passing in single file to cross a rocky hillside called Chuquillusca, which is very steep and dangerous [and covered in] jungle and undergrowth.' Gonzalo Pizarro was at the head of the column, with his cousin Pedro Pizarro and Pedro del Barco marching behind. Gonzalo paused to remove a stone from his boot and instructed Barco to lead on. The path at this point crossed two new bridges over small streams, and the jungle suddenly opened into a mountainside clearing a hundred yards wide. Pedro del Barco led the column across the open space and into the jungle beyond. Pedro Pizarro felt, later, that the new bridges should have made them suspicious, for this proved to be the setting for a native ambush. Manco's men suddenly rolled 'a great quantity of huge boulders' down the open hillside on to the column. Three Spaniards were swept to their deaths. Archers hidden among the trees opened fire simultaneously and killed two others. The front of the column ran forwards but found the path blocked by a rocky outcrop that the natives had reinforced with a breastwork. The rear of the column, which contained Gonzalo Pizarro and Paullu Inca, fled backwards. According to the quipocamayos, the native recorders of oral history who were interrogated three years later, Gonzalo and the other leaders wanted to retreat, thinking that the barrage had destroyed the front of the column. Paullu insisted that it would be a mistake to abandon the vanguard, and his advice was followed, despite a suspicion that he might be luring the Spaniards into a further ambush. Pedro Pizarro felt that the natives botched what could have been a perfect ambush. They should have held their fire until the head of the column was attacking the breastwork, and released the boulders only when the rest of the invaders were crowded on the clearing. As it was, the Spaniards rejoined one another and retreated that night to the place where they had left their horses. They had 'many wounded and many who had grown cowardly', and they sent back to Cuzco for reinforcements. According to the quipocamayos, thirty-six Spaniards were killed at Chuquillusca.*

While waiting for the reinforcements, Gonzalo Pizarro attempted to parley with the Inca. Two full brothers of Manco's sister-queen Cura Ocllo were sent on this embassy. Manco was furious at the treachery of these men, who were called Huaspar and Inquill. He overruled Cura Ocllo's pleas and ordered their execution, declaring, 'It is far better for me to cut off their heads than for them to depart with mine.' Cura Ocllo 'was so distressed by the death of her brothers that she refused ever to move from the place where they had been executed'.

Ten days after the first battle, Gonzalo Pizarro returned with more men to attack the fortification of Chuquillusca, which was some fourteen miles from Vilcabamba itself.* The natives had provided gun embrasures in their new wall, and attempted to fire from these with four or five arquebuses captured from Villadiego. They had not mastered the complexities of fire drill. 'Since they did not know how to ram the arquebuses they could do no damage. They left the ball close to the mouth of the arquebus, and it therefore fell to the ground when it emerged.' A hundred of the best Spanish soldiers – Pedro Pizarro among them – climbed on foot through the thick undergrowth to the top of the mountain. While their compatriots continued a frontal attack on the fort, the upper contingent descended to a 'pampa that lay on the far side of the mountain, where Manco Inca had his residence'.

When the natives saw that this force was outflanking them, runners went to warn Manco, who was still in the fortification. 'When he heard this, three Indians took him in their arms and carried him "á vuelapié", half-running, half-flying, to the river which flows beside this fort.' The Inca dived in and swam across. With the Inca gone, the defenders of the fort vanished into the jungles. The Spanish attackers went hurrying off up the mountainside, thinking that Manco had been caught there. 'Had we realised that he was in the fort he would never have escaped. We Spaniards and our auxiliaries would have found him, had we not all gone climbing up thinking he was there.' Titu Cusi wrote that his father shouted back from the far side of the river: 'I am Manco Inca! I am Manco Inca!' Mancio Sierra de Leguizamo recalled that Manco shouted across 'that he and his Indians had killed two thousand Spaniards before and during the rising, and that he intended to kill them all and regain possession of the land that had belonged to his forefathers'. Gonzalo Pizarro's men 'travelled about for over two months from place to place on his trail', but Manco was hiding with some forest Indians and they never found him. Nevertheless, for all his defiance, he had had a hair's-breadth escape for the second time in two years.

Although the Spaniards failed to capture the Inca himself, they caused havoc in his new settlement, Vilcabamba. They marched back to Cuzco in July 1539 with one of his brothers, Cusi-Rimache, and with his coya

Cura Ocllo, who, in the correct Inca tradition, was also his sister. Titu Cusi described the hardships of this poor creature, who had just witnessed the execution of her two brothers and had worse in store. She and the Spaniards 'reached the town of Pampaconas, where they intended to try to rape my aunt. She would not allow it, but defended herself fiercely throughout, and finally covered her body with filthy matter so that the men who were trying to rape her would be nauseated. She defended herself like this many times during the journey until they reached [Ollantay]tambo.'

While the Inca princess was being held here, there was a report that Manco was willing to negotiate his surrender. This brought the Marquis Pizarro hurrying back in late September from Arequipa, where he was about to found what has since become the second city of Peru. There was an exchange of envoys between the two rulers. Cura Ocllo added a message to one letter to say that she was being well treated. Pizarro went so far as to send Manco an imported pony and some silk with a Negro attendant and two Christian native envoys. But Manco killed the entire party, including even the gift horse.

Pizarro was furious at the rebuff. He felt unequal to launching another punitive expedition, for his active young brother Gonzalo was by now marching towards Quito. The frustrated old Governor therefore vented his anger on the Inca's beloved wife – with whom, according to Cieza, he and his secretary Picado were rumoured to have had intercourse. Cura Ocllo was stripped, tied to a stake, beaten by Cañari Indians, and shot to death with arrows. She told her persecutors: 'Hurry up and put an end to me, so that your appetites can be fully satisfied.' Beyond that she endured her martyrdom in silence and without a movement. Most Spaniards were disgusted by this brutality: 'an act totally unworthy of a sane Christian man'. To heighten the cruelty, Cura Ocllo's body was floated down the Yucay river in a basket, so that it would be found by Manco's men. The Inca was 'grief-stricken and desperate at the death of his wife. He wept and made great mourning for her, for he loved her much.'

Not content with this savage murder, Pizarro went on to execute many native leaders who had surrendered to the Spaniards. Villac Umu and some chiefs protested about the killing of Cura Ocllo and were therefore supposed to have shown rebellious sympathies. Villac Umu had just been captured, in October, in the Condesuyo. Illán Suárez de

Carvajal wrote that the high priest was 'a person of great authority among the natives. If treated well, he should be most influential in making the Inca come out in peace more quickly.' Instead of lenient treatment, Pizarro burned Villac Umu alive. He also burned Tiso, who had been living peacefully in Cuzco for the past nine months. These executions were glossed over in most official records. But the vicar-general of Cuzco, Luis de Morales, revealed that Pizarro murdered sixteen of Manco's commanders, having taken them to Yucay on the pretext of giving them land there. Titu Cusi recorded that, in addition to Villac Umu and Tiso, Pizarro burned 'Taipi, Tanqui Huallpa, Orco Varanca, Atoc Suqui . . . Ozcoc and Curiatao, and many other former captains, so that they could not return to join my father, and so that the Spaniards would be leaving their rear secure.' This treacherous murder without trial of the Inca commanders was one of the most sordid acts of the Conquest.

Manco's second rebellion thus came to an end, apart from Illa Tupac's stand at Huánuco. It was the last effort on a national scale to dislodge the European invaders. Although the Spaniards had little difficulty in snuffing out the isolated rebellions, the natives had shown great courage and resilience in making the attempt.* No one could say that the Inca empire went under without a struggle. And the natives had the consolation that, contrary to all predictions, their Inca was still at large. The jungle-covered mountains of Vilcabamba had saved Manco and earned themselves an unsavoury reputation. Hernando Pizarro told the King that the Spaniards who returned from Vilcabamba were appalled by the difficulties and lack of food. 'They write that great resources are needed to undertake a penetration of that land. It can be done only with very heavy expenditure, and since those who went could not afford this their enthusiasm came to an end.' Manco Inca had survived.

13

PAULLU AND MANCO

WHILE Manco was struggling to liberate Peru and survive in Vilcabamba, his half-brother Paullu was enjoying the lordship of Cuzco. The Spaniards were frankly baffled by Paullu's enthusiastic collaboration. Bishop Valverde told the King: 'At the present time we have great need of a son of Huayna-Capac called Paullu, through whom we rule the native Indians who are at peace and well disposed towards us.' They regarded him as 'a good Indian, sensible and well disciplined', and praised him as 'a good friend and servant of His Majesty'. They were impressed that 'the Indian natives think highly of him, and also because he is courageous in battle' and 'knows much about warfare'. He gained a massive reputation as the leading authority on native affairs. Alonso de Toro said in 1540 that he had 'seen Paullu advise the Governor,

Manco Inca seated in state on his usno throne-dais

Lieutenant and other officers on many occasions', and felt that he gave good advice.

Some Spaniards thought Paullu was too good to be true. Bishop Valverde wrote to the King that Paullu's days must be numbered: 'we believe, and are in fact certain, that the rebellious Inca will either win over Paullu or will kill him because of his friendship towards us'. Other Spaniards questioned Paullu's loyalty. On Gonzalo Pizarro's expedition to Vilcabamba, Captain Villegas voiced his suspicions 'because we do not know what agreements or plots he has made with his brother Manco Inca, to whom he owes stronger loyalties than to us'. And when the Marquis Pizarro was trying to negotiate with Manco from the Yucay valley, many Spaniards suspected Paullu of treacherous dealings with his brother.

Such suspicions flattered Paullu. If he was active at all during this period, he was more probably trying to sabotage the negotiations that threatened his possession of the Inca title. He assumed an air of wronged innocence and argued, doubtless correctly, that he had never indulged in any treachery to the Spaniards. His friends stood by him. As Martín de Salas said: 'I have heard many people say that Paullu was deceitful – but I have also found that no treachery has ever been discovered against him.'

It now remained for Paullu to reap the rewards of devoted service to what proved to be the winning side. He adopted the accepted course for conveying his merits to the King. A series of twenty questions was drawn up, each of them describing outstanding services to the Spanish cause. From 6 to 12 April 1540 twenty friendly witnesses were asked to testify to the accuracy of Paullu's claims. All were heroes of the past years of fighting, and almost all were in their middle or late twenties as was Paullu himself. Paullu's record of service was outstanding, and the witnesses did him full justice. The document that was sent to the King presented a glowing picture of Paullu's collaboration and ended with a plea for royal favours.* This Probanza was heard before the Licenciate Antonio de la Gama. To reward this official for his help, Paullu spent part of 1540 on a treasure-hunting expedition with La Gama to extract gold and silver from the natives of the Collao. The pair were eventually recalled by the jealous council of Cuzco.*

The Probanza made a great impression in Spain, and Paullu soon became a rich man in post-Conquest society. Almagro had given him the Colcampata palace in Cuzco. This was Huascar's old palace, at the upper edge of the city, half-way up the steep hillside leading to Sacsahuaman. From here the Spaniards' Inca ruled over an increasingly hispanicised native aristocracy. His palace looked out across the pink-tiled roofs of the conquistadores' houses that were rapidly occupying the gutted Inca city.

In addition to this town house, Francisco Pizarro gave Paullu a rich repartimiento at the beginning of 1539 to bribe him to join Gonzalo's Vilcabamba campaign. A repartimiento was nominally granted to enable an encomendero to instruct the natives in the Christian faith. Paullu was still an unbaptised heathen. No matter. Pizarro justified the grant to Paullu because 'you are now going in pursuit of Manco Inca Yupanqui, instigator of the war'. The repartimiento was called Hatun Cana. It included the towns of Pichihua and Yauri in the Canas country at the headwaters of the Apurímac, and these contained 922 tribute-paying Indians and a further 4,391 persons who had been resettled in its towns. The repartimiento also embraced some villages around Muyna at the far end of the valley of Cuzco, some settlements taken from Mancio Sierra de Leguizamo in the Condesuyo, and the districts of Asihuana and Surita in the eastern Antisuyo. It brought Paullu an annual rent of 12,000 pesos.* Paullu claimed some Indians at Alca near Arequipa as being his personal mitayos. Paullu also had land on the Copacabana peninsula jutting into Lake Titicaca, and at Episcara in the Jaquijahuana valley just outside Cuzco.*

All this wealth made Paullu a potential prey for Spanish ruffians. It was difficult to make Spaniards respect even this leader of the Indian community. One Spaniard struck Paullu in the street, pulled his hair and insulted him; but he was never punished. Others robbed Paullu's house. A royal decree of October 1541 gave him a Spanish tutor to look after his interests and prevent him from signing dubious contracts. A further decree put a stop to the practice of placing Spanish guards in Paullu's house, for fear that the guards might abuse his wives or belongings. Various property suits were settled in Paullu's favour, and he was given his friend Juan de León as tutor.*

The greedy puppet Inca even begged the King 'to grant him the Indians and lands held by his brother Manco Inca at the time that he rebelled, since he has forfeited them by rebelling against your service'. In 1543 Prince Philip wrote personally to acknowledge Paullu's loyal service, 'for which I thank you greatly and charge you to continue to perform henceforth'. The following year, Paullu's very numerous illegitimate offspring were legitimised *en masse*.* But the crowning glory came in 1545, when Paullu received a coat of arms containing a black eagle rampant, sinople palms, a gold puma, two red snakes, a red imperial fringe, the inscription 'Ave Maria' and eight golden Jerusalem crosses.*

Paullu wore Spanish clothes and learned to copy his name, but he could not speak Spanish nor read and write it. His fine record of collaboration had only one blemish. Although the proud holder of an encomienda, he was still a heathen. During 1541 and 1542 he began to remedy this by receiving instruction in the articles of the Christian faith. The Spanish authorities were understandably excited at the prospect of wholesale conversion that might follow the baptism of the puppet Inca.* His instruction was partly performed by the vicar-general of Cuzco, Luis de Morales, in whose house he lived for five months. Paullu had formerly been a champion of the native religion in Cuzco, celebrating the feast of Raymi in his palace of Colcampata and harbouring the sacred stone idol that had once stood on the holy hill of Huanacauri.* Now, with the zeal of a convert he betrayed to the Christians some mummified remains of 'his father Huayna-Capac and of other uncles and cousins of his to be buried, despite the lamentations of his mother and relatives'. Paullu was finally baptised in 1543 by the Friar Juan Pérez de Arriscado, a Prefect of the Order of St John, and Vicar of the church of Cuzco. He was given the Christian name Cristóbal, after his sponsor Cristóbal Vaca de Castro. Years later, the ex-Governor Vaca de Castro left 600 pesos in his will to 'the sons and heirs of Paullu Inca, who was baptised during my time [of office] and called himself by my name'. Paullu's wife Mama Tocto Ussica was also baptised as Doña Catalina, and following the Inca's lead a rash of native royalty performed the ceremony of conversion. Paullu's mother Añas Collque became Doña Juana, and his sister became Beatriz Huayllas Ñusta. Among the males who adopted sonorous hispanicised titles were 'Don García Cayo Topa, Don Felipe Cari Topa, Don Juan Pascac, Don Juan Sona and many others'. The conversion of Paullu Inca and his relatives was not as successful as the Spaniards had hoped. The natives now 'regarded Manco Inca as their universal lord, and not Don Cristóbal [Paullu] Inca, because he was a friend of the

Christians with whom he always consorted, and because he had been baptised'.

Many natives fled into the forests of Vilcabamba to join Manco there. The Inca returned to Vilcabamba from his retreat in the depths of the forests, and began to create a tranquil native state. His followers built towns and terraces, the Inca religion was revived, and an attempt was made to reproduce in miniature the well-ordered state that had existed throughout Peru only a decade earlier. But the period before the advent of the nightmarish strangers must have seemed infinitely remote.

Although there was now no hope of raising a third rebellion of the exhausted native populations of Spanish Peru, Manco's men continued to harass the Spaniards whenever possible. An expedition under Pedro de Valdivia had left Cuzco in January 1540 to resume the conquest of Chile that had first been attempted by Almagro five years earlier. Valdivia marched farther south than Almagro had done. He learned at Santiago in 1541 that Manco Inca had warned the Chileans of the Spaniards' approach and told them to hide their gold, food and clothing so that the Spaniards would turn back. 'So faithfully did they carry out these instructions that they even destroyed their own clothing, suffering the greatest privations.' A large native uprising ensued in this southern province of the Inca empire. Ten thousand Indians attacked the Spanish settlement of Santiago and burned it completely, killing a number of Spaniards and horses in the process. Valdivia built a strong fort and survived there with great difficulty until 1545.*

One of Manco's tormentors in 1535 had been Diego Maldonado. This man had led an expedition of citizens of Cuzco against Villac Umu at Andahuaylas during the rebellion of 1538–9. Maldonado received the large encomienda of Andahuaylas as a reward, and Manco now attacked it with special violence. The Marquis Pizarro wrote at the beginning of 1541 that Manco was 'going about causing immense damage and slaughter of Christians and natives'. He had made the roads impassable to Spaniards not marching in battle order, and was committing the appalling crime of absolving the natives from their service in the encomiendas. Pizarro determined on a fresh attempt to suppress this gadfly. 'The Inca and his men are marching close to [Huamanga] and are said to be moving in to attack it. I am also informed that he came to the repartimiento of Andahuaylas, which is possessed by Diego Maldonado, citizen of Cuzco, and attacked various Spaniards with a horde of warriors, fought and killed some of them . . . and stole property being transported from Lima to Cuzco. If some solution is not found, his depredations will continue to increase, and the Inca and his followers will be encouraged to commit them. I therefore intend to make war on the Inca . . . and feel that it should be done this summer.' Pizarro intended to raise and equip a hundred men from the chief cities of Peru for this expedition, and the Council of Huamanga ordered all citizens to prepare for the defence of the city; each man was to 'have his cross-bow ready with a dozen bolts'.

While Pizarro was planning a final attack on his adversary, Manco was apparently contemplating yet another exodus from Vilcabamba, a place that had been invaded so easily by Gonzalo Pizarro. He now planned to go to Quito, 'which was fertile country, and where the Spaniards would have less opportunity to harm them, and where they could fortify themselves better'. The Inca left Vilcabamba with all his army and possessions and descended on the area east of Huamanga. But he came to the conclusion that Peru was too full of Spaniards for him to risk this migration, and therefore returned to the dubious security of Vitcos and Vilcabamba.

The proposed annihilation of Manco Inca was soon forgotten. Peru was about to be shaken by events that left Manco a forgotten exile in his mountainous backwater. The followers of Diego de Almagro had never forgiven the execution of their beloved leader by Hernando Pizarro in July 1538. Hernando left for Spain nine months later taking a cargo of treasure for King Charles. But Almagrist sympathisers headed by Diego de Alvarado and Alonso Enríquez de Guzman were determined that he must be punished. Their shrill accusations finally led the Emperor to imprison Hernando indefinitely in the castle La Mota at Medina del Campo.

The Almagrists who remained in Peru were increasingly frustrated. Many had just failed to participate in the ransom of Atahualpa; their subsequent ventures in Chile and various jungle expeditions had been sorry failures; and they had been defeated in the attempt to hold Cuzco. They had no estates and were treated with contempt by Pizarro and his arrogant secretary Antonio Picado. This group of impoverished and bitter men rallied around Almagro's son Diego de Almagro, whom he had had by a native woman of Panama and who was now growing to manhood.

Lima was full of rumours of an impending attempt to assassinate the Governor, but Pizarro took no special precautions. A group of twenty armoured Almagrists headed by Juan de Herrada forced their way into his undefended palace on the morning of Sunday, 26 June 1541. Most of Pizarro's companions fled, but the sixty-three-year-old Marquis strapped on a breastplate and killed one of the attackers before he himself died from a mass of wounds. His half-brother Francisco Martín de Alcántara fell at his side, and the violent Francisco de Chaves was killed by his Almagrist friends when he tried to reason with them. The mangled bodies of Pizarro and his few followers were secretly buried in hurried graves, while the young Diego de Almagro was paraded through the streets and proclaimed as Governor and Captain-General of Peru.*

Bishop Vicente de Valverde came down from Cuzco when he heard of the murder of his friend Pizarro. The bishop escaped on a ship for Panama, but was killed and eaten by cannibalistic natives on the island of Puná – the base from which he and Pizarro had first landed in Peru a decade earlier.*

The deaths of these two leaders, Pizarro and Valverde, brought the first phase of the Conquest to an abrupt end. Francisco Pizarro had been the undisputed leader of the enterprise from the beginning. He was one of those baffling men driven on by ambition, but with no clear purpose or objectives. He was not a religious zealot like Cortès, and was indifferent to the conversion or welfare of the natives. His ambition drove him upwards from obscure beginnings as the illegitimate son of an undistinguished officer, until he amassed a huge fortune and great power. López de Gómara said that Pizarro 'found and acquired more gold and silver than any other of the many Spaniards who crossed to the Indies, more even than any commanders there have been in the history of the world'. Yet Pizarro made no great use of his wealth. He did not dress or live particularly lavishly; and his success came too late in life for him to acquire extravagant interests such as building, women, horses and hunting, the arts or social success. He never married, although he was evidently fond of his Inca princesses and of the children they bore him. He was a simple soldier, illiterate and a poor horseman, but deeply respected and obeyed by the young men he led. Their respect was well placed, for Pizarro was a magnificent leader, infinitely tenacious and determined during the wretched years of exploration before the discovery of Peru, and decisive and coolly courageous during the Conquest. He was calm and diplomatic: he won the backing of the Spanish Crown for his dubious undertaking, established a reasonable relationship with the Incas Atahualpa and Manco, and often cooled the tempers or restrained the excesses of his army of adventurers. But Pizarro could be cruel and callous, and his total lack of administrative experience made him an uncertain Governor.

Far less is known of Pizarro's colleague Valverde, who had little chance to prove himself as ecclesiastical leader of the young colony. It is easy to brand Valverde as a hypocrite because of his meeting with Atahualpa when the Spanish cavalry was poised to deliver its murderous charge, or because of his want of compassion in the deathbed conversion of the Inca. He emerges in better light in the long and articulate letter he wrote to the Emperor in 1539: this was the first humanitarian protest to come from Peru. Had he lived, Valverde might have done good in his capacity as official protector of the Indians.

The young rebel Diego de Almagro and his followers occupied Lima and went on to control Peru for almost a year. At the time of Pizarro's assassination King Charles had already sent an emissary called Cristóbal Vaca de Castro with full powers to settle the affairs of Peru. Vaca de Castro was on the coast of modern Colombia when Pizarro was assassinated. He soon landed in northern Peru and was joined by an army under Alonso de Alvarado, Pizarro's lieutenant-governor of the northeastern settlement of Chachapoyas. Together they advanced into Peru, and their army was swollen by Pizarrists and others whose instinct was to rally to the royal standard. The Almagrist cause was finally crushed for a second time at the battle of Chupas, just outside the young city of Huamanga, on 16 September 1542. A group of Manco's warriors looked on from surrounding hills while the two armies of Spaniards battered each other. They saw them use all the weapons of sixteenth-century war: crossbows, arquebuses, artillery and the classic medieval impact of fully-armoured horsemen charging one another with lowered lances.

With the assassination of Pizarro and the crushing of the native rebellions, the conquistadores were in full control of Peru. The Inca treasures had long since been melted down and distributed. Spanish settlers therefore had to rely on the native population to provide the wealth for which they had come to the Americas. With no authorities imposing restraint, many Spaniards who had been lucky enough to receive encomiendas abused the native inhabitants with exorbitant demands. Luis de Morales, Vicar-General of Cuzco, reported that some Spaniards were branding natives on the face and using them as slaves, in flagrant violation of royal ordinances. Other Indians were forced to provide unpaid service in the houses and fields of Spanish masters. Under the terms of an encomienda grant, the encomendero was forbidden to live among the natives of his repartimiento. The Indians had to carry his tribute to the Spaniard's town house, and the encomendero developed no understanding or sympathy for his charges. Diego Maldonado, holder of the rich encomienda of Andahuaylas, conceived an idea of leading an expedition to explore the South Pacific. The money for his dream had to come from his natives, and he was merciless in extracting tribute from them. He ordered his majordomo to tolerate no delays, made his Indians double their output, and erected storehouses on their land to speed the collection of produce, to such an extent that the curaca of one of Maldonado's holdings, Don Pedro Atahualpa of Urco-Urco and Chuquimatero, made a formal complaint to the Corregidor of Cuzco.* Throughout the 1540s legal documents referred to legacies, transfers and outright sales of quantities of natives who had been entrusted to the care of encomenderos. One encomendero, Melchor Verdugo, described his comfortable life to his mother: 'I am living in a town called Trujillo in which I have my house and a very good repartimiento of Indians that must contain some eight or ten thousand vassals. I believe that there has not been a single year in which they have not given me five or six thousand castellanos income. I am writing all this to you to give you pleasure.'

All Peru came to revolve around the 480 encomenderos. Each of them established a large house, preferably of stone, and filled it with relatives, guests and spongers. The encomenderos kept plenty of horses, Negro slaves and Indian women as household servants, and they generally acquired a Spanish wife who raised her children in the traditions of the mother country. Encomenderos were expected to provide a missionary priest to convert the Indians of the encomienda. But willing priests were hard to find. Some encomenderos kept priests as private chaplains; others appointed their majordomos or stewards as 'lay missionaries'.

The worst offenders were often adventurers newly arrived from

Spain. Violent, greedy, excessive gamblers, but frustrated by having no encomienda grants and no easy conquests, these newcomers formed a white rabble that was most arrogant and vicious towards the natives. Vaca de Castro tried to revive the Incas' fine system of tambo post-houses, chasqui couriers and official storehouses. In the preamble to his Tambo Ordinances of 1543, he admitted that 'I have seen with my own eyes that most of the towns, tambos and settlements of the Indians are deserted and burned, and that the Christian Spanish citizens load the Indians in great quantities with excessive loads and long marches.' Morales complained that Spaniards surrounded themselves with native women 'in the manner of the law of Mahomet' and Cieza de León admitted that 'they seized the natives' wives and daughters for their own uses and committed other atrocities'.

The abuse of Indians was undoubtedly patchy, with wide areas of Peru still untroubled by undue oppression. Many encomenderos probably demanded less tribute than had been provided to the Inca state and religion. An immense volume of official reports and legal documents survives from this period, but there are no diaries or impartial travellers' accounts, no Indian records and almost no local administrative sources. As Lewis Hanke observed, under such disadvantages historians draw conclusions at their peril.* To their great credit, many Spaniards had a strong moral conscience about their new conquests. The King was receptive to any reports from whatever source. As long ago as 1521 the Crown decreed that no one should prevent any useful communication reaching the King. The result was a flood of correspondence, much of which was passionately pro-Indian. Cieza de León wrote that 'His Majesty was informed by many people from many sources of the great oppression the Indians suffered from the Spaniards.'

The written reports were reinforced and dramatised by a group of reformers headed by Bartolomé de las Casas, who returned to Spain from Guatemala in 1539. Las Casas and his followers condemned the institution of the encomienda itself. Miguel de Salamanca deplored a system under which 'all the profit derived from the natives' work goes to those who hold them in encomienda. This is contrary to the well-being of the Indian republic, against all reason and human prudence, against the welfare and service of our Lord and King, contrary to all civil and canon law, against all rules of moral philosophy and theology, and against God's will and his Church.' Shortly after his return, Las

Casas added two further manifestos to his growing output on the subject of native welfare. In one essay he argued that the papal donation had authorised the King of Spain to conquer the Americas to improve the lot of its inhabitants – not to enslave them in encomiendas. The other work, 'Very Brief Account of the Destruction of the Indies', reinforced the reforming arguments with lurid details of Spanish atrocities and exaggerated statistics of their depredations. Rumours of these scandals spread through Europe as quickly as the first reports of the new conquests. Other nations were jealous of Spain's brilliant empire. Atrocities make good reading at any time, and other Europeans eagerly accepted Las Casas's powerful hyperbole, particularly when it was dramatised by the fine but fanciful illustrations of Theodore de Bry. The result was the 'leyenda negra', the legend of excessive Spanish cruelty, that has caused and still causes passionate condemnation or defence of the Spanish performance. Both sides have been able to support their arguments with plenty of contemporary material, partly because it was common in the sixteenth century for political theorists to argue hotly. Hanke quoted the English historian Pelham Box for a more reasoned view of Las Casas's contribution: 'If he [La Casas] exaggerated on details he was right in fundamentals and his truth is not affected by the use hypocritical foreigners made of his works... It is not the least of Spain's glories that she produced Bartolomé de las Casas and actually listened to him, however ineffectively.'

The self-questioning set in motion by the small group of reformers bore fruit remarkably quickly. In 1536 King Charles, pleased at the success of the conquests, had issued a law that confirmed encomienda holdings for two generations. The settlers hoped that this might lead to grants in perpetuity.* But the King then had a complete change of heart. He ordered a council of statesmen, jurists and ecclesiastics to formulate laws for the good government of the new realm and particularly for the protection of its native population. The fruit of their deliberations was the famous New Laws, issued at Barcelona on 20 November 1542.*

The New Laws started with regulations for the establishment of the Council of the Indies. They continued with ordinances concerning the Indians that were so sweeping and so strongly in favour of the Indians that Las Casas himself might well have drafted them.* Spanish authorities were ordered to keep continuous watch for the welfare and protection of the natives. Slavery of Indians was to cease immediately and not be resumed on any pretext whatsoever – not even 'under title of rebellion'. The encomienda system remained, but the most flagrant abuses were curtailed. Titles to encomiendas were to be scrutinised, and no one was to hold excessive numbers of tributary natives. Indians were to be removed from any who abused them, especially from 'those principal persons' involved in the disturbances in Peru. Royal officials and ecclesiastics were to lose encomienda holdings and receive no future awards of Indians. The four Audiencias that controlled the Indies were to assess the exact amount of service and tribute to be paid, and this was to be written in a book in each encomienda. These assessments were to be 'less than [the natives] used to pay in the time of the caciques . . . before they came under our rule'. Finally, no grants of encomienda would in future be made to anyone, and existing encomiendas would revert to the Crown on the death of their present holders. The conduct of explorations was also strictly regulated. The natives, in short, were to enjoy rights almost equal to those of Spaniards.

The New Laws curtailed the existing holdings of all American settlers and deprived them of future hopes and prospects. 'These things were heard by the people over here with great indignation . . . Finally there was a wild tumult with the news flying from one part to another.' A flood of furious protests poured into Spain from Mexico and particularly from Peru, where a civil war was already brewing. King Charles sent a Viceroy, Blasco Núñez Vela, to succeed Pizarro and Vaca de Castro as administrator of Peru. As this royal official approached from the north, a new constellation advanced into Peru from his estates in the Charcas. This was the dashing Gonzalo Pizarro, the last of the four magnificent brothers to remain in Peru, a reckless, vain young man, but the hero of countless native wars and Amazonian exploration. In August 1544 he wrote on behalf of the cities and encomenderos of Peru to protest in detail the clauses of the New Laws.* He pointed out in particular that the ordinance forbidding royal officials to hold encomiendas was unworkable, since many of the original conquistadores and encomenderos also held official posts. The letter ended with a bitter attack on the new Viceroy who had reached Peru a few months earlier. Blasco Núñez Vela was behaving with extreme obtuseness, provoking the Spaniards in Peru in every possible way. He applied the New Laws with unnecessary vigour, imprisoned the ex-Governor Vaca de Castro and killed the Royal Factor Illán Suárez de Carvajal.* King Charles had also appointed four oidores or judges for the Audiencia of Lima, and these were driven in September 1544 to depose the Viceroy and set him on a ship for home.

Gonzalo Pizarro entered Lima the following month, a figure of legendary panache, rich, handsome, and arrayed in a sumptuous outfit of black velvet, gold, plumes and jewels. Settlers from all over Peru flocked to his support. It was a situation familiar to the twentieth century but new to the sixteenth. A home government had legislated too liberally on behalf of colonial natives. The European settlers rebelled, loudly protesting their loyalty to Crown and mother country, but demanding a free hand in the exploitation of the territory they had won and settled. To them the chief inducement to remain in the colonies was the luxury provided by native labour. While an army of settlers rallied to Gonzalo Pizarro in Lima, the deposed Viceroy had landed in northern Peru and made his way to Quito. Gonzalo marched north against him in mid-1545, and royalists and settlers clashed at the battle of Añaquito, on the equator just north of Quito, on 18 January 1546. The Viceroy Blasco Núñez Vela was killed in the battle, and the rebel Gonzalo Pizarro was undisputed master of all Peru.

The details of these turbulent years are of little direct concern in the history of relations between Spaniards and natives. Manco Inca ruled at Vilcabamba and kept himself informed of the progress of the Spaniards' civil war. Paullu continued as puppet Inca in Cuzco, now a Christian, a rich encomendero and recipient of royal favours and colonists' respect. It was the ordinary natives who suffered the brunt of the disturbances, pressed into the service of Christian armies and subjected to colonists whose behaviour was under no governmental or legal control.

When Gonzalo was at the height of his power, many settlers wanted to make an unilateral declaration of independence, with Gonzalo Pizarro as King of Peru. It was even proposed to marry him to an Inca princess, to unite the royal family of Peru with the first family of conquistadores. Gonzalo Pizarro, who had had affairs in many Peruvian cities and who had just had a daughter by one María de Ulloa in Quito, wrote: 'I cannot think of marriage at present: I am wedded to my lances and horses.'

With Gonzalo unmarried, the illegitimate children of the various Pizarro brothers received much attention. The Marquis Francisco

Pizarro had had four children: Francisca and Gonzalo by the princess Inés Huaylas, and Francisco and Juan by the princess Añas who was baptised Angelina Añas Yupanqui.

Both mothers subsequently married prominent Spaniards. Pizarro married Inés to one of his servants, Francisco de Ampuero, and gave them an encomienda in Lima. Ampuero was ambitious and intelligent. He soon became a regidor of Lima, where the couple lived for many years.* Vaca de Castro gave Inés a pension of 6,000 pesos a year, and in 1575 the couple were still 'married and much honoured'. Angelina married Juan de Betanzos in 1542, soon after the death of Pizarro, and her husband learned Quechua and became an expert on native customs. He wrote a chronicle, full of material gleaned from his wife and her royal relatives, and the couple lived in the Carmenca district of Cuzco.

When the Almagrists assassinated Francisco Pizarro, his four young orphans were also in danger. Their fiery aunt Inés Muñoz, widow of Pizarro's half-brother Francisco Martín de Alcántara, was put to sea with the young Francisca and Gonzalo, who had been legitimised by royal decree at their father's request.* The Almagrists secretly told the pilot to abandon them on a desert island, but he landed them instead in northern Peru. Inés Muñoz greeted Vaca de Castro as he marched south, and he installed her and the children in Trujillo during the war of Chupas.* They returned to Lima after Vaca de Castro's victory. Inés Muñoz appointed Antonio de Ribera as their guardian and married him herself. The infant Juan died in 1543, but his brothers Gonzalo and Francisco were placed with an old tutor called Cano. Their uncle Hernando wrote to their uncle Gonzalo: 'My pity for these children of the late Marquis is so great that I can hardly bear to speak of it: the one who died seems the best off.'

When Gonzalo Pizarro was advancing on Lima, the Viceroy wanted to send the orphans to Spain. But Licenciate Agustín de Zárate intervened to stop this, because the nine-year-old Francisca 'was a grown maiden, rich and beautiful, and it would not be decent to send her among sailors and soldiers'. At the height of Gonzalo Pizarro's rebellion his lieutenant Francisco de Carbajal wrote to him that his niece was 'a big girl and beautiful, and I think it is time you arranged a marriage for her'. This sounded interesting, and it occurred to the dashing Gonzalo – a bachelor in his early thirties – to marry his eleven-year-old niece himself. Such a marriage would unite him with the descendant of the great Marquis and of the royal house of the Incas – a valuable match if he ever decided to crown himself king of an independent Peru.*

A second of the Marquis's mestizo children, the nine-year-old Gonzalo, died in 1546.* But his son Francisco was growing up happily in Cuzco. He was at school with another mestizo, the future historian Garcilaso de la Vega, who recalled that Francisco 'was a great friend and rival of mine, for when we were eight or nine years old his uncle Gonzalo Pizarro used to make us run and jump together'.

The new ruler of Peru, Gonzalo Pizarro, also had young children: a son and potential heir-apparent known as Francisquito, and a daughter called Inés. This Inés and her cousin Francisca went up to Cuzco at the beginning of 1547 and received confirmation together. Tomás Vásquez described them as 'charming little ladies' and Juan de Frías wrote that his wife 'loves them so much that she would like to have them with her to bring up. There is someone here who could teach them to read and play the virginals.' Young Francisquito, son of Gonzalo Pizarro and 'a Christian Indian woman' was legitimised by royal decree in August 1544.* He fell into the hands of the Viceroy, but was recovered by his father at the battle of Añaquito, and remained with him. Juan Pizarro had also left a daughter, Isabel, whom he disowned in his will of 1536, but who was brought up with the respect due to the descendant of a Pizarro.

The rebellion of Gonzalo Pizarro was opposed in circumstances very similar to those that overwhelmed Diego de Almagro the Younger. Once again an unassuming royal emissary, the Licenciate Pedro de la Gasca, appeared in the New World and gradually accumulated a royalist army to oppose the rebels. Gasca began by winning over the commander of the rebel fleet, which was patrolling the Pacific off Panama. Gasca brought with him very wide powers from the King. But his most potent weapons were royal decrees completely reversing the most liberal clauses of the New Laws. In June 1545 the Duke of Alva advised King Charles to suspend the New Laws in view of the civil wars, and to content the Spaniards 'wholly by promising perpetuity' of tenure of encomiendas – anything rather than risk a stoppage in the flow of treasure from the Indies.* King Charles forgot his good intentions in the face of such a danger. In October 1545 he formally revoked the clause prohibiting grants of future encomiendas, and in February 1546 he revoked the law that removed Indians from those who had mistreated them or who had been involved in civil strife.* Pedro de la Gasca wrote letters to every potential loyalist in Peru, reminding them of their duty to the King and explaining the revocation of some of the laws. He sailed down the Pacific coast in mid-1547 with a force of loyal Spaniards from the Isthmus and Mexico.

Gonzalo Pizarro was by now enjoying the taste of power. He wrote to his lieutenants: 'Spain's desires are well understood despite her dissimulation. She wishes to enjoy what we have sweated for, and with clean hands to benefit from what we gave our blood to obtain. Now that they have revealed their intentions, I promise to show them that we are men who can defend our own.' But he wrote to the King with an air of wronged innocence: 'I have never in word or deed offended against your royal service nor neglected my duty as a sincere and loyal vassal.' And he shouted to Gasca's emissary: 'Look here, I am to be Governor because we would trust no one else, not even my brother Hernando Pizarro. I do not care a jot for my brother Hernando or my nephews and nieces . . . I must die governing! There is nothing more to be said.'

The quiet ecclesiastic Gasca advanced resolutely into Peru. Settlers flocked to the royal standards, many of them disgusted by the brutality of Gonzalo Pizarro's regime, which had executed 340 Spaniards in its brief duration. In December Gasca formally accused Gonzalo of treason. 'You did not simply intend to appeal against the Ordinances; you wished to usurp the government. . . . Despite the fact that the Viceroy temporarily suspended the execution of those Ordinances, you pursued him to the death, thus manifesting your indifference regarding the Ordinances and your anxiety to rid yourself of the Viceroy who might obstruct your path to power.' Perhaps the exercise of independent power is just as strong a motive for colonial secession in Algeria or Rhodesia today, particularly when coupled with a wish to be free of any restraint in dealing with the native population.

Gonzalo's forces won another sweeping victory over a royalist force at Huarina, on the south-eastern shores of Lake Titicaca, in October 1547. But Gasca advanced steadily southwards along the Inca highway towards Cuzco. Native auxiliaries helped him to bridge the upper Apurímac in May 1548, and warned his men of a Pizarrist ambush. The two armies met on the plain of Jaquijahuana or Sacsahuana a few miles west of Cuzco on 9 April 1548. Forty-five of Pizarro's stalwarts died fighting, but the vast majority of his army ran across the field to the royalist lines, and Gasca lost only one man in the battle. Gonzalo Pizarro was captured and executed next day. It was on this same field that the Incas had defeated the Chancas a century earlier, at the start of their expansion. And it was the fifth time in sixteen years that armies fought at the edge of Cuzco: Huascar against Chalcuchima in 1532; Francisco Pizarro against Quisquis in 1533; Manco against Hernando Pizarro in 1536; Hernando Pizarro against Almagro in 1538; and now Gasca against Gonzalo Pizarro.

Paullu Inca gave a warm greeting to President Gasca when he entered Cuzco. Paullu was as loyal as ever to the Spanish cause in general, but it had not been easy for him to survive a decade of civil wars among his European friends. Only an unerring sense of politics enabled him to change sides whenever necessary. Never for a moment during these turbulent years was Paullu out of favour with the current Spanish master of Cuzco – a fine record even for this master of political opportunism.

Paullu played little part in the short-lived rebellion of Diego de Almagro the Younger, although he may have sent a contingent of native troops to fight on the Almagrist side at Chupas.* He basked in the approval of the victorious Governor Vaca de Castro, who was particularly proud of his own role in the baptism of the Inca. The Viceroy Núñez Vela came to Peru with special instructions to favour Paullu, and Gutiérrez de Santa Clara, a chronicler who was in Cuzco at the time, said that Paullu supported the Viceroy at the start of the colonists' rebellion. But when Gonzalo Pizarro occupied the city, Paullu found no difficulty in switching his allegiance. Gonzalo's lieutenant-governor in Cuzco was Alonso de Toro, an old friend of Paullu and one of the most flattering witnesses to the Probanza of 1540; so Paullu instructed his natives to support the rebel side in patrolling the road to the coast.* His blockade was so effective that the Viceroy in Lima was soon completely ignorant of Pizarro's movements in the mountains. Paullu's cousin Cayo Topa also marched with the rebel army.

But with the arrival and successful advance of Gasca, Paullu shrewdly sensed that the royalist side was again going to triumph. Yet another *volte-face* was called for. Paullu sent messages of support to the royal envoy immediately after his landing in Peru, but succeeded in keeping his overtures secret from the rebels occupying Cuzco.* Paullu's brother-in-law Pedro de Bustinza – the husband of the most important Inca princess Beatriz Huayllas or Quispique – was guarding the royal highway for Gonzalo Pizarro. He had with him a force of natives under Cayo Topa and another chief 'because the caciques and Indians have great respect for them'. Apparently acting on Paullu's instructions, these natives failed to warn Bustinza and his men of the approach of Gasca's vanguard, so that the rebels were captured and executed. Cayo Topa then joined the royal army at Huamanga, in November 1546, and advanced with it through the rest of the campaign.* Paullu, in short, was 'doing much for the President [Gasca] even though he was with the rebels in Cuzco'. When Gasca finally defeated Gonzalo Pizarro and occupied Cuzco, Paullu's prestige was greater than ever: he was 'a man of great valour, intelligence and energy, beloved of the Indians throughout the land'.

Manco Inca did not intervene directly in the civil wars, but he kept himself well informed of their progress. When Vaca de Castro defeated Diego de Almagro the Younger at Chupas, he executed many of the rebels, including Titu Cusi's former guardian Pedro de Oñate. 'The ditch under the scaffold of Huamanga was full of dead bodies. This gave considerable pleasure to the native onlookers, although they were amazed to think that many of the victims had been captains and men holding posts of honour. They took news of all this to their King Inca Manco Yupanqui at Vitcos.'

The young Diego de Almagro fled to Cuzco with one of Pizarro's assassins called Diego Méndez, 'intending to seek refuge with the Inca Manco' who was in sympathy with the Almagrist cause. But Méndez could not resist a final visit to his mistress, and the delay led to the capture of the fugitives. The town council of Cuzco reported that a posse of citizens had galloped off in pursuit of Diego de Almagro and seven or eight of his most important lieutenants. They overtook them in the Yucay valley twenty-five miles from the city and captured them 'with great difficulty'. 'Almagro was going determined to join the Inca, to assemble there any Spaniards of his faction that he was able.' It was probably fortunate for Manco that the young Almagro did not reach him: the Spanish army would certainly have followed in pursuit of such important prey. The captive renegades were taken back to Cuzco. Diego de Almagro was executed, only four years after his father, but Diego Méndez was left alive in the prison of Cuzco.*

Manco Inca was saddened by the deaths first of Almagro and now of his son. He therefore welcomed any fugitives from the defeated Almagrist faction who fled to Vitcos. Before long Diego Méndez himself succeeded in escaping from Cuzco and joining the Inca. He was an important man, half-brother of the dashing Rodrigo Orgóñez who had almost caught Manco in 1537, holder of the rich encomienda of Azángaro, and one of the leading captains of Diego de Almagro the Younger.* Six other Spaniards also escaped to Vitcos in September 1542: Gómez Pérez, Francisco Barba, Miguel Cornejo, one Monroy and two others. Manco was pleased to have these Europeans at his court. He had them instruct his men in the use of captured Spanish weapons. He himself had ridden a horse in the defence of Ollantaytambo and the battle against Captain Villadiego. He appreciated the importance of teaching his men Spanish fighting techniques: only in this way could they hope to resist effectively. The fugitives also taught Manco himself to race a horse and shoot an arquebus.*

The Inca tried to make the fugitives as comfortable as possible on the remote spur of Vitcos. His son Titu Cusi wrote: 'My father ordered that they should have houses in which to live. He had them with him for many days and years, treating them very well and giving them all they needed. He even ordered his own women to prepare their food and drink, took his meals with them, and treated them as if they were his own brothers.'

Manco's presence in Vilcabamba still posed a threat to the security of Cuzco and the road to Lima. The council of Cuzco told the King that they had not dared to send all the citizens to fight at Chupas: two hundred men were left to defend Cuzco, 'which was very necessary, because during these disturbances Manco Inca approached this city, and it was suspected that he would attack it'.

When Vaca de Castro had finished suppressing Almagro's rebellion, he turned his attention to Manco, hoping to achieve his surrender by diplomatic means. Vaca de Castro sent a friendly overture with a present of brocade; Manco sent back parrots and an embassy of 'two of his three principal captains'. Vaca de Castro told them that King Charles had provided safe-conduct and pardon for the Inca, and had ordered that if Manco emerged he be given a good estate and be well treated. The Governor felt that the negotiations were 'progressing with very much heat', largely 'because the Marquis Pizarro and his brothers are gone – for he was afraid of them, both because he had killed Juan Pizarro and for other reasons . . . [Manco's envoys] did not wish to leave before seeing me enter Cuzco, for they are people who look to the valour and reputation of him who governs. When they saw this well fulfilled they became extremely deferential.' Manco requested five specific estates and it was virtually agreed by the end of 1542 that he would emerge if these were provided.* But nothing came of all this. Manco may have been dissuaded by his Almagrist guests, whose lives were endangered if their host surrendered. Or Vaca de Castro, who was busy sponsoring Paullu for baptism, may have lost interest or felt that Manco was asking for too much.

Manco was evidently tempted by the prospect of returning to rule in Cuzco. His chances of doing so improved still further with the arrival in May 1544 of the first Viceroy of Peru, Don Blasco Núñez Vela, and the first four oidores of the Audiencia of Lima. The Viceroy had been sent to enforce the New Laws which promised a new place for the natives in colonial society. Here was another official fresh from Castile who owed no allegiance to the dreaded Pizarros. The seven Almagrist fugitives also took heart at the news of the arrival of the Viceroy: they thought that he might be more leniently inclined towards them than Vaca de Castro had been. Before making a move in this new situation, Manco consulted his Spanish companions. 'Taking Diego Méndez aside, he asked him to explain clearly and without reserve who was the great and powerful captain who had arrived at Lima, whether he was strong enough to defend himself against Gonzalo Pizarro and whether he would remain as universal governor of the kingdom.' Méndez's answers were reassuring on all these points, and it was decided to draft a letter to the Viceroy requesting pardons for Manco himself and for the Spaniards with him, on receipt of which they would emerge peacefully into the Spanish part of Peru. Garcilaso de la Vega said that an embassy went to Lima with such a letter. It was received by the Viceroy and returned with promises of pardon.*

Gonzalo Pizarro, in the meantime, had occupied Cuzco and was moving into open rebellion. He used Manco as a pretext for raising an army and declaring himself its captain-general. The army was to march against the rebellious Inca, and the only possible leader was 'Gonzalo Pizarro, because he is such a great soldier and has so much courage and experience in warfare, and because of the Inca's great fear of him'.

Pizarro, of course, had no intention of moving against Vilcabamba. He arranged for a group of his sympathisers to seize weapons from the battle of Chupas that were still stored in Huamanga, and then set out towards Lima, taking virtually every horse and most of the able-bodied men of Cuzco. The denuded city was soon gripped by fear that Manco's troops were advancing towards it with Méndez as their adviser. This may in fact have happened. Manco's spies would have informed him that Gonzalo Pizarro had led out the bulk of Cuzco's citizens on the start of another civil war. He may have dreamed of another chance to occupy his capital city, and 'as large a force as they could muster' of native troops advanced to the villages close to Cuzco. Gonzalo Pizarro had removed all the horses. The frightened municipal authorities therefore ordered mares to be gathered in the square and assembled every man, including the clergy. The essential was for them to be mounted, for 'there is no fortress for resisting the fury of Indians equal to Spaniards on horseback'. The native troops devastated an area twenty miles from Cuzco, but withdrew without attempting a direct attack on the city. They were possibly only undertaking a punitive raid on Caruarayco, curaca of Cotamarca, who was rumoured to have plotted to assassinate Manco and make himself Inca.*

The Almagrist fugitives were bored and restless in their humid jungle sanctuary. They were encouraged by the favourable reply received from the Viceroy, and further emboldened when they heard that the Viceroy had imprisoned their enemy Vaca de Castro and was about to clash with Gonzalo Pizarro, brother of the man they had murdered. They may also have corresponded with Alonso de Toro in Cuzco, who advised them that their chance of pardon would be immeasurably enhanced if they could dispose of the Inca. They decided to murder Manco, 'for without any doubt Vaca de Castro would pardon them because of this outstanding service and would reward them'. Méndez said that the killing must take place before Manco's troops returned from the raid near Cuzco. An Indian woman servant of Francisco Barba overheard the plotting and tried in vain to warn the Inca.* Manco paid no attention. The Inca was hopeful that he might recover the lordship of Cuzco under the Viceroy, or that his state of Vilcabamba might gain sovereign recognition by assisting the royalists against Pizarro's rebels. But the hopes of both the Inca and his guests were to be dashed in another instance of Spanish treachery.

When the seven fugitives first reached Vitcos, Manco's officers wanted to kill them immediately. It was the Inca who allowed them to remain alive and to stay as his guests, provided they were stripped of their arms. He used to play games with them, and a favourite pastime was throwing horseshoe quoits. They were enjoying such a game when Diego Méndez suddenly produced a hidden dagger. He fell upon the Inca from behind as he was about to throw a horseshoe. Méndez and his companions then repeatedly stabbed the man whose hospitality had saved their lives. The nine-year-old Titu Cusi was with his father at the time, and later reported: 'My father, feeling himself wounded, tried to make some defence, but he was alone and unarmed and there were seven of them with arms. He fell to the ground covered with wounds and they left him for dead. I was only a small boy, but seeing my father treated in this way I wanted to go to him to help him. But they turned furiously upon me and hurled a lance which only just failed to kill me also. I was terrified and fled among some bushes. They searched for me but failed to find me.' Native attendants ran to the bleeding Inca, but he died three days later. His heartbroken subjects embalmed Manco's body and took it from Vitcos to Vilcabamba.

The Spanish assassins ran for their horses and galloped off on the road to Cuzco. They rode all night, but missed the path through the forested hills and camped in a large thatched building. The Indians of Vitcos had meanwhile sent runners to alert the forces marching against Cotamarca. The runners met these returning with their prisoners. Rimachi Yupanqui turned back with a contingent of forest archers and caught the Spaniards on a forest path. Some assassins were dragged from their horses; others retreated into the building, but the Indians piled firewood at the entrances and set fire to it. Any assassins who were not burned to death were speared or shot as they ran out of the flaming building. 'All had to suffer very cruel deaths and some were burned.'

So ended Manco Inca. The heroic warrior who had so frequently confronted Spanish forces and eluded Spanish pursuers fell to Spanish treachery. He was stabbed in the back by men whose lives he had spared and who had enjoyed his hospitality for two years – the same men who had stabbed his enemy Francisco Pizarro. His death was a tragic loss for the natives of Peru. Manco was the only native prince whose royal lineage and stubborn courage enjoyed the respect of Spaniards and Indians alike. With the disgrace of the Pizarros, Manco could certainly have negotiated a return to Spanish-occupied Peru on favourable conditions. But, as an indomitable patriot, he might have preferred to rule his tiny but independent Inca state in Vilcabamba.

The renegade Spaniards undoubtedly planned the murder hoping to win kudos and recognition by this spectacular feat. They timed their attack for a moment when most of Manco's army was menacing Cuzco, in late 1544 or early 1545, and they very nearly made good their escape.

Many Spaniards were ashamed of this second killing of an Inca. The same chroniclers who had embroidered Atahualpa's execution also sought to excuse this squalid treachery and breach of hospitality. Because the killing took place during a game of quoits, they claimed that it was done on an impulse, a flash of sudden anger by a proud Spaniard who had felt insulted by the Inca. But no Spaniard, however volatile, would kill a foreign ruler at whose court he was a fugitive on so slight a pretext. The assassination was not a crime of passion. All the best contemporary sources agreed that it was premeditated murder.†

The Spaniards were fortunate in the timing of Manco's assassination. The Inca was removed just as Gonzalo Pizarro's rebellion was ripening, so that the Indians of Vilcabamba were left leaderless during the three years that the Spaniards were most seriously divided. Manco was the only remaining native leader with the will to organise yet another rising or the stature to inspire it. His state at Vilcabamba was a small affair, a handful of native towns in some secluded mountain valleys – many other native tribes established similar pockets of resistance in other corners of the Andes. But Manco possessed qualities that made him and his state exceptional. He alone had been born an Inca prince, and had acquired during his youth and adolescence the self-assurance of a man certain that his family ruled virtually the entire world. He never became dazzled by the Spaniards' civilisation or by their aura of invincibility: he had campaigned and lived among them in the early days of the Conquest and was fully aware of their weaknesses. Although he tried to adopt the most effective Spanish weapons, he did not ape the Spanish way of life in other respects and never abandoned the Inca religion in favour of Christianity. Under him there was a revival of militant Inca nationalism in Vilcabamba, with great insistence on the minutiae of court procedure and religious observance.

Manco's other great strength lay in his doggedness – an essential quality for survival in a guerrilla chief – combined with the ability to organise and act on a large scale. His campaigns demonstrated this ability, with the massive native levies of 1536 and 1538 as visible proof of his effectiveness as an organiser. Manco's character is more elusive. His enemies, and most particularly partisans of Paullu such as Cristóbal de Molina or the Quipocamayos of Vaca de Castro, tried to maintain that he was cruel and ferocious. They told stories of his cutting down those who were lukewarm to his cause – the very thing of which Paullu had boasted in his own Probanza. They claimed that Manco's threat to exterminate every last native collaborator during the siege of Cuzco did irreparable damage: it forced the yanaconas to side with their Spanish masters. His son Titu Cusi, on the other hand, described his father as a man of generosity and high ideals, slow to take violent action.

Manco lived for three days after the stabbing by his Spanish guests. He had the satisfaction of learning that his murderers had not escaped, and had time to settle the succession on his eldest legitimate son Sayri-Tupac, a boy aged five whose names meant Royal Tobacco – the Incas used tobacco, sayri, as a medicine, and as snuff, but did not smoke it. The young Sayri-Tupac was placed under the regency of an Inca noble called Atoc-Sopa or Pumi-Sopa.

A guerrilla movement depends on the energy of its leadership. The state of Vilcabamba was left sadly weak with the murder of its young ruler – for Manco was only in his early thirties at the time of his assassination. Sayri-Tupac's regents were fortunate that the Spaniards were entirely engrossed in Gonzalo Pizarro's rebellion during the first three years of their rule. Everything that we know about this period indicates that there was a powerful revival of Inca traditionalism. Without an aggressive leader, the Vilcabamba Indians abandoned their raids on the Spanish highways and were glad to be left unmolested in their remote valleys. Bitterness against Manco's murderers made his survivors avoid everything Spanish. There was no further attempt to assimilate Spanish fighting methods, and the regents hoped that geography and the well-tried native weapons would be enough to preserve them in quiet isolation.

14

SAYRI-TUPAC

WITH the settlement of the colonists' rebellion Pedro de la Gasca – like Vaca de Castro before him – would have liked to complete his mission in Peru by pacifying the Vilcabamba natives. He never contemplated doing so by force, although an expedition on the scale of that of 1539 would almost certainly have defeated Manco's feeble successors. In the troubled conditions following the execution of Gonzalo Pizarro, Gasca probably did not dare risk an army of his supporters leaving Cuzco for the remotenesses of Vilcabamba. His men were weary of fighting, and there was nothing to tempt them into the Vilcabamba jungles, which had by now acquired an unsavoury reputation. People had forgotten how nearly successful the expeditions of Rodrigo Orgóñez and Gonzalo Pizarro had been. All they now remembered was that these two most dashing young conquistadores had both failed to capture Manco in his eyrie. Even Cieza de León, writing at this time, dismissed the expedition of 1539 by recording simply that Gonzalo had 'occupied a rocky emin-

The Inca Sayri-Tupac meets the Viceroy Cañete in Lima

ence and destroyed two bridges'. Also of course Gasca was an ecclesiastic who preferred diplomacy to warfare.

Gasca turned to Cayo Topa, who had accompanied him for the past six months, 'to ask him to send two of his servants to persuade the Inca to submit to His Majesty, and to assure him that we wished to receive and treat him well'. These envoys returned in July 1548, accompanied by six messengers from the cautious regents. They brought exotic presents from the jungle city of Vilcabamba: parrots, jungle cats and flutes, and, according to Bernabé Cobo, they brought Paullu 'objects of gold and silver, some exquisite cumbi cloth of the quality they used to weave in the time of their Inca ancestors, and various species of curious birds and animals that are raised in those provinces'. The envoys wished to establish the extent to which Gasca himself was involved in the overture, and to reassure themselves about his attitude to the natives in general.

Gasca's answers to the regents were almost entirely reassuring. He sent back presents appropriate to the ages of the boy Inca and his older tutors, but he also included a threat of violence with his messages of goodwill. 'I sent clothing of various coloured silks – blouses and cloaks – and I sent Sayri-Tupac two casks of preserves, and to Pumi-Sopa, the tutor and regent, I sent two jugs of wine. I sent a very hispanicised Indian called Don Martín with the messengers so that he could persuade them to come out for their own advantage, and could also explain that if they did not come for the benefits, they would be made to come by force.'

This Don Martín was an exceptional figure in post-Conquest society. He was given to Pizarro on his second voyage in 1528, and taken to Spain. He became one of very few Indians to learn fluent Spanish. Pizarro grew fond of the boy and gave him a horse: he fought as a cavalryman during the Conquest and in the Almagrist wars. Don Martín was the only Indian to receive a share of Atahualpa's ransom. He was the only Indian other than Paullu to receive an encomienda from Pizarro: a rich holding near Lima. He was also the first native to marry a Spanish lady, Luisa de Medina, and to receive the ultimate symbols of success, a knighthood and coat of arms. He was a Christian, wore Spanish clothes, and entertained generously in his large house in Lima. He was a faithful follower of the Pizarros and stayed with Gonzalo until the very end. This loyalty betrayed him. On his return from

Vilcabamba, Don Martín's encomienda was summarily confiscated by Gasca; this 'hispanicised Indian' died in Spain when he went to protest to the King.*

The Indian Don Martín returned from Vilcabamba in mid-August with a second group of envoys from the Inca. The regents stated the terms on which Sayri-Tupac would surrender. He would have the land he was now occupying in Vilcabamba together with an adjacent triangle of land contained between the junction of the rivers Apurímac and Abancay and the royal road. He also wanted 'some houses that belonged to his grandfather in Cuzco, and a certain estate and plot of land with some pleasure houses that his grandfather used to have in Jaquijahuana'. Gasca agreed to the Apurímac-Abancay triangle and the various houses, and the messengers returned apparently satisfied by his offer. Gasca explained to the Council of the Indies that the land between the two rivers contained 'only five or six hundred Indians belonging to two citizens, one of whom is Hernando Pizarro' – who was still imprisoned in Spain and whose name was in disgrace with the rebellion of his brother Gonzalo. Gasca had not agreed to Sayri-Tupac keeping Vilcabamba province itself, for the obvious reason that 'if they remain lords of that stronghold they can rebel whenever they wish'.

Don Martín had reported that the young Inca and his tutor seemed most anxious to return to Cuzco on reasonable terms, for they and their followers were in poor health in the heat, damp and relatively low altitude of Vilcabamba. Their enthusiasm was confirmed when the Inca sent one of his officers to take possession of the houses in and near Cuzco, to decorate them and plant crops against his arrival at the next corn harvest. Pumi-Sopa also wrote Gasca an encouraging letter. Preparations for the Inca's arrival were still continuing in 1549. In May of that year Gasca wrote from Lima boasting of the advantages that would result from the Inca's emergence, 'for as long as the natives know that he is in revolt they do not cease to run off to him'. Gasca even went so far as to appoint one Juan Pérez de Guevara to establish the indispensable Spanish garrison in Vilcabamba. Paullu was also making elaborate preparations to go in person to accompany his nephew out of the mountains. He left Cuzco with a great entourage of leading natives, passed through Limatambo above the Apurímac bridge, and reached Huaynacapaco on the high central road into Vilcabamba.* But the reception arrangements suddenly collapsed. Paullu himself grew sick and was forced to return to Cuzco. He died there a few days later, only a few years older than his murdered brother. And, to Gasca's dismay, the regents took fright and Sayri-Tupac remained in Vilcabamba.

Paullu was mourned by both Spaniards and Indians. He was the last of Huayna-Capac's sons to rule in Cuzco, and his ancestry and military record gave him a commanding position in the city's colonial society. 'He died a Christian,' wrote Cristóbal de Molina, 'and a chapel was ordered to be built where he was buried in splendour. There was a service of Spaniards and a mass for him. . . . When the natives learned that he had expired, all the Indian warriors who lived in Cuzco went up to the palace of this Paullu Inca with all their arms, arrows, lances and maces, each with the weapon he used in battle. They surrounded it completely on the tops of all the hills and walls, took possession of it, and gave great cries and lamentations. All the inhabitants of Cuzco also mourned and cried out. But the warriors were the most remarkable: they stayed there, guarding Paullu Inca's palace until he was buried. When asked why these Indian soldiers had assembled there at that time – there were four or five hundred of them – they explained that it was a custom in Cuzco that when the natural lord died, they should assemble there so that no tyrant could occupy the lord's palace and capture or kill his wife and sons, and usurp the city and kingdom. . . . The entire city, both Christians and Indians, wept at the funeral of this chief.'

The grief of some Spaniards was short-lived. A number of opportunists soon approached Gasca to be given portions of the dead Inca's estates. 'But', the President wrote to the Council of the Indies, 'I left these to Don Carlos, the eldest son of Don Paulo, together with the coca plantations and all the rest of his father's property, both because he himself was legitimised by His Majesty and his father had married his mother two days before his death, and also because (although this may be irrelevant) it seemed to me highly inhuman to take the estate from him, as he is the grandson of the ruler of these provinces. It is something that would greatly depress the natives and might even deter Sayri-Tupac from submitting to the rule of His Majesty even though he desired to do so, as he appears to do up to the present. But such reasons are scarcely comprehensible to the [Spanish] inhabitants of this land. Some feel that I am wronging them by not giving them these Indians and taking them from Don Carlos.'

President Gasca wrote in this same letter that he feared Paullu's death might discourage Sayri-Tupac and his followers from emerging from the mountains 'since they now lack the shadow of Don Paulo, whom Sayri-Tupac considered as a father'. His fears were justified. The young Inca stayed in Vilcabamba. Paullu's young son Carlos became the leading native in Cuzco, although Cayo Topa and Titu Atauchi were also senior nobles of the royal house. Cayo Topa acquired a Spanish protector, arranged for his sons to be educated by the Mercedarians, and was himself baptised as Don Diego Cayo. Cieza de León reached Cuzco at this time and eagerly sought out Cayo Topa and 'other orejones who regarded themselves as the nobility. And through the best interpreters to be found, I asked these Incas what people they were, and of what nation.' Titu Atauchi was a nineteen-year-old grandson of Huayna-Capac who had been converted to Christianity, as Don Alonso, and later received various royal favours.*

Gasca also wished to settle the affairs of other descendants of the Incas: the various half-caste children left by the Pizarros. He felt sorry for the Marquis's son Francisco, 'a child of nine or ten who seems well-inclined'. Gasca wished to give the boy his father's encomienda of Yucay, together with the coca plantations of Avisca, and recommended that he be legitimised as his sister had been. He could be left in the care of his mother Angelina, now wife of Juan de Betanzos. The King agreed only that the boy should receive the income from these estates, which were to revert to the Crown. Juan Pizarro's daughter Isabel and Gonzalo's Inés were to be given 6,000 ducats each from their cousin Francisco's income, and sent to Spain. Gasca described Gonzalo's son Francisquito as 'considered to be badly-intentioned; and his father said on several occasions that on his death this boy would remain in his place'. No one wanted another rebellion like that of Diego de Almagro the Younger. The King therefore ordered that Francisquito be exiled to Spain with his sister Inés and cousin Isabel. The children duly set sail from Lima in great secrecy early in 1549.* They were then escorted to their fathers' birthplace, Trujillo de Extremadura, and given the first instalment of their grants.*

A year later the Marquis Pizarro's two surviving children, Francisca and Francisco, followed their cousins to Spain.* The children were in the care of two important citizens: Antonio de Ribera, husband of their aunt Inés Muñoz, and Francisco de Ampuero, husband of Francisca's mother Inés Huayllas. Ribera and Francisca petitioned the Audiencia of Lima for permission to settle Francisca's affairs before sailing for Spain, and she wrote a will to this effect.* The children sailed in April 1551, accompanied by Ampuero, and reached Seville at the end of July. The arrival in Trujillo of these two children of the late Marquis caused excitement among their Pizarro aunts and cousins. Everyone wanted to ensure that the fortunes of these heirs remained within the family.

At Christmas 1551 Francisca and Francisco were taken to see the head of the family, their uncle Hernando, only survivor of the four brothers who had conquered Peru and only legitimate son of Captain Gonzalo Pizarro and his wife Isabel de Vargas. Hernando was now fifty, and his niece was a beautiful girl of seventeen; he had not seen her since she was an infant. Hernando was also a prisoner, who had already spent over ten years incarcerated in the great keep of La Mota at Medina del Campo. When he returned to Spain in 1539 Hernando had been greeted by a storm of protest for his execution of Diego de Almagro in 1538. Hernando's explanations and generous donations to the Crown almost saved him, but his opponents were too vociferous: in 1540 he was imprisoned indefinitely, without formal charges or a trial.

Hernando's confinement was gentle enough. He was in the same prison and apartments that had harboured King Francis I after his capture at Pavia in 1525. He was able to enjoy revenues from his great Peruvian estates, particularly the silver mines of Porco near Potosí. Gonzalo Fernández de Oviedo, who admired Almagro and hated Hernando Pizarro, described him as 'a tall, coarse man with a thick tongue and lips; there was too much flesh at the tip of his nose, which was inflamed'. He described Hernando's imprisonment with envy: 'Pizarro's table and service were sumptuously appointed and attended by many noble gentlemen. Distinguished, important lords often visited him and paid their respects. He was entertained by many forms and varieties of music. He rose at midday, and his bedchamber was richly hung with fine tapestries and canopies. His table service was abundant and magnificent, with as many pieces of gold and silver as a great prince would have. It is known that he heard mass very late: he thought that such laziness towards God and repose in a soft bed enhanced his position and added to his dignity. He never lacked dice or cards to pass the time gambling for high stakes – for money, jewels and horses.' Hernando was not even deprived of feminine companionship. He enjoyed an affair with a noble local girl called Doña Isabel de Mercado, by whom he had a daughter called Francisca Pizarro Mercado.

Hernando Pizarro had in his possession the will written by his brother Francisco in 1539, in which he left his fortune to his two oldest legitimised children, Francisca and Gonzalo. With Gonzalo dead, Francisca was heiress to the Marquis's vast fortune. She was also a beautiful mestiza. Hernando did not hesitate. He sent his nephew Francisco back to live with his aunts in Trujillo; dismissed the heartbroken Isabel de Mercado, who retired to a nunnery; and in mid-1552 himself married his young niece, unperturbed by consanguinity, the thirty-three-year difference in their ages or by his own imprisonment. Francisca moved into the soft bed in La Mota, and for nine years shared her husband's imprisonment. During this time she gave birth to five children of whom three survived childhood: Francisco, Juan and Inés.

Hernando was finally released from prison by Philip II in May 1561 aged sixty and broken in health. He and Francisca returned to Trujillo and immediately began the construction of the magnificent Palacio de la Conquista on the main square. It still stands on the square of Trujillo, but is now occupied by a number of poor families. One corner of the palace is decorated with sculpture. There are heads of the aged Hernando Pizarro, a mean but powerful old man with a long spade beard; of Francisco Pizarro, equally bearded; of the Inca princess Inés in European dress; and of her daughter Francisca with the high cheekbones of her Inca mother. Above is a gigantic carved coat of arms of Francisco Pizarro, containing the chained heads of conquered native rulers.*

Hernando was a bitter old man, whose brilliant career had been ruined by what he considered an unjust imprisonment. All that was left to him was to consolidate his huge fortune. He finally published his brother's will, but took care to suppress clauses that mentioned pious or charitable bequests. He instructed his agents in Peru to press for settlement of all debts and revenues due to him or his wife. He started acquiring property, farms and orchards in and around Trujillo. And he and Francisca even had the audacity to petition the Crown for restitution of 300,000 pesos they claimed that her father had spent in the suppression of Manco's rebellion, and for 20,000 Indian vassals promised to Francisco Pizarro when he was granted his marquisate. The Crown replied by appointing the Fiscal of the Council of the Indies to rebut this exaggerated claim. This Fiscal mounted a probanza in which witnesses testified to every crime, robbery or abuse ever attributed to or imagined against the Pizarros. Hernando promptly produced a counter-enquiry supporting his claims, and similar interrogations were held in Peru. The case dragged on, but the Crown never paid: many decades later it awarded a title to Hernando's descendants in return for dropping the claim.*

The other Pizarro fortunes also stayed in the family. The Marquis's sixteen-year-old son Francisco was married in 1556 to his cousin Inés, only surviving child and heiress of Gonzalo Pizarro. This couple were not remotely as rich as Hernando and Francisca, for most of Gonzalo Pizarro's fortune had been confiscated when he was executed, and the Marquis's young son had never been legitimised. Their efforts to contest Francisca's inheritance were in vain. And the marriage was short-lived, for young Francisco died in 1557; his widow Inés married one of the first conquistadores, Francisco de Hinojosa, two years later.

The settlers' revolt and the unilateral declaration of independence by Gonzalo Pizarro did little to discourage the reform movement which flourished under Bartolomé de las Casas. The New Law that forbade royal officials and religious orders from holding encomiendas had aroused the opposition of those sections of the colonial administration. This and other New Laws were repealed. But others remained on the statute books and many were reiterated, notably those condemning cruelty, slavery and personal service. The reform movement continued to enjoy the support of most of the clergy.

The Spanish Conquest was now championed by an erudite humanist as well qualified as Las Casas himself. Juan Ginés de Sepúlveda produced a learned treatise in 1547 that gave convincing arguments in support of Indian conquests. Sepúlveda was a great authority on Aristotle, and he based his arguments on the ancient Greek concept that some men are slaves by nature. His manuscript had the enthusiastic support of the Council of the Indies, and was submitted to the Council of Castile. Bishop Las Casas returned from his Guatemalan diocese just in time to prevent its publication. The treatise was submitted to two universities who declared that it was based on unsound doctrine. Such was the power of Las Casas and his followers that they succeeded in having all the books by Sepúlveda justifying the Conquest banned throughout Spain and the Indies, although some were published in Rome.*

The debate grew more passionate. Las Casas produced more works arguing that all conquests were unjust and must stop. He said that the only way in which the Spanish Crown could occupy the Americas without endangering the King's hope of salvation was to convert the natives to Christianity, after which they would become loyal subjects of Spain. Armed conquest was manifestly not the most persuasive way to produce such conversions. The King was worried. In April 1549 the Crown issued instructions governing the conduct of expeditions and conquests: fighting was to be only in self-defence, and all food and supplies were to be paid for.* Later that year, the Council of the Indies advised the King to call a meeting of theologians to decide 'the manner in which conquests should be carried on . . . justly and with security of conscience'. And so, on 16 April 1550, Charles ordered that all conquests in the New World cease until the meeting had decided how they should be conducted. This was an extraordinary move. 'Probably never before or since has a mighty emperor – and in 1550 Charles V was the strongest ruler in Europe with an overseas empire besides – in the full tide of his power ordered his conquests to cease until it could be decided whether they were just.'

The great moral debate took place before a gathering of ten members of the Councils of the Indies and of Castile, and four friars, two of whom were avowed opponents of conquests. Las Casas appeared before them at Valladolid in the hot month of August 1550. He had prepared a Latin treatise of 550 pages, which he proceeded to read seriatim for five consecutive days. The judges listened, trying to concentrate in the heat while he expounded a great eulogy of Indian life and achievements. He was passionately convinced that all peoples of the world were men with equal desires and senses. The Indian races were *not* natural slaves: Aristotle had meant simply that there were a few natural slaves in every society.

Sepúlveda then gave his case to the bemused judges. He did not claim that the Spaniards were supermen, although he argued that sixteenth-century Spanish civilisation was the height of human achievement. Nor did he condone colonial wars for territorial expansion – such wars became glorious in later centuries but were repugnant to sixteenth-century Spaniards. At that time men insisted that a war be 'just', and preferred the word 'pacification' to 'conquest'. Sepúlveda argued that by obeying and emulating Spanish masters, Indians could acquire their manners and religion and be dissuaded from their own cruel idolatries. Since Indians were naturally crude it was right that they should serve more refined Spaniards.

Neither Las Casas nor Sepúlveda tried to differentiate between the widely varied races and habits of the Indian peoples. Their arguments were therefore inconclusive, and the judges could reach no firm opinions, even after reconvening in 1551 and being pressed for their judgements during subsequent years. But it was really the saintly fanatic Las Casas who won: for Sepúlveda's books remained unpublished whereas Las Casas continued to pour out his provocative treatises until his death in 1566. Throughout the century, it was Las Casas's extremely pro-Indian views that circulated in Spain and her Indies. They were also current in the rest of Europe, so that this extraordinary example of enlightened moral debate provided fuel for the legends of Spanish brutality. But in Spain itself these arguments established that Indians were men entitled to enjoy the privileges of any citizen: property, civil liberty, human dignity and Christian communion.

Another moral thinker was active in Spain during the 1540s. The great Dominican jurist Francisco de Vitoria was examining the legal implications of the papal donation of the Americas to the Kings of Spain and Portugal. Vitoria's arguments were more muted than those of Las Casas, but his reputation has grown steadily ever since his death in 1546. He is now acknowledged as a founder of international law. Vitoria was the first to assert that the papal grant had no temporal value as a licence to conquer. Neither pope nor emperor was temporal ruler of the whole world. They could not supplant the natural lords of American peoples simply because the Indians refused to recognise papal dominion or to receive the Faith. Vitoria argued that Spaniards were entitled to travel to America, to live there and to preach the Faith. They had a right to form alliances at the invitation of native rulers, and they could depose such rulers if this was essential to save innocent lives threatened by that ruler. They could fight to protect their freedom to preach. But they had no right to fight under any circumstances for personal gain.

Las Casas himself now turned from general humanitarianism to political thinking, producing in 1553 a treatise that concluded that the Spaniards had no right to occupy the lands of native rulers in the Indies.

Unlike Vitoria, Las Casas thought that the papal donation could have temporal validity. But the pope's only purpose was the spread of Christianity, best done through nominated native rulers. Military superiority was certainly no justification for Spanish conquest, and the conquerors had no right to what they had taken from the natives. Everything should be restored to the last penny even if this meant fighting the encomenderos.*

All this theorising was remote from the reality of life in Peru, although it helped inspire continued royal legislation on behalf of the Indians. It also affected the Spanish attitude towards the Inca royal family. There were a series of contradictions here. Most Spaniards despised Indians and regarded puppet Incas as quaint anachronisms from a lost civilisation, nothing more than a temporary expedient. These same conquistadores did nothing to delay the destruction of pre-Conquest life. But the debates on Spain's just title to the Indies and on the rights of the native 'señor natural' enhanced the standing of the Inca nobility. Now that Spain was firmly in control of Peru the authorities wished to conciliate the surviving Incas, from mixed feelings of guilt, legal rectitude and historical sentiment. It was at this time that Spanish chroniclers rewrote the history of the deaths of Atahualpa and Manco, seeking scapegoats or legal justification and avoiding admissions of calculated political expediency. In the initial Conquest the Spaniards had made much of the justice of overthrowing the 'usurper' Atahualpa in favour of the 'legitimate' royal family. This all backfired when the 'legitimate' Manco Inca rebelled, and the Spaniards never really convinced themselves that Paullu's title was equally good. Now that both Manco and Paullu were dead, the Spanish authorities tried to conciliate their respective descendants. Gasca provided for Paullu's son Carlos Inca; but everyone, from the King of Spain down, wanted to win over the defiant sons of Manco Inca in Vilcabamba.

On 19 March 1552, Prince Philip wrote to Sayri-Tupac. He acknowledged that Manco had been provoked into his rebellion and gave his son Sayri-Tupac full pardon for any crimes that might have been committed since his accession. He also promised that the towns of Vilcabamba would not be awarded to any individual Spaniards by the Crown. Unfortunately, the Viceroy who was to convey this letter died before it reached Peru.* The King repeated the terms of this letter to the next Viceroy, Don Andrés Hurtado de Mendoza, Marquis of Cañete,

6 *Pizarro's coat of arms on the corner of Hernando's palace shows Atahualpa surrounded by chained native chiefs*

17 *Busts of Francisco Pizarro and his Inca lady Inés project from the palace wall*

8 *A modern statue of Francisco Pizarro faces the palace built by his brother Hernando in their birthplace, Trujillo de Extremadura, after his release from prison in 1561*

19 *Hernando Pizarro*

20 *Francisco Pizarro*

23 *Rodrigo Orgóñez*

24 *Hernando de Soto*

21 *Emperor Charles V*

22 *Sebastián de Benalcázar*

25 *Pedro de la Gasca*

26 *Cristóbal Vaca de Castro*

27 *The baroque church of Santo Domingo in Cuzco has often crumbled in earthquakes. But the mortarless ashlars of the sun temple Coricancha, on which it rests, have always stood firm*

28 *Llamas walk past polygonal Inca masonry in a typical Cuzco street*

29 *Hatun Rumiyoc, 'the great stone', demonstrates the uncanny precision of Inca stonemasonry. It has no less than twelve corners*

30 *Indian women walk past the mighty walls of Sacsahuaman, below the gate where Juan Pizarro was mortally wounded in May 1536*

31 *Inca terracing flanks the beautiful Yucay valley, later the fief of the Inca Sayri-Tupac and his descendants. The siege of Cuzco began and ended with battles in this valley*

32 *Sacsahuaman, the finest Inca monument: its huge grey stones give an impression of serene invincibility*

3 *The terraces of Ollantaytambo: 'a horrifying sight' to Hernando Pizarro's unsuccessful attackers*

4 *All that remains of the great hall of Colcampata, where Paullu Inca and Carlos Inca once ruled over the native aristocracy of Cuzco*

35 *Thousands of warriors crowded the terraces of Ollantaytambo to repel Hernando Pizarro's attack. Manco Inca himself appeared among them, 'on horseback with a lance in his hand'*

in March 1555. Prince Philip asked him to pursue the negotiation to lure out the Inca.

The Viceroy Cañete wisely decided to enlist the help of the *grande dame* of Cuzco's native society, Doña Beatriz Huayllas Ñusta. 'In Cuzco, where she resided, there was no surviving lord, male or female, as important as she. For this reason the Marquis [Cañete] wrote to her and strongly requested and affectionately begged her to send the message to her nephew, offering her a reward for this. For he understood that Sayri-Tupac would not confide in or trust any other person.' Doña Beatriz was Huayna-Capac's daughter. She had been mistress of the dashing conquistador Mancio Sierra de Leguizamo,* and then married a Spaniard, Pedro de Bustinza, who was betrayed by Cayo Topa and executed as a partisan of Gonzalo Pizarro. She was at this time married to a former tailor called Diego Hernández, a match that she had considered beneath her dignity but had accepted with great reluctance at the insistence of her half-brother Paullu.* Despite these marriages, Beatriz never deigned to learn Spanish.

Beatriz sent a relative called Tarisca who penetrated Vilcabamba after 'making, as best he could, his bridges across the passages and bridges that had been cut'. Tarisca was received by the regency council of elders and military commanders. These prevaricated, sending him back with some of their own men under an elder called Cusi, to check the validity of the overture. They also demanded that any future embassy should include Beatriz's mestizo son Juan Sierra de Leguizamo.

Juan Sierra duly set out for Vilcabamba in 1557. He was accompanied by Juan de Betanzos, who spoke Quechua and had become an expert on native affairs by virtue of his marriage to Atahualpa's sister (and Francisco Pizarro's former mistress) Angelina Yupanqui; and by a Dominican monk called Melchor de los Reyes.* Murúa said that Beatriz's husband Diego Hernández and a Portuguese, Alonso Suárez, also went on the mission. These last two had been sent from Lima by the Viceroy to deliver the royal pardon sent by Prince Philip, but had failed in attempts to penetrate Vilcabamba from Huamanga or Andahuaylas. They therefore joined Sierra's embassy from Cuzco and all five crossed into the native territory by the conventional eastern route across the bridge of Chuquichaca.

The natives were still highly suspicious. Betanzos and Friar Melchor were repeatedly detained, while Sierra was reprimanded by the native commander 'for having come accompanied by Christians. Juan Sierra excused himself, saying that it had been done on the advice and at the command of the Corregidor of Cuzco and [Sayri-Tupac's] aunt Doña Beatriz.' There was clearly indecision and anguished debate on the native side about the reception that should be given to this overture. At one moment a message arrived from the Inca telling the envoys to leave with their presents and dispatches. While they were on their way home, a contradictory message invited them to go to meet the Inca in person. When they were almost at the meeting place, a further message required that Juan Sierra should proceed alone. Further delays occurred, indicating deep dissension born of insecurity among the native regents. The Inca even admitted as much. He finally received Sierra most cordially 'as being his most important relative' but explained that 'he himself was not empowered to make an agreement, since he was not sworn in as ruler and had not received the royal fringe, having not yet come of age'. Sierra and Melchor de los Reyes therefore had to explain the details of their embassy to the native elders and commanders. Melchor also delivered the presents he had brought from the Viceroy: some pieces of velvet and damask and two jugs of silver and gilt. All this was well received, but the natives again asked for time to debate the proposal and to consult their auguries and oracles.

In the meantime Juan Sierra and his companions were asked to accompany two of the Inca's captains to Lima to bargain with the Viceroy for the best possible terms for the Inca. They emerged through Andahuaylas and reached Lima in June 1557. After a week of negotiation Viceroy Cañete decided, in consultation with the Archbishop and oidores, to be generous. He issued a full pardon dated 5 July and offered the Inca substantial estates if he would emerge within six months of that date. The time limit was imposed to avoid a repeat of the embarrassing anticlimax when Sayri-Tupac had failed to materialise in 1549. The return of Juan Sierra and the native commanders with this offer produced another bout of anguished debate at Vilcabamba. 'They tested their auguries through sacrifices of animals and birds of the field by day and by night, and watched the cloud formations to see whether on those days the sun would appear clear and bright or sad and obscured by cloud and mist.' The omens were favourable, but a powerful group of advisers still feared that Sayri-Tupac might be surrendering himself to the fate of Atahualpa, Chalcuchima or Manco at the hands of the Spaniards.

In the end Sayri-Tupac himself brought the matter out of the clouds and made his own decision. He had now finally attained his majority and been crowned with the imperial fringe and the title Manco Capac Pachacuti Yupanqui. His decision was bold and realistic, but also materialistic and unheroic. For Sayri-Tupac decided to accept the Spanish offer, leave Vilcabamba and settle in Spanish-occupied Peru. He told his hesitant followers: 'We have never been so well fortified as now, nor so prepared for war. It is true that here I am lord of all that I could reasonably want, for all the Indians come to serve me here in their mitas. But you must consider that the Sun wishes that I should leave, so that my domain should be increased and because out there I could be the salvation of my family and of you all. And consider how right it is to go to see our neighbours and friends, and how we wish to visit the lands where we all originated and to which our natural desire draws and inclines us so strongly. I tell you, therefore, that I wish to leave, even though I know that it could cost me my life.' Many of Sayri-Tupac's followers decided to accompany him. There followed eight days of drinking parties and elaborate celebrations, 'with much rejoicing, although some of the most ancient captains were sad about the departure'. Sayri-Tupac's decision meant that he abandoned his father's struggle for independence, relinquished the Inca rule that he had just formally inherited, and left the forested hills where he had spent his youth and childhood. His older but illegitimate brother Titu Cusi Yupanqui later claimed that he himself and the other native chiefs had decided to send Sayri-Tupac to test the Spaniards' good faith – and also, perhaps, because he did not fulfil their conception of a tough guerrilla leader. It was significant that when Sayri-Tupac left Vilcabamba, on 7 October 1557, he did not take the imperial fringe with him.

Sayri-Tupac emerged on to the royal road at Andahuaylas in early November and made his progress towards Lima. He travelled in litters, 'like a lord and King of Peru', but these were not of gold as in the days of the empire. Three hundred Vilcabamba warriors, Chunchos and forest Antis escorted and carried him. 'The caciques and Indians of the provinces through which he passed came out along the road to receive him and feast him as best they could. But their festivities were more for lamentation than for enjoyment – for contrasting the misery of the present with the greatness of the past.' The Inca left his young wife at Jauja and proceeded down towards the coast, sending two salvers and a pitcher of gold worth 5,000 pesos to the Viceroy. Cañete sent back some fine clothing for the Inca and his wife, a mule worth 500 pesos and 'a saddle-cloth of black velvet bordered with silver passementerie, with silver stirrups and stirrup-guards'.

Sayri-Tupac reached Lima on 5 January 1558, the first and only Inca to visit the Spanish capital of Peru. The Viceroy sent the entire council to greet the visitor at the edge of the city, and lodged him in the vice-regal palace. Cañete rose as the Inca entered and 'seated him beside himself in the presence of the oidores'. It was just twenty-five years since the last occasion when a Spanish governor had met an independently-crowned Inca; but there had been an irreversible change of fortunes since the isolated soldier Pizarro confronted the proud, victorious Inca Atahualpa. A few days after his arrival, Sayri-Tupac was invited to a banquet in the palace of the Archbishop Jerónimo de Loayza. At the end of the meal a great silver salver was brought in containing the decree confirming the estates being granted by the Viceroy to the Inca. The story goes that the Inca, possibly irritated by the theatricality of this gesture, plucked a thread from the silk tassel of the tablecloth, held it up, and said that this thread compared to the entire cloth in the same proportion that the estates he was now being given compared to his grandfather's empire. His comparison was not unreasonable.*

Although they were only a tiny fragment of the Inca empire, the estates granted to Sayri-Tupac made him a very rich man in colonial society. There was a provision naming the Inca as adelantado or marshal of the Yucay valley. His chief repartimiento was an estate around Oropesa, at the upper end of that lovely valley in which the Incas used to have their country houses.* The Oropesa lands were now cultivated by Spaniards as vineyards and grain fields: they therefore possessed relatively few tribute-paying Indians. But Sayri-Tupac's other encomiendas were far more lucrative. There was Jaquijahuana on the plain north-west of Cuzco – one of Peru's richest encomiendas, recently confiscated from the rebel Francisco Hernández Girón. Two other repartimientos confiscated from that same rebel included the rocky stronghold of Pucará on the road to Titicaca. Sayri-Tupac thus came to possess the scene of Gonzalo Pizarro's defeat in 1548 and of Hernández

Girón's six years later. The Indians on these rich estates paid their new master annual tribute of over 17,000 pesos de oro, the equivalent of 71 kilos of gold: Sayri-Tupac became a territorial magnate with annual rents worth some $150,000 (£60,000). And the Viceroy Cañete, impressed by the royal status of his guest and pleased at his own diplomatic victory, granted these estates not for the normal two lifetimes, but in perpetuity.

From Lima, Sayri-Tupac made a royal progress to Cuzco, again being fêted by the communities along the route. At Huamanga an aged conquistador Miguel Astete presented the Inca with the royal fringe and diadem that had been worn by Atahualpa. The Spaniards wanted their protégé to wear these royal insignia, but he must have been reluctant to wear a trophy snatched from his executed uncle, particularly as Atahualpa had been the bitter enemy of his father Manco.

The natives of Cuzco were pathetically delighted to have the young Inca and his wife in their midst. Bernabé Cobo wrote that 'they organised a magnificent reception at which the Indians came out in their ayllus and clans with their festive inventions. The Inca and his queen entered in their litters, richly adorned with brocade and precious stones. Sayri-Tupac represented his grandfather Huayna-Capac admirably, for the Indians affirmed that he resembled him closely.' According to Garcilaso, the celebrations included bull-fights, 'and cane floats covered in the most costly liveries; I can bear witness to these for I was one of the people pulling them'. Sayri-Tupac was lodged with his aunt Beatriz, now married to the Spanish soldier Diego Hernández. He was taken to see all the architectural glories of his ancestors, and received the obeisance of the native nobility of Cuzco.

One who came to kiss his hands was his mestizo second cousin Garcilaso de la Vega, who was nineteen and only a few years younger than the Inca. Sayri-Tupac produced two small silver jugs of chicha and handed one to Garcilaso to drink the customary Inca toast. He asked his young cousin why he had not joined him in Vilcabamba, and said that he would have preferred him as an envoy to the two Spaniards who had been sent. 'He detained me for a while, questioning me about my life and activities. He then gave me leave to go, commanding me to visit him often. On parting I made him an obeisance in the manner of his Indian ancestors, which pleased him greatly; and he gave me an embrace, with a happy expression on his face.'

Bishop Juan Solano gives papal sanction to the incestuous marriage of Sayri-Tupac and his sister-queen María Cusi Huarcay

Sayri-Tupac's youth and charm were complemented by the beauty of his sister-queen Cusi Huarcay. Garcilaso described the sixteen-year-old coya as a most beautiful woman, but added the somewhat prejudiced comment that 'she would have been much more so had her dark colouring not robbed her of part of her beauty'. The sentimental priest Martín de Murúa recorded a poem that praised Cusi Huarcay as 'drawn by divine hand, a lady of royal grace, perfect in virtue . . . and consecrated as a nymph at the crystal spring of pure chastity'. Despite this perfect virtue, the church was disturbed by the marriage of a full brother and sister. At the request of the Viceroy and the King of Spain, Pope Julius III issued a special dispensation and the Bishop of Cuzco, Juan Solano, was able to give a church marriage to the royal couple.

The Spaniards were jubilant at the success of their diplomacy. They seemed to have undone Manco's rebellious state of Vilcabamba for the price of a couple of estates – one confiscated from a rebel and the other from Francisco Pizarro's son. When Sayri-Tupac left for Cuzco one of the Viceroy's officials wrote that 'the Marquis [Cañete] is very pleased at having succeeded in this undertaking, for he considers that he did a great service to our Lord God and to Your Majesty. In addition to the gain of having [the Inca] peacefully in Your Majesty's service with the Indians he had in there, it is believed that he will become a Christian. And if *he* is one the others will be; for all the Indians adore him.'

The Augustinian Juan de Vivero, who spoke Quechua, undertook the religious instruction of the young Inca, and all went well. Sayri-Tupac went to the monastery church of Santo Domingo, built among the walls of the Inca sun temple Coricancha, and was seen worshipping there with much devotion, 'although some malicious observers said, when they saw him on his knees in front of the Holy Sacrament in the church of Santo Domingo, that he was doing it to worship his father the sun and the bodies of his ancestors who had been kept in that place'. In late 1558 the Inca Sayri-Tupac and his wife were baptised by Bishop Solano, with the conquistador Alonso de Hinojosa as godfather.* The Inca was baptised with a resounding combination of the names of his father and the Viceroy's father: Don Diego Hurtado de Mendoza Inca Manco Capac Yupanqui. His queen was baptised María Manrique and their infant daughter Beatriz Clara.

Soon after being received into the Church, Sayri-Tupac fell ill and decided that he must write a will. A copy of this testament was discovered recently in the Archive of the Indies in Seville. It was a simple document with no mention of the Inca's imperial past or present wealth. Sayri-Tupac made special bequests to his wife María Cusi Huarcay, his sister Inés, and his young cousins Juan Balsa and Juan Sierra de Leguizamo (sons of his aunts, the princesses Juana Marca Chimpu and Beatriz Ñusta). But he left his daughter Beatriz as his main heiress. He mentioned seventeen native captains who had accompanied him from Vilcabamba and left each of them some fields and a cloak and tunic of fine cumbi cloth. He also commemorated his recent conversion by providing for a chapel to be built in the church of Santo Domingo and by requesting that he be buried in that former temple. As it happened, the Inca recovered from his illness and went to live on his estates in the Yucay valley.*

The corregidor of Cuzco at this time was Juan Polo de Ondegardo, a man full of curiosity about the Incas and with intelligent views on the government of Peru. He now scored a further triumph for Spanish control of the country. In 1559 he interrogated an assembly of leading native elders and sent a report of his findings to Archbishop Jerónimo de Loayza to be used in the first ecclesiastical council of Lima.* Probably as a result of this interrogation, Polo learned that many of the mummies of dead Incas were still being concealed and worshipped by the natives. He found a stone idol representing the founder Inca, Manco Capac, richly dressed and adorned, at Membilla near Cuzco. Gonzalo Pizarro had already discovered and burned the mummy of Inca Viracocha at Jaquijahuana – the village where he himself was later hanged. Polo now discovered that Viracocha's ashes were being venerated almost more than his mummy had been. He also found that Inca's huauque, the symbolic effigy made for each dead ruler. But his greatest triumph was to discover intact the mummies of three Incas and four coyas, including those of the great Pachacuti, of Huayna-Capac and of his mother Mama-Ocllo. The young Garcilaso de la Vega went to say goodbye to Polo before leaving for Spain in 1560, and was shown the five bodies. He recalled that 'the bodies were so perfect that they lacked neither hair, eyebrows nor eyelashes. They were in clothes such as they had worn when alive, with llautus on their heads but no other sign of royalty. They were seated in the way Indian men and women usually sit, with their arms crossed over their chests, the right over the left, and their eyes cast down. . . . I remember touching a finger of the hand of Huayna-Capac. It was hard and rigid, like that of a wooden statue. The bodies weighed so little that any Indian could carry them from house to house in his arms or on his shoulders. They carried them wrapped in white sheets through the streets and squares, the Indians falling to their knees and making reverences with groans and tears, and many Spaniards taking off their caps.' The mummies were taken to Viceroy Cañete in Lima and were examined by Father José de Acosta twenty years later, still perfectly embalmed.* The Spaniards rightly regarded the discovery and removal of these sacred bodies as an important step in counteracting a growing revival of the native religion. So the decade of the 1550s closed with Peru in good shape for the conquerors: the last settlers' revolt was over and forgotten, and the Inca was living near Cuzco as a docile Christian.

15

NEGOTIATIONS

The toiling masses of Peru had been reassured and inspired by the continued existence of Manco's tiny Inca state of Vilcabamba. Bartolomé de Vega admitted that 'the hearts of most of them are with the Inca there, and will be until they die.'

When Sayri-Tupac succumbed to Spanish blandishments and made his pathetic progress through occupied Peru, it seemed as though this last vestige of the Inca empire was extinguished. But the Spaniards' diplomatic success was to be short-lived. For in 1561 the young Inca Sayri-Tupac suddenly died on his estates in the Yucay valley. He was still only in his early twenties, and it was less than three years since he had emerged from the Vilcabamba mountains. He was solemnly buried in the monastery of Santo Domingo, the former sun-temple Coricancha.

Sayri-Tupac's death was a double blow to the Spaniards. They were left with no Inca to lead the native community in Cuzco; for the Inca's beautiful widow María Cusi Huarcay was still in her teens and his only

A contrast in arms and armour: Captain Martín Hurtado de Arbieto and an Inca general child was an infant girl, Beatriz Clara Coya. But far more serious was the realisation that they had totally failed to penetrate Vilcabamba. It had been assumed that possession of the Inca automatically brought possession of his dominions. It now became clear that the separate native state was very much alive, and was in fact in more capable hands than at any time since the death of Manco. As soon as Sayri-Tupac left, his older brother Titu Cusi Yupanqui had taken charge of Vilcabamba.

Titu Cusi – whose names meant Magnanimous and Fortunate – was the son of Manco by a wife other than his full coya. His accession appears to have been a further example of the Inca system of selective succession. The Incas never attached as much importance as Europeans to primogeniture and legitimacy in choosing a new ruler. They also looked for ability, and it was customary for a more capable but less 'legitimate' son to supplant the first choice. Normally the aged Inca chose his successor and Titu Cusi claimed that Manco had lived long enough after being stabbed to name him in this way. This claim was untrue: the infant Sayri-Tupac was undoubtedly Manco's legitimate successor and was recognised as such during the twelve years of the regency. But it does seem likely that Sayri-Tupac was found wanting. The regents and military commanders decided that Titu Cusi would better maintain Manco's tradition as a guerrilla leader and implacable enemy of the Spaniards. They therefore allowed Sayri-Tupac to emerge to a life of comfort in Spanish Peru, taking care to retain the royal regalia, the mascapaicha and the crimson fringe, in Vilcabamba. They also used Sayri-Tupac as a guinea-pig, to test Spanish attitudes after the close of the civil wars.*

As to his legitimacy, Titu Cusi explained to the Spanish envoy Diego Rodríguez de Figueroa that it was of secondary importance among them. He was Inca and high priest *de facto*; 'he was in possession and was recognised by the other Incas; they all obeyed him, and if he had not the right they would not obey him'. In this remark, Titu Cusi tacitly admitted that he was ruling by merit rather than strict rule of succession. He had in fact supplanted another younger brother called Tupac Amaru, who was Sayri-Tupac's full brother and son of Manco and his coya. Murúa said that Sayri-Tupac designated this Tupac Amaru as his legitimate successor. But Titu Cusi, 'his bastard brother, because he was older and Tupac Amaru was a boy, usurped the royal litter and sovereignty from him, and introduced himself as ruler. He made Tupac

Amaru a priest and ordered him to remain as custodian of their father's body in Vilcabamba.' Sarmiento de Gamboa said: 'Titu Cusi Yupanqui is not a legitimate son of Manco Inca, but a bastard and apostate. The Indians consider that another son called Tupac Amaru is legitimate; but he is incapable and the Indians call him uti [impotent].'

The Spanish allegations of bastardy and usurpation were made out of frustration at the failure of their policy. Whether they liked it or not, a capable ruler was now in charge of Vilcabamba. The Spanish authorities were uncertain how to deal with this new situation in the native state. Some wanted to repeat the policy of enticing the Inca out of his retreat; others favoured another military expedition, despite the memory of failure by Rodrigo Orgóñez and Gonzalo Pizarro. Spanish policy therefore oscillated during the ensuing years between diplomatic blandishments and military sabre-rattling, depending on the energy of the local or viceregal authorities at any given moment.

The first reaction to Sayri-Tupac's death was to look for a scapegoat. It was obviously suspicious when a native puppet ruler suddenly died at such an early age. The Cañari curaca of Yucay, Francisco Chilche, was imprisoned on suspicion of having poisoned the Inca, since he had been with him at the time of his death. It was all reminiscent of Chalcuchima's arrest on suspicion of poisoning Tupac Huallpa. But Chilche avoided Chalcuchima's fate and was released after a year when nothing could be proved against him.* Another theory was that Sayri-Tupac had been murdered at the instigation of the Vilcabamba militants, to punish his collaboration and make way for Titu Cusi's succession. This was possible, and it might have been done without Titu Cusi's knowledge or approval. Yet another theory was that the Inca had been eliminated by Paullu's son Don Carlos Inca and his friend Alonso Titu Atauchi. The enmity between Manco and Paullu continued between their descendants; and Sayri-Tupac's arrival in Cuzco had undoubtedly eclipsed the prestige of the former leaders of its native community.*

Titu Cusi was probably innocent of his brother's death. He wrote later: 'As soon as I heard of his death I was deeply saddened, thinking that the Spaniards had killed him just as they had my father.' The wise corregidor of Cuzco, Polo de Ondegardo, immediately sent an embassy to assure Titu Cusi that his brother had died a natural death. The new Inca accepted the explanation, and wrote that it had put an end to his suspicions of Spanish treachery. The two envoys who took this reassurance were Juan de Betanzos and a mestizo secretary called Martín Pando. Betanzos, husband of Pizarro's mistress Angelina Añas Yupanqui and an expert on native lore, was eager to repeat his success in the Sayri-Tupac negotiations. But Titu Cusi sent him back empty-handed. As a capable ruler, however, Titu Cusi knew the importance of surrounding himself with good men. He therefore persuaded Martín Pando to remain with him as his secretary, confidant and adviser; which he did for the rest of his life. Pando probably deserved much credit for Titu Cusi's skilful handling of the Spaniards during the ensuing years.

No further overtures were made until the arrival of a new Viceroy, Don Diego López de Zuñiga, Count of Nieva, in April of the following year, 1561. Nieva's men sent to offer terms similar to those proposed by Cañete. Titu Cusi replied that if rewarded with 'some of the large amounts of my father's lands now in the possession of the King', he was quite prepared to be at peace. This vague but promising reply encouraged the royal Treasurer García de Melo. He suggested a tidy and cheap solution to the problem of finding land for this new Inca: Titu Cusi's son Quispe Titu should marry the young daughter and sole heiress of Sayri-Tupac. This would leave the lands around Oropesa in the Yucay valley in Inca hands but would relieve the Spanish authorities of the embarrassment of finding another estate for Titu Cusi. García de Melo travelled to Vilcabamba to deliver this proposal in person, and coupled it with a request that the Inca accept Christian missionaries into his territories. The Inca felt that peace should be negotiated first. The missionaries and other adjuncts of Spanish civilisation could come later.*

The negotiations were interrupted by the death of the Viceroy Count of Nieva in February 1564. This aged libertine collapsed at his notorious love-nest at Surco outside Lima – the victim either of a heart attack or of a jealous husband. Peru's next Spanish ruler was more dedicated to the tasks of government. He was Licenciate Lope García de Castro, a conscientious but indecisive man. He reached Lima in September 1564 to assume office as Governor-General and President of the Council until the next Viceroy should be appointed. He had with him royal instructions to lure Titu Cusi out of Vilcabamba with an adequate income, after making sure that he was the grandson of Huayna-Capac.*

The new President García de Castro accordingly sent García de Melo back to Vilcabamba with letters to the Inca early in 1565. The bait in the negotiation was still the estates inherited by Sayri-Tupac's daughter Beatriz. In his enthusiasm, Castro offered these estates to the heirs of Sayri-Tupac's daughter and Titu Cusi's son in perpetuity. He was also prepared to add 'two towns of Indians that pay no rent' and a number of 'Indians belonging to the Cathedral and the Mercedarian monastery of Cuzco' which had lapsed to the Crown and were 'worth not more than 400 pesos a year'. Something would also be given to Titu Cusi's younger brother Tupac Amaru to ensure that he would not repeat Titu Cusi's manoeuvre of taking charge as soon as his brother departed.

Titu Cusi sent some Indians back to Lima in April 1565 with a plea for an immediate cash payment to cover the costs of emerging. President Castro reported that the Inca had asked 'not to have to come to Lima to spend his fortune and later have to sell his clothes as his brother Sayri-Tupac had done. He says that [Sayri-Tupac] spent over 10,000 pesos in coming to this city, and in two pitchers and a gold vase that he gave the Marquis [Cañete] and which are said to have cost 5,000 pesos. I replied that he did not have to give *me* anything.... And to please him more I sent him some damask and fancy cloth, and clothed his servants at my own expense.'

The Treasurer García de Melo unfortunately left no descriptions of his missions to Vilcabamba. But the authorities in Cuzco now decided to send an embassy of their own, and their envoy, Diego Rodríguez de Figueroa, wrote an admirably detailed report of his journey. From the diplomatic point of view, Rodríguez's mission appeared to make satisfactory progress in the negotiations already in progress. The Indians who had been with Castro in Lima two weeks earlier returned while Rodríguez was actually with Titu Cusi on May 14, bringing the 'eight yards of yellow damask' and letters setting out the Governor's offer. Rodríguez discussed these proposals with the Inca. It was agreed that Titu Cusi should either remain where he was and accept a garrison of Spanish colonists in Vilcabamba, or that he should emerge within a couple of years to settle near Cuzco or Huamanga, leaving only a Spanish corregidor to govern Vilcabamba.

Rodríguez reported this favourable reception to his superiors in Cuzco. One of the foremost lawyers of Peru happened to be in that city at the time. This was Juan de Matienzo, who had arrived in Peru a few years earlier after many years of training and teaching in the Faculty of Law of the University of Valladolid. He was an oidor of the Audiencia of the Charcas and had come to Cuzco to conduct an enquiry into the conduct of its corregidor.* Matienzo held many theories about the government of Peru and its natives, believing that they should be treated with humanity but firmness. He became intrigued by Rodríguez's reports from Vilcabamba, and decided to go in person to conduct a 'summit' meeting with the Inca.

The meeting was duly arranged, and took place in mid-June 1565 at the bridge of Chuquichaca, the point at which a traveller crossed the Urubamba to enter Vilcabamba.* A group of Spaniards, including Rodríguez, García de Melo and a cleric, remained on the native bank as hostages. The Inca then crossed the bridge with his escort and he and Matienzo drew apart to converse together. Each was fully armed: Titu Cusi with his traditional weapons and Matienzo with sword and arquebus. The Inca preferred not to sit, so the two men stood to talk for some three hours.* Titu Cusi launched into a tearful account of the abuses that had forced his father Manco into rebellion, and explained why the poverty of the Vilcabamba area had forced his followers to raid into Spanish-occupied Peru. He then produced two memoranda: one set out the conditions he asked in return for peaceful surrender, the other detailed the wrongs suffered by himself and his father.

He agreed to accept a Spanish corregidor and priests into his territory, and would himself go to live in Cuzco with his son Quispe Titu, who was to be baptised. In return, his son was to marry Beatriz Clara Coya and to inherit her estates, after Titu Cusi himself had extracted 5,000 pesos a year for the rest of his lifetime. The Inca was also to receive the two towns of Cachana and Canora and to retain the towns he was now ruling, as an encomienda. He was to receive a year's rents forthwith, to be granted full pardon, and to enjoy all dignities and titles appropriate to his rank. Titu Cusi assured Matienzo that 'he made this trip with enthusiasm, in order to avoid the many evils that could result from war' and the Judge came away convinced that 'it must be understood that he truly desires peace'. Matienzo was prepared to accept the Inca's terms forthwith, particularly as he was appalled by what he saw of Vilcabamba across the river: the difficulty of launching an attack into the area, the bad climate, the heat.

The negotiations had a military justification for both the Spaniards and the Incas. The native raids on the Cuzco–Jauja road, which had almost dwindled to a halt during Sayri-Tupac's regency, were revived by his brother. There had been one moment of activity under Sayri-Tupac. This occurred after the defeat of the last settlers' rebellion of Francisco Hernández Girón, when a number of Spanish fugitives joined the Vilcabamba natives and helped to plan raids. One major attack on the Apurímac suspension bridge led to a battle with the Huamanga militia.* That burst of activity in 1555 had inspired the Marquis of Cañete to launch the negotiations that resulted in Sayri-Tupac's emergence. Titu Cusi's raiding increased in intensity until frequent attacks were being made near Huamanga and Jauja, and this now spurred García de Castro into action.*

Titu Cusi himself recalled that 'from one sortie I brought back for my household over five hundred Indians from various places'. Bernabé Cobo described how serious the situation became from the Spanish point of view. Titu Cusi, he said, 'set himself to doing the Christians as much harm as he was able. He attacked the Yucay and many other places, took as many Indians as he could capture to Vilcabamba, and killed travellers. As a result there was no safe place in the districts of Cuzco or Huamanga, and no one could travel from place to place without an escort.' Titu Cusi, in his interview with Diego Rodríguez in May 1565, denied that he had molested Spaniards and claimed that 'the first instruction he had given to his people, whenever they made a raid, was to touch neither the churches nor the crosses. . . . He also said that none of the friars, clerics or soldiers in Peru could complain that he had killed one of them, though he could have done so many times. He could easily have killed two Augustinian friars and two other Spaniards who were shut up in a house at Curahuasi, and many others at many times and in many places.' Rodríguez disputed some of the Inca's protestations of innocence and said that he had himself seen ransacked churches, and 'I told him that in Amaibamba I had seen pieces of the cross, and that [the Indians] had cooked a llama with it.'

In 1564, Titu Cusi's activities acquired a far more menacing aspect for the Spaniards. Castro wrote that he suspected the Inca of helping to foment risings in Chile, Tucumán and among the Juries and Diaguitas 'who are almost his neighbours'. But on 5 December 1564 the son of an encomendero from the Jauja area rode breathlessly into Lima with far more dangerous news. The encomendero, Felipe de Segovia Balderábano Briseño, told García de Castro that he had stumbled across a secret native arms factory in a remote thatched building. One of his yanaconas, a master carpenter called Don Cristóbal Callaballauri, betrayed the plot.

The rebellious chiefs had observed that Spaniards were at their most vulnerable during the celebrations of Holy Week. The carpenter revealed that they planned 'to raise the country in rebellion and kill as many Spaniards as possible during the night of next Holy Thursday [1565], at the time when the holy processions are going through the streets. So many Indian warriors are ready, that for every Spaniard over four hundred Indians will enlist to attack you... In this valley [of Jauja] alone they have assembled over thirty thousand pikes and battle-axes, over ten thousand bows with their arrows, and many halberds and swords. All we carpenters of the repartimiento are making pikes and macana battle-axes... The curacas say that we must all without exception simultaneously enter the streets and squares along which the processions are passing, very silently, to attack you with impetuous ferocity, as men determined to kill their enemies or die in the attempt. Do not doubt that they would easily succeed, for you are unprepared and unarmed on that night, flagellating and illuminating one another.'

Castro was terrified by the news of the plotted rebellion. He immediately ordered the corregidor of the district, Captain Juan de la Reinaga, to make a thorough investigation. Reinaga discovered some eight hundred pikes and large quantities of battle-axes and other weapons, as well as huge stores of native food, much of which was pre-cooked ready for immediate consumption. Two thousand more pike poles had already been destroyed. The plotters had clearly been in contact with Titu Cusi. This was revealed in an enquiry undertaken by another encomendero, Gómez de Caravantes, and by the discovery of a map showing escape routes from Jauja to Vilcabamba. García de Castro wrote to the King that the discovery of the pikes 'fills those of us who know about it with horror'. By discovering the plot in time, Castro felt that he had averted a blood-bath, 'for before it would have been pacified, the majority of the natives would have died'. Castro's men arrested a leading curaca, brother of the powerful Don Felipe Guacra Pucara to whom the King had just awarded a coat of arms and a pension. But sympathetic Augustinian friars released the curaca by night and he escaped to plead his case before the frightened Governor.* He claimed that the pikes were being manufactured as a surprise contribution to the Spaniards fighting in Chile. Castro was unconvinced by this ingenious explanation, but 'dissimulated with him, treating him very well until I can catch the other caciques to learn why these pikes were made'.

Castro was right to fear the manufacture of pikes, for this was the best weapon for stopping a charging horse. The Araucanians of southern Chile discovered the pike and were able to resist Spanish invasion of their country with it. In his alarm, Castro wrote to the King that 'there has been much carelessness in this kingdom. The Indians have been allowed to have horses, mares and arquebuses, and many of them know how to ride and shoot an arquebus very well.' He ordered that all horses or Spanish weapons be confiscated from Indians, with their value refunded to the owners.*

The Jauja rising was apparently part of a concerted rebellion with ramifications in Cuzco, the Charcas and northern Peru. The town council and corregidor of Cuzco were afraid that a conspiracy was afoot there.* In Tucumán, south-east of the Charcas, local Indians had surrounded the Spanish Governor Francisco de Aguirre, and it was believed that he had been killed. In northern Peru, in the wild country of the Bracamoros, where the Marañon turns eastwards and drops into the Amazon basin, forest Indians annihilated a new Spanish settlement called Valladolid and killed its Lieutenant-Governor Captain Francisco Mercado.*

The native resurgence was not solely military. Throughout Peru Spanish missionaries were finding an alarming revival of native religion. At the time of the Conquest, religion had been all-pervading in the Inca empire, but it flourished at different levels. At the top was the official religion, closely identified with the Inca himself. There was the worship of the sun, moon, stars and thunder; of the creator god Viracocha; of the Inca as son of the sun; of the mummies and effigies of his royal ancestors, and the holy hills and caves of the Inca creation legends. There was an official priesthood headed by Villac Umu, with temples in the main cities, convents of mamaconas, and a 'tithe' of part of the produce and herds of each agricultural community. Below this imperial religion, many tribes that were absorbed into the Inca empire still clung to their own tribal deities. Pachacamac was such a shrine; Atahualpa humiliated its priest at Cajamarca and encouraged the Spaniards to desecrate it. In Manco's second rebellion of 1538–9, the Inca gained the allegiance of the Conchucos who fought in the name of their god

Catequil. He was thwarted by the Huanca and therefore destroyed their shrine Wari Wilca.

At a lower level, the common people of Peru were embroiled in a world of spirits and superstition. Almost any circumstance – a double ear of maize, a dream, an accident – represented a supernatural manifestation. Each village was surrounded by a mass of huacas: trees, springs, rocks, caves that had magical significance. Each house had its canopas or household deities, some object displayed in a niche in the wall or carefully wrapped in cloths. The Indians collected talismans of unusual objects, like modern schoolboys. They observed propitiatory rituals throughout their daily lives, sprinkling chicha or coca when ploughing, saying prayers when crossing rivers, making sacrifices at appropriate occasions, and always leaving an object on the pile of stones still to be found at the top of every pass. They lived in awe of their sorcerers, the old men who foretold the future by studying the entrails of animals or the clouds, and were terrified of black magic, spells that could induce pain or love in selected victims. The Incas themselves, for all their sophistication, shared these fears: Atahualpa arranged the destruction of everything he had touched to prevent its manipulation by sorcerers; Sayri-Tupac's decision to leave Vilcabamba was delayed for years by the prophecies of his soothsayers.

The Spanish Conquest of Peru was made in the name of Christianity. Pope Alexander VI had conceded the right to conquer the western Americas to the King of Spain on condition that he arranged their conversion to Christianity. The contract between the Spanish Crown and Francisco Pizarro of 26 July 1529 stipulated that Pizarro's expedition would include missionaries nominated by the Crown. Pizarro duly took some missionaries from the religious orders. He had six Dominicans on his first voyage, although only Vicente de Valverde was present at Cajamarca; Benalcázar brought some Mercedarians including Francisco de Bobadilla when he joined Pizarro on the island of Puná; the Franciscans sent friars, including Marcos de Niza and Jodoco Ricke, during the first years of the Conquest. In his grants of encomienda Pizarro also included the obligation to provide a priest in each repartimiento.

The religious orders soon established monasteries in the principal towns, Cuzco, Quito, Lima, Huamanga and others, and the work of conversion started well. But it was thwarted during the 1530s and 1540s by the disruptions of the civil wars, by a lack of suitable priests (many of the finest had gone to Mexico), by the apathy of many encomenderos (who appointed priests more for their ability as tribute-collectors than for missionary zeal), and by the unchristian behaviour of the more outrageous conquistadores. The church leaders were generally exemplary men, deeply concerned with the welfare of the Indians. Vicente de Valverde was a conscientious bishop of Cuzco until his death in 1541. Two years later Fray Jerónimo de Loayza, first bishop of Los Reyes (Lima), entered the city to rule this new see. He did so for many decades, first as bishop then as archbishop, and became a towering figure in Peruvian affairs. A third bishopric, Quito, was first occupied in 1549 by Garcí Díaz Arias, who had been in Pizarro's palace when he was assassinated. The religious orders produced many good men, but the most outstanding were the Dominicans Fray Tomás de San Martín and Fray Domingo de Santo Tomás and the Franciscan Fray Jodoco Ricke, who imported the first wheat seeds and taught European agriculture to the natives of Quito.

The Peruvians took to Christianity remarkably well. They flocked to mass baptisms even if they understood almost nothing of the crude interpreting of Spanish sermons. They found no difficulty in exchanging the official Inca religion for Catholicism. The striking military successes of the Spaniards were visible proof of the success of their god, and the Indians readily believed that divine intervention had saved Hernando Pizarro's men during the siege of Cuzco. The two official religions had many similarities, with priests, convents, temples and a form of tithe. Both provided religious holidays with the celebrations that the natives loved so much – although the Indians could not understand why Christian priests dampened their festivities by discouraging the traditional drinking and dancing. Both also provided paternalistic priests who recorded the births, marriages and deaths of the community, gave individual guidance and heard confession. There were therefore no martyrs during the spread of the Faith in Peru.

Although the official Inca religion was quickly supplanted, the simpler, more fundamental superstitions could not be uprooted so easily. They survived out of sight of Spanish priests. It was almost impossible for Christian authorities to know which natural objects had religious significance, or to ferret out furtive sorcery or hidden talismans. This basic idolatry was mentioned during the first Council of the Peruvian Church, at Lima in 1551; but that gathering was more concerned with settling procedure for instructing, baptising and administering the docile native congregations. Only during Titu Cusi's reign, in 1565, at about the time of the discovery of the pikes at Jauja, did the church become seriously alarmed by the extent of heathen survivals. This was partly because the priests were getting to know their congregations better, and were being made to learn Quechua. But it was also because of a resurgence of native religions, apparently inspired by the Inca state of Vilcabamba.

A messianic movement whose chief festival was called Taqui Onqoy was secretly preached throughout Peru. The movement was run by a mysterious figure called Juan Chocne or Chono and was preached by tarpuntaes or Inca priests. It did not attempt to revive the full Inca religion. But its priests gathered up the remains of ancient huacas and sacrificed maize and chicha to them in the village squares, amid dancing and shouting. Cristóbal de Molina of Cuzco mentioned one Luis de Olivera, curate of Parinacochas, who surprised a performance of this Taqui Onqoy. The Catholic priests moved urgently to suppress these heathen ceremonies; but they had great difficulty with the revived worship of canopas, the piles of apachitas on the mountain-tops, mummies and llama sacrifice. The extirpation of these petty idolatries was a serious concern for the clergy throughout the latter part of the century. The thought of a religious upheaval on top of the plots for rebellion terrified García de Castro. It made him doubly anxious to negotiate with Titu Cusi and to try to achieve the conversion of the Inca: with the Inca a Christian the heathen revival would presumably lose its inspiration.*

Titu Cusi also had grounds for wishing to avoid war. Although the defences, internal communications and militia of his state were well organised, he knew that Vilcabamba was short of manpower and not able to withstand a determined Spanish invasion. He therefore took seriously any sabre-rattling by the Spaniards. In 1564 a number of natives had been mistreated on the encomienda of one Nuño de Mendoza on the Acobamba river and had fled to Vilcabamba. The corregidor of Cuzco, Dr Juan Cuenca, wrote a strong letter to Titu Cusi demanding their return 'or he would launch the cruellest possible war'. Titu Cusi sent a polite reply, but prepared his warriors and outposts and himself went out to observe whether there were any troop movements along the Spanish highway. A raid from which he returned with five hundred natives ensued, and Dr Cuenca sent an abject letter in which, Titu Cusi said, he 'begged me to let bygones be bygones'. In his second interview with Diego Rodríguez on 15 May 1565, Titu Cusi had said that he knew that Martín Hurtado de Arbieto had asked leave to attack Vilcabamba in the time of the Count of Nieva, and had in fact made some incursions towards Amaibamba.* While Rodríguez was with the Inca the native envoys returned from Lima with letters from Castro. One letter threatened that if the Inca did not immediately accept the terms being offered to him, a commander was authorised to force an entry and make war on behalf of the citizens of Cuzco. Other letters reported the arrest of 'the chiefs of Jauja and of the whole kingdom' for making clandestine arms, and mentioned that the municipalities of Cuzco and of Huamanga wished to make war. 'The Inca and his captains received the news bravely. The Inca rose up and declared that they were not afraid of the Spaniards.' But, for all his bold front, the Inca had no wish for the Spaniards to call his bluff. A letter was in fact on its way from Castro to the King at that very time saying: 'I am determined to make the citizens of Cuzco and Huamanga throw him out of there and settle the place that he holds.'

Titu Cusi therefore gave every appearance of compliance in his negotiations with Juan de Matienzo, Diego Rodríguez de Figueroa, and García de Castro. These Spanish authorities, who overestimated the extent of the Inca's power, believed his claims that their diplomacy had dissuaded him from violent action. 'The Inca disbanded a conspiracy that he claimed to have made throughout the kingdom to attack Spaniards and drive them from it. He restored a quantity of Indian fugitives that he was harbouring from [Spanish] repartimientos. And he refrained from launching an attack on the Apurímac bridge to seize a quantity of gold bars that were being transported across it.'

The cornerstone of any agreement between the Spaniards and Incas was to be the marriage of Titu Cusi's son Quispe Titu to Sayri-Tupac's daughter Beatriz, who was about a year younger than her cousin. Beatriz Clara Coya had been born in 1558, shortly after her parents went to live in the Yucay valley. When her father died, her widowed mother María Cusi Huarcay was left very poor. The Viceroy Count of Nieva appointed various Spaniards to administer Sayri-Tupac's great estates on behalf of his heiress; these custodians received large pensions for their work; but they gave nothing to the Inca's widow. In August 1563, when the daughter Beatriz was five, she was entrusted to the care of the Franciscan nuns of the Convent of Santa Clara,* and shortly afterwards her mother María Cusi Huarcay was befriended by one of the richest citizens of Cuzco, Arias Maldonado. The young widow went to live in the Maldonado household and soon felt secure enough to extract her daughter from the convent and bring her to live there as well.

It soon transpired that the Maldonados were anything but disinterested in their hospitality to the impoverished princesses. One of them, probably Arias, had two daughters by the beautiful and spirited María Cusi Huarcay.* But the Maldonados also had designs on the child Beatriz, who would eventually become one of the richest heiresses in Cuzco. Arias Maldonado had a younger brother called Cristóbal who already had an unsavoury reputation as a seducer.* He now decided to betroth the eight-year-old princess to this Cristóbal, and to arrange that the young man should have intercourse with the little girl to give additional force to his claim. Castro wrote to the King in panic. 'I am afraid that he [Arias Maldonado] might marry her to his brother Cristóbal Maldonado, and I think that this has been done. . . . They tell me that he has been intimate with her, but I do not know whether this is true. This man must not be allowed to have the repartimiento that the girl possesses. For his brother has Hernando Pizarro's repartimiento, and they would become so powerful that no one could oppose them in Cuzco.' Castro ordered that the girl be returned to the convent immediately, and instituted an enquiry into the facts of the matter. By February of the following year the worst was known: 'When Arias had her in his power, as is revealed by the trial, he had her married to his brother' in a ceremony of betrothal in 1566.

The Maldonados had gone too far. The rape of this girl princess made a mockery of the delicate negotiations with the Inca. The Archbishop of Lima wrote that this abuse of the Inca's daughter 'has scandalised the Indians, Sayri-Tupac's relatives, and even more the Inca who has succeeded him' Castro arrested the Maldonado brothers for conspiracy to rebel. Beatriz was hurriedly replaced in the Convent of Santa Clara and her estates were entrusted to the custody of Atilano de Anaya, a respected citizen. On 15 February 1567, King Philip issued a decree from Madrid ordering the President of the Audiencia of la Plata to 'find out what happened . . . do justice in it, and report to us'. But the preamble of this decree had already stated: 'We have been informed that a daughter of the Inca, a girl of eight or nine, . . . was said to have been married to Cristóbal Maldonado.'

The prospect of having Quispe Titu marry Beatriz Clara Coya had a strong appeal for Titu Cusi. He was sensitive that his younger brother Tupac Amaru had a stronger claim than he to be Inca, being the son of Manco Inca and his coya. He was most anxious that Quispe Titu should succeed him, and the boy's chances would be greatly enhanced by marriage to the heiress of Manco's legitimate children Sayri-Tupac and Cusi Huarcay.

Titu Cusi therefore pursued the negotiations with President Castro to the point of signing a formal peace treaty on the Acobamba river in Vilcabamba on 24 August 1566. The Inca was accompanied by his captain-general, Yamqui Mayta, his captain Rimachi Yupanqui (the commander who had caught Manco's assassins) and other military leaders, as well as by his secretary and interpreter Martín Pando. The Spaniards present were the treasurer García de Melo, a priest Francisco de las Veredas and Diego Rodríguez de Figueroa.

The terms of the treaty were remarkably favourable to the Inca. Quispe Titu was to marry Beatriz after his conversion to Christianity, and was to inherit her father's estates as a hereditary fief, with Titu Cusi as his sole guardian. The Inca was to receive an annual income of 3,500 pesos from these estates. But, most important, he was to be allowed to remain in Vilcabamba, holding in encomienda 'the Indians that he now holds, which are many, in the place where he now is'. Titu Cusi insisted that certain Indians who had fled to find sanctuary in Vilcabamba should be allowed to remain, after he had compensated their previous encomenderos. Apart from these, neither Spaniards nor Inca were to receive fugitives from one another's lands. In return, Titu Cusi agreed to accept missionaries and a Spanish corregidor, who was to be the sympathetic Diego Rodríguez de Figueroa. He also agreed to cause no further damage to Spanish territory. If he or any of his captains resumed their raiding, 'war might be waged on them immediately without further warning'. This treaty was ratified by President Castro in Lima on 14 October 1566. The document was taken back to the Inca, who 'approved it, consented and signed it'.

President Castro was pleased with his progress and thought that he would soon have a diplomatic triumph to his credit. He sent a copy of the treaty of Acobamba to King Philip in November 1566 and asked that papal dispensation he obtained for the marriage of Quispe Titu to his first cousin Beatriz. He also suggested that the first Corregidor of Vilcabamba at a salary of 1,500 pesos should be Rodríguez de Figueroa. For, 'although he is not of equal social standing [to García de Melo, Rodríguez] performed very well over this affair, for he went in to the Inca among his Indians when they were at war, in the course of which he overcame many difficulties'. Rodríguez de Figueroa therefore returned to Vilcabamba to assume his new office.

On 9 July 1567 the Inca again ratified the treaty of Acobamba. He performed various rites to the sun and 'then placed his hand on the ground and [swore] to keep the peace agreed with García de Melo and Diego Rodríguez de Figueroa for ever...' The Inca then received Rodríguez de Figueroa as Corregidor and Justice and presented him with a staff as a sign of obedience to the Spanish Crown. He declared that 'he placed himself of his own free will, without reward or coercion, under the power and strength of the Kings of Spain'. He even involved his brothers Tupac Amaru, Capac Tupac Yupanqui and Tupac Huallpa in the submission, and swore that if they refused to comply he 'would despatch them by spear thrusts with his own hands'. And if *he* failed to do so, the Spaniards could immediately make war on him at the expense of his son's estates. The new corregidor, Diego Rodríguez de Figueroa, later described how he had performed a ceremony that echoed Pizarro's occupation of Cuzco. 'The Inca took the staff from my hand and kissed it and touched it to his head, declaring that from then onwards for ever more he would give allegiance to the Kings of Castile. As a token of this, I erected a gallows on a hill and took possession of that gallows by making two knife cuts in it.' The gallows was the traditional *picota*, erected 'so that delinquents on whom the said Rodríguez should do justice on behalf of His Majesty could be placed upon it'. Titu Cusi also agreed to 'march with his men and weapons to the service of His Majesty if summoned'.

The ratified treaty of Acobamba was immediately implemented. The first missionaries crossed the Apurímac and entered the Inca's territory with the new corregidor. They were two priests attached to the cathedral of Cuzco, Antonio de Vera and Francisco de las Veredas. They rapidly instructed the ten-year-old Quispe Titu, and baptised him less than a fortnight after the ceremony of ratification. The baptismal note has survived: 'In the church of the town called Carco in the province of Vilcabamba, I, Antonio de Vera, declare that I baptised on 20 July 1567 Don Felipe Quispe Titu, son of Inca Titu Cusi Yupanqui and of Chimbo Oclla Coya. Francisco de las Veredas and Diego Rodríguez de Figueroa were godfathers.' The baptism was an agreed preliminary for the marriage of Quispe Titu to the convent-educated Beatriz. Titu Cusi himself therefore 'brought his son to the church and presented him to receive the holy baptism'. Titu Cusi also made a declaration before Rodríguez de Figueroa to verify the 'certainty' of his own claim to rule, 'and the command and rule that he exercises at present succeeds by direct line to this heir'.

After these two successes, President García de Castro felt that something had really been achieved. In September 1567 he wrote to the King admitting that Titu Cusi's elusive behaviour had worried him. 'He grew so proud that I thought he would never come, for having almost concluded the agreement he went against me.' Now, however, the Inca had accepted Spanish missionaries and a corregidor, had had his son baptised and had performed his formal act of surrender. In no time he would follow his brother into Spanish Peru. Castro boasted: 'If we had had to bring him out by war it could not have been done for 40,000 pesos' and the King wrote in the margin of the letter 'This is well done.' Three months later, the President was still writing 'Your Majesty will appreciate the great service that has been done to Your Majesty by the reduction of the Inca to his service.' King Philip did what was required of him, although it took time. He wrote in August 1568 to his ambassador in Rome telling him to request papal dispensation for the marriage of Felipe Quispe Titu and Beatriz, 'who are children of four brothers and sisters who married while infidels according to their custom'. The King stressed to the Pope the great benefits that would result to Church and State through the marriage of these Inca cousins. On 2 January 1569 King Philip gave formal ratification to the agreement between Licenciate Castro and the Inca Titu Cusi.* The Pope issued the necessary dispensation, and the various documents were taken out to Peru by the next Viceroy, Francisco de Toledo.

Castro's self-congratulation was premature. Titu Cusi never left Vilcabamba, and Diego Rodríguez de Figueroa did not return to take up permanent residence in the native province. By the time Castro relinquished the government of Peru to his successor in late 1569, Titu

Cusi was as far as ever from leaving his refuge. Not that there had been any rupture. The Inca sent a friendly letter of farewell to the man with whom he had been negotiating for five years. He thanked Castro for the favours he had shown him, and said how sorry he was that he was about to relinquish the government of Peru.* A few months later the Inca dictated his famous narrative, and blandly asserted that he had 'ratified completely' the peace of Acobamba. He had in fact survived a decade of negotiation with the Spanish authorities. He had raised their expectations and restrained their armies with annual exercises of diplomatic or religious goodwill. But he was just as firmly established in Vilcabamba in 1570 as he had been when Sayri-Tupac died.

A zealous friar teaching Christianity

16

VILCABAMBA

TITU Cusi's Vilcabamba had been an independent state for over twenty-five years when he became Inca. It was tightly, obsessively dominated by the militaristic and religious traditions of the old Inca empire, and Titu Cusi himself ruled both church and state. He told Rodríguez that he was 'high priest in what we call spiritual affairs', and he clearly took this role seriously.

The Inca religion had three main bases of worship: the sun and celestial bodies, natural landmarks accepted as shrines, the Inca and his ancestors. All of these were revered in Vilcabamba, but 'the principal god they worshipped was Punchao, which represents the day.' Punchao literally means the moment of daybreak, when the first rays of the sun break over the horizon – an instant of clear beauty in the unhazy air of the high Andes. Manco's followers had succeeded in spiriting the most important golden disc of the Punchao away from Cuzco, although other similar images, such as the one that Mancio Sierra de Leguizamo claimed to have gambled away before sunrise, had vanished into the conquistadores' furnaces. The Viceroy Francisco de Toledo and his commissioned historian Pedro Sarmiento de Gamboa, regarded the sun image as part of a theatrical trick that helped the Inca family dazzle its subjects. 'The idol Punchao . . . is the image of the sun that issued the laws of their religion from Cuzco to the entire kingdom . . . ever since the time of the seventh Inca, who established its cult and religion in order better to tyrannise this barbaric land. The province of [Vilcabamba] and its neighbours were preserved through it.' The Incas had always identified themselves with the sun, and the Inca himself, like the Pharaoh, was worshipped as the son of the sun.

Titu Cusi kept his sun image in the innermost depths of his territory, in the jungle city of Vilcabamba. Sarmiento de Gamboa said that the Punchao was the size of a man, and Toledo described it as made of cast gold: 'it has a heart of dough in a golden chalice inside the body of the idol, this dough being of a powder made from the hearts of dead Incas. . . . It is surrounded by a form of golden medallions in order that, when struck by the sun, these should shine in such a way that one could never see the idol itself, but only the reflected brilliance of these medallions.'

Diego Rodríguez reported frequent examples of sun-worship during his visit to Pampaconas in 1565. When the Inca Titu Cusi first arrived on the plateau where his throne was prepared, he gazed towards the sun and performed a mocha, the Inca act of worship similar to our blowing a kiss. All the captains who entered the ruler's presence made a formal reverence to the sun and to the Inca. This obeisance was even performed by a contingent of six or seven hundred forest Indians – the near-naked savages known to the Incas and Spaniards as Antis, because their rain forests lay in the eastern quarter of the empire, the Antisuyo.* And at times during Rodríguez's visit, proceedings would be interrupted for religious observances: 'All the Indians rose and began to worship, and the Indian captains began to make offerings, each for himself, with daggers of bronze or iron in their hands.'

During the Inca empire the rocks, hills, springs and caves around Cuzco were endowed with magical significance as huacas, and arranged in lines called ceques similar to the stations along a Christian pilgrimage route. The natural features of Vilcabamba's turbulent landscape became shrines in a similar pattern. The most important of these shrines was an oracle called Chuquipalta, near Vitcos and Puquiura. The indignant Augustinian hagiographer Antonio de la Calancha wrote – not without a hint of awe – that the site contained 'a temple of the sun, and inside it a white stone above a spring of water. The devil appeared here, and it was the principal mochadero – the common Indian word for their shrines – in those montañas. . . . There was a devil, captain of a legion of devils, inside the white stone called Yurac-rumi, within that sun temple.' He and his band persecuted any who did not believe in the shrine, and 'many died from the horrible abominations that it wrought on them'. The devil 'gave answers from a white stone outcrop, and was visible on various occasions. The stone was above a spring of water, and [the natives] worshipped the water as something divine.'

What is almost certainly the stone of Yurac-rumi (which means simply 'white stone') was discovered by Hiram Bingham at a place in the Vilcabamba valley now known as Ñusta España. It was a great outcrop of white granite, twenty-five feet high and fifty-two feet long. This white rock was covered in complex Inca cuttings: rows of rectangular seats, ten projecting square stones, a cave with niches, a flattened top, platforms on the sides, and artificially-enlarged cracks probably intended to channel offerings of chicha or blood from llama sacrifices. On the eastern and southern sides lay a swamp and pool of eerily dark water, and the entire place, in a wooded ravine, had an other-worldly atmosphere, fitting exactly Calancha's description of the diabolical shrine. 'We were at once impressed', Bingham wrote in his report to the American Antiquarian Society, 'and convinced that this was indeed the sacred spot, the centre of idolatry in the latter part of the Inca rule' (plate 46).

One striking demonstration of Titu Cusi's religious observance occurred during his ratifications of the treaty of Acobamba, when he received Rodríguez de Figueroa as corregidor in July 1567. 'The Inca then stood up. Looking towards where the sun then was, with his arms outstretched and his hands open with humility, in the manner of a reverence, he said: "I swear by thee, O Sun, who art creator of all things and whom I hold to be God and worship; and by thee, Earth, whom I regard as mother from whom is produced all sustenance for the support of man." '

That same treaty of Acobamba provided for the admission of Christianity into Vilcabamba and into Titu Cusi's own life. It even insisted that 'Indians who were idolators should not perform their rites and ceremonies where the priests could see them.' The first priests to enter the remote heathen state were Antonio de Vera and Francisco de las Veredas, and the Inca immediately employed them in the baptism of his son Quispe Titu. Vera remained alone after the baptism and was supported in his work by an annual salary of 600 pesos. This money came from the treasury, since Governor García de Castro had accepted the missionary activity as a royal expense. It was paid in half-yearly instalments, and of the total 600 pesos '400 are for the alms of the said Order and 50 for the wine and wax with which the holy office is celebrated; the remaining 150 are for his food, since under the agreement the Inca does not give him any'. Father Vera was also helped at the outset by Titu Cusi himself. He was permitted to build a church at Carco above the Apurímac at the western edge of Vilcabamba province. 'The Inca gave an image, a tall, large wooden cross next to the door of the church. Father Antonio de Vera preached the holy gospel to the Inca and to the people who live with him.'

The Inca himself had for some time been intrigued by Christianity. Despite his position as chief priest of the Inca religion, Titu Cusi had always been reasonably tolerant of Christianity, unlike some of his more militant followers. Diego Rodríguez, at his first meeting with him, had boldly asked leave 'to say something in praise of God and of our holy Christian religion. The Inca gave permission. I then said many things which were the fruit of study in books I had brought.' Rodríguez on that occasion went too far in exhorting the Inca's captains to adopt Christianity. 'The Inca himself was moved to anger. He spoke very fiercely to me, saying that no Spaniard who had entered his territory had dared to treat of these things and to praise our Lord Jesus Christ, nor had they in the time of his father. It was great insolence on my part, and he was disposed to order me to be killed.' A few days later, in mellower mood, Titu Cusi allowed Rodríguez to go further. Titu Cusi had in fact been christened during his boyhood in Cuzco, and Rodríguez reminded him that 'I had seen in the baptismal book of the principal church (in Cuzco) that the Inca had been baptised and named Diego. He told me that it was true, that he was a Christian, and he confessed it before the Indians. He said that they had poured water over his head, but that he did not remember the name.' On the strength of this Titu Cusi agreed to Rodríguez' request to assemble the Indians who were Christians, and even issued orders for 'a great cross to be made' and 'ordered them to come and hear me preach'. The Inca was apparently seized by a sudden enthusiasm for Christianity, for he 'wrote to the Mercedarian and the Franciscan friars, to ask two of them to come and preach in his territory, and that he would give them the products of the land in exchange for their doctrine. He also wrote to the Judge Matienzo, thanking him for having sent me, for I had made known to them the law of our Lord Jesus Christ.'

That was in 1565. Two years later, inspired by Father Vera and also because Governor Castro 'wrote me many letters begging me to be converted a Christian', Titu Cusi decided to investigate for himself the Spaniards' powerful religion. He began by enquiring of Rodríguez and his secretary Martín Pando 'Who was the leading figure among the ecclesiastics in Cuzco, and which religious order was most respected and most reputable? They told me that the most reputable and most authoritative and flourishing order was that of lord Saint Augustine, and its Prior – I am referring to the monks residing in Cuzco – was the most important figure of all who were in Cuzco. When I learned that this was so, I became greatly enthused about that order and religion above any other, and decided to write many letters to the said Prior asking him to come and baptise me in person, since it would give me great pleasure to be baptised by his own hand, he being such an important person.'

The Prior Juan de Vivero penetrated Vilcabamba in August 1568 accompanied by one of his friars, Marcos García, and by two prominent citizens, Gonzalo Pérez de Vivero and Atilano de Anaya, the man now caring for Beatriz Clara Coya. The Inca travelled up from the city of Vilcabamba to meet them at Huarancalla, above the Vilcabamba river between Lucma and Vitcos. After a fortnight of instruction in the articles of the Faith, Titu Cusi learned the catechism and was baptised by Prior Vivero, with Gonzalo Pérez de Vivero acting as godfather and Angelina Siza Ocllo as godmother. Titu Cusi adopted the name of the then Governor, and became Don Diego de Castro Titu Cusi Yupanqui. One of his wives was baptised at the same time, as Angelina Polan-Quilaco. After a further week of religious instruction, Prior Vivero and the two laymen departed, leaving Friar Marcos García alone in Vilcabamba – Antonio de Vera had left the area when the Augustinians arrived. The Augustinian chronicler Calancha unkindly made no mention of Vera's eighteen months of missionary work before the arrival of the Augustinians.

Marcos García appears to have been a dedicated but stern and inflexible ecclesiastic. He started his mission on the crest of the Inca's Christian enthusiasm. Calancha reported that Titu Cusi used his conversion as an excuse for 'holding a great celebration at finding himself a Catholic', and the Inca himself wrote: 'I informed my Indians of the reason why I had had myself baptised and had brought those people to my land, and the results that men derived from being baptised, and why this father had remained in the land.'

García was sure that the entire province was about to follow its ruler's lead. He remained in Huarancalla until October, teaching and instructing and 'baptising some children with their parents' consent'. He then went across to the western slopes of the Cordillera Vilcabamba, where he spent the four months of the rainy season with the mestizo Martín Pando, proselytising in Carco and the other villages towards Huamanga and the Apurímac – the area already penetrated by Antonio de Vera. He established churches in three villages and crosses in five others, baptised ninety people, and left young men behind to continue the instruction. García then returned to Huarancalla and spent the months of March to September 1569 'baptising and teaching the Indians of the entire region'. Just before he had left for the western area, messengers had reached the Inca from his remote forest tributaries, the primitive Pilcosuni, to tell him that they were prepared to receive Christian missionaries. The Inca wrote to Prior Juan de Vivero: 'I was delighted that these poor people should wish to hear the evangelic law and to follow in my footsteps. I sent to tell them that ecclesiastics from your devout order would go to convert them to the faith of Our Lord.'

Somehow the hoped-for mass conversion never took place. According to Calancha, Titu Cusi had initially been attracted by Marcos García's 'lack of greed for land that contained silver, and his extreme chastity in a province where dissolution held sway'. But García's stern teachings soon lost their appeal. The friar caught some of his boys practising idolatries and 'punished them with ten or a dozen strokes'; their fathers resented this corporal punishment – which the Incas reserved for serious criminal offences – and complained to the Inca; Titu Cusi summoned the priest and reproved him, and Marcos García had to apologise. More serious friction was caused by García's fire-and-brim-stone sermons. He reprimanded the Inca and his baptised subjects for practising polygamy, for having drunken festivities and for continuing to visit the ancient shrines. There was a reaction towards the more lenient ancient religion.

A climax in Titu Cusi's own attitude came when he took to wife a second ñusta who had also been baptised as Angelina. Marriage to two Christian wives was bigamy to García. The second wife, Angelina Llacsa, later admitted that her husband had taken a dislike to Friar Marcos 'because he reprehended him with some liberty'. Calancha was more direct: 'The servant of God reprimanded [the Inca] with apostolic zeal' and Titu Cusi became very angry. He resented religious restraints on his marital adventures just as much as Henry VIII had done forty years earlier.

In September 1569, about a year after the Inca's baptism, another Augustinian called Diego Ortiz went to join Marcos García in the province of Vilcabamba. The Inca was pleased by this new arrival, for Ortiz apparently had a more friendly manner. García himself was by now becoming disillusioned by his thankless task and was anxious to return to Cuzco. Ortiz, by being more permissive, soon became a close confidant of the exuberant Inca, 'to such an extent that [Titu Cusi] made a celebration whenever he visited him, declaring that he loved him like a brother. He gave him presents of birds and of their types of food.' García had established a church at Puquiura, and Ortiz now obtained permission to build one at Huarancalla.

Ortiz's church flourished, and the Indians liked him because he 'taught, cured and clothed them'. He gained a reputation for medical skill by treating highland Indians who were feeling the effects of the change to the hot forests of Vilcabamba, and also by curing the ailments of the Manarí and Pilcosuni jungle Indians.* Poor Marcos García, on the other hand, was still fulminating against the natives' beloved drinking-parties, which he took to be the cause of their 'homicides, sodomies, patricides and cases of incest by the dozen'. He became increasingly unpopular and suspected that the local chiefs were trying to poison him with herbs. He set off for Cuzco, but the Inca was angry at this unauthorised departure and sent five natives to arrest and bring him back. Diego Ortiz hurried over from Huarancalla and joined García in his appearance before the Inca at Puquiura.

After reproving García, Titu Cusi made an expansive offer to the two friars. ' "I want to take you to Vilcabamba, for neither of you has seen that town. You will go with me, for I wish to entertain you." They left next day in the company of the Inca. The fathers had wanted [to make this journey] and had tried to go to Vilcabamba to preach, for it was the largest town. It contained the university of their idolatries and the witch-doctors who were masters of their abominations.'

The journey was made during the rains in early 1570, and part of the road was flooded. Calancha seized on this detail in the contemporary reports and magnified it into an outrage prepared by the natives to humiliate the friars. 'The two ecclesiastics thought that it was a lake, but the Inca told them: "We all have to pass through the midst of this water." O cruel apostate! *He* was travelling in a litter while the two priests were on foot and barefoot.... They slipped and fell, and there was no one to help them stand. They held one another's hands while the sacrilegious [natives] roared with laughter at their expense... They came out on to dry land frozen and covered in mud.'

Titu Cusi arranged quarters for the friars outside the city of Vilcabamba, for he did not wish them to see 'the worship, rites and ceremonies in which he and his captains were engaged every day with the witch-doctors'. The famous *Relación* was prepared while they were all in Vilcabamba. It was originated by Titu Cusi, translated and dictated by Marcos García, written down by the secretary Martín Pando, and witnessed on 6 February 1570 by Diego Ortiz and a number of native chiefs at 'San Salvador de Vilcabamba'. This document gave Titu Cusi's more tactful explanation for his restrictions on the missionaries' activity in the capital city. 'They have not baptised anyone here because the people of this land are still very new in the things that must be known and understood concerning the law and commandments of God. I shall arrange that they learn them little by little.'

Calancha told a sorry story of the friars' tribulations at Vilcabamba. It is worth repeating for its bathos: Calancha in his indignation was quite unaware of the comic aspect of the anecdote. The heathen priests of the city, he said, decided to test the friars' chastity. 'They sought out the most beautiful native women – not mountain women, but ones from the temperate valleys where the women are the fairest and most elegant in the land.... The native women attempted everything that the devil could teach them, employing the greatest wiles known to sensuality and the most dangerous graces known to dissipation. But the holy men defended themselves so valiantly that the women returned crestfallen.' The witch-doctors now tried dressing their sirens in mock monastic habits. Two women went in black habits, and when they were rebuffed, two went in white habits. 'They entered as far as the beds – for Indian houses and their inns or tambos do not have doors. ... This battery of women continued day and night with changes of habit and relays of different Indian women. If the two religious left their house for the country, the women sought them out. And if they came to the town to preach against their dissolute ways and to deplore women wearing monks' habits, the battery did not cease, but invented new sensualities and stirred up terrible temptations.' Calancha admitted another explanation: 'Juana Guerrero, wife of the Inca's secretary Martín Pando, declared that the Inca sent these pairs of Indian women wearing habits as a joke, to make fun of the priesthood and the habit of St Augustine, rather than from any desire to rob them of their chastity.'

There is a far simpler explanation for the masquerading women. Vilcabamba lay in the forest, in the territory of the Campa Indians. The great nineteenth-century traveller Antonio Raimondi wrote that the Campas 'do not go about naked, but are constantly covered with a wide and long sack of cotton cloth woven by the women. This sack reaches to their ankles and is white when new. But it rapidly turns yellowish and also reddish as it is dirtied by the annatto dye with which they paint their faces ... In addition to this sack, many use a sort of capuchin hood to cover their heads ... The women, in general, laugh readily, particularly those of tender age ...' This would explain exactly how the giggling women came to be dressed in monastic garb of the colours of different religious orders.

Vilcabamba, with its witch-doctors and seductive transvestites, was too much for the two friars to tackle. They asked permission to return to their flocks in Huarancalla and Puquiura, and walked back by the flooded road. Once in Puquiura, they determined on a daring confrontation with the native religion. They rounded up their Christian flock and gave each of their boy acolytes a piece of firewood. With a cross at their head they brazenly marched the faithful from Puquiura up to the shrine of Chuquipalta which contained the white rock of Yurac-rumi. They then set fire to the rock temple and burned and exorcised the pagan site. There was uproar among the supporters of the Inca religion. 'The Inca's captains were furious and planned to kill the two ecclesiastics with their spears, thinking nothing of cutting them to pieces. They arrived at the town wishing to give vent to their fury.' The Inca and his wife also hurried to Puquiura, but the Christian community in this part of Vilcabamba was strong enough to save the friars. Their only punishment was that Ortiz was to return to Huarancalla and García was expelled from Vilcabamba for good. The Inca's anger was in fact remarkably short-lived: Ortiz was soon forgiven and restored to favour.

The viceregal authorities approved of the missionaries' work, and it was the royal treasury – not the church – which paid to send vestments and church ornaments into Vilcabamba. An entry in the treasurer's accounts for December 1569 provided a carefully itemised list of ornaments: a damask altar cloth; damask chasuble with gold border, stole, maniple, amices and singlet; tabernacle screen; silver wine jugs, chrismatories, chalice and paten; paintings, candelabra, missals, bells and so forth. These were sent as a reward and encouragement for Diego Ortiz's brave and solitary mission.

The Augustinian Diego Ortiz thus continued to live and preach in the remote valleys of Vilcabamba throughout 1570. He and his predecessors Antonio Vera and Marcos García were the only Europeans to live for long in this forbidding remnant of the Inca empire, apart from the fugitives from the battle of Chupas twenty-five years earlier. Each lived alone, except for the few months during which García and Ortiz were in the area simultaneously, and each bravely taught the beliefs of his religion in the face of militant opposition by many of the Inca commanders. It says much for their missionary zeal that they continued with what must often have seemed a miserable, unrewarding task. But it was thanks to Titu Cusi's personal protection that they were allowed to do so.

Every glimpse that we have of life in Vilcabamba shows that much power lay with a caste of military commanders. This is hardly surprising. The small territory was in a constant state of cold war, threatened at all times with a Spanish invasion. The Inca empire itself had been militaristic, noted more for its conquests, administration and civil engineering than for artistic or cultural achievement. Vilcabamba was a microcosm of that empire and the army and church were naturally the strongest repositories of tradition and reaction. With its shortage of manpower, Vilcabamba could afford only farmers and soldiers: there were few skilled artisans to spare for the arts, for elaborate masonry, weaving or metalworking.

Although most of the military leaders of Manco's rebellions of 1536–7 and 1538–9 had been executed by Francisco Pizarro in the Yucay valley, a new generation of commanders emerged in the guerrilla warfare of the 1540s. These were the men who destroyed Manco's assassins and governed Vilcabamba during Sayri-Tupac's minority. They probably saw in Titu Cusi the most aggressive and capable of Manco's sons. They therefore agreed to the departure of Sayri-Tupac – without the royal fringe – to the easier life of colonial Cuzco, and accepted the relegation of the weaker Tupac Amaru to the seclusion of a temple in Vilcabamba city.

When Diego Rodríguez visited Vilcabamba in 1565, there were endless incidents of petulant hostility from the military. When he first arrived at the bridge of Chuquichaca he hoisted a flag of truce on the eastern bank and waited for some days, plagued by the biting flies that still infest that hot canyon. Titu Cusi's first reply was a letter saying that 'he did not want any Spaniard to enter his territory either in peace or in war . . . if I did his captains would kill me'. Ten native warriors appeared, wearing magnificent uniforms with plume headdresses and masks on their faces. 'They . . . asked if I was the man who had had the audacity to wish to come and speak to the Inca. I said yes. They replied that I could not fail to be much afraid.' In the end they relented and permitted Rodríguez to enter.

After he had swung across the Urubamba in a basket, and was travelling up the Vilcabamba valley with ten of the Inca's men, a hundred native soldiers appeared as an additional escort. They told Rodríguez that had the Inca not 'given orders for me to enter, they would have killed me there and then. Each began to brandish his weapons, calling the Spaniards bearded cowards and thieves. Others said, "May we not kill this little bearded one, to avenge what his brethren have done to us!" I appeased them by saying it was true that when the Spaniards came they did much harm. . . . I then gave them some drink and . . . needles, ornaments, knives and other things, so we made friends.'

At Vitcos, Rodríguez saw the heads of the seven renegades who had killed Manco: these were still exposed as a symbol of resistance to Spain. Whenever he met native military commanders he found them dressed in full Inca regalia. Two orejones who accompanied the Inca carried halberds and were dressed in feathered diadems with much gold and silver ornament; some carried lances decorated with feathers of many colours and others carried battle-axes; and 'all wore masks of different colours'.

There were belligerent incidents even when Rodríguez was with the Inca. Titu Cusi admitted that if any Christians were to remain in Vilcabamba, 'his people might kill them from fear and terror, without his permission to kill'. On one occasion 'a little Indian entered. After making reverences to the sun and the Inca, he came towards me brandishing a lance and raised it with great menace. He began to cry out in Spanish "Get out! get out!" and to threaten me with his lance.' When a letter arrived from Governor Castro in Lima containing threats of violence, even Titu Cusi joined in the demonstrations of defiance. 'He said . . . all he could think of in abuse of the Spaniards and in praise of his Indians. Then all the Indians arose and began to worship . . . Some said they would kill four Spaniards, others five, others six, and others ten. One of them, named Chinchero, said that as I was there why was he not ordered to kill me, for he wanted to stab me with the dagger in his hand. The Inca kept silent, answering nothing, so the Indian went back to his seat.'

There was another such display the following afternoon. The Inca 'began to boast, saying that he could himself kill fifty Spaniards, and that he was going to have all the Spaniards in the kingdom put to death. He took a lance in his hand and a shield, and began to act the valiant man, shouting, "Go at once and bring me all the people that live behind these mountains. I want to go and fight the Spaniards and to kill them all, and I want the wild Indians to eat them." About six or seven hundred Anti Indians then marched up, all with bows and arrows, clubs and battle-axes. They advanced in good order, making reverence to the sun and to the Inca, and took up their positions. The Inca again began to brandish his lance and said he could raise all the Indians in Peru: he had only to give the order and they would fly to arms. Then all those Antis made an offer to the Inca that, if he wished it, they would eat me raw. They said to him, "What are you doing with this little bearded one here who is trying to deceive you? It is better that we should eat him at once." Then two renegade Inca orejones came straight at me with spears in their hands, flourishing their weapons and shouting, "The bearded ones! Our enemies!" I laughed at this, but at the same time commended myself to God. I asked the Inca to have mercy and protect me. And so he rescued me from them and hid me until morning.' Next day the atmosphere had cleared: 'The Inca and all the captains began to laugh heartily at what had happened the day before and asked me what I thought of yesterday's festival. . . . They explained that it was only their fun.'

Spanish writers equated the leading Inca commanders with their own civil and military ranks. They referred to the most important as 'governor' or 'captain-general' and to the military field commander as 'maese del campo'. The governor at the time of Rodríguez's visit was Yamqui Mayta. He marched immediately behind the Inca and sat on his right hand; he had a personal retinue of 'sixty or seventy attendants with silver plates, lances, belts of gold and silver, and the same clothes as were worn by all who came with the Inca'. He embraced Rodríguez in the same way as Titu Cusi. In the negotiations for Titu Cusi's possible departure to Spanish-occupied Peru, it was a condition that Yamqui Mayta would accompany the Inca. Yamqui Mayta died soon after Rodríguez's visit. His successor in the top administrative position was Huallpa Yupanqui, who had been christened Don Pablo – he witnessed Titu Cusi's *Relación* under this name.

Rodríguez mentioned the presence of a 'maese del campo' but did not name him; we know from various sources that this position was held, a few years later, by Curi Paucar. It was a young captain called Paucar who had greated Diego de Almagro on his return from Chile with a moving speech about the provocation that had caused Manco's first rebellion. He revealed his passionate royalism. 'We feel most deeply in our hearts that a natural lord given us by the sun, honoured, loved, revered and obeyed by us, has been treated and insulted like the least among us.' It was Paucar who turned against Almagro and attacked his men at Calca. Paucar was later one of the victors in the battle behind Jauja during the second rebellion. Curi Paucar was among the orejones who accompanied Sayri-Tupac to Cuzco and was mentioned in that Inca's will. This defiant general inspired the military revival in Titu Cusi's Vilcabamba.

Another leading figure, described by Calancha as 'one of Titu Cusi's close friends' was Don Gaspar Sullca Yanac; he also witnessed the Inca's *Relación*.* The mestizo secretary Martín Pando was also a close confidant of the Inca. He naturally acted as intermediary in the various meetings between Rodríguez and Titu Cusi, and wrote down the 1570 *Relación*. When Diego Rodríguez first saw him he was carrying a sword and shield and wearing 'Spanish clothes and a very old cloak'.

It is difficult to estimate how many warriors were being led by these commanders. Rodríguez, an admirably dispassionate observer, mentioned seeing various contingents totalling several hundreds, and it is likely that a sizeable portion of the Inca's forces would have accompanied him and his commanders to the Pampaconas meeting. Francisco de Toledo, in his report to the Council of the Indies of 25 March 1571, estimated that there were five hundred Indian warriors in Vilcabamba, although Yamqui Mayta once told Rodríguez that he had been planning a great raid with '700 Antis and 2,000 other Indians'.

In addition to the mountain Inca troops in their short tunics, Titu Cusi could also call upon various forest tribes, the 'people behind these mountains' of whom he boasted to Rodríguez. These jungle Indians had an unsavoury reputation among Spaniards and Incas alike. Diego Rodríguez described them as 'Antis who eat human flesh' and was alarmed by their threats to eat him raw; Garcilaso mentioned two hundred 'carib' troops sent by Sayri-Tupac to greet Juan Sierra in 1557 – carib was the common term for cannibals.* Even Titu Cusi admitted to the Augustinian Prior Juan de Vivero that the Chunchos 'have only one blemish, which is that they eat human flesh'. The Incas themselves had suffered their worst defeats at the hands of jungle Indians, and the Spaniards soon lost their stomach for ventures into the rain-forests. Thus today, four hundred years later, new tribes are still being contacted by the Brazilian Indian Protection Service and by the missionary organisations working on the Peruvian-Brazilian border not far to the north of Titu Cusi's stronghold.

The names of the forest tribes become hopelessly garbled in Spanish accounts. Juan de Matienzo, the official who met Titu Cusi at the bridge of Chuquichaca, said that the Inca possessed 'much land and many people: for instance the provinces of Vitcos, Manarí, Sinyane, Chucumachai, Niguas, Opatare, Pancormayo in the mountain range that runs towards the north sea, and the provinces of Pilcosuni towards the region of Ruparupa, and of Huarampu and Peati, and of Chiranaba and Ponaba. All these obey him and pay tribute to him.' The names that occur most frequently in other sources are the Pilcosuni and Manarí. The Inca's hold over them was tenuous: in his letter of 24 November 1568, Titu Cusi told the Augustinian Prior that two messengers had recently come to Vilcabamba to say that the Pilcosuni would accept Christian missionaries. But Titu Cusi implied that he had no real power over them. The Pilcosuni lay to the north-west of Vilcabamba: Toledo said that their territory was 'at the back of Jauja, to the east, bordering on the land which Titu Cusi Inca now has in revolt'. Manco himself had visited the Pilcosuni at the invitation of their chiefs immediately after his destruction of the idol of Wari Willca, which was the chief shrine of the Jauja Indians. Baltasar de Ocampo described the Pilcosuni as numerous and very warlike, 'a hill people, in a country of very great mineral wealth [with] valleys where there was a marvellous climate for the cultivation of grains and sugar-cane'.

Titu Cusi had a firmer hold on the Manarí who lived to the east of the Pilcosuni. This was a purely forest tribe. When Ocampo visited Manarí territory he was given typical Amazonian jungle food by the tribe. Matienzo listed the Manarí first among the tribes subject to Titu Cusi, but Toledo, in his report of 1 March 1572, said that the 'Manarices and Anaginis Antis contain many people, some of whom serve the Inca but others of whom have rebelled against him and have come to me.' Both Matienzo and Toledo mentioned the Opatari, whose territory lay east of Vilcabamba, across the Urubamba in the forested hills between the Manarí and Cuzco.

The valleys of the Vilcabamba and Pampaconas rivers, which formed the heart of Titu Cusi's kingdom, offered spectacular changes of scenery. Gleaming through breaks in the clouds were the snows of Salcantay, Soray, Pumasillo and Suerococha, the final peaks of the Andes. In places the forests crept up sheltered valleys almost to the level of the permanent snows. On other, more exposed mountainsides there were desolate pastures where llamas could graze on mosses and long, slippery ichu grass. This was a land of vertical planes, where different levels of climate and vegetation superseded one another like geological strata on the steep valley walls. Below the grazing lands of the high glacial valleys was an area where the Incas' descendants could plant their maize and potatoes in milder conditions. The wind-driven mists from the canyons were fragmented here, and there were dizzying views from hill spurs into the valleys thousands of feet below. Choqquequirau is a ruin on such a spur, jutting out above the Apurímac on the western side of the

Cordillera Vilcabamba. Seen from its terraces, the thunderous river is no more than a silver-grey line creeping along its gorge far below; and looking downstream, the darkly-matted valley walls disappear into a hazy distance like the background of a Leonardo painting. Baltasar de Ocampo described Vitcos – which overlooked the Vilcabamba river on the eastern side of the cordillera – as being 'on a very high mountain, from which the view commanded a great part of the province of Vilcabamba'. And García de Loyola mentioned an Inca fortress that was 'on a high eminence surrounded by rugged crags and jungles, very dangerous to ascend and almost impregnable'. Most visitors to Peru see Machu Picchu, which is perched on a narrow saddle of rock high above a hairpin curve of the Urubamba. The granite sugarloaf of Huayna Picchu towers above the ruin, and the surrounding forested hillsides are often gripped by shrouds of low clammy cloud. Such scenery makes Machu Picchu one of the world's most eerily beautiful ruins.

The valleys below the spurs occupied by the fortresses of Vilcabamba were tightly wooded, not with the tall trees of the Amazonian rainforests, but with smaller, gnarled trees and dense vegetation clinging to the steep slopes. Many of the branches are loaded with gloomy, dripping moss, and the woods are dark and sombre. There are snakes, particularly bushmasters and corals, many creepers, and canes such as the long, springy zigua. Close to the rivers are thickets of *Nicotiana tomentosa*, the tree tobacco of the Andes, and Machu Picchu boasts a giant slipperwort with great flat leaves and a brilliant yellow flower. The terraces of the ruined city are covered in wild strawberries to distract the greedy visitor, and there are many fine orchids in the surrounding forests. With its sweeping range of vegetation, the Vilcabamba area could offer the Inca survivors most of the plants they wanted, from potatoes, quinua and maize on the upper hills, to coca and tropical fruits in the warm canyons. But there were hardships to be endured in return: an altitude lower than the Inca homelands, a climate plagued by a long rainy season with fogs and violent electric storms, and Amazonian heat.

17

TITU CUSI AND CARLOS INCA

THE neo-Inca state of Vilcabamba owed its existence to the presence there of Manco Inca and his sons. A few thousand survivors of the Inca empire clung to its inhospitable valleys as much to be near the descendants of their ancient rulers as to avoid the persecutions of Spanish-occupied Peru. Perhaps the most fascinating aspect of Vilcabamba was Titu Cusi himself. He was the sun around which the small state revolved. All that we know reveals him as a capable and conscientious ruler. He was constantly moving about his kingdom, living without undue ostentation but preserving the ceremonials traditional to his rank. He took his duties as head of the Inca state and religion seriously and participated with enthusiasm in the various festivals that gave so much pleasure to his subjects. The proximity of Spanish Peru did not terrify

Titu Cusi's grandfather, the great Inca Huayna-Capac, last undisputed emperor of Peru him as it had Sayri-Tupac's regents. He maintained cool-headed relations with the Spaniards and was receptive to new influences from them. His prime concern was the well-being of his subjects and the preservation of his state and dynasty.

Matienzo described Titu Cusi in 1565 as 'a man aged thirty-three; very able and somewhat larger than other Indians,' while to Rodríguez de Figueroa he was 'of middle height and with some marks of smallpox on his face; his expression rather severe and manly'. Calancha described him as a fat man. When Rodríguez first saw him he was wearing full ceremonial costume: a multi-coloured feather headdress, a diadem on his forehead and another on his neck, a coloured mask, and a silver plate on his chest; below his tunic he wore garters of feathers, with small wooden pompoms; and he carried a golden lance, dagger and shield. On other less ceremonial occasions he was seen wearing a tunic of blue damask and a cloak of very fine cumbi cloth woven from vicuña wool, or a tunic of crimson velvet with a cloak of the same material.

Something of Titu Cusi's character emerges from the reports of those who met him, and from his own writings. He was an exuberant, emotional man who could switch from sudden anger to jovial good humour. There was the occasion when he allowed his captains to threaten Diego Rodríguez: he himself joined in the bombast 'saying that he could himself kill fifty Spaniards' and brandishing his lance; next day he laughed heartily at the episode and dismissed it as a joke. He treated the Augustinian friars in a similar way: angry with them one moment, playing jokes on them another, and then telling Diego Ortiz that he loved him like a brother.

He was a thoughtful host, ensuring that Diego Rodríguez was well looked after from the time he crossed the bridge of Chuquichaca. Soon after their first meeting, the Inca sent Rodríguez 'a message by the mestizo [Pando to say] that I must rest, for I must be tired, and that on the following day he would do all that I desired'. At one time he gave the Spaniard a cup of chicha to drink, but Rodríguez recalled that: 'I drank a quarter of it, and then began to make faces, and wipe my mouth with a handkerchief. He began to laugh, understanding that I did not know that liquor.' Four years later, he took the two friars to see the city of Vilcabamba to please them, as a friendly gesture after berating Marcos García for his unauthorised attempt at departure.

There was none of the stony reserve or inscrutability of Atahualpa about Titu Cusi. He talked openly to Rodríguez and Matienzo and even, at one point, showed Rodríguez the scars on his leg inflicted by his father's assassins. He showed obvious delight at presents Rodríguez gave him, such as 'half an arroba of crystals and pearls, and seven bracelets of silver' or 'a very good looking-glass, two necklaces of coral beads, and a paper book' or the eight yards of yellow damask that arrived from President García de Castro. Nor was he above throwing himself tearfully at the feet of Judge Matienzo when they met on the bridge of Chuquichaca.

Titu Cusi was remarkably tolerant of the Spaniards, often against the wishes of his more narrowly militant followers. The fact that he permitted Diego Rodríguez to erect a large cross and to preach at Pampaconas was an example of this lenience; so also was his reception and protection of the various friars, even when they had the audacity to censure the natives' festivities or marital customs, and even when they provoked the fury of the Vilcabamba reactionaries by destroying the sanctuary of Chuquipalta. The Inca kept by his side the half-caste secretary, Martín Pando, who marched about in his tattered Spanish cloak; Titu Cusi used to practise European fencing with him. After 1570 the friar Diego Ortiz became his close companion.

Titu Cusi was open-minded towards the Spaniards despite an acute awareness of Spanish duplicity and cruelty. Immediately after meeting Judge Matienzo on the bridge of Chuquichaca he poured out his grievances 'with moving tears'. 'He told how his father and he had been kept in prison by Juan Pizarro, like dogs in a collar, until they should give him a chest full of gold. His mother and one of his sisters had been raped in the valley of Tambo. His father had taken refuge in these mountains and on various occasions [Hernando] Pizarro, [Gonzalo] Pizarro and Orgóñez had come to conquer him, so that many Indians and Spaniards lost their lives. In return for his father's hospitality, seven Spaniards of the Almagro faction had murdered him.' These same injustices formed the bulk of a memorandum handed to Matienzo in 1565 and of the *Relación* dictated to Pando and Marcos García in 1570. Throughout these works, the Inca depicted his father as a wise idealist, a tolerant man, slow to anger but powerful in action. The fact that he bothered to recite the abuses showed that he correctly sensed a strong humanitarian conscience among the higher Spanish officials. And the way in which he described his father Manco revealed the qualities that Titu Cusi himself was trying to emulate.

Titu Cusi's appraisals of the Spaniards were remarkably perceptive in other ways. Apart from making capital of the Christians' feelings of guilt about the Conquest, he had clearly pondered the motives of President Castro and others towards himself. In a revealing passage in his *Relación*, he said that he had originally deduced three motives behind the generous Spanish overtures: irritation at the guerrilla raids from Vilcabamba; King Philip's guilty conscience at having seized so much of Peru from another monarch; and the threat of rebellion in Spanish-occupied Peru inspired by the independent native state of Vilcabamba. He went on to say, however, that he had since decided that the Spaniards' principal motive had probably been the propagation of Christianity.*

Some modern historians have dismissed Titu Cusi's flirtation with Christianity as a diplomatic ruse, the exploitation of a subject on which Spaniards were particularly gullible. The astute Inca is credited with being 'submissive by necessity and Christian by political convenience'. This interpretation is not necessarily correct. In his *Relación* the Inca appeared genuinely convinced that his own conversion had been the Spaniards' prime objective. This impressed him, as did the sincerity and lack of greed of the missionaries, and the obvious success of the Spaniards' god in promoting their interests. Although Titu Cusi never allowed Christianity to supplant the native religion, he personally protected it and allowed it to develop into an alternative religion among his subjects. Titu Cusi's initial enthusiasm for the new religion dwindled when Marcos García tried to impose puritanical austerity. But after García's departure the Inca kept the Christians Diego Ortiz and Martín Pando as his closest confidants. Titu Cusi was thus able to declare in his *Relación* that 'I adopted and have maintained my Christianity up to the present.'

If Titu Cusi respected the Spaniards' Christian piety, he also feared their notorious greed. The Vilcabamba area contained some mineral deposits, notably the silver mines of Huamani and Huamanate and the gold-bearing ravine of Purumata.* One of the qualities that was said to have endeared Marcos García to the Inca at the outset was his 'lack of greed for land that contained silver'. An innocent Spanish prospector called Romero appeared in Vilcabamba in 1570 and asked permission to search for gold. 'The Inca gave him permission, and he discovered rich veins in his search for mines. In a few days he mined quantities of gold. Romero thought that the Inca would be delighted, and brought him the gold in the hope of negotiating a new licence for a period of months during which he could mine much. When the Inca saw the gold he thought that it could arouse greed and attract thousands of Spaniards, so that he would lose his province. He therefore ordered them to kill the Spaniard Romero.' Intercession by Diego Ortiz could not save Romero, who was beheaded and thrown into a river. This was the only Spaniard killed on Titu Cusi's orders. The Inca rightly saw that the lure of mineral wealth was the one magnet that would certainly bring Spaniards swarming into Vilcabamba. In the same way that President Castro had reacted in panic to the discovery of the pikes at Jauja, Titu Cusi recognised the extreme seriousness of Romero's gold.

The Inca also appreciated the dangers of racial tension between Spaniards and Indians. He refused a request by Diego Rodríguez to admit a few Spaniards to Vilcabamba to open trade in coca and timber, on the grounds that 'if the Spaniards lived among them and there was some dispute as a result of which the Spaniards killed one of them or they killed a Spaniard, there would be trouble'.

By the late 1560s Titu Cusi was evidently pursuing a deliberate policy in his negotiations with the Spaniards. There was an exciting possibility that his purpose became more than mere prevarication. He may have been planning nothing less than the survival of Vilcabamba as an independent state, a native enclave related to Spanish Peru in much the same way as Basutoland or Bechuanaland were related to modern South Africa. Had he succeeded, Vilcabamba might now have a seat in the United Nations like Lesotho or Botswana. Titu Cusi's policy during the decade of the 1560s divides into three phases. At the outset he revived his father's raids into the frontier areas of Spanish Peru, encouraging native uprisings such as that at Jauja. In the middle years there were the negotiations, starting with the embassies of García de Melo and culminating in the act of submission to the Spanish Crown. During this period Titu Cusi apparently changed his aims, and decided to negotiate for coexistence, rather than pursue his confrontation. This reappraisal was caused by awareness of the power and permanence of the Spanish occupation. The Inca himself observed this power on sorties in which he participated, and he noted the ease with which the

Jauja rising was extinguished. The change of outlook may also have been the work of Martín Pando, the everpresent mestizo who had arrived with the news of Sayri-Tupac's death and who advised the Inca throughout the decade. Every communication between Titu Cusi and the Spaniards was transcribed and translated by this intelligent secretary, and Pando participated and interpreted at every meeting with a Spanish envoy. During the later years of the 1560s the natives applied the policy of coexistence, permitting missionaries to operate in Vilcabamba and doing nothing to offend the Spanish authorities. Titu Cusi admitted embassies, and corresponded with Spanish governors as a monarch and head of state.

Titu Cusi was highly successful in weathering the renewal of official Spanish interest in Vilcabamba. Every year something was done to maintain Spanish hopes and curb Spanish action. In 1564 there were encouraging letters and meetings with García de Melo; in 1565 there were constructive conversations with Rodríguez de Figueroa and Juan de Matienzo; 1566 saw the signing of the treaty of Acobamba; in 1567 the treaty was ratified, the Inca performed his act of allegiance and Quispe Titu was christened; 1568 saw the baptism of Titu Cusi himself; and in the following two years there was a steady correspondence between the Inca and his friends in the Spanish government and Church.

Titu Cusi's letters contained a skilful blend of proud independence and obsequious submission. They were written, thanks to Martín Pando, in acceptable diplomatic language and generally expressed views that the recipient would have wished to hear. Reviewing his actions in the *Relación* of 1570, the Inca could insist that he had 'ratified completely . . . the peace which I had given my word to keep with the King and his vassals'. 'As your lordship knows, when you sent Diego Rodríguez to me to be corregidor of my land, I received him.' He had kept the peace 'firstly by the reception I gave to the Oidor Licenciate Matienzo at the bridge of Chuquichaca . . . and also by receiving priests into my land to instruct me and my people in the things of God. . . . Another testimony of this peace, and one which entirely confirms it, is the renunciation that I made to your lordship in the name of His Majesty of all my kingdoms and lordships, no more nor less than what my father possessed.'

All this apparent acquiescence was compatible with a policy of preserving Vilcabamba's independence. This could be done only by giving the Spaniards no pretext for invasion. Hence Titu Cusi's orders to his men to cease raiding and to avoid killing Spaniards or harming churches. His acceptance of missionaries and personal adoption of Christianity removed any religious grounds for attack. His reluctance to admit Spanish settlers ensured that no unforeseen incident could form a *casus belli.* And his concealment of Vilcabamba's mineral riches prevented an invasion prompted by greed.

But there was one Spanish requirement with which Titu Cusi could not comply. It was the action by which the Spaniards judged the success or failure of their diplomatic efforts. It was the one part of the treaty of Acobamba that was vaguely worded and to which the Inca did not refer in his *Relación*. This was the departure of Titu Cusi and his brothers from Vilcabamba and their settlement in Spanish Peru. Titu Cusi was not tempted by the comforts of Cuzco as his brother Sayri-Tupac had been. The Inca continued to reside in Vilcabamba; and his presence there inspired the survival of that defiant native state.

While Titu Cusi maintained his proud exile in Vilcabamba, other members of the Inca royal house enjoyed the luxuries of Cuzco and Quito. Some had acquired encomiendas because of their royal ancestry, and all basked in the esteem of natives and sentimental Spaniards.

At the head of this aristocracy was the magnificent figure of Paullu's son and heir Don Carlos Inca. He was about twelve when his father died, and had always enjoyed the full rewards of Paullu's collaboration and adroit political trimming. 'He was raised by his parents just as they themselves lived since they had become Christians: with much lustre, with Spanish tutors and servants, and having for his ornament magnificent horses, harnesses and other finery.' Carlos was the only pure Indian to be educated in the midst of a dashing group of sons of conquistadores, learning with them to ride, hunt and fence. He acquired the social graces of a sixteenth-century gentleman, and a good classical education from tutors such as the priest Pedro Sánchez and canon Juan de Cuéllar 'who taught Latin and grammar to the mestizos, sons of noble and rich men'. The quipocamayos spoke in awe of Don Carlos as 'very well educated, a good scrivener and horseman, charitable, and skilled at arms and music'. Garcilaso de la Vega (himself son of an Inca princess) recalled that he 'knew no Indian who spoke Spanish except for two boys who were my fellow students and who had gone to school and learned to read and write from childhood. One of these was called Don Carlos, son of Paullu Inca.'

The Spaniards saw no advantage in crowning Carlos with the Inca fringe as they had his father. Although he wielded no real power, he was an ornamental figurehead, hereditary leader of the native Peruvians with the landed wealth to support this position. He lived in the Colcampata palace with his mother Catalina Ussica, who had married Paullu in a religious ceremony shortly before his death, and who survived her husband by some thirty years. Colcampata overlooks Cuzco, and Manco's besieging troops captured it intact in their first rush down from Sacsahuaman. The great hall of the palace therefore survived the siege and was still standing in the 1560s. It 'served as an assembly place for rainy days, and it was here that the Indians celebrated and solemnised their festivals'. Carlos held court there, surrounded by a cluster of impoverished native nobles who acted as his courtiers. He entertained many dignitaries who visisted Cuzco, and provided hospitality for Spanish and mestizo citizens. He maintained the chapel of San Cristóbal founded by his father at Colcampata, and himself founded a chapel dedicated to the Virgin of Guadelupe in the Franciscan convent.

Carlos's reward for such exemplary behaviour was to be accepted and honoured in Cuzco society. He played a prominent part in the processions and tournaments that were a regular feature of the city's life. Garcilaso described a procession for the feast of St Mark that started at the Dominican convent (the former temple of Coricancha) and ended at a hermitage alongside Carlos's Colcampata palace. A garlanded bull marched at the head of the procession and was led right up to the high altar of the church. Another procession was held annually to mark the festival of Corpus Christi. The eighty Spanish citizen-encomenderos each provided a lavishly-ornamented float carrying a holy image and carried by that citizen's tributary Indians. The caciques from around Cuzco were allowed to wear 'all the regalia, ornaments and inventions that they used in celebrating their major festivals in the time of the Inca kings. . . . Some came looking like paintings of Hercules: dressed in a puma skin with their heads enclosed in those of the animals, because they prided themselves on being descended from lions. Others wore the wings of a very large bird they call condor on their backs, like wings that are painted on angels . . . Others wore strange devices with their clothes plated with gold and silver. . . . Others came as monsters with hideous masks and the skins of various small animals on their hands as if they had hunted them. They made great gestures, pretending to be mad or foolish, to please their kings in every way: some with grandeur and riches, others with nonsense and bathos.'† The native men and women marched along with their flutes, drums and tambourines, but 'the songs they were singing were in praise of Our Lord God'. Similar masked clowns still perform at the dances and processions that take place in every village of the altiplano during carnival. But in the mid-sixteenth century, the celebrations were watched by the councils of Church and of the city. Prominent among these were Carlos and 'the surviving Incas of royal blood, to honour them and make some demonstration of the fact that that empire had once been theirs'.

Carlos enjoyed other distinctions. He was the only pure Indian to be an alderman of Cuzco.* He was also the only Inca noble to marry a Spanish lady – a rare creature in early colonial Peru, where many landless Spaniards had to settle for native wives. Carlos's wife was Doña María de Esquivel from Trujillo in Spain. 'A very important lady . . . and a person of much Christianity' according to the quipocamayos.

Carlos's ascendancy in Cuzco was interrupted for a short time when his cousin Sayri-Tupac ruled there as crowned Inca between 1558 and 1561. For obvious reasons, Paullu's son Carlos did not take part in the ceremonies to mark the arrival of Manco's son; and Sayri-Tupac was lodged with his aunt Beatriz rather than in Colcampata. Sayri-Tupac's untimely death was therefore a relief for Carlos. The eccentric mestizo writer Felipe Guaman Poma de Ayala echoed rumours that were doubtless current at the time by saying, bluntly, that 'Don Carlos Inca and Don Alonso Atauchi . . . killed Sayritopa Inca by giving him poison, because his emerging from the jungles annoyed them, as did the way in which the whole kingdom honoured and respected him.'

Alonso Titu Atauchi, whom Poma de Ayala accused of complicity in Sayri-Tupac's murder, was an ambitious young man who aspired to a career similar to Paullu's. He was a grandson of Huayna-Capac and son of a general who had been killed on Atahualpa's orders alongside Huascar at Andamarca. His lineage was thus as good or better than that of Paullu's sons, and Sarmiento de Gamboa described him as the only surviving member of the ayllu of his uncle Huascar.* Alonso Titu Atauchi was twenty-one when Francisco Hernández Girón rebelled, and he led a contingent of four thousand native troops to help the royal army. He claimed to have helped defeat Girón at Pucará in October 1554, and asked for favours. The King thought that he had found another Paullu. In October 1555 he awarded Titu Atauchi the resounding hereditary title of perpetual Alcalde Mayor de los Cuatros Suyos, or Lord Mayor of the Four Quarters. Don Alonso was allowed to wear the royal fringe or mascapaicha, to carry a rod of justice at all times, to surround his house with chains as a sign of immunity, and to bear a coat of arms.*

Cuzco during the 1560s was a wild place, and Carlos Inca allowed himself to become involved with its more disreputable elements. The marriages of conquistadores and native ladies produced a generation of spirited mestizos. These had been the classmates of Garcilaso de la Vega, who left Peru for ever in 1560 and often wrote nostalgically of his schooldays with them. These mestizos had now reached manhood. They were proud of their parentage on either side and of the feats of their fathers.

The laws of Spain discriminated against mestizos, depriving them of some rights of Europeans but forbidding them to live in the countryside with the Indians. In 1549 Charles V decreed that mestizos could not hold public office; nor were they allowed to make Indians serve them. They were not even allowed the prestige of bearing arms – the decree forbidding this complained that 'there are many who are better arquebusiers than the Spaniards'. Mestizos were even prevented at times from becoming priests. Such discrimination bred resentment among the proud mestizos of Cuzco. Native-born creoles were also bitter at seeing the best offices and encomiendas given to a steady stream of well-connected opportunists from the mother country, and at the spectacle of the aged libertine Count of Nieva rewarding his favourites. Nieva, for his part, wrote to the King that intermarriage of Europeans and natives should be stopped 'because there are already so many mestizos and mulattos, and they are so badly inclined that one must fear harm and unrest in this country because of their great numbers'.

The trouble that Nieva feared was not long in coming. In 1562 there was an abortive plan to revolt and have four consuls in the Roman manner to rule Peru. In early 1566 it was feared that the corregidor of Cuzco, Diego López de Zúñiga, might rebel, and he was hastily replaced by Jerónimo Costilla. But the most serious mestizo plot was hatched at the end of 1566. The two brothers Arias and Cristóbal Maldonado were in the thick of it – they had just been thwarted in their attempt to marry Cristóbal to Beatriz Clara Coya. So was their cousin Juan Arias Maldonado, the son of Diego Maldonado el Rico, holder of Andahuaylas, the richest encomienda in Peru.* Another mestizo, Pedro del Barco, son of the conquistador of that name, had his Indians at Muyna make 150 feet of artillery fuses, while Juan Maldonado collected forty arquebuses in Cuzco.

Many of the plotters' meetings took place in Carlos Inca's Colcampata palace. There was big talk that the mestizos, who called themselves 'montañeses' would kill the Spaniards and win fat repartimientos. Juan Maldonado, Carlos Inca, Pedro del Barco and their followers were to assassinate the corregidor of Cuzco; others were to rebel in Arequipa and Huamanga; Arias and Cristóbal Maldonado were to stab their enemy President Lope García de Castro, having been introduced into his presence by Melchor de Brizuela, the respected chief constable of Lima. Don Carlos Inca 'was to summon the natives to rise and seize the food supplies', and after the Spanish towns were taken, the plotters would ask Titu Cusi to come to their support.

On 11 January 1567 one of the conspirators, Juan de Nieto, told of the plot in his confession to Juan de Vivero, Prior of the Augustinian monastery and the man who was to baptise Titu Cusi in Vilcabamba the following year. Vivero decided that he must violate the secrecy of the confessional to warn the corregidor Costilla that a plot was afoot. The corregidor immediately sent Atilano de Anaya, who was now guardian of Beatriz Clara Coya, to warn Castro in Lima. Anaya completed the journey in only eight days. Castro immediately arrested Arias and Cristóbal Maldonado, about whom he had already heard incriminating rumours. He also arrested the constable Brizuela as he was in the act of releasing the Maldonado brothers from prison. Meanwhile, the plotters had been discovered in Cuzco and most were arrested by 18 January. Juan Arias Maldonado and seven other mestizos were sent under heavy escort to Lima. The Maldonado brothers and Brizuela were exiled to Spain and their estates confiscated by a triumphant Castro.* Carlos Inca barely escaped the exile that had been imposed on his fellow conspirators. The corregidor Jerónimo Costilla proceeded against him, but the action was referred to President Castro. He in turn elevated it to the Audiencia of Charcas, where it remained on the records as a sword of Damocles over Carlos's head.*

Carlos continued to live in splendour in Cuzco despite this threat. He was much in evidence when a new Viceroy, Francisco de Toledo, made a triumphal entry into Cuzco in 1571 – the first Viceroy to visit the city. Toledo's secretary Antonio Bautista de Salazar recorded the scene, a familiar spectacle throughout colonial regimes. 'A vast crowd of natives came down the side of a hill in front of the spectators. The Incas came in front, followed by the provinces of the four suyos each with its flag and a great number of pennants in a variety of colours. Almost all the Indians had discs of gold or silver on their chests, and breastplates of the same with a great quantity of plumage. They were facing the sun, and I cannot find a comparison for how splendid they looked. When each suyo or tribe arrived before the Viceroy, they made their reverence and mocha in their manner, with a short speech of welcome. Each tribe then went out dancing in its own way, and these dances, though sideshows to the main ceremony, were no mean thing to see and note. They then staged skirmishes, guazabaras or puellas as they call them, with one another.'

Some months after Toledo's arrival, Carlos's Spanish wife at last gave birth to a son after many years of marriage. The Viceroy did Carlos the particular honour of acting as godfather to the infant, who was called Melchor Carlos after one of the Three Kings, because the baptism took place on Epiphany 1572. This made Toledo and Carlos 'compadres' or co-godparents, a very close relationship throughout Latin America to this day. The baptism ceremony took place in the chapel of San Cristóbal, Colcampata, and was attended by a great gathering of native notables. Once again 'there were festivals, rejoicings, fireworks, dances and many newly invented and costly conceits, which they knew how to organise admirably in Cuzco in those days'.

Carlos Inca wrote to King Philip to express his gratitude for Toledo's kindness to him. He wrote that 'for many years I have longed to go to your court to kiss the royal feet and hands of Your Majesty.' And his praise of the new Viceroy was particularly fulsome. 'What a marvellous thing it is that, although those who have governed this kingdom have done so very well, none [has governed] with as much zeal as [Toledo] is now doing. He has come to deal in person with the prosperity of the natives and of this country.' Paullu's son was following him in abject hispanicisation (plate 41).

Another important group of grandsons of Huayna-Capac were the children of Atahualpa. When tied to the stake at Cajamarca, Atahualpa Inca had commended his children to Francisco Pizarro even though they were under Rumiñavi's control in the province of Quito.* Atahualpa's children were captured during the invasion of Quito. But Pizarro had by then established Manco as puppet Inca: he therefore had no political use for these infant descendants of the Quitan house. He decided that the safest place for them was in the protection of Christian monasteries. Five were entrusted to the Dominicans in Cuzco, and three to the Franciscans in Quito, since these were the only religious houses founded by 1534.

There were three boys and two girls in the Cuzco group, and these survived the civil wars under the protection of the Dominicans and particularly the zealous Domingo de Santo Tomás.* It distressed Santo Tomás to see these princes growing up without family or resources. He therefore organised enquiries on their behalf in Cuzco in 1554 and in Lima in 1555, and one of his witnesses was Doña Inés Yupanqui, Pizarro's former love and now wife of the powerful Francisco de Ampuero.* Santo Tomás took these enquiries to Spain and presented them in person to King Philip in Flanders. The eloquence of Santo Tomás and the magic of Atahualpa's name produced the desired result: a pension of 600 pesos a year for two lives to the sons called Diego Illaquita and Francisco Ninancoro.* This Diego, Francisco and their brother Juan Quispe-Tupac lived quietly in Cuzco, and they and their sons were known to various Spanish chroniclers: Garcilaso de la Vega, Guaman Poma de Ayala, Sarmiento de Gamboa and Bernabé Cobo all consulted them.*

Other sons of Atahualpa flourished in Quito. The Inca's sons Carlos and Francisco were educated in the Franciscan monastery of Quito and were baptised in 1548. In the following year President Pedro de la Gasca gave Carlos the encomienda of Conocoto, so that he became the second Inca descendant to join the privileged few who held such grants.* These boys found a champion in the Franciscan friar Francisco de Morales, who wrote to the King in 1552 that young Francisco Tupac-Atauchi was in the College of San Andrés in Quito with 'no means of support: Your Majesty should give him something with which to marry'.

The Crown mentioned him in a decree in 1556, and he was soon receiving pensions of 1,000 pesos a year. Thus endowed, the Auqui (Prince) Francisco did make a good marriage: to the daughter of the curaca of Otavalo. He became a rich man in colonial society and acquired various properties in Quito (including one, close to the quarries, that is still known as Tierras del Auqui), and land in Chillo, Latacunga, Otavalo and Cumbayá, and a fruit orchard in Ambato. By 1576 Francisco was the only person receiving a pension from the royal treasury of Quito. An official report said that 'he maintains a household and is a very quiet and peaceful person; he is content with this favour, for he supports himself with it'. Francisco Tupac-Atauchi was made a director of public works in Quito, and died after 1580, rich and respected, to be buried in a chapel that he had founded in the church of San Francisco.*

Other members of Atahualpa's family survived in the Quito area. When Sebastián de Benalcázar invaded Quito, he sent Captain Diego de Sandoval into the province of Chaparra, far to the north of the Inca empire near Cali in modern Colombia. Sandoval captured a group of Inca fugitives including a sister-wife of Atahualpa who was baptised as Doña Francisca Coya. The natives of Quito venerated her with pathetic passion. One of her serving women later testified that this Francisca Coya became pregnant and 'I asked her by whom, and she told me she was pregnant by her master Captain Sandoval.' She gave birth to a daughter, and her granddaughter married a prominent Spaniard called Vicente de Tamayo: their family spread throughout the colonial period. Another of Atahualpa's wives, Doña Isabel Yaruc Palla, became the love of another of Benalcázar's lieutenants, Diego Lobat. She possessed much influence as the widow of the Inca, but collaborated closely with the conquerors. She betrayed a plotted native rebellion to Pedro de Puelles, Governor of Quito, as a result of which he arrested the curaca of Otavalo and imprisoned him and other conspirators.*

In both Quito and Cuzco, members of the Inca royal family thus used their privileged birth to secure pensions and honours in Spanish colonial society. It is hard to condemn them for profiting from Spanish generosity or for trying to assimilate European ways. But their collaboration was purely selfish, and it lacked the dignity of Titu Cusi's lonely exile in the forested hills of Vilcabamba.

Forced labour: the forbidden practice of porterage

18

OPPRESSION

THE first impression of any sensitive visitor to Peru in the mid-sixteenth century was a terrible decline in the native population. Vicente de Valverde wrote to the King as early as 1539: 'I moved across a good portion of this land and saw terrible destruction in it. Having seen the land before, I could not help feeling great sadness. The sight of such desolation would move anyone to great pity.' Vaca de Castro wrote a few years later: 'There has been and still is a great decline of the Indian natives, which I have seen with my own eyes on the road from Quito to Cuzco.'

The depopulation was most obvious on the coastal plains. Cristóbal de Molina described conditions in three valleys in the 1540s. 'I shall tell you about two provinces that were reputed to have contained 40,000 Indians when the Spaniards entered this country. One was Huaura beside Huarmey, which Almagro took as a repartimiento because of its large population and reputation for being very rich; the other is Chincha, which Hernando Pizarro took, and which also had 40,000

Indians. Today there are not more than four thousand Indians in the two provinces. In the valley of this city [Lima] and in Pachacamac five leagues from here, which was all one entity, there were over twenty-five thousand Indians. It is now almost empty, with scarcely two thousand.'

Chincha, a hundred miles south of Lima, had been one of the most populous valleys of the coast. Cieza de León reported that by 1550 its population had shrunk by five to one, and Bartolomé de Vega wrote that Chincha had only a thousand inhabitants by the 1560s.* A decade later, it was reported that Chincha had only five hundred natives and Pachacamac only one hundred.* Cieza reported many other instances of terrible decline, notably in the Santa, Ica and Nazca valleys and at Paramonga, where 'I believe there are no Indians at all to profit from its fertility', and Garcilaso said that the valleys of Lunahuaná and Huarcu had fallen from 30,000 to 2,000 by the year 1600.*

Some observers made damning comparisons. Fernando de Armellones said that 'we cannot conceal the great paradox that a barbarian, Huayna-Capac, kept such excellent order that the entire country was calm and all were nourished, whereas today we see only infinite deserted villages on all the roads of the kingdom.' Another report concluded that 'it is clear that the government in the past was better and more valuable: for under the Inca's rule the Indians were daily on the increase'.

Other writers were alarmed at the terrible prospect facing Peru. The eminent Jesuit José de Acosta wrote: 'many believe that what remains of the Indians will cease before long' and the Dominican Santo Tomás pleaded to the King that 'unless orders are given to reduce the confusion in the government of this land its natives will come to an end; and once they are finished, Your Majesty's rule over it will cease'. Diego de Robles also stressed the urgency of the problem, 'for if the natives cease, the land is finished. I mean its wealth: for all the gold and silver that comes to Spain is extracted by means of these Indians.' And Rodrigo de Loaisa wrote 'I must advise Your Catholic Majesty that the wretched Indians are being consumed and are dying out. Half have disappeared, and all will come to an end within eight years unless the situation is remedied.'

These horrifying examples of population decline referred to parts of Peru most exposed to Spanish occupation. Many coastal natives retreated into the hills, and in the sierra itself the Indian population had diminished less catastrophically. Spanish officials attempted to estimate the rate of decline in the mountain valleys along the route from Lima to Vilcashuaman: their estimates ranged from 3·75 to 1, to 1·5 to 1.* Modern historians have attempted to estimate the population of the Inca empire at the time of the Conquest, and to calculate the rate of decline. John Rowe took five provinces for which contemporary estimates of population have survived, and deduced from this random selection an average rate of decline of four to one for the whole country during the fifty years after the Conquest.* George Kubler felt that the decline was only two to one – still a shattering figure.* The best contemporary estimates for the population of Peru in the late sixteenth century gave a figure of some 1,800,000. By Rowe's standard of decline, the population at the time of the Conquest would thus have been some seven million.†

What caused this appalling depopulation? The most obvious cause would be disease, for the Peruvians after centuries of isolation had no immunity to European diseases. Contemporary sources were curiously silent about any epidemic during the first decade after the Conquest. There may well have been an attack of hemorrhagic smallpox that spread across country from the Caribbean just before Pizarro's arrival.* But the next recorded epidemic was not until 1546, when Herrera said that an epidemic, possibly typhus or plague, 'spread over the entire land, and people without number died from it'. A contemporary enquiry revealed that in 1549 'there was a great epidemic and mortality among the Indians in the Cuzco district and throughout the Collao and other provinces of Peru'. Miguel de Segura testified that in and around Cuzco 'a great quantity of Indians died from a disease said to be like hay fever'. In that same year the municipal records of Cuzco recorded an epizootic that ravaged the herds of llamas and alpacas. The next major epidemic that was well documented occurred between 1585 and 1591, at a time when the population was beginning to stabilise. The Viceroy Villar wrote that a disease resembling 'smallpox and measles ... destroyed and killed a great quantity of Indians' in Quito, where the mortality was estimated to be 30,000. The epidemic swept down the coast and soon struck Lima, where the native hospital of Santa Ana averaged fifteen deaths a day for months on end. Almost a quarter of the city's population died. Thousands also died in Cuzco and Arequipa.

The disease originated as smallpox and measles with high fever; its victims became almost unrecognisable from virulent pustules, became delirious and often developed severe coughs or ulcerated throats before dying. The town council of Huamanga cut the Apurímac bridge to try to stop its spread, but the epidemic raged for five years.*

Disease was therefore important, but was not the main cause of the sharp decline during the first forty years of Spanish rule. That decline resulted more from profound cultural shock and chaotic administration. Since the death of Huayna-Capac the people of Peru had lived through a numbing series of catastrophes. Their calm, rigidly organised society was shattered in quick succession by a ferocious civil war, a bewildering conquest by foreigners totally alien in race and outlook, two mighty attempts at resistance, and a devastating series of civil wars among the invaders.

In the quarter-century after the outbreak of war between Huascar and Atahualpa, there were only brief lulls during which armies were not on the march in Peru. Thousands of Peruvians died fighting, particularly in the battles of the civil war between Atahualpa and Huascar. Vicente de Valverde estimated that 20,000 were killed during the year of Manco's first rebellion.

In such turbulent times, many natives grew so deeply demoralised that they lost the will to live. This is still a serious threat to primitive peoples who witness the collapse of their way of life – tribes of the Mato Grosso, for example, give up the ghost despite efforts to protect them. A group of aged Inca officials interrogated in the 1570s described this pathetic condition. 'The Indians, seeing themselves dispossessed and robbed . . . allow themselves to die, and do not apply themselves to anything as they did in Inca times.' This same demoralisation led to a sharp decline in the birth-rate, a phenomenon accelerated by population movements and the disruption of the Inca marriage system. A fall in the birth-rate may have been the most important of all factors in producing the population decline – a few acute Spanish officials, including Andrés de Vega and Luis de Monzón, sensed its importance.*

Spaniards rarely wantonly killed natives except in war, but they were nevertheless responsible for much of the decline. Under the Incas the steep valleys of the Andes and the rivers running across the coastal desert had been organised to maintain a large population. The Spaniards, preoccupied with their personal fortunes and embroiled in passionate civil wars, neglected the public works of the Inca regime. Precious irrigation canals were allowed to fall into disrepair, agricultural terraces that used to climb in neat ranks up Andean mountainsides crumbled and became overgrown, roads and bridges that had been built for runners were pitted by heavy horses and wheeled traffic. The Incas maintained great storehouses as insurance against bad harvests. These were dissipated and looted at the outset of the Conquest, and the great llama herds of the Incas were slaughtered, dispersed and never restocked. One official wrote: 'It is said that [early conquistadores] killed great numbers of llamas simply to eat the marrow-fat, and the rest [of the meat] was wasted.' Gutiérrez de Santa Clara described the vast herds assembled in the Jauja valley by President Gasca. Each soldier in his army received a llama every fortnight, and 'in this way all the food, the vegetables, llamas and alpacas that were in that valley and district were totally consumed'.

Cieza de León, who also wrote in the mid-sixteenth century, said that 'when the natives hid their flocks the Spaniards tortured them with cords until they gave them up. They carried off great droves and took them for sale at Lima for next to nothing. When the unfortunate natives went to beg for justice from the Marquis [Pizarro] he turned them away saying that they lied. So they wandered from hill to hill complaining of their ill-treatment.' This cruel plundering inevitably led to local famines. Pascual de Andagoya, a man who had tried to sail to Peru even before Pizarro, wrote a harrowing report to the Emperor in July 1539. He said that bands of Spanish marauders were roaming Peru in the chaos following Almagro's death. 'But what is worse is that the Indians are being totally destroyed and lost. Someone – an official of Your Majesty – told me here that it was not in fighting for one side or the other that fifty thousand souls died in Cuzco. They begged with a cross to be given food for the love of God; and when they were given none they threw the cross on to the ground. One man in Cuzco collected 200,000 hanegas of maize from the Indians and was selling it in the native market. The soldiers and citizens took all the Indians' cloth and food and were selling it in the square at such low prices that a sheep [llama] was sold at half weight. They were killing all the [llamas] they wanted for no greater need than to make tallow candles . . . The Indians are left with nothing to plant, and since they have no cattle and can never obtain any, they cannot fail to die of hunger.'

Although legally free men, the Indians were grossly overworked, underpaid and overtaxed. Under the Incas they had lived in a paternalistic society without money, personal property or writing. It was impossible for them to grasp that they were now regarded by the authorities as free individuals expected to earn money, compete and stand up for their rights, if necessary by written Spanish law. All they knew was that their country had been conquered and parcelled out among the conquerors. Once the rebellions of Manco Inca had failed, the natives passively accepted their lot. Diego de Robles complained that 'the natives of this land are by nature a very subservient, timid people, not free to complain of the abuses they receive from their encomenderos'.

Innumerable writers described the natives' hardships with deep compassion. Francisco de Morales, Franciscan Provincial of Peru, told the King: 'The injuries and injustices that have been and are being done to the poor docile Indians cannot be counted. Everything from the very beginning is injury. Their liberty has been removed; their nobles have lost their nobility, authority and all forms of jurisdiction; [the Spaniards] have taken their pastures and many fine lands, and impose intolerable tributes on them.' The Licenciate Francisco Falcón said that 'generally speaking, they all pay excessive tributes – far more than they are able to pay or than their estates are worth.... They have never given express or tacit consent to such tributes' and they received nothing in return. Baltasar Ramírez echoed this: 'The tributes and taxes they pay... are all endured only with great difficulty and hardship. Nothing is left over for them to have any leisure, to endure times of necessity or illness – as we Spaniards have – or to raise their children. They live in poverty and lack the necessities of life, and never finish paying debts or the balance of their tributes. We can see that they are dying out and being consumed very rapidly.' Bartolomé de Vega began his admirable *Memorial* of 1562: 'The first way in which the Indians of Peru are abused is in the assessments and tribute which the encomenderos claim His Majesty has ordered them to pay... They make them give excessive tributes of the produce of their lands – tributes that some repartimientos are simply incapable of paying. Because of this the Indians flee to the wilds, and wander lost outside their homelands.'

Hernando de Santillán, an official sent out from Spain to the Audiencia of Lima, left a heart-rending account of the Indians' misery. 'Even if it freezes or if their cereals and other foods are dried up and lost, they are forced to pay their tribute in full. They have nothing left over from what they can produce. They live the most wretched and miserable lives of any people on earth. As long as they are healthy they are fully occupied only in working for tribute. Even when they are sick they have no respite, and few survive their first illness, however slight, because of the appalling existence they lead. They sleep on the ground ... and their diet is maize, chilli and vegetables: they never eat meat or anything of substance except some fish if they live on the coast. The only furnishings in their houses are some jars, pots, spindles, looms and other equipment for working. They sleep at night in the clothes they wear by day and scarcely succeed in clothing their children, most of whom are naked. ... They are deeply depressed by their misery and servitude ... and have come to believe that they must continue to work for Spaniards for as long as they or their sons or descendants live, with nothing to enjoy themselves. Because of this they despair; for they ask only for their daily bread and cannot have even that. ... There are no people on earth so hardworking, humble or well-behaved.'

By the end of Gonzalo Pizarro's rebellion, the situation was at its most chaotic. There was no yardstick for measuring the amount owed by each repartimiento or each individual Indian. Francisco Pizarro had awarded great tracts of land to the care of his conquistadores, and these had extracted as much as they could from the natives entrusted to them. 'Under the Christians the only rule has been to use and exploit them according to the greed of each encomendero.' Everyone recognised that the first step to alleviate this situation must be an equitable assessment based on the population and fertility of each repartimiento. The King had already ordered Pizarro and Valverde to carry out such an assessment in July 1536; but Manco's rebellion intervened. Vaca de Castro was ordered to do it in June 1540; but there was the rebellion of Almagro the Younger. The instruction was repeated in chapter 34 of the New Laws of 1542 and in a letter to Viceroy Blasco Núñez Vela of 14 August 1543; but Gonzalo Pizarro rebelled.* When Pedro de la Gasca regained control of Peru, he made a complete overhaul of its encomiendas, dividing the largest, removing repartimientos from rebels and rewarding his own supporters. Gasca closeted himself with Archbishop Jerónimo de Loayza of Lima at the tambo of Huainarima above the Apurímac. They proceeded to award land grants that Gasca's biographer Juan Calvete de Estrella boasted to be worth one million and forty-one thousand pesos rent a year – almost as much as all Atahualpa's ransom gold.* After the rains in 1549 Gasca sent seventy-two inspectors to try to assess the amount that should be paid by the Indians of the various repartimientos. 'He hoped that this assessment would be the salvation of the natives: through it the country could be brought to order and justice, and the Ordinances that had not been repealed would be obeyed.'

Such salvation never came, for Gasca's assessors were generally local encomenderos, and their findings did little to damage their friends' incomes. A few of Gasca's encomienda awards have survived, and they are horrifying documents. The annual levy from the Conchucos, a remote region between the Cordillera Blanca and the upper Marañon, included 2,500 pesos of gold and silver; 400 hanegas (640 bushels) of wheat, 800 hanegas of barley, 200 of maize and 100 of potatoes; 30 llamas; 3 arrobas (76 lb) of tallow fat for candles; 30 pigs aged over 18 months; 300 birds, of which half must be hens; 45 brace of partridges; 1,040 eggs – 'twenty to be delivered every Friday fish day'; 25 loads of salt; 20 willow or alder logs at least 20 to 25 feet long and 100 agave poles; 25 small tubs, 25 plates, 25 wooden bowls, 6 saddles and 20 chopping-blocks; 120 pairs of sandals with their ankle laces; 20 seja palms for lassos; 10 sacks and 4 aprons, and 30 ropes each 30 feet long and all of sisal. Almost all this awesome array had to be delivered to the encomendero's town house in Huánuco, many days distant. In addition to the produce, the repartimiento had to provide a total of eighty people to serve as herdsmen, farm labourers and personal servants. And what was expected of the encomendero in return for all this bounty? The official award gave the answer: 'And so that you, the encomendero, should take these tributes with less worry or scruples of conscience, we charge you to instruct the said natives in the tenets of our Holy Catholic Faith . . .'

An alarming aspect of these assessments was the demand for precious metals and European produce, or for Peruvian foodstuffs from different climates. 'They demand gold and silver from those who have no mines . . . pigs from those who do not raise them, and chickens that do not exist in this country . . . maize, wheat and ají [chilli] from those who have no suitable land for them; wood and kids from those who have none on their land; cotton cloth from mountain Indians who do not pick it; and Indian ploughmen and servants from people who have not been accustomed to provide them . . .' Vega said that natives had to travel far outside their repartimientos to find the necessary goods, and Rodrigo de Loaisa demonstrated how difficult it was for the Indians to acquire enough precious metals.* Diego de Robles recommended simply that 'No Indian should be ordered to pay in tribute anything that he does not have or grow on his land.'

The natives were forced to carry their tribute to the town house of the encomendero, who was rigorously forbidden to live within his repartimiento. This isolation was originally intended to protect the Indians from direct abuse, and to keep the communities of Spaniards grouped together for greater security. But it made the encomenderos callous absentees. And the transportation could create prodigious problems for the natives. Bartolomé de Vega, a sober writer, reported that the Indians of Parinacocha had to carry their tribute over two hundred miles to Cuzco. 'They sometimes have to carry five hundred fanegas [800 bushels] of maize. Fifteen hundred Indians transport it, with three carrying one fanega . . . and they carry all the things required by the assessment: wheat, maize, cloth, bars of silver, etc. Indian men are loaded with it, and so are the women, the pregnant ones with their heads on their swollen bellies and those who have given birth with their babies on top of the loads. . . . Since Peru is such a mountainous country these people climb with their loads up slopes that a horse could not climb. They go sweating up the hillside with their loads, and it is heart-rending to see them. The Indians often take two months to deliver the tribute to their encomendero, including the outward and return journey and the time spent in the Spanish city.'

This journey to transport tribute to an encomendero's house was often an Indian's only contact with Spanish life. Many Indians chose to remain in shacks at the edges of the towns, to work as unskilled labourers. Any agricultural disaster stimulated this pitiful migration, which continues to this day.

The power of the encomenderos did not end with the receipt of great loads of tribute. In theory they had no jurisdiction either civil or criminal. But in practice they wielded unquestioned control of their Indians. In the remote rural areas, nothing but conscience and divine law controlled the conduct of the encomendero towards his charges. Hernando de Santillán described how they filled the vacuum caused by

An encomendero usurps the Inca privilege of being carried in a litter

the overthrow of the Inca and his administrators. 'Each of these encomenderos made himself an Inca. In this way, by virtue of their encomiendas, they enjoyed all the rights, tributes and services that each district made to the Inca – plus those they added to them.' Although the encomenderos themselves did not reside on their encomiendas, they invariably employed one or more Spanish stewards or majordomos to supervise the collection of tribute. These majordomos were the scourge of the Indians. Generally of humble origin, their task was to maximise the tribute yielded by the encomienda. They organised the Indians' farming, establishing local workshops, traded with Spanish merchants, and introduced European crops and techniques. Majordomos were often literate and efficient managers, and they naturally expected to make an additional profit themselves. A humbler form of Spaniard was the estanciero, a peasant who lived among the Indians and supervised their herds and plantations. Encomenderos fortunate enough to possess shares in mines also employed Spanish miners to exploit the mineral, and these miners were skilled and highly paid.*

The Crown hopefully decreed that Indians should pay less tribute than they had under the Incas, and some Spaniards believed that this was the case.* But Francisco Falcón was sure that 'the natives are forced to pay far more tribute than during the Inca era' and he stressed the important differences between the Spanish and Inca systems. Much of the tribute extracted during Inca times was stored for the welfare of the people, particularly during times of crisis. The Indians were supported by the state while at work, and they knew exactly how many days' labour would be required of them each year. Under the Spaniards, there was nothing to show for all the natives' work, and the only benefit was relief from military service in inter-tribal or expansionist wars. The Peruvians had been torn from the shelter of a benevolent, almost socialistic, absolute monarchy into the cruel world of feudal Europe. Because of their handicaps in language, education and race they remained at the base of the feudal structure, and only a tiny handful of Inca and tribal nobility could appreciate the finer aspects of post-Renaissance European civilisation. The mass of Peruvians, the hatunruna of the Inca empire, saw an Inca master replaced by a Spaniard; but they lost heavily on the exchange.

At the time of the Conquest, the Incas had been introducing a hierarchy of centrally appointed administrators. These imperial officials, the famous graduated system of decimal administrators, were supplanting the traditional chiefs of assimilated tribes. The chiefs – whom the Incas called curacas and the Spaniards called by the Caribbean name caciques – were being relegated to the position of honorific figureheads responsible for little more than collecting tribute and organising labour levies from their tribes. With the overthrow and humiliation of the Incas, the delicate balances of their administrative system collapsed, and Manco was unable to reassert them during the two years before his rebellion. In the confusion the natives turned to their traditional tribal chiefs for leadership – it was because the curacas of the Chachapoyas, the Huanca of Jauja, and the tribes of the Charcas refused to accept Inca leadership that the second rebellion of 1538–9 failed.*

The chiefs seized upon the Spanish Conquest as an opportunity to regain complete control of their tribes. At the same time the more enterprising central Inca officials tried to remain in power as regional curacas. Both types of official tried to establish themselves as hereditary chieftains. Damián de la Bandera wrote of the traditional chiefs: 'Under the Incas they had very little jurisdiction or dominion over the Indians. With the entry of the Spaniards and collapse of the Inca's government, they rose up and acquired all the property and privileges that had belonged to the Inca, including civil and criminal jurisdiction that they did not formerly possess. Each became on his own dungheap what the Inca had been in the entire kingdom.' Cieza de León wrote of the second type of curaca: 'When the Spaniards entered, many of the [provincial governors] remained with permanent command in certain provinces. I know some of them, and their power was so well established that their sons have inherited the property of others.' Rumiñavi at Quito and Illa Tupac at Huánuco were examples of such Inca governors who tried to perpetuate their rule. Polo de Ondegardo said of both types: 'Since the Christians entered the land . . . the curacas and their sons seized a more complete and widespread licence than they were given in the past. Each one became an Inca in his district.' Felipe Guaman Poma de Ayala, who as a mestizo experienced the sufferings of the natives, wrote that with the overthrow of the Inca empire 'the existing hierarchy disappeared. Instead, lowly Indians profited by the confusion and created little kingdoms for themselves or became curacas without being entitled to by birth.'

The curacas' takeover was helped by the fragmentation caused by the encomienda system and the mountainous geography that effectively isolated each valley. Most encomienda grants, in fact, referred to particular curacas by name, and the area of the repartimiento was defined as that controlled by the curacas. But in many instances encomiendas cut across Inca provincial boundaries, giving the heads of tiny units a chance to assert themselves as independent lords.*

At the outset of the Conquest the curacas had borne the brunt of Spanish cruelty. They were thought to hold the key to hidden treasure, and were treated as hostages for the good behaviour of their Indians. Diego de Vera told how the encomendero of Bambamarca, near Cajamarca, had 'killed two caciques by fire and with a dog, demanding treasure from them', and Antón Quadrado reported to Gonzalo Pizarro that the curaca of Huambacho had died as a result of a flogging.* Pedro de la Gasca executed Juan de la Torre, who had tortured a Cañari chief for treasure and then hanged and quartered him; Diego de Almagro the Younger claimed that chief Luis de León of Arequipa had hanged himself to escape persecution by Hernando Pizarro.* The dispassionate Pedro Cieza de León reported atrocities against curacas by Gonzalo de los Nidos and by Alonso de Orihuela 'who is still living, in this year

1550'. 'These were awarded certain chiefs and Indians. After having robbed them of all they possessed, they put the chiefs into pits up to their waists and then demanded gold. They had already given all they had and could give no more. So the Spaniards flogged them with whips; then, bringing more earth, they covered them to the shoulders and finally to the mouths. I even believe that a great number of natives were burned to death.' Many curacas also died in the massacre of Cajamarca or leading their troops in the battles of Manco's rebellions; others were executed in reprisals by Morgovejo, Francisco Pizarro and others.

But when the fighting and treasure-hunting ended, the Spaniards came to regard curacas as valuable helpers. Since the encomenderos themselves were not allowed to live in their repartimientos the collection of tribute became the responsibility of the curacas, who were themselves exempt from paying it.* Many curacas used their new autonomy to join the Spaniards in exploiting the natives. Bishop Vicente de Valverde, a sensitive champion of the natives – despite his sinister behaviour during the kidnapping of Atahualpa – was the first to deplore the cruelty of curacas towards their Indians.* Many later writers described similar excesses. The Viceroy Marquis of Cañete complained in 1556: 'The caciques are the ones who take everything from them. The Indians are so subjected to them, that no slaves are as domesticated or serve to such a degree. This arises because they have not begun to appreciate that they are free – at least as far as their caciques are concerned.' Gasca said at the same time, 'not only do the curacas punish them very severely, they also exact excessive work and tribute from them', and Santillán wrote that 'the caciques blatantly rob and fleece the natives'. Even Domingo de Santo Tomás wrote to his mentor Las Casas: 'The Indian natives of this land are very subjected to and obedient to their caciques' and a violent opponent of Las Casas said: 'The natives have no greater god than their own curacas, whom they hold in the greatest fear. The curacas do not leave them property, daughters, wives or freedom, but take all. No man dares complain about them to the law. But if one did complain, woe betide him! For they have a thousand ways to kill and rob them without being discovered.' Juan de Matienzo described the Indians as 'greater slaves than my own Negroes', and Miguel Agia wrote: 'I consider the natives' own caciques to be crueller and more tyrannical than the greatest Spanish tyrant in the world.'

The most common accusation against the curacas was that they extorted far more tribute than was required by the encomendero. 'The curacas are delighted if there are many levies and taxes, because if they have to collect ten, they collect fifteen and keep the balance; and the poor always bear the burden.' Bartolomé de Vega said that no individual Indian 'keeps more than his cacique allows him to keep, and the cacique takes all that he wants or can extract from his Indians under colour of their main tribute'.

Under the Incas the curacas had been responsible for assembling working parties to take a turn, 'mita', at labour for the Inca or his public services. Some curacas continued and abused this practice. Damián de la Bandera said that 'they themselves use the Indians, and they rent them out like beasts, keeping the payment themselves'. This traffic sometimes included women. Rodrigo de Loaisa claimed that 'not only do they take their wives and daughters. And the Indians are so wretched that they do not dare complain.' The outraged Dominican Diego de Vera said: 'I have often seen encomenderos and caciques remove the Indians' sons and daughters. And it is certain that the removal of Indian girls to the house of their encomenderos or of any Spaniard is tantamount to taking them to a house of public prostitution.'

Curacas varied as much as encomenderos. Many, possibly most, took good care of their natives. Santillán knew some curacas who 'make deposits of food to give to the poor or to pay tribute when there was nothing else with which to pay it'. Spanish encomenderos used to place harsh majordomos called sayapayas in their repartimientos to oversee the collection of tribute.* These majordomos sometimes supplanted dead curacas and became the most flagrant oppressors.

The stories of curacas' cruelty may also have been exaggerated, since it obviously eased Spaniards' consciences to be able to write: 'These demons emerged from among the Indians themselves, but . . . treated their own natives with more cruelty than any Spaniard.' Francisco Falcón said that Spaniards used stories of curacas' exploitation as a pretext for robbing the curacas. Far from being a lucrative sinecure, the collection and transmission of heavy tributes could be a thankless task.* Diego de Vera told how the best curacas tried to defend their Indians from Spanish abuse, and Antonio de Zúñiga had had weeping curacas come to him and say: 'Father, I do not understand this justice of yours. The assessment requires that I pay my encomendero so many pesos de oro and so many pieces of cloth every year. But some Indians leave my village to avoid spinning and weaving: they go to other districts and become vagabonds. Because of this the work quota falls on the few who remain in the village.' Licenciate Hernando de Santillán, a senior Spanish official, said that when curacas complained about excessive tribute, the Spaniards 'set fire to some and scorched or completely roasted them . . . they imprisoned others in dark prisons until they hanged themselves in despair . . . and kept them without food until they sent out to fetch the tribute that was lacking'.

The downtrodden natives had a third master in addition to their encomendero and curaca. The most important Spaniard with whom they were in direct contact and who was allowed to live among them was the priest. The conversion of the natives was the only obligation of the encomendero, who often was supposed to pay the doctrinal priest. The most oppressive encomenderos naturally tried to appoint priests of similar character, and before long such priests were milking the natives for an additional tribute.* Bartolomé de Vega reported to the King how this imposition had become officially recognised. 'Every repartimiento of Peru pays tax to the priest who instructs the Indians, in addition to the principal tribute they give the encomendero. The same assessors who signed the principal tax have fixed and assigned this one. The repartimiento of Ilabaya [near Arequipa] gives its priest the following tribute: 25 fanegas of maize [forty bushels] and 12 fanegas of wheat every year; one native sheep [llama] every month; a pig, or a sheep in its stead, every three months; a hen on every meat day; two arreldes (eight pounds) of fish on every fish day; six eggs every day; and also firewood, water, salt, fodder for his mounts, and the necessary personal services. This is the tribute that the Indians give the priest, to whom the encomendero pays three or four hundred pesos for instructing his Indians.' Some dissolute priests had native girls in their houses: the first ecclesiastical Council of Lima ordered that they should have 'no woman whatsoever, but only males, to prepare food'. Felipe Guaman Poma de Ayala said that children were sent for religious instruction from the age of four. 'The curates and parish priests took advantage of this to have concubines at their disposal: as a result they produced dozens of children, increasing the number of mestizos. . . . There are priests who have up to twenty children.' Other zealots maintained private stocks, prisons, chains and whips to punish religious offenders: blasphemers or converted Indians who renegued were given ten days in the stocks, a hundred lashes in public, and had their hair shaved.* Against this, there were many fine, conscientious priests in Peru, and the monastic orders produced the humanitarians who agitated so tirelessly for the welfare of the natives. The Peruvians were initially receptive to Christianity, and priests and friars went about the country baptising, converting, destroying pre-Conquest shrines and mummies, and trying to remould the Indians into zealous Christians. Many ecclesiastical restrictions were unpalatable and unsuitable for the Indians. But in the sixteenth century, as now, the only outsider who cared for the natives and devoted his life to them was often the priest.

The tribulations of the Peruvian natives did not end with the payment of agricultural and manufactured tribute to their encomendero, curaca and priest. Spaniards needed servants and labourers as well as native produce This service was supplied to some extent by the class of Indian called yanaconas. Under the Incas, some yanaconas had been artisans or specialists exempt from ordinary tribute and mita service; others were simply a shiftless proletariat without roots in tribal agricultural communities. The yanaconas advanced themselves with the arrival of the conquering Spaniards. They entered Spanish households as unpaid domestic servants, enjoying in return immunity from tribute and some of the comforts of their Spanish masters.* Their loyalty was tested during Manco's rebellions and most remained selfishly faithful to the invaders. Their numbers swelled and their status improved so that some became respected old retainers in Spanish households, others acquired the skills of Spanish artisans, as tailors, cobblers, smiths, barbers or silversmiths. Rodrigo Loaisa said admiringly that 'these yanaconas have learned our trades and succeeded marvellously in them . . . demonstrating their natural aptitude and ingenuity'.

The yanacona thus rose above the most humble agricultural and labouring work, and Spaniards turned to the docile hatunruna to provide such personal labour in addition to tribute obligations. They employed the men 'to fetch fodder for the horses; to fetch water, wood, etc.; in the kitchen garden and around the kitchen; they kept many Indian women for housework, cooking, nursing the children, accompanying and serving the ladies and their daughters; many Indian men were employed tending cattle, sheep, goats and so forth – for most citizens have taken to raising cattle on the lands of the Indians.' The natives were of course paid for such labour, but their wages were grossly inadequate – a mere six pesos and ten bushels of maize for a year's work. 'This manner of payment is far worse than if they paid them nothing but kept them as branded slaves in their houses: for a master gives his slave food and clothing and cures him when he is ill. But they make an Indian work like a slave and give him no food or clothing or medical attention: for the maize is insufficient to feed him and the salary not enough to clothe him.'

Diego de Robles said that 'in eighteen years over half the natives have expired in many repartimientos on the coastal plain, because of the excessive work given to them by their encomenderos and by people who employ them to work their lands and other kinds of property. It should be ordered that no curaca should hire out any Indian against his will to anybody . . .' Antonio de Zúñiga described the indifference of the authorities, who 'seem to wish that when it comes to work the Indians should be bodies without souls who can work like beasts, but when it comes to food they should be souls without bodies and not eat'. Miguel Agia gave a brutal definition of personal service as 'perpetual service that Indians perform for the Spaniards to whom they are entrusted in encomienda . . . labour without pay or distinction of sex or age, introduced by the force of the sword for the satisfaction of private individuals. In many encomiendas [natives] were not free men, but slaves.' Agia investigated one encomienda in which 140 of the 180 tributary Indians worked with their wives and children in a wide variety of 'personal service' for their encomendero. Even a royal decree that tried to correct the abuses of personal service admitted that the daily wage 'seems so little pay that it is scarcely different from working for nothing'.

Personal service could take many forms. It went far beyond simple jobs on the farm and in the pastures, which the natives found quite tolerable. From the outset of the Conquest, Spanish armies and expeditions had commandeered regiments of native porters, and it was manifest that this abuse contributed directly to the country's depopulation. There were dozens of grandiose attempts to discover eldorados in the forests of the Amazon. Hundreds of Spaniards lost their lives on these desperate adventures; but their native porters perished long before the European masters. 'Some two or three hundred Spaniards go on these expeditions. They take two or three thousand Indians to serve them and carry their food and fodder, all of which is carried on the backs of the poor Indians. . . . Few or no Indians survive, because of lack of food, the immense hardships of the long journeys through wastelands, and from the loads themselves.' 'In our days they are taken heavily loaded, in shackles and dying of hunger. There has been no expedition that has not cost thousands of Indian lives. They take them in this way and leave them there, all dead.' Some expeditions were notoriously bad. Diego de Almagro accused Pedro de Alvarado of killing innumerable natives on his march towards Quito; Cristóbal de Molina and Hernando de Santillán accused Almagro of similar atrocities on *his* expedition to Chile 'from which a hundred leagues of desert were left strewn with dead Indians'; and Almagro's son then accused Hernando Pizarro of allowing his men 'to take Indians in chains to carry what they had pillaged. . . . When Indians grew exhausted, they cut off their heads without untying them from the chains, leaving the roads full of dead bodies, with the utmost cruelty.' Licenciate Salazar de Villasante accused his enemy Melchor Vázquez de Avila of allowing his men to seize peaceful Indians from around Quito to take on the conquest of the Quijos forests. 'They took a thousand Indian men and women to the Quijos by force and keep them there to this day, deprived of their husbands, wives and families. Many of these have died in there from ill-treatment, heavy work and undernourishment; others have been put to dogs, with greyhounds set on to them, and others have committed suicide to escape being captive.' Hernando de Santillán accused the ill-fated Pedro de Ursúa of causing the deaths of all his Indian porters, 'carrying off an entire province and depopulating a Christian town'.

The situation was at its worst during the civil wars, when each army, royalist or rebel, pressed natives into service as it marched across Peru. One of Gonzalo Pizarro's captains, Francisco de Almendra, found artillery at Huamanga in 1543 and, 'loading it on to the backs of the barbarians it was carried by road to Cuzco' – over two hundred miles of hard country. Gonzalo Pizarro later transported this same artillery to Lima. At Parcos he half garrotted and then burned two chiefs alive. The surviving chiefs then rounded up six thousand men, women and girls to transport the army's baggage and artillery towards Lima* Pizarro's adversary Pedro de la Gasca described the situation in 1549: 'A great mass of natives has died as a result of porterage. They take them along, burdened down and exhausted by the sun and rough roads.

They have them tied in chains by day and throw them into stocks at night so that they will not flee. They march along with their loads, fifteen or twenty linked on a chain, with iron collars on their necks. If one falls, all must fall.' Although the natives did not fight, 'all the weight of the war falls upon the Indians'.

Even when there were no armies, Spanish individuals, merchants and encomenderos forced Indians into porterage service during these lawless years. Francisco de Morales complained that 'one of the things that has most destroyed this land is the burdening of the Indians. It is essential that Your Majesty pass an inviolable law that ... under no circumstances whatsoever may any Indian be compelled to carry.' The town council of Jauja admitted in 1534 that excessive porterage 'is to the detriment of His Majesty's orders and of the liberty of the Indian natives. Many of them have died from being loaded, or have left their towns and fled into the mountains'. The council of Quito forbade the transportation of natives outside the cities, even for conquests or expeditions. Domingo de Santo Tomás wrote that the natives 'are burdened by force', and Archbishop Jerónimo de Loayza said that 'they are always treated like hired mules; if they are given something to eat, it is so that they will go for more leagues'. President Pedro de la Gasca told the King that the loading of natives should be abolished. 'There is no part of the country where beasts of burden cannot now penetrate, and it is high time that the unfortunate natives should be spared the performance of such tasks.' Miguel Agia echoed this: 'cargoes should be placed on beasts and not on the frail shoulders of men'. The authorities encouraged the importation of mules for the mountains, and in 1552 even gave a licence to a Levantine, Sebrián de Caritate, to import camels for use on the coastal desert. But the camels were a failure; and a Viceroy was still complaining twenty years later that natives were being made to act as porters, even in flat country.*

Indian porters were theoretically paid for their services. But the natives were generally unaccustomed and indifferent to money. They therefore had to be coerced into working. The pretext for demanding such service from reluctant, 'lazy' natives was a so-called continuation of the Inca 'mita' system. The Inca mita had consisted of controlled labour quotas on public projects or on the lands of the Inca or of the official religion. It had been a reasonably cheerful communal effort in a society that did not use money. Spaniards now claimed that 'all tribute had been remitted in the form of labour' during Inca times and tried to usurp the system for selfish purposes.

There were, as always, many humanitarian Spaniards who brought these abuses to the attention of the King of Spain. One of the New Laws ordered that natives were not to be given excessive loads, and under no circumstances to be loaded against their will or without adequate payment – even in remote areas where there were no beasts of burden. 'Anyone who violates this is to be punished most severely, with no remission in respect of any person whatsoever.' This law was never revoked, but was repeated in June 1549, after the defeat of Gonzalo Pizarro. The King went further. On 22 February 1549 he issued a decree forbidding personal service or labour involving any element of coercion. Indians were supposed to be free subjects and must be properly paid if they were to work for others.* The decree was issued soon after Gonzalo Pizarro's rebellion. When it reached Peru President Gasca told the King that it was too explosive to reveal.* The King repeated it to the second Viceroy, Antonio de Mendoza, who entered Lima in September 1551. He also thought it prudent to delay publication, but it was finally issued in June 1552, when Mendoza, an old man, lay dying. 'All were greatly scandalised by it, . . . were highly indignant and held meetings to discuss it.' The oidores of the Audiencia of Lima had made various unauthorised limitations to the decree when they published it. They allowed Spaniards to employ Indians, provided they paid for the work at a fee fixed by the corregidor. They attempted to institute a system for assessing this wage, and their method was announced in Lima in September 1553 and in Cuzco in November. This coincided with a fresh assessment of encomienda tributes conducted by the notably pro-Indian Licenciate Santillán and the Dominican Santo Tomás. These reduced and sometimes halved the quotas of a number of encomiendas, and were accused of doing so without having visited the repartimientos or consulted the encomenderos.*

Once again, the home government had legislated too liberally in favour of the natives. The settlers regarded the decree on personal service, even watered-down, as an intolerable restraint. They took it seriously enough to rebel, commit treason, fight and die. The citizens of Cuzco revolted under a respected leader, Francisco Hernández Girón, in November 1553. The cities of Huamanga and Arequipa joined the insurrection. For a year the Spaniards of the southern Andes were in revolt, protesting their loyalty to the King but again demanding a freer hand in exploiting the native population. Girón was finally defeated at Pucará, a crag of red rock to the north of Lake Titicaca, in October 1554; he was captured at Huamanga in November and beheaded in December. The royal prohibition on personal service remained in force, and was reiterated in 1563.

The settlers had not embarked on a second hopeless rebellion simply to force the natives to be paid porters. There were far bigger financial enterprises at stake. With the introduction of more beasts of burden, porterage became unimportant compared with two highly lucrative industries that were growing in Peru: coca and silver. Each of these commodities had to be produced under gruelling conditions, and the natives would never undertake them without coercion. The settlers rebelled again in order to be free to force the Indians to perform such labour.

Coca is a bush that grows on the eastern slopes of the Andes. Its leaves, which resemble myrtle or laurel, are mixed with lime and chewed by Peruvian Indians to produce a tiny quantity of cocaine; this mild narcotic deadens the sense of hunger and fatigue. Under the Incas, coca had been one of the privileges of the royal family and priests; with the fall of the empire anyone could buy the leaves, and the habit swept the native population of Peru. Its use has diminished, although any modern traveller standing too close to a native bus risks being hit by a spinachy wad of chewed coca.

By the middle of the sixteenth century coca was used extensively in heathen rites, and was almost worshipped for its magical power as a stimulant. It formed a bond among the natives and was an important obstacle to the spread of Christianity. Because of this, coca was condemned and attacked with passion by ecclesiastics. Diego de Robles declared that 'Coca is a plant that the devil invented for the total destruction of the natives.' Juan Polo de Ondegardo, Martín de Murúa, Cristóbal de Molina and the first Augustinian missionaries all described its use as an offering or fetish in heathen rites,* and it was condemned at the first ecclesiastical council of Lima of 1551.*

Coca plantations lay at the edge of humid forests, thousands of feet below the natural habitat of the Andean Indians. This did not deter Spanish planters and merchants who made huge profits from the coca trade. They forced highland natives to leave their encomiendas and work in the hot plantations. The change of climate was devastating to Indians with lungs enlarged by evolution to breathe thin air. Antonio de Zúñiga wrote to the King: 'Every year among the natives who go to this plant a great number of Your Majesty's vassals perish.' There were also ugly diseases in the plantations. A tiny mosquito-like dipterous insect that lives between 2,500 and 9,500 feet in the Andean foothills carries the destructive 'verruga' or wart disease, in which victims die of eruptive nodules and severe anemia.* Coca workers also caught the dreaded 'mal de los Andes' or uta, which destroys the nose, lips and throat and causes a painful death. Bartolomé de Vega described the native hospital of Cuzco 'where there are normally two hundred Indians with their noses eaten away by the cancer'. Those who escaped the diseases returned to their mountain villages debilitated from the heat and undernourishment: they were easily recognisable, pale, weak and listless. Contemporary authorities estimated that between a third and half of the annual quota of coca-workers died as a result of their five-month service, and 'everybody who lives in Cuzco is well aware of Indians who die in the Andes and of the intolerable hardships they suffer there'. Even King Philip said in a royal decree that coca was 'an illusion of the devil' in whose cultivation 'an infinite number of Indians perish because of the heat and disease where it grows. Going there from a cold climate many die, and others emerge so sick and weak that they never recuperate.'

But the coca trade was too lucrative. When Cieza de León reached Cuzco after Gonzalo Pizarro's defeat, everyone was talking about this wonderful crop. 'There has never been in the whole world a plant or root or any growing thing that bears and yields every year as this does . . . or that is so highly valued.' Some coca plantations were yielding 80,000 pesos a year, and Acosta reckoned that the annual coca traffic to Potosí was worth half a million pesos. Such wealth produced a powerful coca lobby. Its protagonists defended the trade because it produced the only commodity that was highly prized by the natives. They argued that coca alone could inspire Indians to work for reward and to participate in a monetary economy. They also said that the trade was too large to suppress and formed too important a part of the Peruvian economy. A number of serious authors felt that coca might be beneficial to its users: they saw the feats of endurance of Indians whose sense of hunger and pain were deadened by the cocaine.* But the problem was best summarised by Hernando de Santillán who wrote that 'down there [in the coca plantations] there is one disease worse than all the rest: the unrestrained greed of the Spaniards'.

The Viceroy Marquis of Cañete tolerated the coca trade but tried to mitigate its hardships. He decreed that workers should serve only for twenty-four days, and be adequately paid. They must also be given a daily maize ration, even during Sundays, feast days and during the outward and return journeys. Royal decrees of 1560 and 1563 prohibited forced labour in the coca plantations, and a law of 1569 ordered Viceroys to protect the health of coca workers and to try to prevent the use of coca in witchcraft. Coca thus remained part of the Peruvian scene, although its use eventually declined somewhat as the natives' diet improved.

The other great industry that required native labour – whether voluntary or forced – was mining. At the height of Gonzalo Pizarro's rebellion, in April 1545, an Indian yanacona called Diego Gualpa discovered the silver mines of Potosí at the south-eastern edge of the Charcas altiplano. Gualpa climbed a conical hill in search of a native shrine and was thrown to the ground by a high wind. He found himself gripping silver ore and reported the discovery to various sceptical Spaniards at the nearby mining town of Porco. He persuaded Diego de Villaroel, his master's majordomo, to examine the find. Five fabulously rich veins were discovered, and a silver rush was soon in progress.* For a few years the local encomenderos were able to work the silver with natives from the area. The silver was close to the surface of the hill, and the Indians received good wages and were even allowed to prospect on their own account on rest days. The ores were refined in native smelting ovens that used strong winds to whip up the necessary heat. Indians of the altiplano were attracted to Potosí to gain silver to pay their tribute, and because there was a rumour that this high, desolate hill was free of disease.*

Potosí did not keep its attractive reputation for long. By 1550 the mine operators were having to recruit as far afield as Lake Titicaca. Persuasion soon gave way to coercion. Domingo de Santo Tomás was one of the first to draw attention to the horrific side of this remote mine. He wrote to the Council of the Indies in 1550: 'Some four years ago, to complete the perdition of this land, there was discovered a mouth of hell, into which a great mass of people enter every year and are sacrificed by the greed of the Spaniards to their "god". This is your silver mine called Potosí. So that Your Highness can appreciate that it is certainly a mouth of hell . . . I will depict it here. It is a mountain in an extremely cold wasteland, in whose district, for a distance of six leagues [twenty-five miles], no grass grows even for cattle to eat, and there is no wood to burn. They have to fetch food on the backs of Indians or of the llamas they possess. . . . The wretched Indians are sent to this mountain from every repartimiento – 50 from one, 60 from another, 100, 200 or more. No one who knows the meaning of liberty can fail to see how this violates reason and the laws of freedom. For to be thrown by force into the mines is the condition of slaves or of men condemned to severe punishment for grave crimes. It is not the law of free men, which is how Your Highness describes these poor people in your provisions and ordinances.' The letter went on to describe the hardships suffered by the natives and their families on the long, cold march to Potosí, and after they reached that bleak place.

The Crown piously decreed on 25 December 1551 that Indians could be assigned to mines provided they went voluntarily, attracted by good pay. But the natives hated work, and they particularly came to hate the hard work underground in the mines. They wanted to live a simple life, asking only for enough food to eat and some home-made alcohol to drink. They were unambitious and indifferent to money. But there was no one else to work the mines – Negroes could not survive at the high altitude of Potosí – and in the primitive conditions of sixteenth-century mining, hundreds of workers were needed. Royal good intentions and native reluctance were overcome by intense economic pressure. Potosí regularly produced 150,000 to 200,000 pesos' worth of silver a week, and the royal fifth amounted to a million and a half pesos a year. Mule trains loaded with silver crept across the altiplano, over the Andes and down to the Pacific coast to fill the treasure fleets that financed Spain's grandiose ambitions in Europe.*

The Potosí ores could only be smelted by using native wind furnaces, which could operate only during a high wind. The Spaniards were therefore lucky when, in 1559, a Portuguese called Enrique Garcés noticed natives mining vermilion and deduced that there must be deposits of mercury nearby. He opened a mercury mine at Huancavelica, high in the mountains between Huamanga and the coast; in 1563 its Indians revealed another fabulously rich mercury deposit to one Amador de Cabrera. Everyone had been hoping that mercury might be found in Peru, for experiments in Mexico were showing how it could be used in refining silver. There was jubilation at the sensational discovery of such vast deposits. For a time the Potosí ores resisted refining by mercury, but in 1571 Fernández de Velasco finally perfected a method for using Huancavelica mercury at Potosí. The Viceroy Francisco de Toledo convinced himself that the method worked. He explained excitedly that he would make of Huancavelica and Potosí the greatest marriage in the world.*

Native labour at Huancavelica reacted similarly to that at Potosí. For a time the mine was worked by voluntary workers from the neighbourhood. But as the workings grew and the conditions became worse, the authorities had to resort to compulsion. García de Castro said that the surrounding districts must provide small quotas, by force if necessary. A commission of enquiry found the mine operators prepared to pay well and provide reasonable conditions. But when evidence was heard from the Indians themselves it was concluded that only compulsion could make them work.

Why did the Indians try so desperately to avoid working in the mines? The few Spanish writers who penetrated them emerged appalled by what they witnessed. The Cerro Rico of Huancavelica, a mine that made a fortune for its operators, was entered through a great pilastered portal surmounted by the royal arms of Spain and cut into the mountain face. Beyond was a large gallery that gradually gave way to a labyrinth of narrow, twisting paths. There was no attempt to provide ventilation shafts for the deeper workings or to observe elementary safety precautions in cutting new galleries. The roofs were propped by weak green timbers that broke easily. To reach the face, miners had to negotiate staircases, caverns, difficult passages and low tunnels. There was no lighting, only a tallow candle held by each miner that emitted thick black smoke. Antonio Vázquez de Espinosa was horrified by the noise, confusion and intolerable smoke and smell as the workers crowded into the narrow spaces at the tunnels' ends.*

But the Huancavelica mercury mine threatened its miners more directly than through discomfort or the risk of collapse. When the Indians broke the hard, dry ore with their crowbars they were struck by a thick, toxic dust that contained no fewer than four poisons: cinnabar (sulphide of mercury), arsenic, arsenic anhydride, and mercury vapours. These caused severe damage to the throats and lungs of Indians already debilitated by heat, exhaustion and bad diet. It led to a dry cough and fever, and many who volunteered to work longer than the minimum requisite period died of incurable 'mal de mina' in which they coughed up a mixture of blood and mercury. The miners worked underground all day in the heat, wearing only thin shirts and trousers. They emerged suddenly in the evening into the cold of the 12,000-foot elevation and were given cold drinks by their wives. This easily led to pneumonia.* Miguel Agia, who visited Huancavelica, said that 'experience has shown that sending them to such work is sending them to die'.

Conditions were no better at Potosí. Here, too, one entered a fairly spacious tunnel, beyond which the mine descended in a great central shaft. This sank deeper and deeper, reaching a depth of some 750 feet by the end of the century. It was scaled by a series of three crude ladders made of hides, with steps of thong 22 inches apart. One chain of Indians descended on one side while another ant-like column laboured up the other. Rodrigo de Loaisa described the week-long shift that became standard. 'The Indians enter these infernal pits by some leather ropes like staircases. They spend all Monday on this, taking some bags of roasted maize for their sustenance. Once inside, they spend the whole week in there without emerging, working with tallow candles. They are in great danger inside there, for one very small stone that falls [down the shaft] injures or kills anyone it strikes. If twenty healthy Indians enter on Monday, half may emerge crippled on Saturday.' And Alfonso Messia described how the miners descended seven hundred feet 'to where the night is perpetual. It is always necessary to work by candlelight, with the air thick and evil-smelling, enclosed in the bowels of the earth. The ascent and descent are highly dangerous, for they come up loaded with their sack of metal tied to their backs, taking fully four or five hours step by step, and if they make the slightest false step they may fall seven hundred feet. And when they reach the top out of breath, they find for comfort a mine-owner who reprimands them because they did not come quickly enough or because they did not bring enough load, and makes them go down again on the slightest pretext.' The hide ladders often broke through excessive wear, rot and age. The Indians climbing them were supposed to carry standard pallets of a hundred pounds of ore, with a quota that increased to 25 pallets a day, or 12,500 pounds a week.* The mines of Potosí and Huancavelica did not reach full production until after the 1560s. When Gonzalo Pizarro's rebellion ended in 1548 these horrors were only just beginning to afflict the natives. During the first fifteen years of Spanish occupation, the period of untrammelled settler rule, the Indians' hardships came from the encomiendas and from abuse by individual Spaniards. But just as the authorities acted to correct these abuses, the greater menace of the mines grew in intensity.

There is much to be said in mitigation of Spanish behaviour. All the quotations in this chapter were written by Spaniards resident in Peru. There was no lack of champions, and their protests were eloquent and surprisingly effective. All the higher Spanish authorities, from the King downwards, were concerned to protect the natives of Peru from the worst excesses. Some of the remedies they attempted will be described in later chapters. Compared with other colonial regimes the Spaniards were distinguished by their efforts on behalf of the natives; and they did not suffer from sexual or racialist prejudices that corroded the colonial efforts of some northern European countries. The majority of the inhabitants of modern Peru are Quechua-speaking descendants of the Indians of the Inca empire, and there is a flourishing class of mestizos.

Any modern government demands the fruits of a large proportion of its subjects' labour. It could perhaps be shown that the proportion of the year that the Peruvian peasant spent working for himself was similar to that of modern industrial man – although there is no remote comparison in standards of living, and most of the tribute of the Peruvians went to enrich private individuals rather than for collective use by the State. When contemplating the horrors of the silver and mercury mines, it is worth remembering that all sixteenth-century mines, in Europe as in the Americas, were exploited under similar labour conditions. The sixteenth century was a hard time, and its cruelties were by no means peculiar to colonial America.

Many of the problems of colonial Peru stemmed from the complete incompatibility of its natives and the Spanish conquerors. Miguel Agia gave an excellent comparison, one that can be applied to Peruvian Indians to this day. 'The Spaniard and Indian are diametrically opposed. The Indian is by nature without greed and the Spaniard is extremely greedy, the Indian phlegmatic and the Spaniard excitable, the Indian humble and the Spaniard arrogant, the Indian deliberate in all he does and the Spaniard quick in all he wants, the one liking to order and the other hating to serve.'

19

EXPERIMENTS IN GOVERNMENT

THE decade of the 1560s during which Titu Cusi performed his prodigies of diplomatic evasion was a time of political experiment and expectant uncertainty in Spanish Peru. With the defeat of the last settlers' rebellion of Francisco Hernández Girón, Peru was finally at peace. The Spaniards had by now established a remarkably stable society in Peru. They continued to found municipalities, each on the traditional Spanish town plan. Plenty of Spanish women sailed across, from the earliest days of the Conquest. These ladies married the richest Spanish settlers and ensured that daily life in the colony was almost identical to that in the mother country. Spanish children born in Peru were reared as true Spaniards. The new municipalities had all the customary ingredients: large houses in which encomenderos lived surrounded by

A native alcalde tied to a picota and beaten for failing to produce his quota of mita labourers their families and servants, plenty of priests, nuns and friars, a prosperous merchant class, Negro and Morisco slaves and freemen, and all the necessary professionals – doctors; lawyers and the lesser procuradores; notaries to satisfy the Spanish passion for written records – and the artisans: tailors, masons and carpenters, barber-surgeons, shoemakers, saddlers, silversmiths and jewellers, smiths who also produced arms and armour and ironworkers who doubled as veterinarians.

With the exception of Cuzco and Quito, the municipalities in Peru were new creations. Most lay along the coast or on the roads that connected the mountains with the nearest outlet on to the Pacific. The centre of gravity of Peru descended from the Andes to the sea coast, where it has remained to this day. By 1560 there were some eight thousand Spaniards in Peru. Of these a third or more were involved in encomiendas or lived in the houses of the 480 encomenderos; a quarter had independent occupations; a third or more were rootless idlers who gambled, bullied Indians, filled expeditions into the jungles, or pestered Viceroys for sinecures. Most Indians remained unaccultured in the mountains, remote from these strange new communities that ruled their country.*

The authorities now had to evolve a system of government for the country. There was little precedent for a European colonisation of a remote and totally alien race. The first experiments in colonial rule were further complicated by the special circumstances of Peru. The country had clearly enjoyed admirably stable government before the Conquest, and the authorities now wished to preserve what was left of the large native population. But Peru had witnessed two revolts by dissatisfied colonists and was dangerously remote from the mother country. It also possessed silver mines that were rapidly becoming a prop of the entire Spanish empire. The Crown wished simultaneously to preserve its purity of Christian conscience and foster the Peruvian natives; but also to maintain its own revenues and reward the Spanish colonists – clearly incompatible aims. In this time of experiment, Peru seethed with political theories. An avalanche of memorials, essays and treatises advised the King how to administer the country, and the standard of political thought and recommendation was remarkably searching and idealistic.

Many political theorists wanted to find a place for the Inca nobility in the administration of Peru. The King felt compassion for the descendants of his fellow monarchs and remorse for the treatment of

Atahualpa and Manco Inca. Medieval Spain had a tradition of respect for the 'señor natural', the wise, legitimate, beloved ruler; and Francisco de Vitoria had stressed the legal rights of native kings.* Writers such as Luis de Morales, Cristóbal de Molina and Bartolomé de Vega painted pathetic pictures of the impoverished Inca nobility, dispossessed of the luxuries they had enjoyed in complete security only a few years previously, and now living without income amid the splendours of the defeated empire.

In fact the situation of the Inca royalty had improved by the 1560s. The only sons of Huayna-Capac to survive the native civil war and the Spanish Conquest – Manco and Paullu – were both dead, and their eldest legitimate sons Sayri-Tupac and Carlos Inca possessed rich encomiendas as a result of fits of generosity by Charles V, Gasca and Cañete. Some princesses had married conquistadores who held encomiendas, and others also received pensions from the Spanish Crown. This privileged group, which should have given inspiration to the natives of Peru, was thus enriched by tribute from encomiendas and embroiled in Spanish society. Instead of attempting to champion the native cause, the royal Inca descendants used the prestige of their lineage to obtain personal concessions. They were as eager for titles, coats of arms, fine Spanish clothes and unearned income, and as indifferent towards the proletariat, as any Spanish hidalgo.

Beneath the few lucky princes and princesses who received encomienda tributes, there were several hundred Inca nobles living in Cuzco. These traced their royal descent to one or other of the eleven Inca monarchs, and preserved the lineage or ayllu of their Inca ancestor. The lands around Cuzco that had been reserved to the Inca or the sun religion had been parcelled out among these royal descendants. Each thus had a small agricultural holding, and they were further exempted from paying tribute.* But these nobles were, like many curacas, often obsessed with 'dressing in silks or having horses, drinking much Castilian wine or having Spanish friends'. Because of their noble origins, any who became Christian were given the title 'Don', which was a distinguished mark of aristocratic parentage among Spaniards. But few natives learned Spanish, and even the nobles remained pitiful figures in post-Conquest society.

Some Spaniards were impressed by the tractability of the native nobility. Luis de Morales wrote that many Indians 'are able and of very sound judgement and intelligence. These are being wasted tending horses or guarding cattle.' The Spanish authorities therefore toyed with ways in which they could harness native administrative ability outside the oppressive encomendero–curaca system. Francisco Pizarro and Vicente de Valverde had tried to preserve native communities wherever they found them, particularly towns like Jauja, with 100,000 inhabitants, or Vilcashuaman with 40,000.* Their efforts were completely frustrated by the anarchy of the rebellions and civil wars. Vaca de Castro hoped to create new native settlements, but his only foundation was a town of Chachapoyas tribesmen established far from their homeland in the province of Vilcas.* Native self-rule could only operate outside the tribal curaca system if new settlements were founded. The priests wanted isolated natives gathered into such communities to ease the task of converting them. Their prompting led to a royal decree in 1549 that required the authorities to found more native towns in Peru. Each town was to have native 'alcaldes ordinarios, magistrates to judge civil actions, and also annual aldermen elected by the Indians from among themselves' to see to the good of the community.

Gasca wrote to the King after his triumphant return to Spain and suggested that native magistrates should be empowered to hear the lowest civil suits between natives, and to arrest delinquent Spaniards and escort them to the Spanish corregidor in each Spanish city.* The King in turn asked the new Viceroy, Marquis of Cañete, to consider such appointments* and Cañete had the corregidor of Cuzco, the formidable native expert Licenciate Juan Polo de Ondegardo, arrange such a system for the twenty thousand Indians living in and near the city. Polo divided Cuzco into four wards or parishes. The corregidor nominated eight candidates in each parish, from which the natives elected three, one of which was chosen by the corregidor to be alcalde.* Each alcalde administered justice among the natives but could not sentence to death or mutilation. They performed well, and it was soon reported that 'it has been seen by experience in Cuzco that the four alcaldes who have been appointed from the natives administer affairs so well that no robber or delinquent remains hidden from them, and they rapidly settle the suits that arise among them concerning land, water supplies, pastures, and other things'.

Cañete had already appointed the well-born Alonso Titu Atauchi as perpetual Alcalde Mayor de los Cuatros Suyos in Cuzco, with a special rod of justice. He tried a similar experiment in Quito, naming Atahualpa's cousin Don Mateo Yupanqui as Alguacil Mayor de los Naturales, Chief Constable of the Natives, with a staff of office and an annual salary of a hundred pesos. Don Mateo administered very efficiently for eighteen years and won the admiration of the Spanish corregidor. In 1563 the Audiencia of Quito decided to institute native alcaldes similar to those of Cuzco, with powers to 'arrest Spaniards who misbehave in their districts and take them prisoner to the corregidor'. Another worthy, Don Diego de Figueroa y Cajamarca, grandson of Huayna-Capac's captain-general and curaca of the Inca settlers in Quito, became one of these alcaldes. He taught native and Spanish children in his parish, built a fine church, founded settlements alongside Quito and rose to become the Alcalde Mayor of the entire province, to be obeyed by all its curacas in judicial affairs.* In the new coastal city of Guayaquil, a native curaca called Don Pedro Zambiza bravely helped repel an attack by Francis Drake. His reward was to be named 'Capitán y Alcalde Mayor de los Naturales' in Guayaquil, and he eventually succeeded Don Diego de Figueroa as Alcalde Mayor of Quito.* Cañete had also organised the first complete native community, with citizens, alcaldes, justice and council, at La Magdalena de Chacalea in the valley of Lima.

These tentative steps in native self-government were an exciting experiment in colonial rule. For a while political theorists in Peru were discussing bolder moves. The Audiencia of Lima in 1561 instructed its regional inspectors to reduce natives into new towns in which they would govern their own affairs.* Cañete died in 1561, and royal instructions encouraged his successor the Conde de Nieva to continue native appointments. Licenciate Juan de Matienzo suggested that a suitably hispanicised Indian from another district should be installed in each province for a two-year term as native corregidor and justice.* Archbishop Loayza of Lima urged that native alcaldes and councillors should be appointed everywhere to serve under Spanish corregidores, and Polo de Ondegardo and José de Acosta both felt the need for such officials.* The Spaniards had occupied a country that they knew to have been brilliantly administered. They could have entrusted much of its government to native officials, selecting candidates with enough authority to break the collusive rule of encomenderos and curacas. Such boldness would have put the government of Peru centuries ahead of other colonial regimes, and given substance to the official claims that Indians were equal subjects of the Spanish king.

Unfortunately, despite the successes of the first native officials, Peru's governors hesitated to go further. Cañete had given only the very humblest civil jurisdiction to his native administrators, and he laid down a forbidding list of conditions for the candidates: they must be of noble descent, be Christian, speak Spanish, be worthy, capable, just, virtuous, clean, of good appearance, and know some law. But he told the King that he hoped to revive the native population by founding more self-governing native communities.* Cañete's successor Nieva was lackadaisical about native affairs. He tried establishing two native alcaldes for the yanaconas of Lima, despite local protests. He also conceived new officials called native judges who would judge the smaller civil suits and be allowed to arrest Spaniards provided they immediately handed them over to Spanish authorities. The innovation did not take root.*

The next governor of Peru, García de Castro, was well aware that the King keenly wanted the natives reduced into new towns and settlements. But his efforts to make them leave their scattered dwellings foundered completely: the natives refused to move and Castro was too indecisive to force them.* García de Castro sent Dr Gregorio González de Cuenca to settle the administration of some thirty repartimientos around Trujillo on the north coast, and his detailed ordinances allowed the natives there a degree of self-government in their villages. President Castro also tried installing two Indian alcaldes to hear native suits in each encomienda repartimiento. He specified that curacas could not serve in this capacity; but his alcaldes did not have enough authority or official support to make any impression on those already entrenched in encomienda government.* Perhaps the natives were too cowed and demoralised by a quarter of a century of Spanish occupation to produce enough acceptable candidates. More probably the curacas were too firmly in control of their tribesmen. The Spanish governors of Peru did not have the courage to surrender sufficient power from Spanish to native hands. The hesitant moves towards native self-government therefore never realised their full potential.

An older tradition of Spanish colonialism was the appointment of Spanish protectors of the Indians. Before Pizarro even sailed towards

Peru, his and Almagro's partner Hernando de Luque was instructed to protect the natives of the undiscovered territories. The cleric Luque never went to Peru, but his instructions were repeated to Vicente de Valverde in 1536, and he was officially appointed 'protector of those provinces' in April 1538.* Valverde welcomed the appointment. He had been appalled by the devastation he saw when he travelled inland after Manco's rebellion, and wrote to the King about his appointment: 'It is essential to defend these people from the jaws of the wolves who threaten them. I believe that if someone is not specifically concerned to defend them the country will be depopulated. The Indians have been overjoyed ... to know that Your Majesty has sent someone specially to protect and defend them.' Valverde's responsibilities as protector were vast; but his powers of enforcement were negligible. The protector was empowered to send inspectors anywhere – but had few to send. He could hold enquiries into cases of abuse; could recommend to the civil governor that an individual be deprived of his Indians or given corporal punishment; or could himself punish offenders with up to 50 pesos' fine or ten days' imprisonment. Valverde tried some cases in Cuzco. Records have survived of two instances of abuse of female servants by Spanish encomenderos; Valverde issued fines and imprisonment, and appeals by the Spaniards to Francisco Pizarro failed to reverse the sentences. But one lone bishop could not restrain a legion of cruel conquistadores.*

Valverde was succeeded as Protector of the Indians by Bishop Loayza of Lima, and other leading ecclesiastics held the position in other parts of the Indies. One of the most notable was the saintly Juan de Valle, whose bishopric lay just north of the limit of the Inca empire. King Philip wrote to Valle in 1557: 'We have learned of the care you have taken in the conversion of the natives of that province and in their good treatment; how you have striven and continue to strive to protect and defend them and to prevent them being harmed; and the persecution and hardship you yourself endure for this. We are most pleased .. and charge you to continue what you have been doing to care for these people.' But these well-meaning bishops had no real power to enforce good treatment. One of Valle's chaplains told how he had arranged a written tribute assessment and forbidden porterage, and how he hopefully 'issued many orders to the Spaniards on how to save their consciences regarding the Indians ... how to behave in future and in the restitution they must make to them'. The bishops did what they could to exhort and shame their congregations, but the duty of protecting the natives devolved in practice on the civil authorities.

The machinery of central government was tenuous. The Viceroy in Lima presided over an audiencia with four oidores, judge-councillors, and a fiscal or prosecutor. The Audiencia of Lima was founded in 1542; in 1559 Philip II created the Audiencia of La Plata or the Charcas, in 1563 that of Quito, and in 1565 that of Chile. The presidents of these three subsidiary audiencias acted as local governors and captain-generals, and each audiencia had four or five oidores and a fiscal.

Royal rule extended into the provinces by means of the corregidor, an official who lived in each Spanish municipality and presided over its council. The corregidor was originally an unpaid, honorary official, generally a leading encomendero. Cañete reduced the number of corregidores to twelve, gave them an adequate salary, and tried to appoint well-educated, competent men such as Juan Polo de Ondegardo. Each Spanish town also had its municipal council, cabildo, headed by alcaldes (mayors or justices) and containing regidores (aldermen). This municipal government was entirely in the hands of encomenderos: holders of encomiendas had to live in towns and were naturally the richest citizens. The only presence of central government or legal protection in the Peruvian countryside came from an annual visit by an alcalde and a regidor from the nearest town. These made rapid summary hearings and delivered sentences. The cabildos occasionally legislated on behalf of the Indians, if only to keep them alive as a willing work force; and the visiting magistrates sometimes acted to protect natives from rapacious Spaniards. The corregidores were also instructed to protect the natives and to punish offences against them as severely as if they had been done to Spaniards. But the corregidores were responsible for vast tracts of mountainous countryside and were kept too busy in their towns to administer beyond. And it was rare for these local officials to punish their fellow encomenderos. The natives, therefore, were effectively unprotected by any royal authorities. Most were probably unaware of the existence of other powers beyond their immediate masters: encomendero, curaca and priest.

At the close of the civil wars, this lack of protective authority was obvious to any sympathetic observer. It was clear that the natives were at the mercy of a group of predatory masters. If they were completely ignored and abandoned, their numbers would decline and they would never embrace Christianity spontaneously. One solution would have been to restore real authority to native administrators. But most political theorists shrank from that bold solution and favoured instead the concept of increased protection of the Indians as a vulnerable inferior social order. The natives had been so abused during the anarchic years of the Conquest and civil wars that they seemed incapable of asserting their rights as free citizens. The law already discriminated against natives by making them liable to encomienda tribute and all its attendant horrors; Spanish vagabonds who entered Peru had no such obligations. The natives were never given a chance to prove themselves as free citizens, and it was asking far too much to expect them to survive the shocks of the Conquest, to acclimatise to a competitive monetary society and a written legal code in a foreign language, and to ward off aggressive armed members of the conquering race. Well-intentioned Spaniards therefore concluded that the natives must be defended by special legislation, with Spanish protectors assigned to enforce their collective rights. But such special legislation meant that natives lost their equality as free subjects of the King of Spain. They sank instead to the status of legal minors. Gone were the fine ideals of the New Laws, and the hope that the Peruvians would advance from being regimented subjects of the Inca to becoming free Christian citizens of a European monarchy.

Soon after the defeat of Gonzalo Pizarro, the Audiencia of Lima asked the King for the right to appoint a protector in each town 'for the better treatment and protection of the Incians'. Gasca insisted on the corregidor's duty to protect the natives, and so did Cañete.* Cañete went further, and tried installing provincial corregidores with wide protective powers. But all these good intentions ignored the problem of finding capable and conscientious Spaniards. To do the job properly, a protector had to be compassionate with his charges, adept in Spanish law, aggressive and inflexible towards his fellow countrymen, and an altruist unconcerned with financial reward. Few such paragons had made their way to sixteenth-century Peru. Someone tried installing a protector in repartimientos that had lapsed to the Crown, but Juan de Matienzo told the King in 1562 that these had been removed 'because of the robbery, ill-treatment and fraud they have done to the Indians'. Cañete soon removed his provincial corregidores, saying that they were thieves.* And Viceroy Francisco Toledo later removed all regional protectors other than the Archbishop, 'because each and every one of them was robbing and milking the Indians and oppressing them in dozens of lawsuits'.

Despite these ominous portents, Governor García de Castro convinced himself that some new form of authority was needed to administer and safeguard the natives. His solution was the 'corregidor de indios', a Spanish official who would administer the natives, a counterpart to the Spanish corregidor who exercised central authority in the Spanish cities. García de Castro pressed ahead with his plans in spite of the scepticism of the religious authorities. He explained to the King how the new officials could prevent sedition among the natives, could oversee their reduction into larger towns, administer justice among them and stop the priests doing so, regulate the collection of tribute by the curacas, maintain a census of the country, and protect the Indians from various abuses. The new corregidores de indios were to be paid by an annual charge of two silver tomines from every Indian tributary whom they found to register.* The Governor's brainchild was to be given wide powers at the expense of the existing local authorities.

Poor García de Castro, a well-meaning but ineffectual man, cannot have foreseen the storm of opposition that would be aroused by his new officials. The most vocal opponents were the ecclesiastics, who stood to lose the jurisdiction that had effectively been usurped by the priests in each native community. Archbishop Loayza argued that the Governor could never find enough worthy laymen for the new posts: the task of Indian protection should be left to the priests supported by a central inspector. Francisco de Morales, the Franciscan Provincial, condemned officials who 'enrich themselves and their relatives and friends at the expense of the blood of those innocents'; Rodrigo de Loaisa said that the corregidores made fortunes trading with their charges, and encouraged their vices; and Reginaldo de Lizárraga said that the prior of his Dominican convent had refused absolution to the new officials.* The ecclesiastics supported their violent verbal attack with a campaign of non-cooperation with the corregidores de indios. They refused, for instance, to produce their baptism and marriage records to help the new officials find elusive tribute-payers.

The church authorities roused the natives themselves against the new corregidores de indios, even though the supposed function of the new officials was to provide added protection to the downtrodden. The curacas joined the opposition. The Licenciate Francisco Falcón helped them prepare a petition presented to the Audiencia of Lima in July 1565. This sensibly urged the gradual increase of native self-government through Indian alcaldes and regidores. It also claimed that no new official was needed to control the docile Indians, and that the salaries of the corregidores de indios should in any case be paid by the encomenderos rather than by the Indians themselves. The petition was to be forwarded to Spain, and Archbishop Loayza urged the curacas to collect money for a financial sweetener to go with it. The curacas' arguments were plausible but their motives were suspect: for the new officials threatened to curtail the power of the curacas as much as that of the priests.*

Further opposition came from other threatened authorities: the cabildos of the Spanish cities (which were composed of encomenderos); the existing corregidores, who were reluctant to surrender their jurisdiction over the natives; and even the relatively impartial audiencias. The Audiencia of the Charcas sent García de Castro a well-reasoned memorandum. It stressed how the new office would give rise to additional suffering for the poor natives, by demanding yet another levy, payable not in kind but in silver. So great had been the outcry among both Indians and encomenderos in the Charcas that the Audiencia had had to suspend this levy. García de Castro argued that any corruption by his new officials would immediately be exposed by disgruntled priests, encomenderos and curacas, and could then be severely punished. But the Audiencia of the Charcas warned of the risk of the corregidor de indios joining the other three exploiters in a conspiracy of silence.* It also felt that the new authorities would be impotent without a settled base such as the existing corregidores enjoyed in the towns: the natives should therefore be reduced into larger communities before the creation of these new posts.

Governor García de Castro changed the terms of operation of his corregidores de indios to meet some of these objections. But he continued to believe in the need for the new officials to fill the administrative vacuum in rural Peru. He was convinced that the ordinary natives would never be protected and the fledgling native alcaldes could never survive bullying by curacas without help from this new authority. He therefore began the introduction of corregidores de indios, experimentally at first in the Lima area, and he continued for years to press the King for an edict confirming his new creation.

While García de Castro and others were pondering ways to administer the natives, Peru had for many years been racked with heated debate about the future of the encomienda system itself. The Crown regarded the original encomienda grants as temporary expedients to ensure the occupation of newly-won territory and to reward its conquerors. Encomenderos had no title to land and no jurisdiction over the natives in their repartimientos. They simply enjoyed receipt of heavy tributes during their own lives and those of their immediate heirs. The recipients were frustrated by the limitations in these awards, and became increasingly alarmed at the prospect of losing their entire source of income at the end of the second lifetime. They longed to have their encomienda grants made perpetual.

The struggle for perpetuity had been waged sporadically since before the conquest of Peru. For a time during the 1530s it looked as though the Crown might concede perpetuity, but the royal conscience swung strongly against the encomenderos in the New Laws of 1542. These declared that the entire system would soon cease and no further encomiendas would be granted. As soon as Gasca had defeated Gonzalo Pizarro and rewarded dozens of loyalists with encomiendas confiscated from rebels, the Peruvian settlers organised a fresh lobby for perpetuity. In 1549 they sent the Dominican Provincial in Peru, Fray Tomás de San Martín, and one of the most respected first conquistadores, Captain Jerónimo de Aliaga, to request perpetuity from King Charles I. The Council of the Indies heard the petition in 1550. Most members favoured perpetuity, but an opposition headed by Las Casas, San Martín (who had fallen under the spell of Las Casas and changed sides), and Gasca delayed the decision.

The arguments in favour of perpetuity were always similar, and they had a persuasive ring to them. A perpetual grant would give the encomendero a continuing interest in the welfare and improvement of his repartimiento. He would invest capital in the land, teach his Indians better farming methods, nurture and protect the natives, and stimulate economic activity in the area. With grants of limited duration, the encomenderos were 'mercenaries and not farmers' who 'tried only to drink the sweat of the Indians before departing'. But if the grants were perpetual they would become a stable, loyal aristocracy rich enough to defend the colony for the Crown.

At the beginning of 1554, at the height of the rebellion of Francisco Hernández Girón, loyal encomenderos met in Lima to prepare a second petition. They requested perpetuity, permission to reside in their repartimientos, and civil and criminal jurisdiction in the second instance – which would allow them to hear complaints or appeals by Indians against curacas. They chose Antonio de Ribera as their emissary, and he reached Charles I in Brussels in late 1555. His request was reinforced by the most potent possible sweetener: a promise of 7,600,000 pesos if the requests were granted. The 467 encomenderos of Peru were then receiving gross annual tribute of some 1,200,000 pesos, from which they had to provide priests. Their offer therefore represented seven years' purchase: a cheap price for properties with steady earnings and no foreseeable risk. How they proposed to borrow or scrape together this great sum was never fully discussed.

Charles I abdicated in January 1556, and Philip II became King of a bankrupt country. He was naturally strongly tempted by the cash offer and wrote in September to the Council of the Indies that he had decided to accept. He planned to create a creole aristocracy, selling titles and coats of arms, and to give the hereditary encomenderos full local jurisdiction subject only to appeal to central tribunals. The Council was appalled. To its credit it sent back a vigorous condemnation of the proposal. It warned the King that perpetuity would favour a few hundred encomenderos at the expense of thousands of other Spaniards now settled in Peru; and it would mean perpetual slavery for the natives. The grant of sweeping jurisdiction would be contrary to the original papal donation, and it would be folly to give such powers to a group of turbulent settlers, many of whom had recently rebelled. Such determined opposition to the decision of the new King was a brave action by the Council. Philip had allowed his need for money to overshadow the humanitarian arguments that had so impressed his father.

Opposition was also gathering momentum in Peru. A group of curacas met in Lima in 1559 and appointed Domingo de Santo Tomás and Bartolomé de las Casas – both of whom were then in Spain – to represent them in a counter-proposal to the King. They offered a magnificent bribe: 100,000 ducats *more* than the highest offer made by the encomenderos. But they asked for revolutionary concessions. Encomiendas were to revert to the Crown on the death of the present holder, and no further awards of Indian land were to be made to Spaniards. Encomenderos and their families were to be forbidden to enter their repartimientos on any pretext whatsoever. The tributes of individual Indians were to be related to their ability to pay, and tributes on Crown repartimientos were to be halved. Native lords were to be granted a number of privileges and concessions, and the natives were to have representative assemblies to comment on important issues concerning them. This petition was presented to the King in 1560, and Philip took it seriously enough to ask his representatives in Peru to consider the proposals and send their views.

King Philip had already sent a commission to Peru to investigate the entire question of perpetuity. He secretly authorised his three commissioners to grant perpetuity if they saw fit, but said that he preferred them 'to send us a very complete and detailed report with your recommendations'. The champions of the two parties, Antonio de Ribera and Domingo de Santo Tomás, also returned to Peru in 1560.* The debate on perpetuity continued throughout the decade of the 1560s in an orgy of rhetoric, abuse, machination and idealism. The King and his commissioners invited comment on the subject from any dignitaries in the Viceroyalty; the result was an avalanche of treatises.* During 1562 the commissioners sent pairs of emissaries to hear the views of Spanish citizens and native chiefs in the provinces of Peru. The emissaries to the central highlands were both well qualified: the ardently pro-Indian Domingo de Santo Tomás and the more sober Polo de Ondegardo, who was now an oidor of the Audiencia of La Plata.

The results of these enquiries were predictable. Santo Tomás held meetings of curacas in various towns along his route and reported that 'in all of them the Indians, after having fully comprehended the matter, were emphatic that they wished to be directly subject to Your Majesty [rather than to encomenderos]. They give ample reasons for this – their arguments could not have been better advanced by academics.' Another group of curacas met near Lima in January 1562 and nominated some liberal ecclesiastics and royal officials to press the native case: that both Crown and natives would benefit immeasurably if the encomenderos were pensioned off and the natives placed directly under royal administration.* Meanwhile the clergy of Peru launched into such violent sermons condemning perpetuity that the Viceroy Conde de Nieva had to summon the heads of the religious orders to say that he would severely punish any priests who continued to excite the natives. Santo Tomás returned from his mission to the sierra convinced that all the natives wanted to be incorporated under the Crown, and that the encomenderos could not possibly pay the sums promised by Ribera. The Viceroy was sceptical and suspicious: he was sure that the priests were plotting to win Peru for the Papacy in collusion with the rapacious curacas.*

Opposition to perpetuity also came from the mass of ordinary Spaniards whose fathers had not won encomienda awards. Some of these were merchants and artisans, others newly arrived adventurers known to the authorities as vagabonds, and others were mestizos uncertain whether they belonged to Spanish or native society. Some such citizens of Cuzco wrote to the commissioners explaining that the Peruvian economy depended on mobility of native labour, for farming, mining and trade, and that the encomenderos were restricting its movement. Other unprivileged Spaniards and mestizos planned pathetically inept rebellions.

A middle view on perpetuity came from some royal officials, notably Juan Polo de Ondegardo, Juan de Matienzo and Hernando de Santillán. Polo had attempted to tell his meetings of natives that there could be benefits from perpetuity, and he did the research for a report on past and present practices of curacas in raising tribute, mita labour and the administration of law and religion. Polo, Santillán and Matienzo supported perpetuity of encomiendas provided no jurisdiction passed to the encomenderos. Matienzo was the most highly qualified juridical author in Peru during the century. He now wrote his famous *Gobierno del Perú,* the first full-length attempt at a political programme for the colony. Matienzo was sympathetic to the Indians, anxious to protect them from the most flagrant abuse and to train them towards Christian civilisation. But he was paternalistic towards the natives and pessimistic about their uncompetitive character; and he was convinced of the rectitude of the Spanish Conquest.

Both Matienzo and Polo stressed the historical fact that the Incas had conquered Peru and ruled – 'tyrannised' – it for only a few generations. This fact had been known since the beginning of the Spanish Conquest and was often repeated in enquiries into pre-Conquest life and interrogations of aged Inca officials. Only now did Spanish theorists console themselves and justify their Conquest by arguing that they had simply supplanted other equally unwelcome conquerors.*

The encomenderos themselves continued to press their full demands. The commissioners wrote to the Spanish cities in March 1561, inviting them to debate perpetuity in their cabildos and send delegates to negotiate terms with the commissioners in Lima. By mid-1562, eight cities had responded, offering between them 3,938,000 pesos over eight years as the price for perpetuity with jurisdiction.*

The commissioners heard all this evidence and produced their report and recommendations. They were disparaging about the natives, whom they considered stupid and without capacity or intelligence. 'They have no means of governing themselves and no other identity or will. They would have no idea what to do tomorrow unless they were given orders.' The curacas would take charge if the encomenderos were completely removed, and 'it is not inconceivable that the natives would receive greater harm and oppression from them than from the encomenderos'. The commissioners favoured the reduction of Indians into larger communities, as was done by the Incas 'whose government in human affairs we understand to have been so very good'. The natives in the new communities 'should have their own government – alcaldes and regidores in the form of a council – with some laws and ordinances for good government. These alcaldes should exercise civil and criminal justice in the first instance.' Suspicious of clergy and curacas, admitting the misbehaviour of many encomenderos, and wanting to swell the royal exchequer, the commissioners recommended a three-part compromise. A third of the encomiendas were to be sold in perpetuity with jurisdiction in the second instance to the most deserving encomenderos, who would thus provide a loyal aristocracy. A third were to lapse to the Crown on the expiry of the existing two-life tenure, provided the curacas paid as they had promised. The remaining third were to lapse to the Crown, but then be reassigned for one lifetime to deserving Spaniards.

Their work done, the commissioners sailed home to Spain. Unfortunately, while they were on the Atlantic some secret papers of the Viceroy Nieva fell into the hands of a Peruvian resident sailing to Spain in another ship. The documents revealed that the commissioners had conspired with the Viceroy in extensive illegal activity during their stay in Peru. Together they had sold lapsed encomiendas, seized and sold

Crown lands, sold confirmation of encomiendas to their existing holders, sold offices and sold justice. They made such vast illegal fortunes that the group was obliged to organise a network of intermediaries to buy contraband and smuggle this back to Spain through three pliant sea-captains. Such massive corruption was remarkable even in sixteenth-century Spain. One commissioner had died in Peru, but the other two were arrested as they landed: one received a long prison sentence and the other was barred from public office for a period of years. The Crown at once sent Licenciate Lope García de Castro to arrest the Viceroy Conde de Nieva. But García de Castro did not reach Peru until October 1564, and his quarry had died suddenly the previous February.

Nieva was a notorious libertine who had already silenced priests who dared to criticise his immorality. Licenciate Monzón reported to the King that the repartimiento of Surco near Lima 'has been assigned for the recreation of certain ladies, from which great dissolution results, both for them and for any third parties who intervene'. Nieva's taste in women was for the highest-born ladies to be found in Peru, and many of these were married. On the night of his death he had gone out at midnight in his carriage and was brought back dying at two in the morning. The oidores circulated an official report that the Viceroy had died of apoplexy, and Archbishop Loayza repeated this to the King.* But gossip was rife that he had been stabbed by a jealous husband, or 'killed one night by sacks of sand when he was caught climbing a ladder on to a balcony', or from 'a tumour on his private parts'. Whatever Nieva's fate, the report signed by him and the commissioners was discredited by the revelations about their misconduct. Their recommendations for the future of encomiendas were therefore ignored.

The Crown did not accept the flamboyant offers of the encomenderos or the curacas. The ranks of the encomenderos were steadily being thinned by deaths of the second occupants, and they were losing their financial pre-eminence: the fortunes of mine operators and coca traders far surpassed the agricultural tributes of the ordinary encomenderos. The strength of the encomendero lobby thus gradually declined, but the question of perpetuity remained unresolved, and, with it, the future government of Peru.

Juan de Matienzo's *Gobierno del Perú* was sent to Spain in late 1567 and read with interest by members of the Council of Castile, since its author was an eminent lawyer. Matienzo's policy towards the natives was colonial: kindly but realistically harsh. At the same time one Luis Sánchez presented a more stridently liberal appeal to the President of the Council of Castile. He demonstrated that the Peruvian Indians were the worst afflicted of any in the Americas and deplored the failure of local administrators to enforce royal laws.* The King had also received the report of his discredited commissioners and many other conflicting political theories.

In 1568 King Philip finally chose a new Viceroy to succeed the Count of Nieva. He selected Don Francisco de Toledo, fifty-three-year-old brother of the Count of Oropesa and a man who thought that his career of minor public service was at an end. Before embarking for his distant appointment, Toledo asked the King to hold a special Junta to discuss the affairs of the Indies, and particularly the future of encomiendas. The special meeting, known as the Junta Magna, met in the house of the most powerful man in Spain, Cardinal Diego de Espinosa, President of the Council of Castile. The secret deliberations of the Junta have not been published, but it evidently debated the various issues that would determine the future of Peru and its inhabitants: the machinery of administration, conversion of the Indians, future of encomiendas, assessment of native tribute, role of native aristocracy, reduction of Indians into larger communities, and provision of labour to work the all-important mines.* The new Viceroy Francisco de Toledo was instructed to implement the wishes of the Junta on these crucial points of policy, but was left wide discretion about the way he did so.

Toledo proved to be a man of outstanding ability, one of the world's great colonial administrators. He landed at Trujillo and marched down the coastal desert to reach Lima on 30 November 1569. There were great ceremonies in which Licenciate García de Castro and the Audiencia of Lima handed over the government, and the Viceroy was received by Archbishop Jerónimo de Loayza, the clergy and citizens. During his viceroyalty, Francisco de Toledo established the pattern of Spanish colonial rule that was to last until the nineteenth century. He also settled the fate of the Peruvian natives.

The three exploiters: curaca, encomendero and priest

20

TOLEDO'S SOLUTIONS

During the twelve years of his viceroyalty Francisco de Toledo tried to settle the future of Peru's native population, dictating a massive body of legislation covering most aspects of native administration. Toledo was a tireless official, and he insisted on seeing everything for himself before deciding on the remedy. He was honest and honourable but cold and unfeeling; a man with the brain and energy of a fine lawyer but the temperament of an ascetic. His commanding presence won respect and obedience, and during his term the office of Viceroy gained power and prestige. But Toledo's autocratic methods made enemies among the clergy and audiencias of Peru, and he had few friends in the Council of the Indies in Spain. Even King Philip was cool towards this outstanding public servant, although it was the King who kept Toledo in office despite opposition from the Council of the Indies. Toledo constantly tried to further the interests of the Spanish Crown and Church, but in so doing he tried to remain strictly within the limits of legal and ecclesiastical custom.

Toledo's attitude towards the Peruvian natives was closely inspired by the proposals in Juan de Matienzo's *Gobierno del Perú*. The natives were to be protected from abuse, but it was paternalistic protection by a superior civilisation. The Peruvians had failed to win equality in colonial society. Toledo now brushed aside any pretence that Indians enjoyed the rights of free citizens. His legislation defined their position as that of a docile proletariat, inferior subjects, but people with certain legal rights that should be protected by the authorities.

Although Toledo's statutes legislated in great detail on many aspects of native life – too much detail to be easily enforceable – he is remembered for settling three fundamental problems: the reduction of natives into towns, the organisation of the mines, and the establishment of forced labour on a national scale.

The reduction of the natives into larger towns and cities had long been seen as a panacea for Peru's problems. As early as 1503, before the discovery of Mexico or Peru, a royal instruction had ordered that 'it is necessary that the Indians be gathered into towns in which they would live together'. The same order was constantly repeated to successive colonial governors; and reduction into towns was recommended by innumerable political theorists, official or amateur, who submitted their ideas to the King. Philip II reiterated the order to García de Castro in 1565* and it formed one of the main instructions to the new Viceroy Toledo from the Junta Magna.

Official enthusiasm for reductions arose from a feeling that the natives would be more manageable if they were easily accessible in a town. Spaniards also believed that civilisation could be achieved only by living in a city. In the words of the royal decree: 'It is something very convenient and necessary for the increase of the Indians, so that they could be better instructed in the articles of our Holy Catholic Faith and would not wander scattered and missing in the wilds, living bestially and worshipping their idols.'

Toledo accepted the challenge of a task that had daunted his predecessors. In two sets of instructions in 1571 he appointed 'visitadores': inspectors who were to examine every corner of Peru to organise the reduction of its inhabitants. The inspectors were to inform encomenderos, curacas and natives of their mission. They were to conduct a census and investigate a long list of questions about the history and administration of each area. They were to enquire whether the Indians were being abused by their encomenderos or curacas, and to inflict public punishment and restitution if there had been excesses. They were to investigate the region's capacity to pay tribute, reassess its tribute as appropriate, and fix the daily wage for labourers 'so that the Indians should have more adequate salaries and food than they have been given up to now'.

'But the chief reason for this general inspection is . . . that Indians who live apart and scattered should be reduced into towns arranged with orderly planning in healthy places of good climate.' The inspectors were to try to enlist the co-operation of encomenderos and curacas; were to choose a site for the new town; draw out the streets and square on an orderly grid; provide for a church, town hall, prison, chief's house, and dwelling for each Indian family with a door on to the street for easy supervision; arrange for the Indians to move into the new town within a short period; and see that their former houses were then destroyed.*

The Spaniards had no difficulty in justifying the reductions. Toledo claimed, as usual, that it was being done in the natives' interest. 'They have been left amid their idolatries, drunkenness and concubines in order that the Spaniards and their caciques could tyrannise and exploit them more easily, with no guardian [to protect] their pay [or ensure] justice and good treatment.' 'To learn to be Christians they must first learn to be men, and be introduced to government and the politic and reasonable way of life . . . It is not possible to convert these Indians nor to make them live in a civilised way without removing them from their hideaways.' The Spaniards were distressed at being unable to control what went on in remote native farms. The Jesuit Bartolomé Hernández wrote that they lived 'in tiny huts, very dirty and dark, where they gathered together and slept like pigs. They got drunk there, and fathers became involved with their daughters or brothers with sisters, with no restraint of any kind. The remedy is that these Indians should be reduced into towns. And the Viceroy [Toledo] has taken this solution very much to heart and in a very Christian manner.'

The Indians saw it all very differently. Peru is a vertical country, a washboard of steep hillsides and sharp valleys. Riding across country one is for ever climbing slowly out of one valley and then zigzagging equally slowly down into the next. High on the sides of the valleys are tiny native farms whose fields seem fixed to the mountain with the adhesion of wet rags. The farms prove to be meagre thatched huts surrounded by a few terraces or sloping fields of maize and potatoes. The Indians living there produce just enough food to survive. They sleep together on the mud floor, brew their own chicha and sugar alcohol, and surround themselves with guinea-pigs, chickens and dogs. They live in poverty near the subsistence level. But they are happy and hospitable, untroubled in their isolation by the wants or ambitions of materialistic society. They are and were deeply conservative peasants closely bound to every rock and spring in their little world. To them Toledo's neat towns represented the end of liberty, uprooting from traditional holy places, and greater exposure to the forced labour and exploitation of Spanish society.

The natives offered Toledo 800,000 pesos if he would desist from the resettlement programme. But the implacable Viceroy brushed this aside: to him it confirmed their desire to remain near their shrines, to continue ancient religious practices and to cohabit 'with sisters and even daughters'. He knew that they preferred to be 'scattered and dispersed, living in mountains and canyons where only their chiefs took any notice of them'. So the reductions went ahead, despite the scepticism of Spaniards and the violent opposition of the natives. There were emotional scenes as weeping women were torn from their pathetic little huts to be transplanted.

The resettlement was a colossal undertaking, involving a million and a half people. It has never been thoroughly studied, but it evidently took place, at any rate in southern Peru where the Viceroy himself was nearby to inspire action. Diego Dávila Brizeño, corregidor of Huarochiri, wrote that he had reduced over two hundred villages of the Yauyos 'into thirty-nine towns in which they live at present'. The natives tried to destroy the first settlements, but Dávila Brizeño completed their resettlement, 'building fine temples and hospitals, destroying all the ancient towns, and bringing the Indians to live in the settlements. The chief curaca Don Sebastián, a man who is fluent in our Spanish language and is very sensible, assisted me greatly in this.' Andrés de Vega reduced the Indians of Jauja and wrote that 'the houses that are being built since the reductions are small and rectangular, imitating those of Castile – for previously there used to be round huts. The materials used are adobe and clay or fieldstone and clay, with the roofs of thatch and alder wood. Only churches and some houses belonging to the town council or to [important] individuals have tile roofs.' The inspectors of the Condesuyo, Licentiates Mexia and Herrera, reported in 1571 that they had founded forty-seven towns, at times reducing as many as eighteen villages to form one town. Juan de Matienzo told the King of the reduction of the natives of Villanueva de la Plata from eleven native villages, and noted that houses had been built for two native alcaldes, two caciques and a scrivener.* Licentiate Juan Maldonado de Buendía reported that he had created twenty-two new towns from 226 settlements in the districts of Moquegua and Arica on the south coast of Peru. 'I left nominated alcaldes and regidores in the most important towns, with ordinances and limited jurisdiction so that they could govern themselves. I divided the tribute among the poor Indians so that each individual should know what he had to pay, and so that the caciques could not rob their property under colour of tribute. In the course of my inspection over four hundred adult Indians were catechised and baptised: even though it is so long since this country was discovered, they were living as infidels because the villages were so dispersed. Over five hundred Indians who were living in sin were married. The old men and widows, whom the caciques did not spare, are now free from vexation. And the encomenderos have been deprived of the empires in which they acted as lords of their Indians.'

These were the ideal Spanish views of the reductions. Some Spanish administrators were less happy about these new settlements. Luis de Monzón admitted that, among the Rucanas living near Huamanga, 'some Indians have been reduced one or two leagues [four to eight miles] from their previous homes. They complain constantly, saying that their fields are in their former villages, and it causes them much effort to go and cultivate them.' Antonio de Chaves y de Guevara was more outspoken about the effect on the province of Huamanga as a whole. 'The towns that have recently been settled are for the most part not permanent. Because of the distress involved in moving to different climates and unhealthy locations far from their fields, many villages have been repopulated in their original locations with the permission of the governors and corregidores of the district.'

Although some reductions failed, most were successful. These tidy new settlements played an important part in causing memories of Inca administration and religion to fade, and in making the Peruvian natives more susceptible to Spanish rule. Baltasar Ramírez was full of admiration for the way in which Toldeo's energy had accomplished this great migration in only two years. He called it 'a marvellous work, of very great benefit both temporal and spiritual'.

Toledo himself regarded the reductions as an essential prelude to the true conversion of the natives. While the Indians were scattered in isolated farms and villages the priest on each encomienda had an impossible task. Toledo told the King: 'I found that the doctrine being given to the natives by these curates was insipid and done *propter formam* – as is evident by the state of the natives' Christianity.' The Church itself was becoming alarmed and self-critical. Many Church leaders appreciated that although Christians had conquered Peru, Christianity had not really won the souls and consciences (or in modern parlance the hearts and minds) of the Indians. The subversive revival of native religion in the mid-1560s demonstrated an urgent need to make the Church more effective. A second ecclesiastical council was held in Lima in 1567 and much of its business dealt with the suppression of heathen practices. Priests were told to eradicate 'the innumerable superstitions, ceremonies and diabolical rites of the Indians'. They were to stamp out drunkenness and all native rites, arrest witch-doctors, and discover and destroy shrines and talismans.

During the last third of the sixteenth century and the start of the seventeenth, the Church launched an aggressive campaign to eradicate any spiritual opposition. A synod at Quito in 1570 instructed curates to attack any 'ministers of the devil who obstruct the spread of our Christian religion. There are famous witch-doctors who practice these offices in a pact with the devil and amid many superstitions. Some guard the huacas and converse with the devil; others act as priests, hear the Indians' confessions and preach the superstitions of the devil.' Throughout these years, any leaders of the native religion were rooted out and flogged, placed in the stocks or imprisoned. But these witch-hunts rarely if ever ended in burning or execution, as in the later colonies of North America. It was just as important for the priests to destroy the countless holy objects venerated by the natives. They went about this task with apostolic fervour: Francisco de Ávila boasted that he himself burned over thirty thousand idols and three thousand mummies during his missionary career.* Even such destruction did not eliminate all heathen beliefs. Juan Meléndez complained that 'they return like dogs to the vomit of their ancient idolatries'. The Andes are full of superstition to this day: almost every market, cemetery and native house has a few weird objects with magical significance.

The Church's crusade against heathenism produced an important literary by-product. The early Quechua dictionaries were written to help priests in their missionary work, the first by Juan de Betanzos, the husband of an Inca princess, and another published at Valladolid in 1560 by Domingo de Santo Tomás, the champion of Atahualpa's sons. The best Quechua dictionary was produced by Diego González Holguin; Luis Bertonio and Diego de Torres Rubio wrote vocabularies for the Aymara language spoken around Lake Titicaca; and Luis de Valdivia wrote one for the Indians of Chile.* The missionaries were also helped by a series of thorough studies of the native religions they were fighting. All the chroniclers writing about the Incas naturally studied their religion, but manuals were now produced specifically to help suppress idolatry. The authors of these valuable studies included Polo de Ondegardo, Juan de Betanzos, the first Augustinians, Cristóbal de Molina of Cuzco, Francisco de Ávila, Cristóbal Carrillo de Albornoz, Hernando de Avendaño, José de Arriaga, and Pedro de Quiroga. Most of our knowledge of Indian religion is derived from these anti-heathen reports.

The authorities knew that the spread of Christianity was being hindered by Spanish oppression of the Indians. The King often instructed his colonial governors to protect the native population. Francisco de Toledo tried to do his duty in this respect. This severe Viceroy, the scourge of the Incas, issued copious legislation on Indian affairs. As an upright man with a deep sense of legality he hated corrupt oppression. But his protection of the Indians was done without compassion: it was condescending solicitude by an enlightened superior. 'I am informed that the Indians are not free, as a result of their weakness and imbecility and the great awe they have towards Spaniards . . . it is therefore my duty as their protector to see that they are not cheated in their work.' Toledo therefore laid down wage-scales and minimum subsistence allowances for different categories of native worker. An ordinary labourer was to receive a peso a month, half a fanega (three quarters of a bushel) of maize, and some land to grow more food; other workers, domestic servants, shepherds, old men or youths received smaller amounts of pay and rations. Toledo also insisted that daily labourers be paid promptly, without reductions or withholdings, in proper silver at the end of each week.

Toledo had no sympathy for the natives' traditional oppressors, the encomenderos, curacas and priests. He complained to the King in his report of March 1572 of 'the scandal perpetrated on the natives by the first conquistadores, and their oppression and cruelty'. He agreed that Indians granted in encomienda should be liable to no personal service – only tribute.

Toledo was in two minds about extending encomienda rights into perpetuity. The cabildo of Cuzco, which consisted only of encomenderos, offered to purchase perpetuity on 150 encomiendas, but Toledo advised the King to sell it only to a few of the best settlers in each community. Himself the son of a feudal house, Toledo hoped to create a landed aristocracy in each Peruvian town; in the same way he relied on garrisoned fortresses such as Carlos Inca's Colcampata to dominate native areas. In fact no decision was ever reached regarding perpetuity. The Cortes (parliament) of Castile recommended in 1573 and 1574 that perpetuity should be sold for a sum based on the annual tribute income of each encomienda. The King took no action beyond calling another special commission: the Junta de la Contaduría Mayor of 1579. This produced a recommendation similar to that of the discredited commissioners of 1562; but King Philip again did nothing. A new formula was proposed in 1584, and the then Viceroy of Peru wrote in 1586 in favour of general perpetuity for all encomiendas. Philip II, exasperated by so much contradictory advice, suspended all further discussion of perpetuity in 1592.

In the meantime the encomiendas were declining in importance: their loads of native produce and hard-won tribute were insignificant beside the fortunes being made in mining or coca. Many encomiendas lapsed to the Crown with the death of the second holder. The new oppressor of the Indians became the corregidor, the royal official appointed to administer Crown holdings. Corregidores were appointed for only a few years: they had to intensify the pace and ingenuity of exploitation in order to make fortunes during their brief terms of office. Some of the encomendero families succeeded during this period in acquiring title-deeds to native land. Property won in this obscure and illegal fashion formed the nucleus of the haciendas of many of the present leading families of Peru.*

Francisco de Toledo was even more critical of oppressive curacas than he was of encomenderos. He followed Juan de Matienzo in a bitter condemnation. 'When these caciques are bad they inspire all the multitude in imitation of their way of life, which is commonly so bad that it must be seen to be believed.... They hold drunken revels and authorise all the evils that take place at them. They inspire the worship of idols and of the mummies of the dead, and the keeping of concubines – not one, but fifty or a hundred. They are responsible for robberies and tyranny suffered by these wretches. For if an Indian pays four or six pesos of tribute, these rob him of forty pesos in addition. They burden the unfortunate poor with tribute but exempt the rich and the relatives of their mistresses and petty chieftains, all of whom they maintain as the instruments of their evil-doings, which include the killing of Indians who do not comply with their wishes.'

Since he regarded curacas as iniquitous petty tyrants, Toledo naturally legislated to curb their powers. He even tried to prevent curacas from becoming officials in his new towns. But he acknowledged that 'the natives cannot be governed except by using the curacas as instruments of the executive, both temporal and spiritual', and he relied on curacas to assemble tribute and labour quotas. Toledo forbade curacas or other natives to possess Spanish weapons, to play dice, give banquets or presents to Spaniards, buy Spanish goods above a certain value, or travel on horses or litters unless they were infirm or the chiefs of a province. But he confirmed the immunity of curacas and their eldest sons from paying tribute or performing personal service, and he established special colleges for the sons of curacas, so that they survived as an educated native aristocracy. He also imitated the Inca system of succession in caciquedoms: an inefficient or disloyal curaca could be desposed in favour of his son, and a younger son could be chosen to succeed if he was more promising than the eldest.

Toledo was at first unsure how to protect the Indians from their oppressors. The Augustinian Prior Juan de Vivero, who had baptised Titu Cusi, wrote to the King in 1572 recommending a protector-general for all Peru, who would accompany the Viceroy and have lieutenants in each city. He would be responsible 'for the defence of these miserable Indians, for there is no one to look out for them. Any who hold this office must be highly favoured by Your Majesty and your ministers: for they will have to suffer martyrdom every hour if they do what they ought.' Toledo tried to follow this advice. He appointed one Baltasar de la Cruz as defensor general in 1575, having previously made this man defender of the Indians of Potosí and seen that he was conscientious. This defensor general was the mouthpiece of the natives, presenting petitions and pressing lawsuits on their behalf, 'for the Indians as legal minors cannot plead'; he reported weekly to the Viceroy and had lieutenants in the cities; his duties were the usual ones – ensuring that correct wages were paid and that encomenderos, curacas and priests did not impose illegal levies or extort personal service.

The defensor general was the figurehead of native protection. At a lower level García de Castro's corregidores de indios were beginning to function. The inspectors of Toledo's visita general administered justice alongside the corregidores de indios and saw the problems they encountered. Toledo had at first been suspicious of the corregidores de indios, feeling that there must be some justification in the opposition to them by ecclesiastics and curacas. But the findings of the visita general and his own observations changed Toledo's mind. He saw that such officials were needed for areas outside the control of the Spanish corregidores in the few European cities. He recommended to the King in November 1573 that they be granted royal approval. This came in a decree of 27 February 1575, and Toledo instituted the new officials in a series of lengthy ordinances. He changed their name to 'juez de naturales', natives' judge, and greatly reduced the number, to correspond to 71 provinces into which he was dividing Peru. He wisely ordered that these native judges should be paid by the encomenderos and not by the natives themselves. He insisted that these new officials could not be encomenderos and could not hold other office; they had to be over twenty-six years old and of good character and social position; they presumably had to be altruistically sympathetic to the natives. Such paragons were almost impossible to find among the Spaniards in Peru.*

In 1578, Toledo lost patience with the protectors of Indians 'because each and every one of them was robbing and cheating the Indians and oppressing them in dozens of lawsuits'. He created instead a new form of protector in each city, and a defender and procurator to look after the legal needs of the Indians in each audiencia. These new officers were to be subject to 'residencia' – a critical review by another official at the end of their term of office, something that could lead to disgrace or an ugly legal battle. The duties of the officials created or refurbished by Toledo were defined in a series of ordinances, in 1574, 1575, 1579 and 1580. The final set of orders covered all aspects of native protection and included eleven new clauses that tried to defend the Indians from stray Spaniards who ventured among them, and from mestizos, encomenderos, priests and caciques.* When he left Peru in 1582, Toledo wrote a memorial to the King reviewing his administration. In this he defended his new protective officials: 'I beg Your Majesty to see fit to preserve the corregidores de indios because of the evident good that results to the Indians... They are proving invaluable in eliminating the greater part of the legal interests of the lawyers and audiencias, the licence of the clergy and friars, the profits and commerce of Spaniards, the dominion and lordship of the encomenderos, and the power and tyranny of the caciques.' It was a high-minded appeal from a colonial governor. King Philip responded by approving Toledo's ordinances in July 1584 and recommending them as a basis for the future rule of Peru.*

Contemporary Spanish authors such as the Jesuit José de Acosta or the Dominican Rodrigo de Loaisa approved of the institution of these protective corregidores de indios but doubted whether suitable incumbents could ever be found. It was here that human nature defeated the good intentions of legislation. The Spaniards were far ahead of other colonial nations in trying to find idealists to devote themselves to the just administration and protection of the inhabitants of some remote region. Many doubtless performed their duties conscientiously, but their efforts went unrecorded. But others tried to exploit their brief appointments by milking the Indians through monopolistic trading and other extortions. The protection of the Indians therefore continued to concern viceroys of Peru throughout the colonial era. But the senior authorities were at least nominally on the side of the natives and were trying to help them.

At a much lower level the natives were allowed to govern themselves in the new settlements. Toledo's most respected advisers, notably Archbishop Loayza and the political theorists Polo de Ondegardo, Juan de Matienzo and José de Acosta had all stressed the need for more native alcaldes and regidores to extend the authority of the new corregidores de indios. Matienzo suggested that a hispanicised Indian from another province should be installed in each area, with judicial powers, the title of corregidor and a two-year term of office. Toledo did not attempt anything as daring as this. But he did confirm the creation of a series of native officials to administer his new settlements. The native alcalde could enforce justice at the humblest level. He had a staff of office with a cross engraved on its silver tip; and he sat in judgement in the public square, but with no written record. He could hear criminal actions, but could sentence only up to one peso fine or twenty strokes – no punishment involving death, mutilation or bloodshed. He could also hear civil suits between Indians involving not more than thirty pesos.

In the ordinances instituting these new officials, Francisco de Toledo was more concerned to enforce his puritanical prejudices than to concede much authority. The native officials were repeatedly ordered to suppress dubious practices that were 'so offensive to Our Lord God, and caused by diabolical persuasion'. These included premarital sex, concubinage, incest and drunkenness. The ordinances were obsessed with protecting single Indian girls above the age of ten from the attentions of possible violators – including the lay and ecclesiastical authorities and their own brothers and fathers. Toledo was equally keen to curb the activities of women of questionable virtue, who were apparently operating in the tambos or travelling about with the curacas. The native alcaldes were also responsible for maintaining hospitals (which were financed by contributions from the Indians), for policing the streets and organising the markets. In all this, Toledo hardly mentioned the purpose announced in the title and preamble to the ordinances: 'the government of the Indians with the intention of preventing the wrongs and oppression received from their encomenderos.'

In the places where native-ruled communities were allowed to develop they became successful communes. Miguel Agia wrote that in those in Peru the natives 'increase and multiply in abundance. They have enough to eat and wear and are fairly rich, with enough to pay their tributes. The clothes they wear are better and cheaper than usual, being made in their own workshops.' But all too often the humble Indian who tried to function in one of the elected positions was crushed by his curaca, priest or corregidor. Felipe Poma de Ayala, a mestizo who had lived in communities with native officials, said that 'the first [native] alcaldes were not obeyed or respected by the Indians, who called them "michoc quilliscachi", meaning "judges empowered to spy and tattle". The Indians generally name youths to be alcalde, so that they could be ignored.'

All Toledo's efforts on behalf of the Indians were nullified and overwhelmed by the voracious demands of the mines. The Viceroy attempted in vain to reconcile two contradictory forces: the protection of Indians and the desperate need for labour in the mines. As a conscientious public servant, Toledo had to ensure that Peruvian treasure continued to flow to Spain. Reluctant, stubborn Indians had to be made to mine the mercury and silver – by force if necessary.

Soon after reaching Peru, Toledo summoned a special Junta to debate the morality of compelling Indians to work in the mines. This commission, headed by Archbishop Loayza, contained the former governor García de Castro, six magistrates and eight eminent ecclesiastics. It concluded unanimously in October 1570 that mining was in the public interest: compulsion could therefore be tolerated without scruples of conscience.

The Viceroy paused at Jauja on his visita general, and sent his legal crony Dr Loarte to inspect the great mercury mines of Huancavelica. These had been growing rapidly during the past decade, and it was now clear that the daily wages offered by the mine-owners could no longer induce Indians to mine mercury. Toledo therefore undertook the repugnant task of forcing the provinces around Huancavelica to provide quotas of conscript labour to serve in the mines. His justification was the mandate given him by the Junta of learned men in Lima. He wrote to the King that 'although this was something previously condemned on other occasions in this country and forbidden by decrees from Your Majesty, and although priests and bishops have greatly stressed the damage and hardship it caused the natives, no single vote failed to approve it'. With his conscience eased by this vote, Toledo legalised the use of forced labour in the public service. The Incas had formerly obliged Indians to work for the common good in a system called mita. The word mita was now revived to describe forced labour in the mines – for the common good of the 'mother country' on the far side of the world.

Toledo ordained that conscripts for Huancavelica should be raised from two hundred miles of mountainous country that embraced Jauja and Huamanga. Three thousand Indians were to serve a mita of one month, or two months during the rainy season from January to April when surface work was impossible. This represented about one seventh of the adult male population of the area. Toledo fixed the daily wage at one real or tomín of good silver – a tomín in Jauja at that time purchased twenty eggs or six avocado pears or three pounds of bread. This payment was supplemented by ten pounds of meat and three-eighths of a bushel of maize per man per month. Toledo tried to ensure that the mineworkers were paid promptly in good silver, were compensated for time spent travelling to and from the mine, and worked only during daylight hours.* Toledo thus raised the mitayos' daily wage and caused an outcry among the mine-owners. But he did not raise it sufficiently to leave the natives with a surplus after they had paid their tributes.

So the mitayos were rounded up and marched off to serve their quotas. Some tended great braziers of liquefying quicksilver, but most worked below ground amid poisonous dust and fumes from tallow candles. Antonio Vázquez de Espinosa never forgot the sight of three or four thousand of them 'working away diligently, breaking up the hard flinty ore with their picks and hammers. When they have filled their little sacks, the poor wretches climb, loaded down with ore, up the ladders or rigging, some of which were like masts and others like cables, but so difficult and distressing that a man empty-handed can hardly get up them. That is how they work in this mine, amid many candles, loud noise and great confusion.'

The ecclesiastics who voted in favour of compulsion began to have serious doubts about what they had done. In July 1571 Archbishop Jerónimo de Loayza, now aware of the implications of the policy he had endorsed, formally retracted his vote. Abundant evidence of native hardship and death in the mercury mines was reaching Lima, and the other ecclesiastics on the 1570 Junta followed Loayza in his retraction. They now insisted that although the work itself was worthy, Indians must be persuaded to do it – not forced. Nothing happened, and on 11 March 1575 the ecclesiastics who had served on the Junta signed a collective letter. They denied having approved the phrase 'compel and force' and were shocked to find that 'the Indians are compelled to work the mines; they are taken to work in them by force with the authority of the law; they complain passionately about this, and it causes them great distress and harm'. A week later seven doctors from the young University of San Marcos wrote a collective letter telling the

Cruel punishments in the mines

King that Indians had for four years been compelled by force to work in the mines. 'We assume that Your Majesty has not been informed of this, since you have not ordered that it be remedied. For it is so contrary to divine and natural law that free men should be forced and compelled to such excessive labour, so prejudicial to their health and lives.' The Archbishop also wrote on the same day protesting that the Indians were being paid in poor silver and treated disgracefully in the mines. 'They are, in short, deprived of their freedom. The ordinances are being observed only when it comes to "compelling and forcing" them – words that I myself signed in the ordinances, acting without consideration and very culpably. Here we have done everything possible to have the situation remedied, but it continues as always. For the love of God, let Your Highness order their revocation!' This was a courageous retraction and admission of error. But it was too late. Toledo had acted with the moral approval of the Junta, and the compulsory mita had received legal sanction. Once enforced it had become far too valuable to be withdrawn.

It was at the time of Toledo's arrival that Fernández de Velasco discovered a new process that allowed Huancavelica's mercury to be used in refining the stubborn ores of Potosí. Toledo was jubilant at the potential of this 'greatest marriage in the world'. He therefore hurried across the Bolivian altiplano towards Potosí as soon as he had finished his inspection and reorganisation of Cuzco.

Potosí was already becoming a huge city whose population of 150,000 made it one of the largest in Christendom, smaller only than London, Paris and Seville and equal to Naples and Milan. It was a proud, flamboyant place, a boom town full of exuberant and extrovert mining millionaires. Charles I had given Potosí the title of Imperial City, with a suitably boastful legend on its coat of arms: 'I am rich Potosí, treasure of the world and envy of kings.' Philip II sent a shield inscribed: 'For a powerful emperor or a wise king, this lofty mountain of silver could conquer the whole world.'

Potosí celebrated the arrival of its first viceregal visitor with five days of extravagant celebrations. Toledo called a meeting of miners to consider the construction of water-driven mills to grind the ore so that it could be refined by quicksilver. Four rich miners volunteered to build a lake to trap summer rains at their own expense. The miners then spent three million pesos in the construction of a system of thirty-two lakes, a ten-mile artificial sluiceway, eighteen dams and hundreds of waterwheels: a remarkable feat of engineering that guaranteed the power to grind a steady flow of silver.* By the end of the sixteenth century, the boom city of Potosí had all the trappings of a Klondike or Las Vegas: fourteen dance halls, thirty-six gambling houses, seven or eight hundred professional gamblers, one theatre, a hundred and twenty prostitutes, and dozens of magnificent baroque churches.

All this wild wealth depended on the labour of thousands of Indians: the mine operators demanded 4,500 men working at any one time. To supply this constant need Francisco de Toledo organised a gigantic mita, which he regulated in a characteristically voluminous series of ordinances at La Plata in February 1574.* Toledo designated an area covering much of southern Peru, Bolivia and north-western Argentina. Sixteen highland provinces stretching from Cuzco, six hundred miles as the crow flies from Potosí, to Tarija in modern Argentina were forced to provide mitayos. There were some 95,000 eligible males aged between eighteen and fifty in this area – only curacas, curacas' sons and the infirm were exempt. In Toledo's system, a seventh of the eligible men were called up at any one time; but in practice harassed curacas often had to send men back to Potosí more frequently than for one year in seven. Under this system there were some 13,500 mitayos at Potosí at any time, of whom 4,500 worked underground for a week while the other two-thirds worked on the surface or rested.

The proclamation of a mita was sent to the curacas in each community two months before the men were due. The curaca was then legally liable to provide his quota of men, even if it meant hiring expensive casual labour to fill gaps caused by depopulation. Some Indians were able to buy their way out of mita service. But most were sent off under the escort of a captain-general of the mita. Alfonso Messia described such a contingent from Chuquito, the district along the west bank of Lake Titicaca. 'From the province of Chuquito 2,200 Indians depart every year for the full mita. All normally go with their wives and children. I have twice seen them, and can report that there must be seven thousand souls. Each Indian takes at least eight or ten llamas and a few alpaca to eat. On these they transport their food, maize and chuño [dehydrated potato], sleeping rugs and straw pallets to protect them from the cold, which is severe, for they always sleep on the ground. All this cattle normally exceeds thirty thousand head . . . and it is all worth over three hundred thousand pesos. They normally take two months over the journey of over a hundred leagues to Potosí, since the cattle and children cannot walk faster. Out of all this community and wealth removed from the province of Chuquito, only some two thousand people return: of the other five thousand some die and others stay at Potosí or the nearby valleys because they have no cattle for the return journey.' Messia showed that such Indians actually made a net financial loss in going to Potosí. Their income from seventeen weeks' labour during their six months at Potosí was 46 pesos; but they needed 100 pesos to support themselves and pay tribute during the ten months from the time they left Chuquito until they returned.*

Once at Potosí the wretched mitayo endured 'four months of excessive work in the mines, working twelve hours a day, descending sixty or a hundred estados [350–600 feet] to perpetual darkness where it is always necessary to work with candles, with the air thick and evil-smelling, trapped in the bowels of the earth'. Toledo's mine ordinances decreed that Indians were to work from an hour and a half after sunrise until sunset, with an hour's break at midday. But as the mine shafts sank deeper into the mountain even these hard hours were ignored:

the mitayos were forced to remain below ground for six days at a stretch. As dawn rose over the bleak altiplano each Monday morning, the 'capitanes de la mita' appeared at the foot of the mountain with their contingents of labourers. These men clambered down to the face and worked there in groups of three, two hacking at the ore for the duration of a tallow candle while the third rested. At night they slept in the mine gallery. Conditions grew steadily more intolerable. The mine operators imposed heavy weekly quotas on each mitayo and fined Indians a day's wage for every 100 pounds lacking from this quota.* The mines were emptied every Sunday and on thirteen important saints' days a year. On these days the Indians were traditionally allowed to prospect for themselves, and they managed to resist all attempts to stop this privilege. But even this did not earn the mitayo enough to subsist and return to his village: many had to stay in Potosí and work as free labourers, 'mingas', at three times the fixed wage for mitayos. Some of these mingas grew accustomed to the climate, pay and exuberant conditions of Potosí and stayed on indefinitely. There were some forty thousand mingas living in Potosí throughout the seventeenth century. Women were in short supply in this high, remote boom town. Although some of the native women were available at a price to anyone, many Indians turned to homosexual practices and resented Spanish wrath about this.*

It was not long before the great mines of Potosí and Huancavelica began to disrupt the entire pattern of settlement in the central and southern Andes. Indians hated the hard work in unnatural surroundings and the enforced absence from their own homes. The most obvious way to escape this obligation was to flee from the 'provincias obligadas' to become vagrants in other parts of Peru. Such evasions became a serious exodus: Diego Muñoz de Cuéllar estimated that the population in the affected provinces was halved in the forty years after Toledo's legalisation of the mita.* By the middle of the seventeenth century the population of the sixteen affected provinces was reduced by over three-quarters. The Indians who remained behind had to furnish the same contingents of mitayos as their forebears had under Toledo, and the daily wage of a mitayo at Potosí remained unchanged for two hundred years. With widespread evasions, and more frequent service demanded of those who remained, agriculture suffered irreparably in these areas. And by a curious irony the amount of royal tribute being evaded by

Indians who fled from royal repartimientos in the Charcas and Collao actually came to exceed the royal fifth brought in by the silver mines.

Francisco de Toledo, a conscientious servant of the Crown and a man who wished to protect the Indians, is remembered for legalising the mine mita. He was only admitting the inevitable: Peru was conquered for precious metals, and once the silver and quicksilver mines had been discovered they had to be exploited. A genuine and remarkable solicitude for native welfare was thus overwhelmed by the insatiable demand for silver. The population of the Inca empire had been accustomed to communal work for the public good and it continued to labour, but for its Spanish conquerors. Some four-fifths of the men within the former empire became subject to forced labour. Of these, ninety per cent had to fulfil agricultural levies on encomiendas and some forty per cent had additional labour obligations, either in the mines or in textile mills established along the coast.

Despite all this exploitation the native population survived with its language and many of its customs unchanged. It declined to less than a million by the end of the colonial era in the early nineteenth century, but has since grown to more than the highest estimates of the population of the Inca empire. The Quechua-speaking Indians of Peru now form the majority of its population – a very different situation from that of the natives of former colonies in North America. This survival is largely due to the tenacity of the Peruvians, to the inhospitable altitude of much of Peru, and to the fact that Spain did not permit immigration from anywhere but the mother country. But it was also due to a genuine concern for native welfare throughout the colonial Church and government. However much the Spaniards exploited or abused the Peruvian natives, they left the Indians on their lands, and they stopped short of the slavery or extreme racial prejudice that has occurred in other European colonies.

21

THE INCA PROBLEM

THE Viceroy Francisco de Toledo reached Peru at the end of 1569 with instructions from the King and Junta Magna to resolve the country's native problems at all levels. He naturally concerned himself with the Inca survivors: Titu Cusi in Vilcabamba and the various royal nobles in Cuzco. Toledo started with an open mind toward the Incas. He wrote to the King in February 1570 that he intended to continue García de Castro's policy of negotiating with Titu Cusi, hoping to lure him into Spanish Peru with suitable estates. He was confident that 'the Inca's Indians and captains, now that he is baptised, are supported by the hope that the agreement confirmed by Your Majesty will be ratified'. He began by being similarly well disposed towards the Inca nobility of Cuzco. One of his first actions on

The Viceroy Don Francisco de Toledo

reaching Cuzco early in 1571 was to act as godfather to the infant son of Carlos Inca.

But Toledo became increasingly disturbed by the threat to Spanish authority posed by these survivors of the former monarchy. It was hard to find any moral excuse for attacking Vilcabamba. Although the Inca descendants had lost real power they were still venerated by the natives and by sentimental Spaniards. Both Las Casas and Vitoria had stressed the importance of the 'señor natural', the legitimate native lord. Las Casas had produced various treatises disposing of Spain's rights to be in the Indies. The bishop went so far as to praise the native resistance in Vilcabamba and to recommend that for the sake of his soul's salvation the King restore all Peru to Titu Cusi.* Some Spaniards had excused Pizarro's Conquest by depicting it as the overthrow of the usurper Atahualpa in favour of the legitimate Huascar and Manco. But the defence of 'legitimacy' could not be used to justify suppression of Vilcabamba, which was clearly a peaceful native state ruled by the son of the very ruler restored by the Spaniards.

Toledo therefore began to question Inca rule itself. It had been well known since the early days of the Conquest that the Incas' expansion out of Cuzco occurred only two or three generations before the Spaniards' arrival. The clear-headed Toledo saw that this late expansion could serve as propaganda against corrosive doubts about Spain's title to Peru. If the Incas could be shown to be late-comers and conquerors instead of long-established utopian rulers, the moralising of Las Casas's disciples would be deflated.

Before Toledo left for Peru, the King had warned him against scandalous priests who 'wished to concern themselves with justice and the lordship of the Indies, on the pretext of protecting the Indians'. Toledo now wrote that 'the books of the fanatic and virulent Bishop of Chiapa [Las Casas] served as the spearhead of the attack on Spanish rule in America'. He collected many copies of Las Casas's works and withdrew them from circulation. He also undertook a quasi-scientific enquiry to undermine the Incas' reputation.

Francisco de Toledo left Lima in October 1570 for the first extensive tour of highland Peru undertaken by any Viceroy. He decided to hold a series of interrogations along his route in which native chiefs would be invited to testify about Inca rule. The first enquiry took place at Jauja on 20 November and was followed by others at Huamanga, the tambos of Vilcas, Pina, Limatambo and Mayo along the royal road, and in Cuzco and the Yucay valley. The questions at these interrogations dealt with the situation of the area *before* the arrival of the Incas: had each community been independent? did it pay tribute to any higher authority? did it elect its own rulers or were these hereditary? did it have other forms of government? Other questions asked the name of the Inca who had first occupied the area and whether he had done so by force, whether he had imposed curacas of his own choosing, and whether the curacas succeeded by hereditary right or were replaced by the Inca.

The questions were tendentious and dealt only with subjects about which the Viceroy wished to prove a point. Most of the witnesses were aged Indians who had once lived under the Incas. These old men were perhaps inclined to give answers they thought the interrogators wished to hear. But apart from these serious limitations, the enquiries were honest and were to some extent genuine attempts to discover the truth as an aid to the Viceroy in formulating his native policy.*

The replies were detailed and had a truthful ring to them. Witnesses generally replied that the pre-Inca communities in the central mountains had been independent and had chosen leaders only in time of war. They described the occupation of the area by Tupac Yupanqui, who appeared with a powerful Inca army and accepted the peaceful submission of the local leaders. The first witness at Jauja, Alonso Pomaguala, gave a detailed description of such a surrender by his own great-grandfather. The witnesses also declared that the Incas replaced incompetent curacas, but preferred to select successors to dead curacas from among their sons. Such sons were often groomed by an upbringing at the court in Cuzco.

The first set of *Informaciones* were sent to King Philip from Cuzco in March 1571, accompanied by a genealogy of the Inca kings that showed how few had ruled before Pizarro's arrival. Toledo was pleased with the results. He told the King that the testimonies of these native chiefs showed that the Incas had no hereditary right to rule Peru and were legitimate rulers of only a tiny corner of it.* One of Toledo's chaplains wrote a jubilant letter at this time. 'The influence of Father Casas was so great, and he gave the Emperor and the theologians such serious doubts that His Majesty wished to leave this kingdom to the Inca usurpers. . . . This Father preached and wrote with great passion and effect, confirming that false deed – Inca rule – so well that few men did not believe him. He endorsed it with his exemplary life, his authority as a bishop, the authority given him when His Majesty installed him in the Council of the Indies so many years ago, and with his canes and venerable age, for he must have been ninety when he died. I myself was one of those who believed him most. It seemed to me the greatest crime to remove their sovereignty from these Incas; until I saw the contrary here in Peru.' This priest was convinced that Las Casas had known nothing of Inca tyranny, 'which fact Your Excellency [Toledo] is making abundantly clear with great authority in the investigations you are conducting'.

Toledo now delved deeper in his interrogations. He devised new questions for enquiries held near Cuzco in May and June 1571. The earlier *Informaciones* had proved that the Incas were recent conquerors. Toledo now sought to show that their reputation for moral rectitude also had blemishes by Christian standards. He asked whether the Incas sacrificed human beings, practised anthropophagy or 'crimes against nature'. Other questions asked whether the Incas had made their subjects work simply to prevent their natural tendencies to idleness, and whether the native character required firm government and guidance. The witnesses answered, as was already well known, that the Incas did not tolerate cannibalism or sodomy, but that they did sacrifice human beings on rare occasions. Regarding the native character, the witnesses felt that Indians were naturally lazy and docile and therefore *did* require the imposition of work and firm leadership.

A third enquiry, in January 1572, examined the earliest origins of the Inca tribe. The inquisitors interrogated survivors of the tribes who had occupied the valley of Cuzco before the Incas had arrived there some two hundred years earlier. The answers revealed a surprisingly bitter hatred of the Incas among some members of these minority groups in the Inca heartland.*

At the same time as his *Informaciones*, Viceroy Toledo had commissioned a thorough history of the Incas from his companion on the visita general, Captain Pedro Sarmiento de Gamboa. This remarkable man distinguished himself as a cosmographer (he was investigated by the Inquisition for his studies of the universe), a geographer and explorer, a military commander, historian, and naval commander against Francis Drake. Sarmiento's work, the *Historia Indica*, was carefully researched, partly from the same informants who had testified to the

Informaciones but also from studies by earlier writers such as Cieza de León or Juan de Betanzos. It was completed by the beginning of 1572, although never published until 1906.* The manuscript of the *Historia Indica* was sent to Spain with four 'paños' or draperies on which were painted 'the busts of the Incas, with medallions of their wives and ayllos in the borders, and the history of what happened during the reign of each Inca'. It is an extraordinary fact that the authors who made the most powerful arguments in favour of Spanish rule – Sepúlveda, Matienzo and Sarmiento de Gamboa – never had their works published in Spain in the sixteenth century; whereas everything Las Casas wrote was published.

Toledo felt that his purpose had been fulfilled with the compilation of Sarmiento's history, the painted paños, and the *Informaciones*. He had revealed blemishes in the Incas' reputation for goodness, and had demonstrated that the Incas were late arrivals on the Andean scene. White propagandists in South Africa take similar pains to show that the first Dutch settlers were moving inland from the Cape before the Bantu from the north had penetrated that far. The purpose of such an exercise is never entirely clear. Two wrongs do not make a right. Toledo had no more intention of handing Peru back to its pre-Inca communities than white South Africans have of reinstating the original Bushmen.

Toledo felt that he must confront the Incas themselves with his findings. He summoned all the known descendants of Huayna-Capac – of which there were many – to meetings on 14 January and 22 and 29 February 1572. They were shown the paños and the interpreter read the texts in a loud voice. 'They were then told by an interpreter . . . that the Incas who had been lords of this kingdom and whom they called "kings" should not be described thus and had not been its kings but its usurpers. At this the Incas, their relatives and grandchildren grew angry with him. To calm them [Toledo] replied that they should not be surprised at this, for the King of Castile had many kingdoms of other peoples that he had won by force of arms, which he had also taken from them as had the Incas.'

The most entertaining reaction to this literary presentation came from Sayri-Tupac's widow María Cusi Huarcay, whom Toledo had recently remarried against her will to an obscure soldier called Juan Fernández Coronel.* 'When she saw that an [illegitimate] daughter of Paullu Inca called Doña Juana was painted above Doña María herself, she was furious and said: "How can it be tolerated that [Paullu] the father of Don Carlos, and [Carlos] and his sister who is a bastard, are in a more prominent position than my father [Manco], my brother [Titu Cusi] and I myself, who are legitimate?" She therefore went with other Inca relatives to complain to the Viceroy. He replied: "Do you not see, Doña María, that Don Carlos and his father served the King, and your father and brother have been usurpers and have always been shut up in the montaña?" To which she answered: "Because you say that my father and brother have been usurpers does not mean they were. If they went into retreat, it was because they were not given enough support as befits rulers of this kingdom."'

Francisco de Toledo had in fact been worrying about María Cusi Huarcay's brother Titu Cusi. His instructions on taking office in 1569 were to conclude the negotiations begun by his predecessor. As he saw the situation at that time, the Inca and his men would be eagerly looking forward to emerging from Vilcabamba. The only problem seemed to be that Toledo had no estates available with which to endow them.* This optimism did not last long. The Viceroy exchanged various letters with the Inca during his first year in office. Nothing came of them. As the months passed, and as Toledo reviewed his predecessor's chaotic negotiations with Titu Cusi, he became increasingly dissatisfied with the diplomatic approach to Vilcabamba. Soon after reaching Cuzco in February 1571, Toledo dictated a massive series of reports for the King.* He revealed that he did not share Castro's awe of the strength of Vilcabamba. Titu Cusi had only some five hundred troops, and his chief defences lay in 'the slopes and density of the jungle on the Cuzco side, and the great river of Mayomarca [Apurímac] towards the royal road from Lima to Cuzco'.

Having hinted that a military solution might be possible, the Viceroy sent the King some letters from the Inca and copies of his replies. He still felt that Titu Cusi might emerge peacefully if his son were married to the heiress Beatriz and if he were granted an annual income of 2,000 pesos. Toledo concluded that 'this obstacle there is inconvenient, for the natives continually turn their eyes to him. In my opinion, it is no solution to bring him down from a trifling fortress of almost uninhabitable savannahs where he is with so few Indians, only to place him in the midst of the 200,000 Indians of this province. If it has to be done, it would be better to bring him down to Lima, where the number of Spaniards always greatly outweighs that of the Indians.'

By mid-1571 Viceroy Toledo's attitude towards Vilcabamba was thus hardening. He was encouraged by the first *Informaciones* which demonstrated the short duration of Inca conquest. He was no longer sure that patient diplomacy was the only solution for this tiresome enclave, and was already convinced that Vilcabamba could not be allowed to survive as a separate state.

The situation in Vilcabamba changed dramatically at about this time – but the change was not known to Toledo or the Spanish authorities. The Inca Titu Cusi was at Puquiura beside Vitcos, visiting the shrine 'where Diego Méndez killed his father Manco Inca'. 'He remained there all day, mourning the death of his father with heathen rites and impudent superstitions. At the end of the day he started fencing (which he had learned to do in the Spanish manner) with his secretary Martín Pando. He sweated heavily and caught a chill. He ended it all by drinking too much wine and chicha, became drunk, and woke with a pain in his side, a thick tongue – he was a very fat man – and an upset stomach. Everything was vomiting, shouting and drunkenness.

'The Inca passed the night bleeding from the mouth and nose. . . . In the morning he complained of his chest, where the pain was tormenting him. His secretary Pando and another close friend, Don Gaspar Sullca Yanac, beat up the white of an egg with sulphur in a bowl. They thought this would be an effective remedy to restrain the flows of blood. When this was given to the Inca to drink, he refused it, saying "Do not give me something that will make me die." But when the Inca saw that his two closest companions were giving him the remedy . . . he said: "Give me the drink. I am very fond of Martín Pando, and he would not give me anything that would harm me." He drank the concoction. At that moment the sickness rose to its final pitch; and he expired.'

The Indians were stunned by the sudden death of their sovereign. In their frenzy of grief many were avid for revenge and sought scapegoats. It seemed as though yet another Inca had been killed by the Spaniards. The only Spaniard in the vicinity was the friar Diego Ortiz, who had a great reputation for medical skill and might therefore have administered the fatal medicine. Titu Cusi's wife Angelina Llacsa later testified that another wife had hysterically incited the crowd against the foreigners. 'As soon as he expired, one of the Inca's wives called Angelina

Quilaco screamed out [and] moved by some malign spirit . . . she ran out and shouted to the captains to seize the friar: for he had killed the Inca and given him poison with Martín Pando, the mestizo who was [the Inca's] secretary.'

A group of militant captains led by Guandopa responded to these cries. Martín Pando was seized and killed immediately. But Diego Ortiz suffered a protracted martyrdom. His hands were tied behind his back with such violence that a bone was dislocated, and he was stripped and left in the open all night wearing only some scraps of cloth. Some Indians thought that the missionary might be able to work a miracle: they ordered him to say a mass to revive the dead Inca. Ortiz was taken to the church of Puquiura and donned his vestments. He started to say a mass, slowly, with much devotion, to the mounting impatience of the crowd. Angelina Llacsa had remained with the body of her husband all night, but she emerged now and saw Ortiz trying to say his mass. He had to explain that God would not respond to such an appeal: it was God's will that Titu Cusi must die. The natives were disillusioned and angry. They tied Ortiz to a cross and beat him; others mixed a foul concoction and forced him to drink it.

Even the militant captains hesitated to kill the Spanish priest. They decided to take him to Vilcabamba for a verdict by the new Inca Tupac Amaru – for it was Titu Cusi's brother Tupac Amaru and not his son Quispe Titu who emerged as the next ruler. He was led by a rope passed through a hole driven behind his jaw. On the first day of the march it rained heavily and the path was flooded: the friar, weakened by exposure and ill-treatment, slipped and fell in the mud, just as he and García had done on this same road during the previous rainy season. Ortiz had a miserable journey, barefoot and half naked, and sleeping one night in a cave with water dripping from the roof. He endured the abuse with Christian fortitude: according to the coya Angelina Llacsa, 'he raised his eyes to heaven and with great humility asked God to pardon them for their sins; the Indians jeered at him for this'.

The three-day ordeal was in vain. The cortège met Tupac Amaru at Marcanay, a few miles before Vilcabamba city, but the new Inca refused to see his brother's priest. Tupac Amaru was a staunch upholder of the native religion, and with his accession Vilcabamba rejected Christianity. The native captains, presumably obeying the Inca's instructions, struck Ortiz 'on the back of the neck with a mace and thus finished killing

36 *Don Francisco de Toledo, Viceroy of Peru, 1569–81*

37 *Philip II, King of Spain, 1556–98*

38 *Conscripted mine workers had to carry their quotas of ore up hundreds of feet of rope ladders: Theodore de Bry's dramatic but fanciful vision of the great silver mines of Potosí*

39 *The Jesuits were proud that the descendants of their saints were united with the Inca royal family. A painting in the Compañía, Cuzco, depicts the marriage in Madrid in 1611 of the Inca heiress Ana María Lorenza Coya de Loyola to Juan Enríquez de Borja. The bride and groom stand on the right. Her parents, the hero Martín García de Loyola and the Inca princess Beatriz Clara Coya, stand on the left. Between them are the Jesuit saints: Ignatius de Loyola, great-great-uncle of the bride, and Francisco de Borja, grandfather of the groom. In the distance are seated the bride's grandfather and great-uncle: the Incas Diego Sayri-Tupac and Tupac Amaru*

40 *Bartolomé de las Casas*

S. C. R. M.

41 *Don Carlos Inca began his letter to King Philip in 1571: 'For many years I have longed to go to your court to kiss the royal feet and hands of Your Majesty . . .' and ended 'Your humblest and most faithful vassal, Don Carlos Ynga.'*

42 The terraces and buildings of Machu Picchu, an Inca city never discovered by the Spaniards. Beyond lie the sugarloaf of Huayna Picchu and the mist-shrouded hills of Vilcabamba

43 The principal temple of Machu Picchu, built of fine coursed masonry

44 *Hiram Bingham, 1912*

45 *Gene Savoy, 1966*

46 *Two of Bingham's men stand beside the great white rock of Yurac-rumi, shortly after he had discovered it and identified it as Chuquipalta, the chief shrine of Vilcabamba*

47 *The state of Vilcabamba was defended by precipitous hillsides and mighty rivers. A view across the Apurimac towards Choqquequirau. The ruins lie in a saddle of the ridge, top centre of this picture*

48 *The royal Inca highway crossed the Apurimac on a famous suspension bridge. The bridge was repeatedly rebuilt until the late nineteenth century, the date of this drawing. It was immortalised by Thornton Wilder as the Bridge of San Luis Rey, but has now been replaced by a road bridge at a lower level*

49 *Modern inhabitants of Vilcabamba: an isolated farm close to the ruins of Choqquequirau*

50 *The descendants of the Incas dressed in Spanish finery but preserved some of their ancestors' roy regalia. Don Marcos Chiguan Topa Inca was royal standard-bearer for the natives of Cuzco in t early eighteenth century. He wears the Inca royal fringe and mascapaicha, and behind him is t coat of arms awarded to Paullu Inca. But he was dismissed by Sayri-Tupac's descendant, the Marq of Oropesa, for cruelty to the Indians in the fief in the Yucay valley*

him. Once dead, they laid him out on the road and the captains made all the Indian men and women pass over him and step on the body. They then made a deep, narrow hole and placed him in it, head down and with his feet upwards, and placed a palm spear in his caecum [*sic*]. They threw earth over him and sprinkled saltpetre and other things over it, according to their rites and ceremonies.' It was felt that God would be less likely to rescue Ortiz if he were buried head downwards. The reactionaries then tried to eradicate all signs of Christianity: they destroyed the churches completely, crushed the metal ornaments, and made coca pouches from the vestments. But many were frightened by the audacity of attacking the Christian religion: their fears were heightened when an Inca palace burned down, someone saw a large snake, and the arm of a man who struck Ortiz withered.†

Diego Ortiz was killed in a xenophobic reaction that rejected all Titu Cusi's attempts at accommodation with Spanish Peru. When Titu Cusi became Inca, he had relegated his young brother Tupac Amaru to the priesthood: 'he ordered him to stay guarding the body of their father in Vilcabamba, where Manco Inca was buried, and he remained there.' The militant captains now decided that this legitimate son of Manco Inca, a man in his late twenties, should succeed as Inca instead of Titu Cusi's teenage heir Quispe Titu.† Tupac Amaru (whose name meant Royal Serpent) was the choice of the native military and religious authorities. During his reign both these establishments attempted a return to the rituals of the Inca empire and tried to forget the Spanish Conquest.

It has often been said that Tupac Amaru had been supplanted because he was simple-minded or impotent. This came from a chance remark by Sarmiento de Gamboa, who did not know of Titu Cusi's death. He reported that the natives consider that another son called Tupac Amaru is legitimate; but he is incapable and the Indians call him uti [impotent]'. In his ecclesiastical office as guardian of Manco's body Tupac Amaru lived among the holy women, the acllas and mamaconas. One romantic Spanish historian, writing in 1925, pondered a situation where a semi-divine prince was installed in a nunnery of frustrated women. His imagination ran riot. He assumed that the holy women's vows of chastity would not apply in this instance. 'There would have resulted a carnal intercourse in which the erotic would have been intermingled with the sacred – an intercourse that would end by aggravating the condition of uti, of simple-mindedness or softness of the prince, so that he would remain far removed from power.'

Whether or not Tupac Amaru was feeble, his accession marked the triumph of the hard reactionaries in Vilcabamba. Christians and Spaniards were blamed for Titu Cusi's death. The supporters of the Inca religion had their revenge for the destruction of the shrine of Chuquipalta. Spaniards were killed, native Christians persecuted, and the frontiers of Vilcabamba closed against the outer world. Titu Cusi's attempts at coexistence were forgotten. There was to be no more flowery correspondence between Vilcabamba and Lima. Titu Cusi's successor was set on a collision course with the equally tough new Viceroy.

Toledo made some final attempts at diplomacy. On 2 January 1569 King Philip despatched his ratification of the Treaty of Acobamba between Titu Cusi and Lope García de Castro. This now reached Cuzco together with papal dispensation for the marriage of Quispe Titu to his cousin Beatriz Clara Coya. On 20 July 1571 the Viceroy asked Gabriel de Oviedo, Prior of the Dominican monastery in Cuzco, to undertake an embassy to deliver these documents to the Inca. The Spaniards did not know that Titu Cusi had just died and Tupac Amaru and his generals had won power.* Oviedo began his embassy by approaching Vilcabamba from the west across the Apurímac, downstream from the Limatambo crossing. The river here swirls through a stupendous canyon with walls of six thousand feet on either side, densely forested on the shaded eastern bank. Oviedo sent four native chiefs across the river, but they never returned. The prior sent two further natives across three weeks later, but prudently remained behind himself. One of the second group of envoys eventually returned, wounded in the head, hands and stomach. Oviedo himself descended into the hot cleft of the valley at the beginning of October, but there was no sign of life on the far bank and no means of crossing the turbulent river. The Prior therefore returned to Cuzco, rather lamely, on 18 October.

The long silence from Vilcabamba caused mounting concern in Cuzco. This was heightened by the rejection of Gabriel de Oviedo's embassy. In January 1572 Prior Juan de Vivero, who had baptised Titu Cusi, wrote to the King that 'the Inca has gone sour for I know not what reason. He would not come out or write or send messengers for some time past, even though serious people have been sent to deal with him regarding his own and his son's affairs.'

Toledo was less patient than the priors. He was a stickler for legality and could not forget that the Inca had formally surrendered himself and his lands to the Spanish crown. He wrote that 'Titu Cusi Inca appears to have been afraid to emerge and come down here [to Cuzco]. He has broken the agreement made with him, without properly understanding the affair at the time.' It may in fact have been the Viceroy's understanding that was at fault: the treaty of Acobamba did not stipulate that Titu Cusi must leave Vilcabamba. Nevertheless, Toledo was beginning to feel that he must invade Vilcabamba. The second series of *Informaciones* and Sarmiento's history convinced Toledo himself (even if they made little impression on anyone else) that the Incas had no moral, historical or legal right to rule Peru. The Viceroy hesitated to use force only because this was contrary to the King's instructions. When he forwarded the history and paños to the King in March 1572, he wrote: 'Your Majesty will appreciate that this affair must be terminated once and for all, in such a way that the final result is that [Titu Cusi] is held quietly – either by peaceful means or else this debate must be ended by war... Your Majesty should determine and order whether or not war should be waged on him... for it can be done very justly, as is evident from the history and paintings I am sending you.'

Francisco de Toledo made a final attempt at diplomacy, even though his patience was worn very thin. He wrote to the Inca – still thought to be Titu Cusi – and deplored the treatment given to Oviedo's envoys. He apologised for not having made greater efforts to meet Titu Cusi during the year since his arrival in Cuzco. The letter ended with an order and a threat. 'If you have faith and loyalty to the service of God and of my lord the King as you have said you have, you must demonstrate this by actually coming out with [the envoys] and by listening to what they have to tell you on behalf of His Majesty the King and of me. And if not, we shall certainly be disabused of any illusions. The necessary future arrangements can then be made.'

The man who volunteered to take this letter was Atilano de Anaya, a respected citizen of Cuzco. He had been caring for Beatriz Clara Coya since the Maldonados' attempted rape, and had accompanied Prior Juan de Vivero to Vilcabamba to baptise Titu Cusi in 1568. Oviedo described him as 'the majordomo or agent in Cuzco of the Inca, with whom he had business relationships'. Calancha called him a 'friend and correspondent of the Inca' and he was 'a grave gentleman with an affable manner and versed in the language of the Indians' according to Baltasar de Ocampo.

Anaya left after the rains in March 1572 and chose to enter Vilcabamba by the orthodox route across the bridge of Chuquichaca. Arriving there, he had a shouted conversation with some of the Inca's troops, 'he from this bank of the river and they from the other. They promised him passage in the name of the Inca, provided he were alone. He complied, leaving his escort on the bank of the river when he crossed to the far side. The Indians made him a hut on a small hill and brought him some food to eat. They told him on behalf of the Inca that he should wait there for three days without passing forward, and that they would provide him with what was necessary . . . He spent the day sitting on an outcrop of rock beside his hut, in such a way that he was seen by Diego, a Negro in his service whom the Indians had not allowed to cross with him.'

The garrison, under its captains Paucar Inca and Curi Paucar, asked whether the loads of silver and presents that Atilano de Anaya had with him represented dues from the Yucay estates of Quispe Titu's betrothed Beatriz – the Treaty of Acobamba said that Titu Cusi was to receive tribute from these lands, but the Spaniards had failed to pay it for some years. The native commanders were afraid of a renewal of diplomacy between the Spaniards and Vilcabamba. They knew that the death of Titu Cusi and accession of Tupac Amaru could not be concealed from Anaya. They wanted to exclude all Spanish influence from Vilcabamba. They therefore panicked and 'that night they killed him with their lances, dragged him out and threw him down a ravine. The Negro heard the noise of this. When day came and he saw neither his master in his former position nor any Indians, he crossed the bridge and went in the hut, but did not find his master's bed or clothing. Following a path he saw Anaya dead and thrown into a small gulley. He went up to him and verified it. Fearing that they would do the same to him he crossed back over the bridge, took the road for Cuzco and came all along it giving the news of the death of his lord, especially in Amaibamba which is the town closest to where the murder took place. The local mission priest sent Indians for [the body], which they brought and he buried.

'Nobody in Cuzco was prepared to believe the Negro's story of the death of Anaya. For although he went first in a tearful state to give the news to [Anaya's] wife, she told him to go to the devil. She said he was a cunning liar, and that he had fled and left his master in order not to serve him. The Negro went to Dr Gabriel de Loarte, the alcalde of the court, and told him what he had seen. But the dead man's wife sent to tell [Loarte] that she could not believe it and begged him to give no credit to the Negro, who was a liar and should be arrested. The alcalde did this. But after two days he released him because the curate of Amaibamba gave the news that he had had the dead man fetched and buried.'

To the Viceroy in his belligerent frame of mind, the news of Anaya's murder seemed an imperative call to arms. Here was solid, defiant provocation. Anaya was the first Spaniard known to have been killed by the Vilcabamba regime since the renegades who assassinated Manco over a quarter of a century earlier. And Anaya was no ordinary victim. He was a prominent citizen, a friend and business associate of the Inca, and a sympathetic expert on native affairs. Worse still, he had been murdered while on an official embassy bearing grants and letters from the Pope, the King and the Viceroy. 'The Indians had [behaved] like barbarians without honour and had broken the inviolate law observed by all nations of the world regarding ambassadors. [Toledo] determined to punish once and for all the Inca Titu Cusi and those with him, and to pacify and reduce that province to the service of His Majesty.'

Toledo did not hesitate for a moment. He knew that he must now invade Vilcabamba. 'Once he was certain [of Anaya's death] the Viceroy took great pains to learn the strength of the Inca in every way.' Everyone who knew Vilcabamba was closely questioned. 'They informed him of the roughness of the road, of the bad passages and mountains where the Indians kept supplies of boulders which they rolled on to those who passed – for it was with these that they had defeated captains in the past. They told him that the Anti Indians might possibly be allied to the Inca, and also the Opataries, those of the provinces of the Manarí, Pilcosuni and Momorí, and the Satis and Zapacaties and others adjacent to these with whom the Inca was in communication.' A census was taken of all the military men then in Cuzco, and an inventory of the arms and ammunition. The city authorities were identified with the decision. 'With the advice and agreement of the most prudent and intelligent people in Cuzco, and with the votes of the city council, Toledo resolved to be rid of that robbers' den and scarecrow bogey, and to launch total war on the Inca as an apostate, prevaricator, homicide, rebel and tyrant.... He ordered the proclamation of a war of fire and blood' on Palm Sunday, 14 April 1572.

22

THE VILCABAMBA CAMPAIGN

HAVING decided to attack Vilcabamba, Francisco de Toledo moved with characteristic efficiency. He knew that the bridge of Chuquichaca was the best means of entry into the remote province. He therefore sent a force under Governor Juan Alvárez Maldonado to rebuild and hold this key point. Alvárez Maldonado left Cuzco on Quasimodo or Low Sunday, two weeks after the campaign was announced, taking ten Spaniards, including Huayna-Capac's grandson Captain Juan Balsa, legitimate son of the coya Doña Juana Marca Chimpu.* Maldonado's men found a native contingent at the bridge, but 'with four shots from our small field guns and the arquebuses of the soldiers, the Peruvians were routed and obliged to retreat to their camp. Our men then occupied the bridge, a matter of no small importance for the royal force.'

The capture of Tupac Amaru by Martín García de Loyola

The Spaniards rebuilt the strategic bridge and 'guarded it with the greatest vigilance' from mid-April until late May. The natives observed this and 'realised that [the Spaniards] must be awaiting fresh troops to enter Vilcabamba'. They anxiously began to prepare supplies of food and missiles for 'the war they already foresaw'. They also sent contingents of magnificently-accoutred troops, wearing all the traditional plumes and metal chest discs. These demanded Spanish intentions and offered to bear messages to the Inca – whom they still pretended to be Titu Cusi Yupanqui. This was doubtless done to give the impression of a powerful, stable state in Vilcabamba.

Toledo in the meantime had been preparing an overwhelmingly strong expeditionary force. He mobilised all who enjoyed the benefits of encomiendas. 'I ordered all the citizens who were of the right age and disposition to go on the campaign, in person and at their own expense. I made the infirm who enjoyed Indian revenues, or women or children [who held encomiendas] pay for one, two or more soldiers according to their income.' The resulting force was a galaxy of 250 distinguished Spaniards and professional soldiers 'all of much lustre and valour, well equipped with arms and clothing, brave and gallant men', but often totally inexperienced in warfare. They marched to the bridge under the direction of Dr Gabriel de Loarte, chief justice of the Audiencia of Lima, and Dr Pedro Gutiérrez, Toledo's chaplain and later a member of the Supreme Council of the Indies.

As the expedition crossed the bridge of Chuquichaca into hostile country, Dr Loarte announced Toledo's nominations for the command. The general in charge was to be Martín Hurtado de Arbieto, magistrate of Cuzco and veteran of the civil wars against Gonzalo Pizarro and Francisco Hernández Girón. Under him were a number of captains, including Martín de Meneses, the Portuguese Antonio Pereyra and Martín García de Loyola. The last was an ambitious Biscayan, a knight of Calatrava who had campaigned in Europe and accompanied Toledo to Peru as the captain of his viceregal guard. He was leading a select contingent of 'twenty-eight outstanding soldiers, sons of citizens and conquistadores of this kingdom'. The captain of artillery was Ordoño de Valencia; sargento mayor was Captain Antón de Gatos; and the officer in charge of Commissariat was Captain Julián de Humarán. Also with the expedition, as 'consultants in warfare', were three aged original conquistadores, Mancio Sierra de Leguizamo, Alonso de Mesa and Hernando Solano.

The expedition also had a large force of native auxiliaries. Don Francisco Cayo Topa was in command of 1,500 warriors from the tribes around Cuzco, and Don Francisco Chilche – the chief who had been suspected of poisoning Sayri-Tupac in 1560 – led five hundred of his Cañari tribesmen, who were as eager as ever to avenge the Incas' massacre of their tribe.*

Toledo wanted to be certain that the Inca could not escape out of Vilcabamba to the south or west. A second force of seventy men was therefore to enter from Abancay, march down the left bank of the Apurímac, cross that great river and climb into Vilcabamba 'through dense montaña and precipitous paths'. These were led by Gaspar Arias de Sotelo, 'one of the most important men in the kingdom' who was to assume overall command if Hurtado de Arbieto died. A third force of fifty citizens of Huamanga under Luis de Toledo Pimental marched into the valley of Mayomarca (Apurímac). It was to occupy the passage of Cusambi to prevent the Inca escaping through the country of the Pilcosuni to the north-west.*

The main force crossed the bridge of Chuquichaca 'with no impediment whatsoever' and marched up the valley of the river now called Vilcabamba. Some twenty miles upstream the valley closes in, with steep forested walls. Here, some thirteen miles before reaching 'Vitcos and Puquiura, there is a bad, steep passage in dense jungle, difficult to cross, called Quinua Racay and Coyao-chaca'. The Inca commanders, according to Murúa, 'decided that this was an opportune place to defeat and destroy the Spaniards, for the difficulty and steepness of the ground were in their favour'. They had strewn the paths with palm thorns and arranged barriers of creepers to entangle the Spaniards.

Martín García de Loyola was leading the vanguard of fifty Spaniards and some Indian allies. As he marched along at the head of the column, 'an Inca captain called Hualpa suddenly leaped out of the forest and, before anyone saw him, grabbed our captain in such an embrace that he could not reach his weapons. The object was to hurl him down the ravine. He would have been dashed to pieces and hurled into the river' by the Indian, who was 'a man of such great physique and strength that he seemed a half-giant'. While they grappled beside the precipice like Sherlock Holmes and Moriarty, 'an Indian servant of the captain, called Corrillo . . . drew Loyola's sword from the scabbard'. 'He struck a slashing blow at [Hualpa's] legs so that he toppled, and followed this with another slash across the shoulders which opened them, so that he fell there dead . . . Corrillo thus took the life of a half-giant with two slashes, and saved his captain's life.' Baltasar de Ocampo recalled in 1610 that 'to this day the place where this happened is called "Loyola's leap" '. But Loyola himself preferred not to remember how ingloriously his life had been saved in this embarrassing episode. In his otherwise boastful petition to the King for favours, he referred to the battle of Coyao-chaca simply as being the first 'hand-to-hand' engagement with the Indians.*

The battle of Coyao-chaca lasted for two and a half hours in the afternoon of the third day of Pentecost, probably the first of June. 'The location was very favourable to the Indians, for their enemies could march only in single file as the path was very narrow. On either side were high mountains between which flowed the great river . . . The Indians were in ambush in different places on the upper slope. Others were on the slope below with lances on which to catch those who fell; and in case any should elude their grasp, they had [posted] Indian archers on the far bank.'

The Indians 'advanced with their lances, maces and arrows with as much spirit, brio and determination as the most experienced, valiant and disciplined soldiers of Flanders . . . They began by sounding a great blast of tarquis, which are their form of bugle. Hardly were these heard before the Indians were among [the Spaniards] . . . They threw themselves into the mouths of the arquebuses, unafraid of the harm these could do them, simply to come to grips.' But their reckless bravery was futile, for the natives were fighting with their traditional maces, stones, lances and arrows, and the Spaniards were using firearms. At the height of the battle an arquebus-shot 'struck a valiant Indian called Parinango, general of the Cayambis, and he fell dead; and with him fell Maras Inca, another captain, and many brave Indians'. The Inca generals ordered a retreat, which was done in good order. As they retired, the natives used their most successful weapon; boulders rolled down the hillside. These killed two Spaniards called Ribadeneyra and Pérez, who were 'buried on the path itself, with two crosses placed over them, for they could find no more level place'. But, as García de Loyola boasted, the Spaniards 'made them retreat with the loss of five captains and other leading Indians' – a loss the tiny state of Vilcabamba could ill afford.

The cautious Spanish commander spent three days reconnoitring routes through the precipitous montaña around Coyao-chaca. His scouts finally found a path that was free of ambushes, and the expedition lumbered forward with all its baggage. It emerged into the valley of Puquiura 'where the Inca had his lodgings [at Vitcos], and where there had been a church in which the Augustinian fathers officiated, and where Titu Cusi Yupanqui died'. To their great delight – for they were already short of food – the Spaniards found maize on the cob ready to be eaten, and many llamas. The expedition had thus gained its first objective: the town and palace of Vitcos, perched high above the little valley of Puquiura. But Vitcos had fallen to Orgóñez in 1537 and to Gonzalo Pizarro in 1539 and the state of Vilcabamba had survived. The Indians now as then retreated into the jungle valley of Vilcabamba itself, hoping to repeat their former escape. If the Inca Tupac Amaru could avoid capture, he could later revive the native state.

The Spaniards knew where to pursue. They continued up the valley of the Vitcos river to its source and over the watershed at an elevation of 12,000 feet. As the road dipped down on the far side, they 'encountered ninety-seven Castilian cows that the Incas kept there, and Castilian sheep and pigs'. The 'maese de campo' Juan Alvárez Maldonado was thrilled by this addition to his dwindling commissariat. 'He shouted "Gather up everything, for it is mine!" and then fell off his horse into a swamp.' When the American explorer Hiram Bingham first penetrated here in 1911, he recalled 'the smooth, marshy bottom of an old glaciated valley, in which one of our mules got thoroughly mired while nibbling the succulent grasses that covered that treacherous bog'. The marsh must have been well camouflaged to have deceived both a seasoned conquistador and an Andean mule: perhaps it was the same infernal bog in which the Indians made the priests García and Ortiz wade and slip on their way to Vilcabamba in 1570.

Hurtado de Arbieto decided to rest on the far side of the watershed at Pampaconas, 'a very cold place' at an elevation of 10,000 feet. 'The expedition stopped for thirteen days, for many soldiers and Indians had fallen sick from a form of measles.' The gallant Spaniards were exhausted by the wild country through which they had travelled, and their general paused 'so that they could rest and cure the sick, and make further enquiries about the road, which was not known by those on the expedition'. While they were at Pampaconas, an Indian prisoner called Canchari stole a Spanish sword and cape and tried to escape to report to the Inca; he was caught and promptly hanged as a warning to the other prisoners.*

On Monday 16 June 1572 the expedition marched out of Pampaconas and plunged into the deep forested valley of the river now called Pampaconas or Concevidayoc. General Hurtado de Arbieto reported to the Viceroy that they set out 'with their weapons and blankets and food for ten days, in accordance with the orders sent by Your Excellency. On that same day Arias de Sotelo reached that place, with the men that Your Excellency ordered him to bring in via Cusambi and Carco . . . He remained there to guard the passes and I, Martín Hurtado de Arbieto, took the direct road for the native forts. Because the road was very wild we could not reach Huayna Pucará, the first new fort they had built, until the Friday.'

The expedition 'went through montañas and ravines with excessive difficulty for all. On the road they found sacrificed guinea-pigs in three or four places: it is common for the Indians to do this in times of war, famine or pestilence to placate their deities and to divine what will happen. They reached a pass called Chuquillusca, which is a cleft outcrop of rock running for a long distance along the course of the turbulent river. It was scarcely possible to march around this. The soldiers and allied Indian warriors were forced to pass it on all fours and holding on to one another, with great difficulty and danger.' A tough Portuguese soldier called Pascual Suárez heaved one of the expedition's small bronze artillery pieces around this rock on his back – a remarkable feat that impressed his companions.

It was near Chuquillusca that Manco Inca had almost destroyed Gonzalo Pizarro's force thirty-three years before, but Tupac Amaru's men did not attempt an ambush here. They did, however, scout along in the forests near the Spanish line of march, 'making a great din, shouting and firing arrows and boulders in each difficult place'. As Hurtado de Arbieto reported, 'some Indians came out in ambushes along the road, but we made them flee with the artillery and arquebuses'. But whenever the Cañari were lured away from the protection of the Spanish arquebuses, 'they returned wounded by enemy lance thrusts: for although the Cañari are so skilful in the exercise of lances, as is well known, the enemy were more experienced'.

On the day after the passage of Chuquillusca, an Inca captain called Puma Inca defected to the Spanish column. He claimed to be a close confidant of the Inca Tupac Amaru and his nephew Quispe Titu, and said that these wanted to come out peacefully to the Spaniards. He declared that these Incas 'were in no way responsible for the death of Atilano de Anaya'. This crime was the work of Curi Paucar who, with the orejón commanders Colla Topa and Paucar Inca, were determined to 'continue the war and resist until death'. Puma Inca then turned to more direct treachery: he told the Spaniards about 'a fort called Huayna Pucará, made a sketch of it, and showed how it could be taken without danger to the Spaniards'.

The expedition spent the night at a place called Anonay, keeping careful watch against a surprise attack. Next day, Friday 20 June, they marched nine miles to the plain of Panti Pampa and came in sight of the fort of Huayna Pucará. The day was spent in reconnaissance and debate about the attack 'which was expected to be dangerous'. Puma Inca explained exactly how the fort could be outflanked. This was not good enough for the excited and voluble Spaniards. 'There were so many different opinions among the captains and citizens that they almost came to blows. For all or most of them were important and selected men, rich and powerful, with much property and wealth, and they were serving at their own expense.' Hurtado de Arbieto himself finally arrived and calmed his agitated officers.

General Arbieto described the position to Toledo: 'For three-quarters of a league [three miles] before the fort, the Inca's Indians had fortified some narrow defiles with many boulders, and had built the fort itself on a knife-edge ridge at the far end. [It consisted] of a wall 200 yards long and 2 wide, crenellated as a defence against arquebus fire, and with four small towers.' Murúa said that the fort's wall was 'of fieldstone and clay and very thick, with many piles of stones ready to hurl or shoot from slings'. Arbieto said that for 'an arquebus-shot's distance in front of it they had placed many palm stakes painted with [poisoned] herbs, with a single narrow entrance through which only one man at a time could enter the fort'. But the Indians' chief defence lay in the approaches to the fort, 'where the road along which [the Spaniards] had to march ran around a half crescent, very narrow, with great rock outcrops and jungle, and a deep, turbulent river running beside the course of the road. It was all most dangerous and frightening to have to pass this and fight the enemy who would be in the heights, which form a steep escarpment above this [long] stretch. . . . All along this knife-edge were heaps of boulders, and above and behind these were enormous rocks with levers to send them tumbling down.' Any Spaniards who survived this deadly avalanche would be picked off by five hundred Chuncho jungle archers who, according to the traitor Puma Inca, were stationed across the river. The Indians clearly hoped to repeat the successes achieved by Manco's army during the first rebellion, and almost achieved by Manco against Gonzalo Pizarro's expedition. But their preparation of the site for this ambush was elaborate and too obvious; and its exact disposition was betrayed by Puma Inca.

General Hurtado de Arbieto decided that the boulders must be captured before his army dare advance along the foot of the slope. He ordered Juan Alvárez Maldonado and Martín García de Loyola to try to climb up through the dense jungle. They took their own contingents and fifty selected arquebusiers, who were protected by twenty-five shield-bearers and fifty Cañari and other native troops. García de Loyola said that they also took one piece of artillery, and 'climbed through the jungle in a place where it appeared impossible to do so'. They left in the dark, at six in the morning of Saturday 21 June, clambering up the mountain through the dense black undergrowth, a world of gnarled trees, mosses and creepers, rotting stumps and dead leaves, and a thick secondary growth of thorny bushes. Visibility in these forested hills is only a few feet, and the Indians had not been maintaining an adequate watch on the Spanish camp. The Spaniards thus achieved complete surprise. They gained the summit in the early afternoon and 'revealed themselves to the enemy who were below, very well drawn up according to their method of fighting'.

Instead of attempting another attack, uphill, against the Spanish firearms, the Indians 'gradually retreated into the fort of Huayna Pucará, abandoning the stones and boulders they had prepared to destroy the Spaniards'. The arquebusiers on the hilltop fired their guns as a signal of success. Hurtado de Arbieto then advanced below the line of abandoned boulders with the main body of the expedition. The sight demoralised the outnumbered natives. Arbieto's Spaniards 'shouted "Santiago!" and attacked the fort. After a good volley of gunfire it was taken. The Indians defended it for a while with spirit and bravery', but 'when the artillery fire began, and seeing that the heights had been taken from them, their general Colla Topa and captains Caspina and Sutic abandoned the fort. Tupac Amaru and Titu Cusi's son Don Felipe Quispe Titu had left the fort for Vilcabamba the previous day, saying that they would await the Christians there.'†

On the following day a scouting party of thirteen picked soldiers and the Cañari chief Francisco Chilche pressed ahead to the natives' second fort Machu Pucará 'where Manco Inca defeated Gonzalo Pizarro'. When the main force arrived a native contingent suddenly attacked with much shouting. This caused confusion among the rich citizens of the invading expedition. The arquebusiers tried to light their wicks. 'In the commotion a servant of Don Jerónimo de Figueroa, nephew of Viceroy Francisco de Toledo, set fire to a suit of padded armour he was wearing. Had he not jumped into a stream that ran past there he would certainly have been roasted.' But the attack came to nothing. To their surprise the Spaniards found Machu Pucará abandoned. Hurtado de Arbieto wrote, almost disappointed, that 'some Indians said that if they lost Huayna Pucará, the first fort, they would not dare to await the Spaniards in the other. But we thought nevertheless that we would find them in it, as its defences are so good, or failing it in the old fort of Vilcabamba.'

The expedition spent the night at Marcanay, the place where Diego Ortiz had been martyred, only a few miles from the hidden city of Vilcabamba itself. The men were delighted to find stocks of maize and tropical foods such as 'plantains, axials, yucas and guavas' which they ate eagerly 'for they were hungry and short of provisions'. They had almost reached their goal: the last free Inca city.

'Next morning, the feast of St John the Baptist, Tuesday 24 June 1572, General Martín Hurtado de Arbieto ordered the entire expedition placed in order by companies with their captains, and the Indian allies also with their generals Don Francisco Chilche and Don Francisco Cayo Topa . . . They marched off taking the artillery, and at ten o'clock they marched into the city of Vilcabamba, all on foot, for it is in the most wild and rugged country in no way suitable for horses.' Pedro Sarmiento de Gamboa, as secretary and ensign of the expedition, planted the royal standard in the main square of Manco's capital and took formal possession with seven of the captains acting as witnesses.*

Hurtado de Arbieto reported that his men 'found Vilcabamba deserted with some four hundred houses intact, and [the natives'] shrines and idolatries in the form in which they had had them here before this city was captured. We found the houses of the Incas burned.' Martín de Murúa confirmed this: 'The entire town was found to be sacked, so effectively that if the Spaniards and [their] Indians had done it, it could not have been worse. All the Indian men and women had fled and hidden in the jungle, taking whatever they could. They burned the rest of the maize and food that was in the storehouses, so that when the expedition arrived it was still smoking. The temple of the sun, where there was a principal idol, was burned. [The Indians] had done the same when Gonzalo Pizarro and Villacastín entered the city, and the lack of food had forced [Pizarro's expedition] to return and leave the country in [the Indians'] power. [The Indians] expected on this occasion that when the Spaniards found no food or anything on which to live, they would turn back and leave the land and would not remain to settle it.'

Murúa went on to give a description of Vilcabamba that confirmed its low, tropical location. 'The climate is such that bees make honeycombs like those of Spain in the boards of the houses, and the maize is harvested three times a year. The crops are helped by the good disposition of the land and the waters with which they irrigate it. There are axials in the greatest abundance, coca, sweet canes to make sugar, cassava, sweet potatoes and cotton. The town has, or rather had, a location half a league wide, rather like the plan of Cuzco, but covering a long distance in length. In it are raised parrots, hens, ducks, local rabbits, turkeys, pheasants, curassows, macaws, and a thousand other species of birds of different vivid colours. There are a great many guavas, pecans, peanuts, lucumas, papayas, pineapples, avocado pears and other fruit trees. The houses and sheds were covered in good thatch. The Incas had a palace on different levels, covered in roof tiles, and with the entire palace painted with a great variety of paintings in their style – something well worth seeing. It had a square large enough to accommodate a good number of people, where they used to celebrate and even raced horses. The doors of the palace were of very fragrant cedar, of which there is plenty in that country, and the garrets were of the same wood. The Incas therefore enjoyed scarcely less of the luxuries, greatness and splendour of Cuzco in that distant or, rather, exiled land. For the

Indians brought them whatever they could get from outside for their contentment and pleasure. And they enjoyed life there.'

The thing that the Spaniards feared most had happened again: the Incas had vanished into the jungles and eluded them as effectively as Manco had done in 1537 and 1539. The only living people in Vilcabamba were some of the Indians who had been with Atilano de Anaya; and at the foot of some rocks were found the bodies of Gabriel de Oviedo's envoys of the previous year. Some of the inhabitants of Vilcabamba soon began to wander back from the bush, and Arbieto went out of his way to feed them and treat them well. But the Incas and their commanders had all fled.

The Spaniards had always thought that the Incas' most likely escape route would be to the north-west, to the territory of the Zapacati and Pilcosuni Indians. It was to block this route that Toledo had sent contingents into Vilcabamba from Huamanga and from Abancay. Hurtado de Arbieto now learned from the returning natives that 'Tupac Amaru and Don Felipe Quispe Titu, with some eighty conscripted Indians, besides the captains and Indians who killed Anaya, had left Vilcabamba the day before [our arrival], fleeing towards Zapacati, whither they had sent some clothing and food a few days earlier.' The Spanish general wrote that the Incas did so 'at the risk of their lives, for they cannot survive there and it is no country for them'.

Immediately after his arrival, Arbieto sent parties into the jungle in pursuit of the elusive Incas. A hand-picked force went northwards in the direction of the Pilcosuni and scaled a mountain called Ututo in pursuit of the prince Quispe Titu. This contingent included all the young men who were related to Inca princesses – Juan Balsa, Pedro Bustinza and Pedro de Orúe. 'They climbed this mountain with incredible difficulty, with no water or food beyond what they had taken from Vilcabamba. They found in that jungle a great number of extremely dangerous rattlesnakes. But after six days Captain Juan Balsa . . . came upon the place where Quispe Titu Yupanqui was with his wife, who was pregnant, and with eleven Indian men and women who were serving them: the rest had vanished.' They returned in only two days and delivered Titu Cusi's son to General Arbieto in the Inca's own palace. 'And there they despoiled Quispe Titu of all his baggage and clothing, leaving him and his wife without even a change of clothes in their prison and with none of their table service.

Because of this they even suffered hunger and cold, although it is hot country.'

Captain Martín de Meneses was also sent into the jungle in search of the Inca. Meneses apparently travelled north-eastwards across the mountains towards the next valley, now called San Miguel. His advance party, which included 'Francisco de Camargo and his company, with Alonso de Carbajal and other soldiers, advanced more than ten or twelve leagues ahead of their captain, to a village of Sati Indians called Simaponeto, where they found a large river. They learned there that Huallpa Yupanqui, the general of the Incas, was marching nearby and was carrying the idol Punchao. Ten of the soldiers of this advance contingent adventured ahead. They crossed the river in pursuit of Huallpa Yupanqui, on a raft of three logs with great danger to their lives. They caught up with him and seized many of his men. And they took from him the idol called Punchao, which caused the Indians to realise the deception in which they had been held.' But although this group captured the holy image of the sun, they failed to capture Huallpa Yupanqui, who continued his march into the depths of the forest. Captain Meneses returned to Vilcabamba with 'the golden idol of the sun, much silver, gold and precious stones and emeralds, and much ancient cloth. All of this was said to be worth over a million, but it was divided among the Spaniards and Indian allies, and two priests who went on the expedition even enjoyed their share.' Martín de Murúa disapproved of this looting of the last few treasures left to the Incas.

Another expedition under Captain Antonio Pereyra moved towards a village of the Panquis. Neither Murúa nor Salazar mentioned that Martín García de Loyola was in this contingent, but that ambitious young man presented a different picture in his petitions to the King in 1572 and 1576: 'I and my company went in pursuit until we reached the town of Panquis, with great hardship because of the density of the jungles. I captured there two brothers of Tupac Amaru and one of his daughters and four nephews, and the Captain Curi Paucar, the principal aggressor in that war, together with a great quantity of Indians and captains.' Two important orejón captains captured in this haul were Colla Topa and Paucar Inca, but the biggest catch was the militant leader Curi Paucar, 'the traitor, the cruellest of all the Incas' captains, the man who was always most insistent in maintaining the war and not surrendering . . . and who had been the principal cause of the death of

Atilano de Anaya'. On the journey back to Vilcabamba one of Curi Paucar's small sons was being carried, but the boy was bitten by a snake and died within hours. So the Spaniards, acting on information from the natives in Vilcabamba, easily rounded up most of the Inca's military commanders. They even captured two venerated relics: the mummified bodies of Manco Inca and Titu Cusi Yupanqui.

But the most important figure was still missing. The Inca Tupac Amaru himself was at large with his commander-in-chief and governor Hualpa Yupanqui. The dashing Martín García de Loyola volunteered to lead a force deep into the northern jungles in pursuit of the Inca. He took forty hand-picked soldiers, including some who had experience of jungle conditions, and descended the 'river Masahuay of the Manarís, who are Chuncho Indians'. They descended this river 'for forty leagues [170 miles] from Vilcabamba' to a place that Loyola called the landing-stage of the Guambos. 'While they were camped there, in a jungle of tall forest and mangroves, he and his soldiers noticed, at midday, five Chuncho Indians in the water ... Captain García de Loyola considered how he could catch some of those Indians to gain information about the Inca: for no one else could know more about his whereabouts.' Most of the soldiers (including Captain Loyola) were too frightened to attempt an attack on these forest Indians. But Murúa's friends Gabriel de Loarte, Pedro de Orúe, Juan Balsa and three others finally volunteered to cross the river in pairs on some balsa rafts. On the far side, smoke appeared in the forest, where the Indians were preparing a meal in a long hut with twenty doorways. After some hesitation, the bravest Spaniards suddenly charged into this long-house. They managed to grab five of the seven Chunchos inside; but two escaped without even pausing to snatch up their bows and arrows. The successful Spaniards found that the hut was being used as an Inca storehouse, stuffed with thirty loads of the finest Inca and Spanish cloths and plumage, 'and, above all, with quantities of gold and silver vases and tableware of the Inca'. The battered and weary Spaniards 'ate very peacefully, most contented with the capture they had made'.

Apart from the treasure, the Chuncho captives proved to be extremely valuable informants. They were 'Manarí Indians, auxiliaries of the Inca, whom he had sent to find his general Huallpa Yupanqui and the rest of the warriors who were with him'. They told García de Loyola that Tupac Amaru was at 'a place called Momorí in the territory of the Manarís, confident that the Christians would not pursue him as far as there because of the density of the country and the difficulty of descending the river, which was full of cataracts, currents and rapids'. But Loyola did not hesitate. 'I made five rafts and sailed down the river with some soldiers of my company, with great danger to our persons. On various occasions we saved our lives by swimming.' They were plunging into the deep rain-forest of the Amazon, a land infested with insects, where the ground and trees and creepers are alive with columns of ants and termites, where travellers are plagued by the bites of ticks, jiggers and borrachudos and by the attentions of determined sweat bees. The Spaniards were by now sailing down the river they knew as Simaponte.* The pursuers survived the discomforts and dangers of the turbulent descent. 'We finally reached Momorí where I learned that the Inca had heard of my expedition and retreated further inland.'

García de Loyola continued the pursuit, striking inland through the forest. 'At the passage of a turbulent river, a native chief and many Indians came out for battle. By using the utmost skill and cunning that I possessed, I made friends with these and reduced them to the service of Your Majesty.' García de Loyola had persuaded his captive Chunchos to fetch the Manarí chief Ispaca, and managed to win over this suspicious chief with a mixture of blandishments and veiled threats. Loyola 'made him a speech, persuading him to say where Tupac Amaru was. To persuade him the more, he offered him some of the Inca's own clothing and Castilian feathers . . . But Ispaca would not accept them, saying that it would be the height of treason to his lord.' But the chief did betray the Inca's movements: 'Five days previously he had left that place, had embarked in canoes and gone to the Pilcosuni, another inland province. But Tupac Amaru's wife was frightened and depressed because she was about to give birth. Because he loved her so much he was helping her to bear her burden and was waiting for her, travelling in short stages.'

García de Loyola did not hesitate for a moment. He left five soldiers to guard the booty and to arrange the forward dispatch of supplies of food from the Inca's stores. 'He took the cacique [Ispaca] and left immediately that night in search of Tupac Amaru. . . . He entered the jungle with thirty-six soldiers, along the path that the Inca had taken. Behind them went their food: ten loads of maize, five of peanuts, three of sweet potatoes and eight of cassava.' It was as well for Loyola that he moved so fast. 'For on that same day Tupac Amaru had had a great debate with his wife, begging her to enter the canoe so that they could travel by water. But she was terrified of trusting herself on that open water [and refused]. Had they entered the canoe and embarked on the open water it would have been impossible to catch them, for they had taken food and stores to cross to the far side.'

The Manarí chief Ispaca also told García de Loyola where he could capture the Inca's commander-in-chief Huallpa Yupanqui. 'He was discovered in a jungle so dense and wild that it would have been impossible to have found him without their advice. Loyola captured him by travelling by night with his soldiers, marching through the dense jungle with torches. The capture of this last military commander was an important victory, for he was evidently trying to join his lord. But the Inca himself was still at large. The Manarí Indians again helped the Europeans, showing them the path followed by Tupac Amaru. García de Loyola wrote: 'I learned that Tupac Amaru Inca was retreating into the densest part of the Manarí jungle. We travelled into it on foot and unshod, with little food or provisions, for we had lost these on the river.'

The Spanish column had now marched for almost fifty miles along the dark forest paths in pursuit of their quarry. 'At nine o'clock one night two mestizo soldiers who were in front, named Francisco de Chávez and Francisco de la Peña, saw a camp fire burning in the distance. They approached it cautiously until they arrived [at a place] where the Inca Tupac Amaru and his wife were warming themselves. They came out upon them, but in order not to agitate them they showed much courtesy, telling them that their nephew Quispe Titu was safe in Vilcabamba and was being treated well; that no abuse or ill-treatment would be done to them; and that their cousins Juan Balsa and Pedro Bustinza were travelling with [the Spaniards]. Because Francisco de Chávez was the first to reach the Inca, and also because he took some rich Inca vases, he became known as "Chávez Amaru". While this was happening, Captain Martín García de Loyola arrived with Gabriel de Loarte and the rest of the soldiers, and they arrested the Inca. They spent that night with great caution and on guard, and returned towards Vilcabamba in the morning.' According to Calancha, Tupac Amaru 'preferred to trust in those who were searching for him than to hide in the jungles into which they were pursuing him. He therefore surrendered to the Spaniards.'

García de Loyola was justifiably elated: 'I captured the Inca with all the Indians he was leading, with his governor and other captains, wives and children.' He was, in the words of his friend Antonio Bautista de Salazar, 'considerably pleased with the successful outcome of this expedition and its booty'. The Spaniards in Vilcabamba were equally overjoyed, and a messenger was sent galloping back to Cuzco to tell the Viceroy. Toledo had been worried by the possibility of the Inca's eluding his men as Manco had done. So when the horseman rode in late one night with the victory dispatches, Toledo ordered religious ceremonies and days of celebration of 'the successful end to this conquest'. The captive Inca was by no means feeble-minded or impotent. Murúa said that Tupac Amaru was 'affable, well-disposed and discreet, eloquent and intelligent'.

The wild native state of Vilcabamba that had inspired awe among Spaniards for thirty-five years proved to be a pathetic bogey. Its few hundred defenders neglected to cut the bridge of Chuquichaca and watched impotently while the heavily-armed Spanish column marched in. When they did fight the natives were brave enough, but their fighting techniques had not developed since Manco's rebellion. They still relied on hand-to-hand fighting, but failed to press home any attacks on the Spaniards after Coyao-chaca. They made little use of the arrows of their forest allies, and their obvious boulder trap at Huayna Pucará was easily outflanked by the invaders. Their ultimate defence was to imitate Manco's flight into the jungles. But whenever contingents of Spaniards pursued on foot through the forests they surrendered without the slightest resistance. It was a sorry end to the last independent vestige of the great Inca empire.

23

THE ELIMINATION OF THE INCAS

THE triumphant Vilcabamba expedition marched back over the Panticalla pass early in September 1572, with the leading Incas as its captives. The native governor and commander-in-chief, Tupac Amaru's uncle Huallpa Yupanqui, did not complete the journey. He had contracted a disease during his flight into the jungle and started internal bleeding in his stomach. 'The illness gripped him and he died one league before reaching Cuzco. So he did not witness the tragedy and sadness that was being prepared in it for his nephew Tupac Amaru.'

The other Incas were with the column when it reached Cuzco on St Matthew's day, 21 September. It paused at the Carmenca gate to form into marching order. Juan Alvárez Maldonado 'chained Tupac Amaru and his captains together. The Inca was wearing a mantle and

Indian women and orejones weep at the execution of Tupac Amaru tunic of crimson velvet, and his shoes were of local wool in various colours. The crown or headdress called mascapaicha was on his head, with a fringe over his forehead, these being the royal insignia of the Inca.' 'For the day of the triumphal entry of the captains with their prisoners, the town council arranged a parade of all the troops who were in the city . . . This was very splendid, but the soldiers of the Conquest looked better in the battle and jungle uniforms they were wearing. . . . Each brought some notable captive. . . . The last was Captain Loyola with the Inca Tupac Amaru, held by a chain of gold round his neck', while Quispe Titu had a chain of silver.

The Viceroy Francisco de Toledo was staying in the finest house in Cuzco, which had belonged to one of the richest citizens, Diego de Silva y Guzman.* The Viceroy 'watched what he wished from a window without being seen', but Loyola was well aware that he was there. He ordered all the prisoners to remove their headbands as they passed the window, and Tupac Amaru to remove his royal fringe. They refused to do this, but instead touched their llautus and bowed towards the unseen presence. Two chroniclers said that Loyola let go of the gold chain and struck his royal prisoner for not obeying.* After Toledo had 'savoured the pleasure of conquest' he ordered that the prisoners be taken to be confined in the 'grand and majestic buildings' of the Carmenca palace. This house had recently been confiscated from Paullu's son Carlos Inca so that Toledo could convert it into a fortress. When Tupac Amaru was imprisoned there, it was being used as a barracks for the city guard, with the Viceroy's uncle Luis de Toledo as its castellan.

The victors of Vilcabamba brought with them the mummified bodies of the two Incas who had died in that province, Manco and Titu Cusi. They also brought the famous image of the sun, Punchao, which was the embodiment of the Inca royal family's descent from the sun. The Spaniards now wished to complete the humiliation of the native religion by persuading Tupac Amaru to follow his brothers in a conversion to Christianity. The task of indoctrinating him was given to two Mercedarians, Melchor Fernández and Gabriel Alvárez de la Carrera. Both were famous for their mastery of Quechua, which 'they spoke so well that they excelled the Incas themselves' and, as teachers, they were so 'adept in their office that they fed the Inca, as it were, with a spoon'. They were reinforced by other notable ecclesiastics, such as the Jesuits under Alonso de Barzana who 'did not leave his side by night or by day', and Cristóbal de Molina, who was in charge of the hospital of Our Lady of the Remedies and was an expert on the native religion.* The Dominican Prior Gabriel de Oviedo also claimed that he himself, with two monks of his order, had had a hand in the instruction of the Inca and his companions.

This formidable battery of proselytisers achieved rapid success. 'It was remarkable, especially as regards Tupac Amaru, to find that heathens who had never been taught the things of our Holy Catholic Faith should have shown such intelligence in understanding it. In three days they knew all that was necessary to enable them to be baptised. Not only did they achieve such good results, but they pressed us to teach them more each day.' The catechism took place in the prison, and the Inca was baptised with the Christian name Pedro.*

The Inca's eagerness to acquire Christianity may have sprung from a feeling that its efficacy was proved by the Spaniards' successes. But he probably hoped that by embracing Christianity he could mitigate his captors' severity towards himself. The Jesuit Antonio de Vega wrote that the Inca 'did not leave [the monks] for a moment, for he said that his consolation and salvation rested with them'. And the native prisoners told Gabriel de Oviedo that 'though they were to be killed, they wished to be Christians and to die Christians'. For Tupac Amaru and his companions soon became aware that their lives were in danger. They were desperate to grasp at anything that might save them: Christianity promised eternal salvation, and, in the present circumstances, it might provide immediate physical salvation as well.

The Viceroy Francisco de Toledo pressed his prosecution of the captives with great urgency. He had reached a clear conviction that the last vestiges of the Inca empire must be eliminated. His interrogations of aged Indians and Sarmiento de Gamboa's commissioned history had reinforced his view that the Incas had been recent conquerors with no better right to rule Peru than the Spaniards now had. He had no time for sentimental admiration of the Inca way of life, or for guilt about its overthrow. He was rapidly integrating the native population into colonial society, establishing a semi-feudal relationship between creoles and natives that has largely survived into the mid-twentieth century. There was no place in all this for autonomous Inca enclaves; nor should there be any figureheads to inspire the native population. Toledo therefore determined in a few days what Elizabeth of England was taking nineteen years to decide: two monarchs of different religions could not exist in the same country.

Having reached this conclusion, Toledo knew that he must strike swiftly and implacably. The people of Cuzco were delighted to be relieved of the bogey of Vilcabamba, and there were great festivities to celebrate the victory. The city council wrote admiringly to the Viceroy: 'This last conquest that Your Excellency has made in the montaña of Vilcabamba . . . has been [against] one of the most effective nuisances that could be imagined. That country was so rough and dense . . . that [its conquest] was coming to appear impossible to all the Viceroys and governors who have discussed or studied it, particularly in view of the lack of success enjoyed by Spaniards on the two or three occasions that they attacked it before now.'

No one in Cuzco had any sympathy for the militant native generals who had tried to lead the defence of Vilcabamba. Martín García de Loyola described Curi Paucar as 'the principal aggressor in that war'. Curi Paucar and his fellow commanders were accused of what amounted to 'war guilt' and were given a summary trial at which nothing appears to have been said in their defence. Toledo's prosecuting magistrate Dr Gabriel de Loarte rapidly sentenced the captains to be hanged. 'The sentence was carried out. The captains were taken out to be executed along the public streets, with the town crier proclaiming their crimes. Three died ordinary deaths and two at the end of the gallows. [The three] had caught the disease "la chepetonada" in prison and were all at their last gasp. Notwithstanding their illnesses they were brought out in blankets to comply with the letter of the law, since they had been murderers. But the other two, Curi Paucar and the Huanca Indian, paid on behalf of them all on the gallows, and were hanged.'

But the Viceroy was not content with the blood of these military commanders. He was determined to execute the Inca himself. During the three days in which Tupac Amaru was being instructed, catechised and baptised by his captors, he was also on trial for his life. He was charged, in general terms, with having ruled a state that had launched raids into Spanish Peru, and in which heathen practices had been tolerated. More specifically he was accused of all the murders that had taken place in and around Vilcabamba during the past eighteen months. Toledo himself summarised the charges against the Inca: 'After the death of Titu Cusi, his son Quispe Titu and his brother Tupac Amaru committed many crimes, causing the deaths and kidnapping of Indians and Spaniards. In particular, Titu Cusi during his lifetime killed a foreign [mining prospector] Romero; and the said Tupac Amaru and Quispe Titu killed a friar of the Order of Saint Augustine [Diego Ortiz] who was teaching the doctrine there in accordance with the agreement [of Acobamba], and Martín Pando. They seized and killed friendly Indians who were bearing messages to them. And finally they seized and killed Atilano de Anaya, whom I sent with letters and peace proposals, and they took and killed many of those accompanying him, placing that entire province under arms.'

The trial of the Inca was hurried and it was manifestly unjust. Tupac Amaru could not be held responsible for raids that had ceased many years earlier. His brief reign had seen a revival of the heathen religion, but this was hardly a capital crime. His brother had admitted Christian missionaries, and it was not necessarily Tupac Amaru's fault that they had not been more successful. The critical part of the Inca's trial concerned the various murders; but although he was held responsible for them, there was no serious attempt to convict him personally of any one murder. Subsequent investigations revealed that it was the lynch mob at Puquiura that killed the mestizo Martín Pando immediately after Titu Cusi's death. These same interrogations did implicate Tupac Amaru in the death of Diego Ortiz – but only to the extent that he had refused to see him at Marcanay and had thus condoned his martyrdom. It was never shown that Tupac Amaru actually ordered the death of the wretched missionary.† The mining prospector Romero was killed on the orders of the previous Inca, Titu Cusi. And the murder of the ambassador Atilano de Anaya – by far the most serious offence in Toledo's eyes – was the work of troops guarding the Chuquichaca bridge. It took place on the day after Anaya's crossing: too soon for instructions to have been sought from Tupac Amaru. Antonio de Vega, a close companion of Tupac Amaru's confessor Alonso de Barzana, wrote that the messengers 'were killed on the road by the captains of the Inca without his knowing anything about it, as was clearly established'. Bernabé Cobo said that Anaya 'was killed by the garrison Indians who were guarding that passage. The killers went to Vilcabamba and told the Inca that they had found some Christians hidden near the bridge of Chuquisaca who were spying out the crossing in order to enter and kill him, and they had taken their lives. The Inca was astonished by what had happened, and regretted that they had not first informed him.'

But the Viceroy and his minion Dr Loarte were not really concerned with proving the guilt of the Inca. Toledo had decided to eliminate Tupac Amaru, and he knew that he must act quickly before a wave of sympathy could save the captive ruler. And so, after only three days of sham trial Dr Loarte sentenced the Inca Tupac Amaru to death. He spared him the indignity of the gallows, and decreed that he be beheaded.

The sentence of death caused an immediate furore in Cuzco. Everyone had been delighted by the triumph in Vilcabamba; but they were stunned by Toledo's threat to execute the defeated Inca after a rushed, cursory trial. It seemed impossible, in the civilised climate of the 1570s, to repeat Pizarro's execution of Atahualpa – done in such very different circumstances almost forty years previously. 'The Spaniards who were in the city, both ecclesiastics and laity, . . . never imagined that the sentence would be executed.' Nevertheless, a group of ecclesiastics hastily organised an appeal to bring about a stay of execution or a full pardon. According to the Jesuit Antonio de Vega, 'the Rector Fr. Luis López and Fr. Alonso de Barzana were certain and satisfied of the innocence of the Inca Tupac Amaru. They did everything possible to prevent his being beheaded. They informed the Viceroy of the truth, presented him with positive witnesses of the Inca's innocence, and begged him on their knees and with tears on many occasions. Seeing that they were accomplishing nothing, they persuaded the Prelates of the Orders and other notable ecclesiastics to do the same.' Baltasar de Ocampo and Reginaldo de Lizárraga listed these religious leaders: 'Fr. Gonzalo de Mendoza, Provincial of the Order of Our Lady of Mercy; Fr. Francisco Corrol, Prior of St Augustine in this city; Fr. Gabriel de Oviedo, Prior of Santo Domingo; Fr. Francisco Velez, Guardian of San Francisco; Fr. Gerónimo de Villa Carrillo, Provincial of San Francisco; Fr. Gonzalo Ballastero, Vicar Provincial of the Order of Mercy; and Father Luis López, Rector of the Company of Jesus, all went to the Viceroy. They went down on their knees and begged to show mercy and spare the life of the Inca.'

The most important plea for mercy came from Fr. Agustín de la Coruña, Bishop of Popayán in southern Colombia and one of the most respected ecclesiastics of his age. Coruña had been one of the first twelve Augustinian friars to enter Mexico and had consistently fought for better treatment of American natives. He travelled south to be Toledo's ecclesiastical adviser on his visitation. Antonio de Vega described him as 'a perfect man, considered by all to be saintly'. Bishop Coruña and the other ecclesiastics 'all begged the Viceroy on their knees, with great emotion, tears and fervour' to spare the Inca, 'for he was innocent and should not die the death that was planned for him. He should be sent to Spain to His Majesty. But the Viceroy resolutely refused, and closed the door on appeals and supplications in this case.'

Tupac Amaru was taken out of his prison of Colcampata to be led down the steep hill to the main square of Cuzco where the scaffold had been erected. His hands were tied and there was a rope at his neck. He was riding 'a street mule with trappings of black velvet, and he himself was completely covered in mourning. So many natives congregated at the death of their King and lord that people who were present say that it was only possible to push through the streets and squares with the greatest difficulty. Since there was no more room on the ground, the Indians climbed the walls and roofs of the houses; and even the many large hills that are visible from the city were full of Indians. Many gentlemen and leading people accompanied him, and there was a great crowd of ecclesiastics.' Gabriel de Oviedo and Baltasar de Ocampo also recalled the gigantic crowd: 'The open spaces, roofs and windows in the parishes of Carmenca and San Cristóbal were so crowded with spectators, that had an orange been thrown down it could not have reached the ground anywhere – so densely were the people packed.' The condemned man was guarded on his way to the scaffold by four hundred Cañari Indians with lances in their hands, and was 'surrounded by the guard and halberdiers of the Viceroy'. 'Juan de Soto, chief officer of the court, was sent on horseback with a pole to clear the way, galloping furiously and riding down all kinds of people.'

There was a moving episode during this calvary. As the prisoner was led down the main street from Colcampata, his sister María Cusi Huarcay suddenly appeared at a window. This fearless woman 'raised her voice with great sobbing and cried out to him: "Whither are you going my brother, prince and sole King of the four suyos?" She tried to press forward but the ecclesiastics prevented her. He remained most grave and humble.' Most Spaniards shared her grief. 'The balconies were packed with people, women and leading ladies who were moved by pity and were weeping in sympathy for him, seeing an unfortunate young man being led to be killed. It was true to say that no one, of quality or otherwise, did not deplore his death.'

Tupac Amaru reached the square and mounted the black-draped scaffold with Bishop Agustín de la Coruña. 'As the multitude of Indians, who completely filled the square, saw that lamentable spectacle [and knew] that their lord and Inca was to die there, they deafened the skies, making them reverberate with their cries and wailing. His relatives, who were near him, celebrated that sad tragedy with tears and sobbing.'

Before being executed, Tupac Amaru made an extraordinary speech from the scaffold. An impartial witness, Toledo's own paymaster Antonio de Salazar, said that 'the Inca Tupac Amaru raised both hands, making a sign similar to that which the Indians used to make to their lords'. Garcilaso described the gesture: 'The Inca raised his right hand, with the hand open, and placed it to the right of his ear. He gradually lowered it from there until it rested on his right thigh.' 'Turning to face towards the direction where there was the greatest number of chiefs, he said in a loud voice in his own language: "Oiari guaichic!" This was a man almost on the point of death. He could scarcely be heard by those immediately alongside him. But the cries, lamentations and shouting ceased instantly. I believe that all were holding their very breath, and everything became as silent as if there had not been a single living thing in the square: such was the authority and sovereignty of the Incas, and the sway they held over their subjects.

'What he said in his language was in summary as follows, as witnessed in a declaration by the interpreters that were on the scaffold, and by others who were nearby and from whom testimonies have been taken. "Lords, you are here from all the four suyos. Be it known to you that I am a Christian, they have baptised me and I wish to die under the law of God. And I have to die. All that I and my ancestors the Incas have told you up to now – that you should worship the sun Punchao and the huacas, idols, stones, rivers, mountains and vilcas – is completely false. When we told you that we were entering in to speak to the sun, that it advised you to do what we told you, and that it spoke, this was false. It did not speak, we alone did: for it is an object of gold and cannot speak. My brother Titu Cusi told me that whenever I wished to tell the Indians to do something, I should enter alone to the idol Punchao;

... afterwards I should come out and tell the Indians that it had spoken to me, and that it said whatever I wished to tell them." '

This remarkable denunciation of the Inca religion seemed too good to be true from the Spanish point of view. The Viceroy Toledo himself wrote to the Cardinal of Sigüenza a month after the event, expressing his surprise and satisfaction. 'The Inca Tupac Amaru made a confession on the scaffold which, from what is understood, was the most advantageous thing that could have occurred for the conversion of these peoples.'

'The Inca then received consolation from the fathers who were at his side, and taking leave of all, he placed his head on the block like a lamb. The executioner [a Cañari] then came forward.' 'He bound his eyes, held him on the dais' 'and taking the hair in his left hand, severed the head with a cutlass at one blow, and held it high for all to see. As the head was severed the bells of the cathedral began to toll, and were followed by those of all the monasteries and parish churches in the city. The execution caused the greatest sorrow and brought tears to the eyes of all.'

Tupac Amaru's body was carried to the house of his sister Doña María Cusi Huarcay, Sayri-Tupac's widow. On the following day, after mass, 'the body of the Inca was interred in the high chapel of the cathedral, the services being performed by the chapter. Pontifical mass was said by Bishop Agustín de la Coruña, the epistle was read by Canon Juan de Vera, and the gospel by Canon Esteban de Villalom. All the ecclesiastics of the city attended the funeral and each said his vigils and joined in the singing of the mass in the presence of the corpse.... There was now a universal feeling of sorrow; and the masses were sung with an organ as for a lord and Inca. On the ninth day all the funeral honours were repeated, the religious coming to join in the vigils and masses of their own accord. From which it may be inferred that the Inca is with our Lord God.'

Toledo ordered that the head of the executed Inca be placed on a pole. But Juan Sierra de Leguizamo observed many natives crowding round it by night to mourn and worship their Inca. As Toledo himself wrote, 'his head could not be allowed to remain on the pike for more than two days, since no punishment could have sufficed to [prevent] the adoration they made towards it, or the cries and wailings of the ten or fifteen thousand caciques and Indians who were present in the square when his head was cut off and his confession heard.' The head was taken down and buried with the body.

Thus ended Tupac Amaru, the last of Manco's sons, the last crowned ruler of Peru, and the last Inca. The sequence of events surrounding his death was achingly familiar. The summary trial, the dignified behaviour of the victim on the scaffold, the exaggerated pomp of the funeral services, the passive despair of the native population, and the subsequent self-reproach of the Spaniards were all sadly reminiscent of the execution of Atahualpa thirty-nine years earlier. The deaths of the two rulers, uncle and nephew, symbolise the start and conclusion of the Conquest of Peru. The stroke that decapitated Tupac Amaru was the final blow of the Spanish Conquest of Peru, almost exactly forty years after the first violence of the ambush in the square of Cajamarca.

Francisco de Toledo felt that he must complete the humiliation of the Incas by destroying their most holy religious relics. The mummified bodies of Manco Inca and Titu Cusi were brought from Vilcabamba and secretly burned in the ancient fortress called Quispi-huaman. The other great trophy was the image of the sun, Punchao, the ultimate prize that had obsessed Spaniards since the start of the Conquest. Various sun images were captured in the course of the Spanish occupation, but the last was the one captured from the general Huallpa Yupanqui in the forests near Vilcabamba. This was the object that Tupac Amaru denounced and exposed in his speech from the scaffold. The Punchao was quite small: the idol of cast gold weighed 6 marks 6 ounces and its silver frame weighed 3½ marks, a total of only 5½ pounds. Toledo wrote that 'it has a heart of a dough in a golden chalice inside the body of the idol, this dough being of a powder made from the hearts of dead Incas. . . . It is surrounded by a form of golden medallions in order that, when struck by the sun these should shine in such a way that one could never see the idol itself, but only the reflected brilliance of these medallions. The soldiers cut them off to make up their shares of treasure.' The Viceroy sent the empty Punchao, stripped of its patens, to King Philip and recommended that 'in view of the power the devil exercised through it, and the damage it has done since the time of the seventh Inca . . . it certainly seems to me an object that Your Majesty could appropriately send to His Holiness.' The famous relic has never been found: it may still exist in some Spanish palace or in the Vatican.*

The Viceroy's attempt to stamp out the cult of the Incas went beyond the destruction of the Inca himself and the holy relics. Within weeks of proclaiming his Vilcabamba expedition, Toledo had told the King that he proposed, 'now that those who are in Vilcabamba are finished, to consume the seed of these Incas in this kingdom'. He explained how he would do this. 'It would be sufficient to punish all the Incas [for being] involved in this plot for rebellion. There would be about three hundred who have kinship relations stemming from the Incas, and who preserve their memory and ayllos.' Any means was sufficient for Toledo in his present aggressive mood. The rooting out of the Inca family was a logical development of the undermining of their divine right by the *Informaciones.* Toledo rightly saw that the Inca myth would be an inspiration to any rebellious Indians over the coming centuries – it in fact was. But it is alarming to see a man so keen on legality state his intention and then find a judicial means to achieve it.

The first victim of the campaign to consume the seed of the Incas was the startled Carlos Inca. In May 1572 he and his brother Felipe were suddenly arrested and put on trial. It was a cruel blow for Carlos, the man who prided himself on his Spanish manners and Spanish wife, who regarded Toledo as his compadre, the godfather of his son Melchor, who had written to King Philip: 'for many years I have longed to go to your court to kiss the royal feet and hands of Your Majesty'. It was only a matter of weeks since the Viceroy had been so flattering about Carlos's loyalty, when he presented the paños. Now Carlos was deprived of all his wealth and estates, including even the palace of Colcampata, and was unceremoniously imprisoned. Toledo for his part was delighted with the acquisition of this palace. He wrote enthusiastically to the King that 'it possesses the largest and best location one could imagine for keeping this town and every house in it dominated by a force of artillery'. He soon started turning Colcampata into a Spanish-style fortress complete with crenellations, artillery and a garrison. He himself dictated detailed memoranda about the guard and the maintenance of the fortress. He wished to imitate the fortresses with which Spain dominated its Flemish and African possessions, and even arranged to move the anti-Inca Cañari and Chachapoyas Indians living in Cuzco into the parish of San Cristóbal in the Colcampata ward.

Also arrested with Carlos were three leading men of royal blood – Don Alonso Titu Atauchi, Don Diego Cayo and Don Agustín Conde

Mayta – and two caciques, Pedro Guambotongo and Francisco Tuyru-Hualpa. These were important men, second only to Carlos Inca in lineage and equally enthusiastic in co-operating with the Spaniards and aping their ways. Alonso Titu Atauchi was the man honoured by Cañete with the title of hereditary Lord Mayor of the Four Suyos, and the recipient of royal privileges. Equally imposing was Don Diego Cayo, a man of seventy who had for thirty years been a highly respected figure in Cuzco society, and whose father had been 'second person of Huayna-Capac' and close confidant of Huascar. Both men had been leading witnesses to Toledo's *Informaciones* and to Sarmiento de Gamboa's *History*. They must have been as startled and incensed as Carlos Inca by Toledo's sudden and unjustified attack.*

The accusations against these men were very vague. They had 'made an alliance with those of Vilcabamba and prevented their coming out peacefully'. They were further accused of holding secret meetings, of preparing arms, and of communicating with the Vilcabamba Inca. Their judge was the fierce Dr Loarte, acting directly for the Viceroy, who had assumed the powers of commander-in-chief during this time of war. One witness against Carlos was Manco's daughter María Cusi Huarcay. She could not resist this opportunity for revenge on Paullu's hispanicised son. She accused him of being in contact with Vilcabamba, although it was highly unlikely that any Inca there would have anything to do with this collaborator – it was probably María herself who kept in touch with her brothers in their retreat. The chances of the accused were not enhanced by the fact that the proceedings were translated by the mestizo interpreter Gonzalo Gómez Jiménez. This young man was accused of homosexuality a few years later in a case involving some of Toledo's servants. He then admitted that he had sometimes falsified the translation of evidence in an attempt to please his master.* The trial of the Inca nobles was interrupted for a time so that the court could prosecute bigger game – Tupac Amaru – but when it resumed Carlos and his colleagues were found guilty and their goods were confiscated. Toledo was pleased to have their wealth, but he was also determined to be rid of the Incas themselves. When he signed their sentence on 12 November he added a clause banishing the more important of the accused from Peru. They were to be sent to exile in Mexico with a recommendation to its Viceroy to provide for them for life. Two Vilcabamba princelings were exiled with them: Tupac Amaru's three-year-old son Martín and

Titu Cusi's fifteen-year-old son Felipe Quispe Titu, together with various ladies and other relatives. The prisoners, escorted by Alonso de Carbajal, left Cuzco at the beginning of 1573 and reached Lima on 18 March.*

Toledo and Loarte had gone too far with this sham trial and heavy sentences. They soon had to face a powerful opposition, more effective than that of the ecclesiastics who made the eleventh-hour attempt to save Tupac Amaru. As early as December 1572, three oidores of the Audiencia of Lima reprimanded Loarte for not having allowed the accused Incas an appeal. Carlos and his colleagues soon became the subject of a legal and political struggle between the Audiencia of Lima and the Viceroy, who was continuing his visitation in the Charcas. Toledo was furious when he heard of this opposition. He dictated one provision from Potosí on 31 January 1573, another from La Plata on 2 June, and a third from Quilaquila on 10 November. In all of them he stressed that he had acted against the Incas in his capacity as wartime commander-in-chief: the affair was therefore beyond the jurisdiction of the Audiencia of Lima, and the accused must be dispatched from Lima to their perpetual exile. The authorities in Lima were openly defiant. One lawyer produced a princeling, presumably Tupac Amaru's young son Martín, during an important church service and told the oldest oidor, in a sarcastic voice loud enough for all to hear, that *this* was the Inca threatening to raise the country behind him. The entire performance was intended, according to Toledo's secretary Diego López de Herrera, 'to make a mockery of the Viceroy's provision'.

In late April, a number of petitions were sent to King Philip, including some from the disinherited Incas themselves in which they accused Loarte of irregularities in their trial and of purloining their confiscated estates. They pleaded their innocence and begged to be allowed to appear before the King himself. They argued, with reason, that they would never have communicated with the Vilcabamba Inca because of the enmity between their respective fathers. They also complained that they were starving in Lima.

These petitions reached the Escorial in December 1573, and the King immediately ordered a full enquiry into Loarte's conduct. The resident judge in Lima, Licenciate Sánchez Paredes, drew up a questionnaire, and more than fifty witnesses were called upon to answer. Roberto Levillier discovered the text of the proceedings in the Archive of the Indies: it consisted of over three thousand manuscript pages. Most witnesses agreed that there had been instances of distorted interpreting and of intimidation of witnesses. Dr Loarte made a spirited defence. He admitted that when Don García Inguill Topa would not reveal contacts between the Cuzco and Vilcabamba Incas, about which he was supposed to be aware, 'it was necessary to threaten torture . . . as a result of which he declared the truth . . . This is something permitted in law in such serious cases or in crimes of treason.' But Dr Loarte's arguments failed. When the examining magistrate sent all the passionate accusations and counter-accusations to Spain, Loarte was given a series of heavy fines. He had judged on flimsy evidence, used dubious legal methods, and sentenced with undue harshness.

Early in 1574 the King reversed the sentences against the Inca nobles. Toledo was furious. He had determined to rid Cuzco of its Incas and was convinced that their return to power there would present a lasting threat to Spanish rule. He wrote to a friend that 'the royal despatch concerning the sentence I gave against Don Carlos, Don Felipe and other rebel Incas from Vilcabamba could do more harm and damage in this kingdom than any other'. Toledo also received a mild rebuke from King Philip, who wrote that, although pleased with the Vilcabamba campaign, 'some things about the execution would have been better omitted'.

The chief beneficiary of this royal clemency was Carlos Inca. He returned to Cuzco to be restored as its wealthiest native noble. There was no fear of Carlos ever causing 'harm and damage' to anything Spanish. All his energies were devoted to the care of his estates. He had many of his illegitimate brothers working as virtual labourers on his farms. They, in revenge, seized on Carlos's exile to petition for a share of their father's property, arguing that *all* Paullu's children had been legitimised by royal decree. But the court found that Paullu's will had left his wealth only to Carlos and Felipe, sons of the wife he married in a Christian ceremony.* So Carlos continued to dominate his illegitimate brothers. He even managed to acquire a piece of land alongside Colcampata that had been left to his legitimate brother Felipe. He lived for eight years after his restoration, greedily amassing fields, terraces and pastures around Cuzco. Carlos wrote his will in February 1582. In it he complained about some acres confiscated by Toledo and for which he had never been compensated. He left something to his aged mother, Paullu's widow Doña Catalina. But he made no mention whatsoever of his royal Inca ancestry or of his native Peruvian roots: Carlos's will was indistinguishable from that of any Spanish property-owner.

The other prominent exiles, Don Alonso Titu Atauchi, Don Diego Cayo and Don Agustín Conde Mayta, presumably returned to Cuzco at the same time as Carlos Inca, as did Carlos's brother Felipe. Cayo and Conde Mayta had never left the highlands, since Toledo exiled them to Cajamarca and Huamachuco in the northern sierra.* Martín de Murúa wrote that Titu Atauchi died of a fever soon after his return to Cuzco from his two years in Lima. His son, Alonso Titu Atauchi, last possessor of the title Alcalde Mayor de los Cuatros Suyos, died in Potosí in 1610.*

Garcilaso de la Vega has caused confusion by dramatising the story of the Inca exiles. He wrote that thirty-six were exiled to Lima and that 'in little over two years thirty-five had died, including two children' because of the change of climate. This was nonsense, a fabrication intended to show that Carlos Inca's son Melchor Carlos was the only survivor of the royal house. In fact, as far as is known, none of the exiles died within two years of leaving Cuzco, and they were cared for in Lima by the many enemies of Toledo who championed their cause.

Not content with this exaggeration, Garcilaso confused Toledo's attack on the Inca family with Castro's earlier punishment of the mestizo plotters of 1567. Garcilaso met two of the mestizo rebels, Juan Arias Maldonado and Pedro del Barco, when they were in Spain. The story they told him grew in his imagination into a mass exile of 'all the mestizos who were found in Cuzco aged over twenty and able to carry arms'. He accused Toledo of having given them 'a long and painful death, which was to exile them to various parts of the New World'. Sir Clements Markham interpreted this remark to mean that thousands of mestizos, 'those bright and happy lads', were sent 'to perish in the swamps of Darien or the frozen wilds of southern Chile'. The four mestizos who had inspired all this hyperbole were in fact living comfortably enough in Spain. The Viceroy Toledo even wrote to the President of the Council of the Indies to appeal for leniency towards Garcilaso's friend Juan Arias Maldonado.* As a result Juan Arias was allowed to return to Peru to enjoy his father's great inheritance. He stayed with Garcilaso before leaving and persuaded the historian to part with all his white linen, some fancy taffeta and a horse. This was to be repaid handsomely when Maldonado reached Peru. But Garcilaso was never repaid: within three days of landing in northern Peru, Juan Arias Maldonado died 'from sheer joy at finding himself in his native land'. His cousin Cristóbal Maldonado also petitioned to be allowed to return to Peru. He was eventually given permission, and his reappearance caused acute embarrassment to the girl he had raped years before: Beatriz Clara Coya.

The mystique of the Incas was too strong to be rooted out by one determined Viceroy. But although the remaining Incas regained their property, and the sentimental memory of the imperial past continued to flourish, the Inca family had lost its power for good.

24

THE INCA SURVIVORS

ALTHOUGH the Incas' independence was gone for ever, the descendants of the various branches of the royal family survived into the seventeenth and eighteenth centuries. It is fascinating to record the varying fortunes of the successors of Huayna-Capac's children Atahualpa, Manco, Paullu and Inés Huayllas. Many of their heirs sought wealth or social prominence, but it was the descendants of Manco's legitimate children Sayri-Tupac and Tupac Amaru who were eventually the most successful.

Atahualpa's grandson was the first royal descendant to try to improve his finances by a trip to Spain. Alonso Atahualpa, son of the Auqui Francisco Tupac-Atauchi, thought he might benefit by appealing to the generosity of King Philip, fellow monarch of his grandfather Atahualpa. The King was the fountainhead of patronage throughout the

Don Melchor Carlos Inca

Spanish empire, and Alonso hoped that remorse over Pizarro's execution of Atahualpa would now make him generous. Alonso was given permission to sail to the mother country in 1585.* Once in Spain, he presented suitable petitions, fathered an illegitimate daughter, and rapidly spent his father's carefully accumulated fortune in a pathetic attempt to make an impression at the Spanish court. Nothing happened; and in 1586 Alonso Atahualpa asked permission to return to Quito. Unknown to him, the President of the Council of the Indies told the King that he saw no harm in letting Alonso return to Peru, 'but I do consider that very great harm could result from giving him any favours – if indeed there is any obligation to do so – for this would spur on other grandchildren and great-grandchildren of the former rulers of that land to come and beg for similar favours'. The King agreed with this very shrewd advice and Alonso received nothing. But he failed to leave Spain. Two years later he was still there, by now imprisoned for debt in the public prison of Madrid. Before the Council of the Indies could intervene, Alonso Atahualpa died in prison. He left a wretched list of debts, the scrapings of a sad improvident who had borrowed from anyone taken in by his boasts and promises. His creditors included various women, a cordwainer and a silversmith whose names he could not remember, the tailor who had made him some suits, and the innkeeper from whom he had been renting a room. The Council of the Indies gave a hundred reales for a pauper's burial of this grandson of Atahualpa, but it ordered that his debts be paid from his father's estate in Quito.

Alonso Atahualpa died a bachelor, but left an illegitimate son and daughter in Quito. His son Carlos Atahualpa Inca became alcalde mayor of the natives of Quito. His daughter Mencía married a Spaniard called Ullua and inherited her grandfather's estates.* Mencía in turn had a daughter called Doña Bartola Atahualpa Inca, but this lady apparently died childless, which caused the end of Atahualpa's direct line.*

One of the greatest fortunes possessed by the Inca nobility was the hereditary estate of Sayri-Tupac. This passed, as we have seen, to his heiress Beatriz, who was in the care of her mother María until the nasty episode with Cristóbal Maldonado caused her removal to a convent. Toledo disposed of these ladies less brutally but more effectively than he had their male cousins: he simply married them to men of his own choosing.

Sayri-Tupac's lovely young widow María Cusi Huarcay was never allowed to enjoy the income from her husband's estates – these were administered by various Spaniards who received handsome fees from them. Arias Maldonado, who sheltered her and probably became her lover, was exiled to Spain. Toledo therefore married María against her wishes to an obscure man called Juan Fernández Coronel, by whom she bore a son and daughter. Her bold protests about her poor position on the paños, her moving cry to her brother Tupac Amaru, and her vindictive testimonials against Carlos Inca showed her to be a woman of powerful character. The last we hear of her is in 1586, when she wrote to the then Viceroy asking permission to return to her beloved Vilcabamba. She tried to tempt the Viceroy by promising to reveal gold and silver mines there, but the Viceroy would not allow her to return on any pretext.*

Toledo had silenced the mother by marrying her to one of his cronies. He now had to dispose of the daughter, the much sought-after heiress Beatriz. He asked the abbess of the Convent of Santa Clara whether the girl wished to take vows. Both the abbess and Beatriz herself thought that she should marry. There was now no question of her marrying her cousin Felipe Quispe Titu, an obscure exile in Lima, although Papal dispensation for their union had recently reached Peru.

Toledo decided instead to offer Beatriz to a most improbable man: the hero of the moment, Martín García de Loyola. Toledo had promised an annual income of 1,000 pesos to the man who could capture the Inca in Vilcabamba. When he learned that it was a captain of his own guard, and that the capture had been so dashing, he raised the reward to 1,500 pesos annually. García de Loyola was a Spanish hidalgo, great-nephew of St Ignatius founder of the Jesuits, and an ambitious man. He hesitated to marry the native princess; but changed his mind when Toledo itemised her fortune. The two were betrothed in late 1572 within weeks of the execution of Tupac Amaru, Loyola's captive and Beatriz's uncle. Loyola even asked, with singular lack of taste, to be allowed to add Tupac Amaru's head to his personal coat of arms; Toledo wrote to the King that he had granted permission; but the Council of the Indies refused the grisly request. Toledo immediately assigned all Beatriz's estates to García de Loyola, who enjoyed their revenues during some eighteen years that he was engaged to the princess. Toledo also helped his favourite to rise in the colonial service. Loyola became Governor of Potosí in 1579 and prepared to marry Beatriz, who was by then aged twenty.*

Since Loyola was already enjoying his fiancée's fortune he was in no great hurry to marry her. But his attempt to do so in 1580 was thwarted by the reappearance of a sinister figure from her past. The banished rapist Cristóbal Maldonado decided to try to restore his fortunes by asserting his rights over the girl. He petitioned the King for leave to return to Peru to contest the validity of the marriage between Beatriz and Martín García de Loyola. He brazenly insisted that 'in 1565 he was himself betrothed to Doña Beatriz Coya, daughter of the Inca, *in facie ecclesiae*, with the consent of her mother and tutor and in their presence. And he consummated the marriage, having marital relations with her in peaceful possession.'

According to this Maldonado, Viceroy Toledo 'having cut off the head of her uncle the Inca and exiled or destroyed all her lineage, sent persons to persuade her to marry Don Martín García de Loyola, a captain of his guard. The Viceroy himself appealed to her in person and ordered her to do it. But she said she could not marry and did not wish to because she was already married to Cristóbal de Maldonado, and by divine law women could not have two husbands.' Toledo had declared her 'free of the marriage she had contracted, because she had been forced into it as a child of only eight years when it took place, and because it was wrong to say that she had had intercourse with her husband'. Toledo produced witnesses to confirm all this, but 'they testified only to her age and to nothing else. Next day a decree appeared declaring her free. She was immediately removed from the convent and betrothed to Martín García de Loyola in the presence of the Viceroy'.

Maldonado was allowed to return to Peru to press his shabby claim. The Provisor of the archdiocese of Lima ordered that Beatriz be returned to a convent until a decision was reached about her marriage. Maldonado was surprisingly successful. An ecclesiastic appointed to investigate the case took a violent dislike to García de Loyola and found against him. The case dragged on. It ruined Beatriz's proposed marriage and affected the career of the ambitious García de Loyola. He was appointed Governor of Rio de la Plata in 1581, but had to delay his departure because the ecclesiastical case against his marriage was unsettled. Five years later the matter was still hanging fire, and he was given a temporary appointment as corregidor of Huamanga and Huancavelica. In 1587 he was made general of the Spanish fleet sent against the corsair Thomas Cavendish.* It was not until the end of the decade that the couple were finally judged free to marry – by which time the Inca princess was over thirty and her husband was forty.

In April 1592 King Philip confirmed Martín García de Loyola in the highly important post of Governor and Captain-General of Chile. He left to take up his appointment in September, and was shortly joined by Beatriz. The leading heiress of the Inca royal house thus found herself a consort in governing the southern section of her ancestors' empire. The Governor's lady gave birth to a daughter, Ana María Lorenza García Sayri Tupac de Loyola, at Concepción in Chile in 1593.

One of the reasons for the appointment of García de Loyola in Chile was his reputation as an Indian fighter. A war had for many years been raging against the stubborn Araucanian tribe in southern Chile. The same tribe had successfully resisted the armies of Beatriz's great-grandfather, the Inca Huayna-Capac. The Araucanians were the first South American Indians to adopt the Spaniards' horses, and they became brilliant horsemen. Martín García de Loyola rode out against them wearing his habit of Calatrava and accompanied by fifty Spanish troops and two hundred native auxiliaries. They camped at Curalava, not far from the city of Imperial. In a dawn attack on 23 December 1598 three hundred mounted Araucanians swept through the camp and annihilated the entire Spanish force. They continued for many decades to use the skull of Governor García de Loyola as a ceremonial drinking vessel.*

The Governor's widow and baby daughter were brought back to Lima by a sympathetic Viceroy who contemplated sending them to Spain. But in March 1600 Beatriz fell ill and wrote her will. She made bequests to 'four girls and a page' and arranged that 'twenty-four poor Indian men and women shall be clothed, in men's cloak and tunic and women's shawl and shift which is their dress, and should carry lighted torches before my body on the day [of my funeral]'. She left the bulk of her great inheritance to her infant daughter Ana. Beatriz died aged forty-two on 21 March 1600 and was buried in the Monastery of Santo Domingo in Lima.* She died greatly respected and honoured – not as the heiress of Sayri-Tupac and Manco Inca, but rather as the widow of a rapidly rising colonial administrator and hero.

Paullu's grandson Melchor Carlos was only three when Toledo's persecution of his father was reversed by royal decree. He was therefore raised in as much luxury as his father Carlos had been. He received a thorough Spanish education in the Jesuit College of San Borja in Cuzco but never acquired his father's classical learning. Melchor Carlos was seized instead by his father's passion for Spanish pomp and ceremonial. He himself wrote that he was 'accustomed to behave with the decorum of a gentleman, mixing and conversing with others who were in the city – its [Spanish] citizens and leading people. I normally kept three or four horses and as many mules in my stables.' Melchor Carlos adored riding in pageants in Cuzco, and earned from Don Alvaro Ruiz de Navamuel y los Rios, the secretary of the Governor of Lima and one of the most elegant men then in Peru, the accolade of being 'one of the finest horsemen there are in this land'.

Anxious to do the right thing, Melchor Carlos founded a chapel of San Francisco and endowed both this and his father's creation of Nuestra Señora de Guadelupe. All this estimable activity earned him the title of perpetual Alderman and Royal Standard-bearer of the Incas of Cuzco, a purely honorific post. The royal Inca ayllus of Cuzco now had an elected honorary body called The Twenty-four. This met in assembly every year (presided over by Spanish officials concerned with Indian affairs) and elected one of themselves to be 'Alferez Real de los Incas' (Royal Standard-bearer of the Incas). The person so elected stood beside the Spanish standard-bearer at the feast of Santiago, and was permitted on that occasion to wear the royal fringe.*

Melchor Carlos inherited his father's estates in 1582, when he was eleven. The inheritance was not a foregone conclusion: Paullu's many encomiendas had now been enjoyed for two lives and were due to revert to the Crown. Melchor immediately petitioned for an extension for his own lifetime. The Viceroy Martín Enríque de Almansa was sympathetic and wrote to the King in favour of the petition on 14 March 1582. He also restored the Colcampata palace to Melchor.

Melchor had been given heady doses of the pleasures of a Spanish gentleman and his father died when he was an adolescent. Free from parental control, the young prince began to sow his wild oats among the native ladies of Cuzco. He had his first illegitimate child at twenty, by an Indian woman called Doña Catalina Quispe Sisa Cháves, and continued to have children by her over the subsequent years. They received resounding names: Juan Melchor Carlos Inca, Juana Yupanqui

Coya, Carlos Inca, and Melchora Clara Coya. But a native wife would not have been worthy of Melchor Carlos: he had to marry a Spanish lady as his father had done. His mother helped him find a suitable bride, and in 1595 he married Leonor Arias Carrasco, the heiress of Pedro Alonso Carrasco, a Knight of Santiago, and granddaughter of two distinguished conquistadores, Pedro Alonso Carrasco and Alonso Pérez.*

Everything was now set for Melchor to hold court over Cuzco as his father and grandfather had done. But things went wrong. The young Inca fell under the influence of the worst elements in Cuzco, a city full of unemployed adventurers and Spanish vagabonds. Melchor Carlos often appeared before the city's justices, but always pleaded his Inca lineage to win pardon. The Viceroy Luis de Velasco, Marquis of Salinas, became increasingly disturbed by the irresponsibility of the man who should have been the best agent for keeping the native population docile. Velasco wrote to the King in 1599 that Melchor Carlos was spending heavily from his inheritance 'with soldiers who could better be described as vagabonds and profligates. He mixes with them and they excite him, as he is boisterous by nature and has pretensions as a descendant of the rulers of this land.' The Viceroy wanted to send Melchor to Spain – anything to rid himself of a rich young man, always in the thick of disturbances in the largest native city, and for ever vaunting his Inca ancestry. But he asked that the King treat Melchor kindly if he went. Velasco was therefore understandably delighted when at the end of 1600 he received a letter from Melchor Carlos actually asking permission to go to Spain. Velasco wrote urging Melchor to come at once to Lima, but told the King: 'With this stratagem I am arranging to remove him from Cuzco if I can, so that there would be no need to do it by orders and violence.'

Matters came to a head in May 1601, when Melchor Carlos's father-in-law Pedro Alonso Carrasco was arrested in Cuzco on charges of conspiracy. A fellow conspirator, García de Solis Portocarrero, was arrested simultaneously in Huamanga and beheaded. 'It was said that Don Melchor Carlos Inca was in league with them' but nothing was proved against him. The scandal made the authorities more determined than ever that Melchor Carlos must be banished to Spain – the country that his father Carlos and grandfather Paullu had so ardently wanted to visit. And so 'he crossed to Spain on His Majesty's orders, having been given a very handsome grant to assist his passage'. He left Cuzco with many of his estates mortgaged, and ran up further debts during a month spent in Lima before sailing.* The mestizo chronicler Felipe Guaman Poma de Ayala made a pen sketch of him at this time. Melchor was shown as a perfect Spanish gentleman. He even sported a pointed beard and moustache, a symbol of the Spanish blood acquired from his mother – Amerindians grow little facial hair, and a moustache is a sign of European blood in South America to this day (see page 457).*

Melchor Carlos sailed on the treasure fleet of 1602 and reached Spain at the end of the year. He at once became friendly with the historian Garcilaso de la Vega, the outstanding product of a union between the Inca royal family and a Spanish conquistador. Garcilaso wrote that Melchor 'came to Spain last year, in 1602, to see the court and to receive the favours to be extended to him for the services of his grandfather [Paullu] in the conquest and pacification of Peru and afterwards in the civil wars. But however great these favours may be, they ought to be greater.'

The next annual fleet, that of 1603, carried the young orphan Ana María Coya de Loyola. The young girl arrived at Valladolid when Melchor was already there. But as daughter of the Governor of Chile she immediately moved into the most exalted court society. The King entrusted her to the care of a distant cousin of her father, Don Juan de Borja y Castro, Conde de Mayalde y de Ficalho, a son of the great Jesuit St Francisco de Borja y Aragón, Duke of Gandía. Don Juan had been Spanish ambassador in Portugal and Germany and served on various leading councils of state. He lived in great luxury in Madrid, with a superb library and art collection, and liked to surround himself with poets and musicians. When he died in 1606, his widow established the thirteen-year-old Ana María with a personal household that included her own servants, nurses, a steward and doctor.

In 1611 the King chose a husband for the rich orphan, who was now eighteen. His choice was her protector's nephew Juan Enríquez de Borja y Almansa. This gentleman was a widower, twenty-two years older than his bride, with a daughter who later became a nun. He had gained a considerable reputation as naval commander in charge of the Havana fleet fighting Dutch corsairs. Ana María brought a great inheritance, with Sayri-Tupac's many repartimientos – which had been granted in perpetuity, unlike Paullu's – and the special pension of

1,500 pesos a year awarded to her father for the capture of her great-uncle. The bridegroom was also very rich: he immediately gave his wife a bracelet of fifty-eight diamonds and a diamond necklace worth 24,000 reales. The couple lived in splendid surroundings in Madrid, with Moorish slaves, large stables and a magnificent library. María also possessed much gold, silver and jewellery and also 'seven cloths painted with figures'—possibly Toledo's paños.

The fleet that took Ana María to Spain also carried a petition dated 16 April 1603 in which many lesser descendants of the Incas asked to be exempted from paying tribute. These nobles empowered four mestizos of royal blood who were then in Spain to present the petition. The four included Melchor Carlos, Alonso de Mesa (son of a famous conquistador) and Garcilaso de la Vega, who commented that 'they write full of confidence that as soon as his Catholic Majesty knows of our grievances he will redress them and will confer many favours on the petitioners because they are the descendants of kings'. They also sent another set of paintings of the busts of the Inca emperors and 'a royal genealogical tree painted on a yard and a half of white China silk'. Nothing came of this attempt. Garcilaso himself was too busy to submit the petition even though it had been addressed to him. He forwarded it to Melchor Carlos, but was indignant when that mestizo 'refused to present the papers, so as not to admit that there were so many others of royal blood. He thought that if he did, he himself would be deprived of a large part of the favours he was claiming and hoped to receive.'

Melchor Carlos was indeed busy begging. Soon after reaching Spain he settled in his mother's birthplace Trujillo de Extremadura. He wrote a memorial called *Ascendencia de don Melchor Carlos Inca*, a massive work intended to establish his claim to the Inca title and to ask for royal favours because of his grandfather Paullu's services to the Crown. Philip III acquiesced to the extent of granting a perpetual annual pension of 7,500 ducats, assistance in bringing Melchor's Spanish wife from Cuzco to Spain, and a recommendation that he receive the habit of a Knight of Santiago. But the King shrewdly refused to renew Melchor's repartimientos in Peru, and forbade him to return to his native land.* Encouraged by this success, Melchor continued to pester the King with further petitions. In 1604 he produced a further 'solicitud', again asking for the habit of Santiago and for a Spanish title. In a later petition he begged to be allowed to return to Peru, and suggested that he be given the title of Admiral – a startling request from a man whose only sight of the sea had been on the journey to Spain.* The King simply continued his policy of transplanting the troublesome Inca from Peru to Spain. A provision of 1605 authorised the royal officials in Cuzco to grant 6,000 ducats for the transportation of Melchor's wife Leonor Arias Carrasco and his household. A further 8,000 ducats were given for the same purpose in 1606; but the poor woman, abandoned by her unfaithful husband, died in 1607 before leaving Peru.

The machinery was now in motion for Melchor's entry into the order of Santiago. Enquiries were held in Madrid and Trujillo with twenty-five Spanish witnesses, most of whom had lived in Peru. The attempt was successful, and Melchor became the first man with native Peruvian blood to become a Knight of Santiago. His father Carlos and grandfather Paullu would have shared his pride in the honour.*

Melchor Carlos became an increasingly pathetic figure in Spain. A weak, pretentious man, he longed to impress Spaniards with his wealth and social importance. He overspent his royal pension in extravagant living, with servants, horses and possessions beyond his means. Spanish society was indifferent, and Melchor diminished the prestige of his name with his bombast and debts. His only successes were the acquisition of the coveted knighthood of Santiago, a coat of arms, an appointment as a royal 'gentilhombre de boca', and some amorous conquests among ladies of Madrid. He already had illegitimate children in Cuzco, and now produced a daughter by one married lady and a son by a lady called María de Silva. Melchor suddenly had hopes of starting a dynasty. The infant was legitimised, and a legal marriage with María de Silva was hurriedly authorised in October 1610. It was too late. Before he could undergo the canonical ceremony Melchor Carlos died, while travelling through Alcalá de Henares, on 4 October 1610, aged thirty-nine. His will boasted that he had kept nine servants, one dwarf and one slave; but he made no mention of his first wife or of his Inca ancestry. The infant heir died in 1611, and the widow María continued the tiresome business of petitioning the King for favours or fighting legal actions against Melchor's illegitimate children over his royal pension. Garcilaso wrote that Melchor Carlos died of melancholy.*

Melchor's illegitimate son Don Juan Melchor Carlos Inca, born in Cuzco in 1592, was brought to Spain by a Spaniard. He soon imitated his father in applying for admission to the order of Santiago, and obtained the cherished honour in 1627. This was his only achievement.* Juan died soon after, without even producing a child to perpetuate this branch of Paullu's succession. It seems fitting that the descendants of men who tried so hard to ape Spanish ways should have dissipated their inherited fortunes in the pursuit of Spanish honours and luxuries.

Paullu left a vast progeny in addition to the two legitimate sons born to his wife Catalina Ussica, whom he married two days before his death. These illegitimate sons received nothing in Paullu's will, even though he had had them all legitimised by royal decree in 1544. Their claim that Paullu's inheritance should be shared among them was rejected, but the sympathetic Audiencia of Lima did at last grant them some royal privileges and immunity from paying tribute.* Few had any wealth beyond the occasional field of maize or potatoes, but most were married and raising young families, and the freedom from tribute and service allowed them to live contentedly. Most resided in the Carmenca district, on the slopes below Sacsahuaman and surrounding the palace of their envied half-brother Carlos.*

Some of Paullu's illegitimate sons were prominent figures in colonial Cuzco, acting as witnesses to Sarmiento de Gamboa's enquiries, or helping Bernabé Cobo gather information for his history.* One lady by whom Paullu had a son was Magdalena Antay, daughter of Alonso Titu Atauchi, who was in turn the son of one of Huayna-Capac's chief counsellors. This union of the houses of Paullu and of Titu Atauchi produced a strong line of Inca nobles. The families with royal blood and privileges intermarried throughout the colonial period, and were obsessed with the importance of their lineages and titles. Many served among the honorary Twenty-four native councillors of Cuzco, and also in the Battalion of Noble Incas that was created for them in the seventeenth and eighteenth centuries. They were firm supporters of Spanish rule. The line descending from Paullu and Titu Atauchi acquired the family name Sahuaraura, and inherited Paullu's aptitude for backing the winning side. Two Sahuaraura, a curate and a sergeant, helped to betray and destroy the eighteenth-century native rebellion led by Tupac Amaru's descendant José Gabriel Condorcanqui. But the son of the sergeant Sahuaraura, who lived in the early days of the Peruvian Republic, changed sides. He wrote a book full of passionate praise of the native precursors of emancipation – including the man betrayed by his father.*

The richest of all the Inca descendants was Francisca Pizarro. She lived in great luxury in the palace that she and Hernando had built in Trujillo de Extremadura after emerging from the prison of La Mota. In 1578 an aged, blind and infirm Hernando Pizarro set about organising his succession, closely assisted by his middle-aged wife Francisca. Between June and August the old man dictated a series of wills, provisions and codicils to dispose of his fortune. There was a staggering catalogue of precious objects, silver, jewels and cash, forty-four estates in Spain, houses, vineyards, and a series of encomiendas in Peru, including the great silver mines of Porco. The avaricious couple wrote to their agent in Peru – Francisca's half-brother Martín de Ampuero* – and urged him to press the collection of every debt and conclusion of every lawsuit. They then formed their estate into a mayorazgo, and entailed it on their second son Juan, having decided that their firstborn Francisco was a wastrel.* Everything about the settlement miscarried. A church and hospital that Hernando endowed for the benefit of his soul were never built. The chosen heir Juan died without succession and his brother Francisco dissipated the entail. And Hernando's widow Francisca remarried three years later to an impoverished widower called Pedro Arias Portocarrero.*

Juan Fernando Pizarro y Sarmiento, the grandson of Hernando and Francisca, decided in 1622 to petition the King for the marquisate originally awarded to his great-grandfather Francisco Pizarro eight-five years before. He produced a verbose plea to support his case. The King was still embarrassed by the huge claim launched by Hernando and Francisca Pizarro when they emerged from La Mota in 1560. He was therefore delighted to grant the marquisate on condition that the claim be dropped – an excellent bargain from the Crown's point of view, since the claim was well founded. Juan Fernando agreed and chose the eloquent title Marqués de la Conquista.*

When the first Marquis of the Conquest died in 1646 there were no survivors of the legitimate descendants of Hernando and Francisca Pizarro. The title passed to a grandson of the illegitimate daughter born to Hernando Pizarro in the prison of La Mota, the product of his liaison with Isabel de Mercado *before* his marriage to his niece Francisca. Six years later the title returned to the illegitimate daughter of Francisco, heir of Hernando and Francisca – at least this lady did have some trace of descent from the original Marquis Pizarro and the Inca princess

Inés. Her succession held the title until the death of the fifth Marquis in 1736. There was no obvious successor and a gigantic lawsuit resulted. Dozens of claimants appealed for the title, some of them descended from the Pizarro family before America had even been discovered. The marquisate eventually passed back to the Orellana Pizarros, the family descended from Hernando Pizarro's union with Isabel de Mercado in the soft bed of the castle of La Mota. The title of Marqués de la Conquista has continued to this day. And the palace built by Hernando and Francisca still stands on the square of Trujillo, a monument to the family that left that small town to make such a stupendous conquest.

Sayri-Tupac's granddaughter Ana María Lorenza de Loyola lived in the very highest circles of the Spanish court. Her husband Juan Enríquez de Borja y Almansa was rich and powerful. She herself, as heiress to Beatriz Clara Coya, owned title to the estates in the Yucay valley that had been granted in perpetuity to Sayri-Tupac.

During Toledo's resettlement programme Beatriz's natives in the Yucay valley had been reduced into villages corresponding to the modern towns of Oropesa, Guayllabamba, Urubamba, Maras and Yucay. But Toledo seized four of these villages in September 1572 in the purge of Inca property that followed the capture of Tupac Amaru. Martín García de Loyola naturally contested this confiscation after he married Beatriz. The case dragged on, but in 1610 a tribunal found entirely in favour of Loyola's heiress Ana María, and the Indian tributaries were returned to her. She then launched a claim against the Crown for forty years' revenue from this part of her estates. After some bargaining Ana María accepted a pension of 10,000 ducats, the creation of a semi-autonomous fief of her Yucay estates, and the title of Marquesa de Santiago de Oropesa, the chief town on her land. It was an extraordinary grant: the first noble title to be awarded in Peru and the only hereditary fief, something that was becoming a medieval anachronism in Spain itself.

King Philip III was exceptionally generous in his grant of the Oropesa marquisate and fief to the young orphan Ana María. Despite the protests of the Viceroy, he issued decrees in 1614 that formally transferred sovereignty over the four towns in the Yucay valley. He even surrendered full jurisdiction, with the right to impose a death sentence, mutilation, exile, enslavement or withdrawal of liberty. The Spanish authorities in Peru were aghast at the donation of such autonomy to the heiress of the Incas. But it went ahead, and sovereignty was transferred in a series of picturesque ceremonies. The Marquesa appointed her cousin Martín Fernández Coronel Inca to take possession in her name from the corregidor.* The keys to the public gaol were handed over amid pomp and music in the presence of a large crowd of Indians and Spaniards. Fernández Coronel released the prisoners and set up a symbolic new scaffold. The entire party then marched upstream with Fernández Coronel making symbolic acts of possession on either side of the road. The marquisate stretched for some twelve miles, the distance from Maras to Yucay, and covered over a hundred square miles.

Ana María and her husband Juan decided that they should visit Peru to see the country and consolidate their new marquisate. They sailed in 1615 in the company of a new viceroy, Juan's cousin Francisco de Borja y Aragon, Prince of Esquilache. The arrival of this exalted couple caused a sensation. Spanish creoles were flattered to receive the first Peruvian marquesa. But Ana María's return meant even more to the Indians: to them she was the sole heiress of Sayri-Tupac and the great Manco Inca, and they were passionately proud to welcome her and her grandee husband. At first the couple stayed in Lima, where the young Ana María bore three sons. They then moved with their young family to take up residence in the Yucay valley. They lived for seven years in a large house in Santiago de Oropesa that had once belonged to Melchor Carlos Inca.

The Jesuits were particularly pleased by the arrival of this fine couple, for it represented a union of two saints who had helped found the Order. Ana María's father was great-nephew of St Ignatius de Loyola, and her husband was grandson of St Francisco de Borja. Years later a painting was commissioned to commemorate this union of Jesuit saints and Inca kings. To the left stand Martín García de Loyola and his wife Beatriz Clara Coya, a dark-skinned beauty dressed as an Inca lady. To the right are their daughter Ana María Lorenza Ñusta and her husband Juan de Borja. Behind the two couples are their illustrious forebears: the Incas Sayri-Tupac and Tupac Amaru and Saints Ignatius de Loyola and Francisco de Borja. The painting hangs, dusty and badly-lit, in a chapel of the pink baroque church of the Compañía on the main square of Cuzco.

In 1627 Juan and Ana María left the beautiful Yucay valley, with its clear air, its canyon walls ribbed with Inca terraces, the swift Vilcanota river, rich fields of maize and quinua and views of shining snowy peaks. They returned to the luxury of the court at Madrid, where Juan became a member of the Supreme Council of War and Junta de Armadas. But on 7 December 1630 Ana María died, aged only thirty-six. She was buried in the church of San Juan opposite her house, and her heartbroken husband survived her by only four years.

The Marquisate of Oropesa was inherited by Ana María's eldest son, a grandee whose full name was: Don Juan-Francisco-Gaspar Ignacio Enríquez de Almansa Inga y Loyola, Marqués de Santiago de Oropesa, Señor de la Casa, Solar y Palacio de Loyola en Azpeitia, Comendador Mayor de Alcañiz en la Orden de Calatrava, y de Villanueva de la Fuente en la Orden de Santiago, and (after 1642) seventh Marqués de Alcañices, Conde y Señor de Almansa y de las Villas de Belver, Cabreros del Monte, Villavellid, Codesal y Ayo.* This second Marquis married twice. He was at first restrained from marrying Ana, daughter of the Duke of Albuquerque, but succeeded in winning her hand when he was nineteen. They had five children, but only one daughter survived childhood and married the Duke of Híjar. Juan was widowed and married again to Juana Teresa, daughter of the Duke of Frías, an elegant and beautiful lady who had already survived marriages to two distinguished men. Their daughter Teresa-Dominga became third Marquesa de Santiago de Oropesa. She married the Duke of Medina de Ríoseco and their son Pascual and daughter María succeeded as fourth and fifth holders of the title.

The fief in the Yucay valley continued as a repository of Inca tradition and Spanish pomp. A fine series of portraits in the Cuzco Archaeological Museum shows eighteenth-century Inca nobles and their ladies dressed in a sumptuous blend of Spanish dress and Inca ornament. One of these portraits is of Don Marcos Chiguan Topa, cacique of Huayllabamba, aide-de-camp and standard-bearer to the Marquis of Oropesa, royal standard-bearer of Cuzco and one of the council of Twenty-four Inca nobles who played an honorific role in the government of the city. Chiguan Topa is shown as a magnificent figure wearing a velvet suit with voluminous lace sleeves, holding the royal standard, wearing a royal Inca diadem and standing in front of an Inca coat of arms similar to that awarded by Charles V to Paullu.* But the fourth Marquis Pascual heard an alarming report that this Chiguan Topa had been 'exacting extortions, violence and tyrannies on the Indians of Huaylla-bamba, to the point of committing the cruel act of branding them on the buttocks like mules'. He therefore empowered the Rector of the Jesuits in Cuzco to depose Chiguan Topa from his caciquedom and other titles. It was one of the last acts of Ana María's descendants.

The fourth Marquis Pascual died without heirs in 1739 and his sister María died childless two years later. The fief in the Yucay valley reverted to the Crown. Sayri-Tupac's direct line came to an end. His descendants had become established in the highest ranks of Spanish society, and produced generals, grandees and ladies-in-waiting to serve the mother country of the Conquest.*

Sayri-Tupac had surrendered to the Spaniards, and his succession was rich and powerful as a result. The descendants of Manco's other sons were less privileged. Titu Cusi had hoped that his son Quispe Titu might succeed him to found a dynasty ruling an independent native Vilcabamba. But the young man's fate was to be led in chains to Cuzco and sentenced to exile by the Viceroy Toledo. Quispe Titu never returned to Cuzco, and survived for only six years in Lima where he died when still in his early twenties. He acknowledged in his will that he was about to have a child by one Doña Francisca Soco, and mentioned that he had a sister called Beatriz Chimbo Aca living in Cuzco. Dr Temple, who discovered his will, wondered whether Quispe Titu's child might have survived to found a succession living in obscurity until the present.* Titu Cusi had died before receiving the estates promised to him in return for leaving Vilcabamba. His son Quispe Titu therefore left only 300 pesos, in marked contrast to the wealth of the heirs of Incas who collaborated and received Spanish bounty.

Tupac Amaru's young son Martín also died a few years after being sent to Lima. He was only six or seven, and his death largely inspired the reports of wholesale destruction of the Inca exiles. But Tupac Amaru's daughter Juana Pinca Huaco, who remained in Cuzco, was to produce a succession that gave birth to the most distinguished descendant of the Incas. She married in 1590 Felipe Condorcanqui, curaca of Tinta to the south of Cuzco.* Their descendants enjoyed the privileges of curacas and received Spanish educations. But their great-great-grandson, José Gabriel Condorcanqui Tupac Amaru, a well-educated, elegant curaca, led a great revolt of the natives of the southern Andes that culminated in the siege of Cuzco in 1781. This and the other native rebellions of the eighteenth century are outside the scope of this book. But it is fitting that a direct descendant of Tupac Amaru should have won a place in Peru's pantheon as a gallant precursor of independence from Spain.

Francisco de Toledo failed in his stated intention of destroying the descendants of the Incas. His execution of Tupac Amaru made a martyr of that last ruling Inca: almost every native leader who rebelled against Spanish rule during the eighteenth century added the tragic name Tupac Amaru to his own. Toledo did not even succeed in humiliating the Inca nobility in Cuzco. Its senior members, the heirs of Paullu and Sayri-Tupac, had their fortunes restored and were rewarded by Spanish kings more sentimental than the hard-headed viceroy. The lesser nobility lived through the colonial era in relative comfort, most of them enjoying the benefits of curaca status, and all clinging proudly to the vestiges of their imperial past. In a sense Toledo was right to feel that this Inca nobility should not enjoy special privileges because of its links with the defeated empire. But in the flush of victory over Vilcabamba he overestimated the importance and potential danger of these representatives of the *ancien régime*: in the event most of them imitated the Spaniards and inspired their native followers to do so. Toledo also seriously misjudged public opinion, both among the natives and his own compatriots. It was presumptuous of him to imagine that he could discredit the Inca empire through his interrogations, or obliterate its charisma by dispossessing its leading representatives. The Inca name retained its magic, and the few royal descendants who survived into the seventeenth and eighteenth centuries revelled in their ancestry, even if they had long since lost any real power.

25

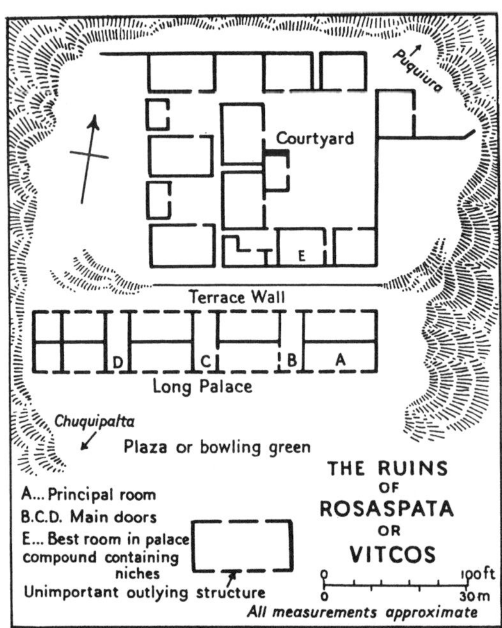

THE SEARCH FOR VILCABAMBA

Every visitor who sees the mist-shrouded crags around Machu Picchu must be curious about the history of this remote province. Everyone knows that the last Incas lived among these forbidding hills. It is often thought that the lost city of Vilcabamba will be found here somewhere – a place never visited by Spaniards, a treasure-house of Inca riches.

When Hurtado de Arbieto occupied the city of Vilcabamba on 24 June 1572, he tried to calm its bewildered population with a show of leniency. A few weeks later, when Martín García de Loyola marched out with his Inca prisoner, most of the men in the expedition wanted to return to the comforts of Cuzco. General Arbieto wrote to Toledo for further instructions, and said that he had not yet decided whether to take the natives 'back up the valley' to new settlements he planned in

After Hiram Bingham's plan of Vitcos

the Vitcos valley. The Viceroy replied on 30 July with a grant of the province of Vilcabamba for two lifetimes to Hurtado de Arbieto as its 'Governor, captain-general and chief justice'. He told the new Governor to do 'no oppression' to the inhabitants, 'for the Indians of that province are now exhausted by the hardships of the past war, and demoralised by being defeated. It would be unjust to add any further affliction. They should instead be favoured in everything. Also, since there might be a shortage of food because of the damage normally caused by wars, do not yet attempt to collect them into towns: allow them to remain in their houses as they were before they were conquered.'

The Spaniards were interested in Vilcabamba in three ways: as a new province and site for a Spanish city; as the scene of the martyrdom of the saintly Diego Ortiz; and as a possible source of wealth. Toledo was naturally determined to leave a Spanish settlement in Vilcabamba. Following his instructions, the new Governor Hurtado de Arbieto offered encomiendas to any deserving Spaniards who chose to remain, and divided 1,500 Indian tributaries among them. He led the expeditionary force back to the Vitcos valley at the beginning of September, together with the captive Incas and many of the former inhabitants of Vilcabamba. On 4 September 1572 he founded a Spanish city called San Francisco de la Vitoria in the valley of Hoyara, on the Vitcos (modern Vilcabamba) river roughly half way between the bridge of Chuquichaca and the valley of Vitcos–Puquiura. He himself remained there while his captains made their triumphal entry into Cuzco.*

Vilcabamba province was not a happy place in the years after its conquest. Toledo's benign instructions were ignored. 'The Indians themselves burned their towns and the Spaniards dismantled them. Thousands of Indians of all ages and sexes found their deaths. All was lamentation and sighs, and everything was death, famine and destruction.' The Governor Hurtado de Arbieto abused his wide mandate as Governor and Chief Justice. A later Viceroy heard that he was forcing Indians to provide personal service, had no fixed tribute assessment, and was condoning gross mistreatment of the natives. The Viceroy Count of Villar sent Antonio Pereyra – Arbieto's former captain – to investigate these disturbing allegations. Pereyra found that they were very true: Hurtado de Arbieto had placed guards on the bridge of Chuquichaca so that Indians could not escape from his fief. He had seized all the mines himself and was forcing the natives to work in unhealthy sugar plantations. He had imposed illegal rates of tribute, lowered the tribute-paying age to sixteen, taxed normally exempt old men between fifty and seventy at half tribute, insisted that women be hired out to work for twelve days a month at only half a real per day, and arranged for unpaid Indians to guard encomenderos' cattle. Villar tried to put a stop to these excesses.* Hurtado de Arbieto died while his oppression was being investigated, and the Viceroy did not allow his young son to succeed as Governor for a second lifetime.*

To Antonio de la Calancha, the cruelties of the Spanish conquerors represented divine retribution for the killing of the holy martyr Diego Ortiz.* Immediately after the Spaniards entered Vilcabamba, they started searching for the priest's body. It was soon discovered in its pit-like grave, under the roots of a large tree near Marcanay. The head was smashed by a club, and the body had five arrow wounds; but although it had been dead for some fourteen months, Ortiz's body was still dry and 'did not smell'. It was moved, in a procession with much burning of wax, to be buried under the altar of the new town of San Francisco de la Vitoria, 'many leagues away'. Six months later the holy relic was moved again, to the Augustinian monastery in Cuzco, where it received much veneration and was credited with a number of miracles. The Augustinians were naturally avid to substantiate every last detail of the death of the missionary. Fray Gerónimo Núñez, Prior of their convent in Cuzco, held an enquiry in September 1582 and interrogated many members of the Vilcabamba expedition as well as Licenciate García de Melo, the first envoy to Titu Cusi. Various natives, including Titu Cusi's wife Angelina Llacsa, were also questioned about the martyr's death.* The details furnished by these witnesses formed the basis of a cult: the devout were encouraged to reflect on every blow, cut or injury suffered by Ortiz. The Augustinians hoped that Diego Ortiz might be canonised because of his martyrdom. The application was finally rejected on the grounds that a saint must suffer martyrdom solely because of his Christian faith: Ortiz had been lynched because of his alleged complicity in the poisoning of Titu Cusi.*

Vilcabamba had more material wealth to offer than the details of its martyr's death. The Spaniards soon began to locate mines in the area – the very danger that prompted Titu Cusi to kill the prospector Romero. In 1586 Sayri-Tupac's widow María Cusi Huarcay, now in her fifties, wrote to the Viceroy offering to reveal a number of quicksilver, gold and silver mines in Vilcabamba. She spoke particularly of some rich gold mines called Usanbi on a jungle river near the town of Vilcabamba. She proposed to return to her native province with her cousin Jorge de Mesa, son of a famous conquistador and an Inca princess, and six or seven Inca relatives. The Viceroy may have been tempted by the offer; but he felt that the risk of allowing the fiery María back to Vilcabamba was too explosive, and therefore refused permission. She also mentioned the silver mines of Huamaní and Huamanate, and these were soon discovered by the Spaniards. They lay on the watershed between the Vitcos and Pampaconas rivers, close to the tarn of Oncoy where the Vilcabamba expedition had found herds of cattle.*

The discovery of silver mines caused the predictable flurry of excitement. The Viceroy reported that the new mines appeared to be even richer than Potosí. Their mining village soon outgrew any other community in Vilcabamba. Baltasar de Ocampo went on behalf of the citizens to obtain permission to move the Spanish town called Vilcabamba closer to the mines. The town of San Francisco de la Vitoria de Vilcabamba was formally moved to the bleak hills between Puquiura and Pampaconas, and acquired a large church run by the Mercedarians.*

For a time, towards the end of the sixteenth century, the silver mines flourished. The Spaniards allocated 480 mitayos from Andahuaylas, Chumbivilcas and Abancay – areas that conveniently fell between those subject to Potosí and to Huancavelica – to work Vilcabamba's silver mines. But the deposits did not live up to expectations. Alfonso Messia and others argued that the Vilcabamba mines were 'worked to very little or no result' and a later Viceroy transferred one quota of mitayos to work elsewhere. Antonio Vázquez de Espinosa and Martín de Murúa both reported that the mines were still active in the early seventeenth century, but their days were numbered.* As its mines became exhausted Vilcabamba lost its glamour. A description of Peru from the early seventeenth century described Vilcabamba as a place 'where there are some poor silver mines, from which they extract 500 bars of silver each year'. Almost the only Spaniards still interested in the place were itinerant Spanish traders who abused and swindled its Indians.

There was a similar moment of prosperity in Vilcabamba's other industries: sugar and coca. One Spaniard, Toribio de Bustamante, was making ten thousand pesos a year from sugar and was able to endow the churches of Cuzco with ornaments of Vilcabamba cedar wood.* African slaves were imported to work in the hot sugar factories in the valleys near the bridge of Chuquichaca. But the Negroes revolted, and the sugar plantations, like the mines, soon became too unprofitable to be worth working. The Mercedarians abandoned their great church at the mining town of Vilcabamba, leaving even 'the ornaments, chalices, bells and images', because their congregation had dwindled too drastically with the final removal of the mitayos. By the time Baltasar de Ocampo was writing, forty years after Hurtado de Arbieto's invasion, Vilcabamba had had its moment of Spanish exploitation and was in rapid decline. By the eighteenth century only two tiny villages remained at San Francisco de la Vitoria and San Juan de Lucma. Cosmé Bueno wrote in 1768 that in Vilcabamba 'there has remained only the memory of the retreat of the last Inca, and the town of San Francisco de la Vitoria, in former times a populous city with rich mines from which much profit was extracted. The remains of the Inca's palace [Vitcos] can still be seen, where the Indians cruelly killed the Venerable Father Diego Ortiz.'

This same Cosmé Bueno reported the first stirrings of a new attraction in Vilcabamba: the search for the lost refuge of its last Incas. 'A few years ago some people, attracted by a tradition that there was an ancient town called Choqquequirau, crossed the Apurímac on rafts and penetrated the montaña. They found a deserted place built of quarried stone, covered in woods and very hot. Sumptuous houses and palaces were recognised.' Choqquequirau is on the Apurímac side of Vilcabamba, on the saddle of a spur projecting some five thousand feet above the river's deep canyon. This overgrown eyrie seized the imagination of historical romantics for a century and a half. It was mentioned by the historian Pablo José Oricaín in 1790 and was searched for treasure by one Señor Tejada, a wealthy landowner who owned those steep hillsides in the early days of the Peruvian Republic.*

The lure of a lost city attracted the first serious visitor, the French Comte de Sartiges, to Choqquequirau in 1834. Sartiges chose an excessively difficult route to reach his goal: he left the Cuzco road on the Vilcabamba side of the Apurímac and climbed a high pass between the glaciers of Soray and Salcantay. From here Sartiges dropped to the hacienda of Huadquiña on the Urubamba a few miles downstream from Machu Picchu: he passed close to that great ruin without suspecting its existence. Fifteen Indians were sent ahead to start cutting a path from Huadquiña towards Choqquequirau. But beyond this Sartiges and his men had to cut their way on foot through high grass and bamboo. They suffered from acute thirst one day and all ended by running for half an hour towards the first glacial stream they met. The flies and mosquitoes were so terrible that they had to sleep in the midst of a ring of smoking brushwood, covered from head to foot in thick blankets. On the Apurímac side the explorers had a difficult descent through the steep montaña below the glaciers of Mount Yanama. Finally, on the fifth day from Huadquiña, the group came down on to the spur of Choqquequirau and spent a week clearing some of the undergrowth from the site. Sartiges was disappointed not to find many Inca objects, but he admired the almost Egyptian look of a screen of trapezoidal arches that masks the rocky end of the promontory. He recalled that one group of envoys to Sayri-Tupac had crossed the Apurímac and met the Inca in a town on the right bank. His conclusion was that the last Incas' residence was Choqquequirau rather than somewhere called Vilcabamba.*

Choqquequirau's next visitor was another Frenchman, Monsieur Angrand, who hacked his way to the remote ruin in 1847. Angrand followed the same difficult route as Sartiges, approaching Choqquequirau from behind, down the densely-wooded slopes below Mount Yanama. Now that the history of the Conquest was becoming popular throughout Europe, the attraction of the 'lost city' became more intense. Angrand was lured to Choqquequirau by a tradition that 'immense treasures were buried among the ruins when the last survivors of the race of the sun retired to this savage asylum'. Angrand measured the buildings in the ruin, and noticed a series of curious stone rings set into the inner wall of a long house on the central square. The rings are still there to this day, looking like moorings on some old stone jetty. They were apparently used to tether something, and Angrand concluded, not unreasonably, that the only animals to need such solid rings would be pumas. The French explorer repeated a local legend that Choqquequirau was the last refuge of Tupac Amaru.

This identification of Choqquequirau as the last refuge was firmly established throughout the second half of the nineteenth century. Its romantic reputation was enhanced by some unsuccessful attempts to reach it. Another Frenchman called Grandidier was driven back by difficult conditions in 1858; a Peruvian called Gastelú claimed that he had travelled all along this precipitous side of the Apurímac; another Peruvian called Sámanez Ocampo questioned Gastelú's honesty, and then claimed that he himself had spent five months at Choqquequirau, without volunteering any details of his visit.*

The great Peruvian geographers Antonio Raimondi and Mariano Paz Soldan both endorsed the view that this was the last refuge. Raimondi, an indefatigable traveller who probably visited more of Peru than any other man before or since, investigated the modern Vilcabamba (Vitcos) valley in 1865, but located none of the Inca ruins in the area. Raimondi thought that Choqquequirau was Vilcabamba because Calancha said that the last refuge was 'two long days' march' from Puquiura, and this distance would include Choqquequirau. Another great traveller who was keenly interested in Inca antiquities, the Frenchman Charles Wiener, also subscribed to the view that Manco Inca's last retreat was located in this ruin above the Apurímac.*

Interest in Choqquequirau reached its peak in the first decade of the twentieth century. J. J. Núñez, prefect of the Province of Apurímac, raised thousands of dollars and led a massive treasure-hunting expedition towards the ruin. He succeeded in crossing the Apurímac because an aged Chinaman had the courage to swim across the turbulent river and attach a line on the far bank. The prefect's party then built a bridge and spent three months of hard work cutting twelve miles of zigzag path up the dense undergrowth of the towering mountainside below the ruin. They cleared and searched Choqquequirau thoroughly, but left without making any dramatic discoveries. Shortly after this, in February 1909, a young American called Hiram Bingham was encouraged by Prefect Núñez to visit the site. He made the giddy descent to the Apurímac, crossed the new bridge and spent a couple of days sketching and photographing the legendary ruin.* This was Bingham's first taste of lost Inca cities, his first introduction to the mysteries of Vilcabamba.

I myself visited Choqquequirau a few years ago. It takes a day to ride down the crumbling side of the Apurímac canyon, down the left bank, which is open, dry and occasionally cultivated. Across the river the climate is totally different: the mountainside appears as an uninhabited tapestry of dark green vegetation. The bridge crossed by Bingham has long since been washed away, but there are now some strands of telegraph wire. I tied myself to a curved piece of wood that slid along these wires, with my feet dangling above the swirling grey waters, and hauled across with my arms for 250 feet along the swaying wires. There are two huts on the hillside of Choqquequirau, one a few thousand feet above the other. I slept at both during the two-day ascent, sharing the mud floor with the poor but extremely kind Indian owners. We then cut the mountain vegetation to reach the ruins.

The central plaza of Choqquequirau perches on a narrow saddle of land projecting over the river, but there are terraces, staircases and buildings hidden in the dark wooded slopes above and below. Choqquequirau is overgrown again, but the Inca buildings rise above the high grasses and bushes. The houses are large rectangular structures of fieldstone set in clay. They have all the familiar features of late Inca work: trapezoidal doors and niches, bosses and rings to hold down the thatch, a steep pitch to the roofs, and an upper half-storey under the eaves reached by lines of steps projecting from the end walls. The long building with the stone rings for Angrand's 'menagerie of pumas' is there, and so is a flight of stepped boulders leading to an upper plaza and groups of houses. The Egyptian-looking feature that intrigued Sartiges is a fifteen-foot-high wall with recessed niches that screens the rocks at the outer end of the plaza: Sartiges smashed through one blind niche in the hope of finding treasure behind. But Choqquequirau's greatest glory is its view. The mighty Apurímac is no more than a silver ribbon piercing its great cleft thousands of feet below. As one looks along the canyon, an endless procession of cliffs, waterfalls, tumbling forested hillsides and dazzling snow-capped peaks fades into the far distance.

In 1909, the year of Bingham's visit, Choqquequirau started to lose its identity as the refuge of the last Incas. The Peruvian historian Carlos A. Romero had recently studied the newly-discovered chronicles by Titu Cusi and Baltasar de Ocampo. He concluded that 'this tradition, so generally accepted, is totally without foundation'. He recommended, instead, that the Inca city Vitcos be sought near the village of Puquiura: Choqquequirau was only an outpost of the Vilcabamba state.*

Hiram Bingham returned to the United States fascinated by his fortuitous glimpse of the edge of Vilcabamba. His friend Edward S. Harkness, who was a collector of Peruvian documents, offered to finance a geologist to accompany him on another expedition. At a Yale class reunion Bingham spoke about his plan to return to Peru; his rich classmates were soon offering to finance other experts for what became the Yale Peruvian Expedition of 1911. In Lima Carlos Romero showed Bingham the paragraphs in Calancha relating to Vitcos and in particular to the nearby shrine at Chuquipalta. This was Yurac-rumi, the white stone above a spring of water, the oracle that the Augustinians García and Ortiz impudently destroyed after their return to Vitcos from Vilcabamba in 1570.

Hiram Bingham had all the necessary qualities for finding Inca ruins: he was full of enthusiasm and curiosity, was brave and tough, and was something of a mountaineer and a historian. He was also phenomenally lucky. The Urubamba valley was just experiencing its small share of the great Amazonian rubber boom – a time when Malaya had not yet supplanted the Amazon basin as the source of the world's rubber. This prosperity justified the cutting of a trail in 1895 under the sheer granite cliffs that had always closed the Urubamba valley below Ollantaytambo. The new road in turn caused a revival of the sugar plantations along the Urubamba and Vilcabamba valleys, and another road had just been cut past the most difficult stretch of the lower Vilcabamba. Bingham and his expedition left Cuzco with a good mule train in July 1911. They marched hopefully down the new Urubamba road into an area in which no one was aware of any Inca ruins.

Bingham was struck by the soaring beauty of the region he was penetrating. His caravan made its way beneath the gigantic precipices of the Urubamba canyon into a world of savage contrasts. The seething waters of the mountain rivers, the granite cliffs and sparkling snowy peaks and glaciers reminded him of the grandeur of the Rockies. But the tropical vegetation that clings to the steep hillsides or cascades over the rocky outcrops, and the mists that shroud the sugarloaf hills were like the most stupendous views of Hawaii. A few days after leaving Cuzco, Bingham's mule train camped between the road and the Urubamba river. This unusual behaviour aroused the curiosity of one Melchor Arteaga, owner of a nearby hut. When he was told the purpose of the expedition, Arteaga mentioned some ruins on the hillside across the river. Bingham's companions chose not to pursue this tenuous lead; but Hiram Bingham himself felt that he should investigate this first hint of an Inca ruin.

Bingham set out on the morning of 24 July 1911 with Arteaga and a

Peruvian sergeant. They crept across the plunging rapids of the Urubamba on a spindly bridge of logs fastened to boulders, and then clambered up a rough path through the jungle on the far side. They paused for lunch with two Indians who had made themselves a farm on some ancient artificial terraces two thousand feet above the river. Bingham left the comfort of their hut, unenthusiastic at the prospect of more climbing in the humid afternoon heat. But just around a promontory he came upon his first sight of a magnificent flight of stone terraces, a hundred of them, climbing for almost a thousand feet up the hillside. These terraces had been roughly cleared by the Indians. But it was in the deep jungle above that Bingham made his breathtaking discovery. There, amid the dark trees and undergrowth, he saw building after building, a holy cave, and a three-sided temple whose granite ashlars were cut with all the beauty and precision of the finest buildings at Cuzco or Ollantaytambo. Bingham left an unforgettable account of his excitement that afternoon, the dreamlike experience of seeing archaeological wonders, of finding each successive treasure of the lost city on that sharp forested ridge. On his first attempt he had discovered Machu Picchu, the most famous ruin in South America.*

Bingham left his companions to map the ruins of Machu Picchu and himself continued down the Urubamba to Chauillay, the confluence of the Vilcabamba river and site of the bridge of Chuquichaca. He made some fruitless chases after rumoured ruins, and then moved up the Vilcabamba river to the existing villages of Lucma and Puquiura. Carlos Romero had told Bingham that this was the probable location of Vitcos, since these place-names occurred in the narratives of Calancha, Ocampo and, particularly, in the newly discovered *Relación* of Titu Cusi Yupanqui. Another new road had been cut below the great cliffs that close the lower part of the Vilcabamba valley – the ravine from which Martin García de Loyola had almost been hurled at the battle of Coyaochaca. Just upstream from the sugar plantation of Paltaybamba Bingham saw an Inca ruin called Huayara among the sugar fields beside the river – evidently the valley that Ocampo called Hoyara, where the Spaniards first resettled the Indians brought up from Vilcabamba.* He moved on to Lucma, the village where Diego Rodríguez slept on his way to Vitcos in May 1565,* and offered its sub-prefect payment for any ruins he could reveal.

The Indians led him to the small village of Puquiura, in a snug valley three miles upstream from Lucma. Bingham did not know the passage in which Murúa said that Puquiura was 'where the Inca had his lodgings, where there had been a church where the Augustinian fathers officiated, and where Titu Cusi Yupanqui died'; but Romero and Bingham strongly suspected that Vitcos must be close to Puquiura. Just beyond the town he came upon a ruin, but it proved to be that of a Spanish ore-crushing mill from the late sixteenth century.* Above the ruined mill a wooded hillside called Rosaspata juts out into the Puquiura valley, and Bingham's guides led him up this hill. At the top, on the type of ridge where the Incas loved to build, Bingham found a ruin he immediately suspected to be Vitcos.

The location of this ruin of Rosaspata was similar to that of Choqquequirau: a knoll at the tip of a spur, with a flatter saddle behind, and a superb view of snow-capped mountains and deep beautiful valleys. There was a compound of fourteen rectangular late-Inca houses on the knoll, arranged in a square, with large and small courtyards. Below and behind this group a long building lay across the ridge, facing backwards on to a level open space on the saddle. This 'long palace' was a fine building, 245 feet long by 32 feet wide, and with fifteen doors along each of its long sides. To Bingham's delight, the doors were of excellent workmanship, with re-entrants and ashlars of white marble cut in the inimitable Inca manner. Each lintel was a solid block of marble six to eight feet long. Bingham remembered Ocampo's description of Vitcos: 'on a very high mountain whence the view commanded a great part of the province of Vilcabamba. Here there was an extensive level space with very sumptuous and majestic buildings erected with great skill and art; all the lintels of the doors being of marble and elaborately cut.' There seemed no doubt that this was Vitcos, the place where Rodrigo Orgóñez captured the boy Titu Cusi, where Diego Méndez stabbed Manco Inca, where Rodríguez saw the heads of the Spanish regicides, and where Titu Cusi prayed before descending to die in Puquiura.*

Hiram Bingham discovered Rosaspata on 8 August 1911, only two weeks after finding Machu Picchu. He was almost certain that it was the ruin of Vitcos, but wanted more positive proof. He remembered Calancha's remark that 'close by Vitcos in a village called Chuquipalta is a temple of the sun and inside it a white stone above a spring of water'. The discovery of such a shrine would prove that Rosaspata was indeed Vitcos. Bingham's native guides mentioned a spring near the ruins, and on the following day he went to investigate, climbing the far side of the hill. He saw a white boulder carved by the Incas, and he was shown a small spring. But it was only when Bingham plunged up a watercourse, past Inca terracing and into thick woods that he suddenly came upon his goal: a gigantic white granite boulder covered in Inca carving, overlooking an eerie pool and surrounded by the ruins of an Inca temple. This was a quiet, dark place, full of mystery, with the great white rock reflected in its pool of black water. Bingham and his companions 'were at once . . . convinced that this was indeed the sacred spot, the centre of idolatry in the latter part of the Inca rule' (plate 46).

The white rock of Chuquipalta was 52 feet long, 30 feet wide and 25 feet high, and lapped by dark water on two sides. It had been flattened on top and its fissures had been channelled to carry sacrificial liquids. There were rows of square bosses projecting from the south side; ten larger square projections on the north; and seats and platforms wherever the shape of the rock permitted. Once again, Bingham's curiosity, luck and determination had revealed an important and legendary site.* As he gazed at his discovery, Bingham could easily imagine the Indian pilgrims bringing their offerings to the temple, the blood of sacrificial llamas trickling down the elaborate channels, the appearances of the devil manifested in the dark water, and the pronouncements of the priests and Inca during the ceremonies at this oracle. He also recalled the reckless effrontery of the priests García and Ortiz, who marched their Christian congregation up from Puquiura with each boy carrying a piece of firewood, and exorcised the shrine of Chuquipalta in a great conflagration of its thatched temples. The finding of this sanctuary brilliantly confirmed that Rosaspata was the palace of Vitcos.

Hiram Bingham was not content with his remarkable discoveries of Machu Picchu, Vitcos and Chuquipalta. His enquiries on the Urubamba and at Lucma had produced a vague report of a ruined Inca city lying deep in the lowland jungle far beyond Puquiura. One native thought that this mysterious forest ruin was ancient Vilcabamba itself. Others warned Bingham that the ruin lay in a region of dangerous forest Indians, and that the area was dominated by a ferocious planter who was most inhospitable to strangers. Bingham had wasted much time pursuing non-existent ruins. But he determined to press on towards this archaeological eldorado.

Bingham also wanted to inspect a village called Vilcabamba that lay some fifteen miles beyond Puquiura. He had no difficulty riding up to this place. Vilcabamba proved to be a village of sixty sturdy Spanish houses with a massive old church and belfry. It lay near some old mine-workings, amid pastures and glaciated valleys, near the source of the modern Vilcabamba river at an altitude of 11,750 feet. The residents confirmed what Bingham suspected: this was the Spanish mining town of San Francisco de la Vitoria de Vilcabamba. It was the community that Baltasar de Ocampo had helped to found near the silver mines at the end of the sixteenth century. There were no Inca ruins, and this Vilcabamba was clearly not Tupac Amaru's sanctuary, the place the Spaniards called Vilcabamba the Old.

An aged Indian at this village of Vilcabamba confirmed the rumours that there was an Inca ruin in the forests to the northwest. He said it was near the sugar plantation of Concevidayoc in the wild Pampaconas valley. Bingham and his companion Professor Harry Foote were again warned of the dangers that lay ahead. But they decided to proceed, to venture into the Pampaconas valley, an area not shown on any map and not even penetrated by the ubiquitous Raimondi.

The Americans' first day's march took them out of the Vilcabamba valley and across a desolate watershed to that of the Pampaconas. They reached the village called Pampaconas: a collection of small huts on a grassy hillside at an elevation of ten thousand feet. They managed to recruit some reluctant porters at Pampaconas, and plunged into the unexplored tropical valley beyond. It took Bingham and Foote three days to reach Concevidayoc, days during which they descended over seven thousand feet into the close-forested canyon. At the tiny clearing of San Fernando they had to abandon their mules and proceed on foot. At another clearing called Vista Alegre there was a breathtaking view down the matted green slopes of the lower Pampaconas. The path was extremely difficult. The explorers often had to crawl beneath the undergrowth on those steep slopes. They finally approached the mysterious plantation of Concevidayoc on 15 August 1911.

The fears about Concevidayoc were unfounded: arrival there was almost an anticlimax. Its owner Saavedra – of whom they had heard such fearsome reports – proved to be a gentle man, owner of a tiny sugar plantation and a primitive house. A few docile Campa Indians appeared, wearing the grubby gowns that Ortiz might once have imagined to be mock monastic habits. Bingham was delighted to see that Saavedra was using fine Inca pottery in his hut. Saavedra and the Campa Indians confirmed that there were indeed some ruins at Espíritu Pampa (the Plain of the Spirits) on the river valley below.

Bingham's Indians from Pampaconas spent two days clearing a path to Espíritu Pampa under the direction of Saavedra's son. The Americans followed this path to a clearing that contained more Campa huts, and after half an hour's scramble beyond they reached a small plain beside a tributary of the Pampaconas. Here, at a terrace clearing called Eromboni Pampa, they saw the first vestige of an Inca building: the ruined walls of a long rectangular structure with twelve doors in each long wall. Other buildings with well-preserved walls lay just beyond the clearing, in a dense curtain of vines, creepers and thickets. One structure had a rounded end and deep niches in the walls. There were an Inca bridge and watercourses. Other rectangular buildings of field-stone set in clay looked unmistakably Inca, with characteristic niches, gables, lintels and bosses. On the following day Bingham's Indians and the local Campa cleared more of the dense tangled jungle below this group of buildings. To their surprise, they revealed a pair of important buildings of careful construction, with rows of small symmetrical niches in the walls. There were potsherds and the remains of Inca aryballi. But Bingham's attention was caught by a pile of crude curved red roofing-tiles, a dozen of them, irregular in shape but clearly imitating the curved tiles that roofed the more important Spanish buildings of sixteenth-century Peru.

Bingham's highland Indians were growing restless in the jungle and his little expedition was running short of food. He was forced to turn back after this cursory glimpse of the ruins of Espíritu Pampa, but was reasonably satisfied that no other Inca buildings lay in the wild curtain of forest beyond the clearing. In a report in the *American Anthropologist* of 1914, Bingham noted that these ruins of Espíritu Pampa were unique in lying so low in the Amazon jungle, at an elevation of only 3,300 feet. Bingham was in no doubt about their Inca origin, and concluded that they were probably built by Manco Inca's followers after the Spanish Conquest. He remembered that Titu Cusi had come to meet Diego Rodríguez at Pampaconas in 1565, and saw 'no reason why the ruins of Espíritu Pampa are not those of the residence of the Inca Titu Cusi Yupanqui in 1565'. And so, only one week after finding Vitcos and Chuquipalta and less than a month after finding Machu Picchu, the brave and determined Hiram Bingham had discovered another intriguing Inca ruin.

Bingham led expeditions back to the Vilcabamba region in 1912 and 1915, to clear the ruins he had discovered and to make further explorations and scientific studies in the area. Parties led by specialists on these Yale expeditions made some gruelling marches – down the wild San Miguel river, across the watershed to the Apurímac crossing, through the high pass between Mounts Soray and Salcantay, and up the densely forested Aobamba valley near Machu Picchu. They found many small Inca ruins in the hills near Machu Picchu, and traces of Inca roads and buildings at various places along the Cordillera. But nothing discovered by these two great expeditions could compare in brilliance or importance with the places found during Bingham's first extraordinary month in the area.†

The expedition of 1912 cleared the vegetation from Machu Picchu. The local authorities pressed Indians into service, and the American expedition had between a dozen and forty men working at most times during the year. The Indians were as reluctant to work for Bingham as they had ever been to work for Spaniards, even though his pay and conditions were relatively good. It took ten days to build a path up the snake-infested hillside to Machu Picchu, and months of labour to hack the dense vegetation from the ruins.

The city that emerged was a place of magical beauty. It contains many buildings in the finest interlocking Inca masonry. But it is Machu Picchu's remarkable unity and state of preservation that are so satisfying to a visitor. Here, standing intact to the roof line, are the houses, temples and buildings of a complete Inca city. The house groups are set amid banks of tidy agricultural terraces, and Machu Picchu is bound together by a web of paths and hundreds of stairways. Its location is fantastic, with the city clinging to the upper slope and crest of a narrow ridge. The sheer sugarloaf of Huayna Picchu rises like a rhinoceros horn at the end of the spur, and the Urubamba roars in a tight hairpin bend around the site, trapped in a green canyon hundreds of feet below. Steep forested hills rise all around Machu Picchu, and its mystery is heightened by ghostly wisps of low cloud that cling to these humid mountains.

Hiram Bingham now had to decide which ruin could have been the lost city of Vilcabamba. There was no doubt that the truncated hill above Puquiura was Vitcos. Excavations on the site revealed a number of rusted European objects among the Inca remains: horseshoe nails, a buckle, a pair of scissors, some bridle ornaments and three jews' harps. These could have belonged to Diego Méndez and the renegades who murdered Manco; they could have come from booty captured by Manco's men; or have been dropped by sixteenth-century treasure hunters. Calancha wrote that the city of Vilcabamba lay two long days' march from Vitcos. But this measurement could apply equally well to Choqquequirau, Machu Picchu or Espíritu Pampa, three ruins roughly equidistant from the hub of Vitcos.

Hiram Bingham came to identify his most glamorous discovery, Machu Picchu, as the lost city of Vilcabamba. In his earlier books – *Inca Land*, written in 1922, and *Machu Picchu, a Citadel of the Incas* of 1930 – he was becoming increasingly sure that Machu Picchu was the right place. By 1951 Bingham was convinced of the identification. He confidently asserted in *Lost City of the Incas* that no one disputed that Machu Picchu was the site of ancient Vilcabamba.

Bingham advanced a variety of reasons for his attribution. In the excavations at Machu Picchu the skeletal remains were studied by Dr George F. Eaton, Curator of Osteology in the Peabody Museum, Boston. Eaton claimed to have established the sex of 135 skeletons and found that three-quarters of these were female. He also found that many skulls in tombs near Machu Picchu had been trepanned: holes had been cut into the skulls for surgical or magical purposes, a common practice throughout pre-Columbian Peru. No trepanned skulls were found inside Machu Picchu itself. Bingham jumped to some extraordinary conclusions from all this. He assumed that the trepanned skulls must have belonged to warriors wounded in battle, and that the absence of such skulls inside Machu Picchu proved that 'robust males of the warrior type' were not admitted to the city. To Bingham's fertile imagination these skeletal remains thus showed that Machu Picchu was inhabited by holy women and the 'effeminate men' who worshipped with them. Having conjured up a vision of holy women, it was a short step to the conclusion that the female skeletons were from the mid-sixteenth century, that they belonged to the priestesses among whom the 'effeminate' Tupac Amaru was cloistered, and that Machu Picchu was therefore Vilcabamba, described by Calancha as 'the university of their idolatries'.

Although the preponderance of apparently female skeletons is interesting, it does not prove that these were of mamaconas, nor that they were from post-Conquest times. Trepanning is generally thought to have been for magical or medical purposes, and is not associated with battle wounds of warriors. The absence of trepanned skulls in Machu Picchu therefore does not help identify the ruin as a religious centre.

Bingham supported his attribution of Machu Picchu with other tenuous arguments. One object excavated at Machu Picchu was a hollow tube apparently used for inhalation. The substance inhaled was probably a narcotic such as the yellow seed of the indigenous huilca tree. If so, Bingham thought that this one tube could explain the derivation of the name Vilcabamba: plain (pampa) of Huilca.

After Pizarro murdered Manco's wife Cura Ocllo in the Yucay valley in 1539 he floated her body down the river. The Yucay–Urubamba runs below Machu Picchu and on to the bridge of Chuquichaca. Because Machu Picchu is the largest of the many ruins near the river, Bingham reasoned that it might be Vilcabamba, the destination of this tragic flotsam. But he forgot that the Spaniards did not know about Machu Picchu, which is, in any case, too high above the Urubamba for recovery of a body floating on it.

Bingham's favourite chronicle source was the seventeenth-century hagiographer Antonio de la Calancha. Calancha wrote that during the journey from Puquiura to Vilcabamba the priests García and Ortiz had been forced to flounder in the swamp called Ungacacha. Bingham was convinced that the identification of Machu Picchu as Vilcabamba would he enhanced if he could find a road between Puquiura and his discovery and if there was a lake called Ungacacha on it. He succeeded, with characteristic tenacity, in making the difficult journey across high roadless tundra and into the dense valleys above his ruin. He asked his native guides the names of every lonely tarn. One was called Yana Cocha or Black Lake, and to Bingham's eager ears this sounded, in a high wind, sufficiently like Ungacacha to encourage his belief that he was on the former road to Vilcabamba.

Bingham's final argument, and the one that seemed to clinch the identification of Machu Picchu as Vilcabamba, was its size. With about a hundred houses it was the most important ruin in the area; and Calancha had described Vilcabamba as the largest city in the province.

It seemed logical that when Manco Inca sought refuge from Spanish horsemen he would have chosen this easily-defensible city. Machu Picchu had clearly been an important town in the Inca empire, with enough buildings of superb masonry to make it a worthy royal capital. Manco's forces occupying Vitcos must have known about Machu Picchu and the other Inca settlements in its vicinity, even if these had been abandoned by the time of the Spanish Conquest.

Bingham's arguments were reinforced by his great reputation as an explorer. Machu Picchu was identified as Vilcabamba. For over fifty years the majority of learned opinion thought that Machu Picchu was the last refuge of Manco Inca and his sons.

Bingham's 1915 expedition explored the hillsides along the west bank of the Urubamba, and located and cleared a number of other Inca communities within a few miles of Machu Picchu.* The ruins cleared by these Yale expeditions were abandoned to vegetation during the following years, until in 1934 Dr Luis E. Valcárcel again cleared Machu Picchu itself as part of the celebration of the fourth centenary of the Conquest. The famous ruin has been kept well trodden by the feet of thousands of visitors ever since.

In 1940 and 1941 another important American expedition penetrated the area and worked on the string of Inca towns on the slopes between Ollantaytambo and Machu Picchu. This was the Wenner Gren Scientific Expedition to Hispanic America. It deployed as many as nine hundred workmen and cleared, mapped and photographed the ruins first seen by Bingham. It went on to discover the towns of Wiñay Wayna and Inti Pata, each of which had spectacular systems of terracing greater even than those of Machu Picchu. Paul Fejos, the leader of this fine expedition, made no rash surmises about the location of Vilcabamba. But he made some observations that have a bearing on the identity of Machu Picchu. He concluded that none of the towns in the area – including Machu Picchu – had been seriously built with a view to defence: none had walls or fortifications such as are found in other parts of the Inca empire, and Machu Picchu's location on a narrow ridge had no special defensive significance. None of the towns in this area showed signs of Spanish occupation or looting (which made them a precious source of Inca archaeology). The existence of so many settlements and terraces made this side of the Cordillera Vilcabamba one of the most densely populated regions of Inca Peru. Machu Picchu was the last of a string of towns that included Pisac, Yucay, Maras, Ollantaytambo and Inti Pata. All were distinguished by flights of superb agricultural terraces, perhaps designed to provide special crops and tropical luxuries for the court in Cuzco.

Some optimists continued to hope that the forests of Vilcabamba might conceal more ruins. Not everyone was satisfied that the 'lost city' of Vilcabamba had been discovered at Machu Picchu. In August 1963 an American expedition dropped by parachute deep into the unexplored Cordillera Vilcabamba some thirty miles north of Espíritu Pampa. It succeeded in crossing the wild hills between the Apurímac and Urubamba, but found no traces of 'the fabled mountain redoubt of the last Incas'. In July 1964 another American explorer, Gene Savoy, decided to return to Espíritu Pampa, the site that Hiram Bingham had investigated for a couple of days in August 1911. In three expeditions during 1964 and 1965 Savoy and his Andean Explorers' Club revealed that extensive ruins lay buried in the jungle beyond the clearing reached by Bingham. As Savoy's men hacked through the extremely dense undergrowth they found the remains of fifty or sixty buildings and almost three hundred houses, all thickly shrouded with mosses, lichen, creepers and damp vegetation. The rain forest has taken control of the site, with great trees towering a hundred feet and more above the crumbling ruins. Savoy's first expedition in July 1964 established that there was a city buried in those jungles. His second expedition, in September–October 1964, explored the streams that create the alluvial plain of Espíritu Pampa; it found traces of Inca roads crossing the peaks of the Marcacocha–Picchacocha range that separate the Concevidayoc (Pampaconas) from the Apurímac and Spanish-occupied Peru beyond. The third expedition spent six weeks from November 1964 to mid-January 1965 exploring the San Miguel river and mapping the ruins of Espíritu Pampa. Detailed surveying was impossible beneath the curtain of the rain forest, and it required extensive cutting and clearing simply to measure the walls of each building.

Savoy's discoveries convinced him that Espíritu Pampa was an important city and could have been Vilcabamba. He found that the main ruins lay some seven hundred yards north-east of Bingham's buildings of Eromboni Pampa. There was a temple with twenty-four doors and 230 feet long, and a 'sunken palace' nearly three hundred feet long. Most of the house groups were built on platforms, presumably as

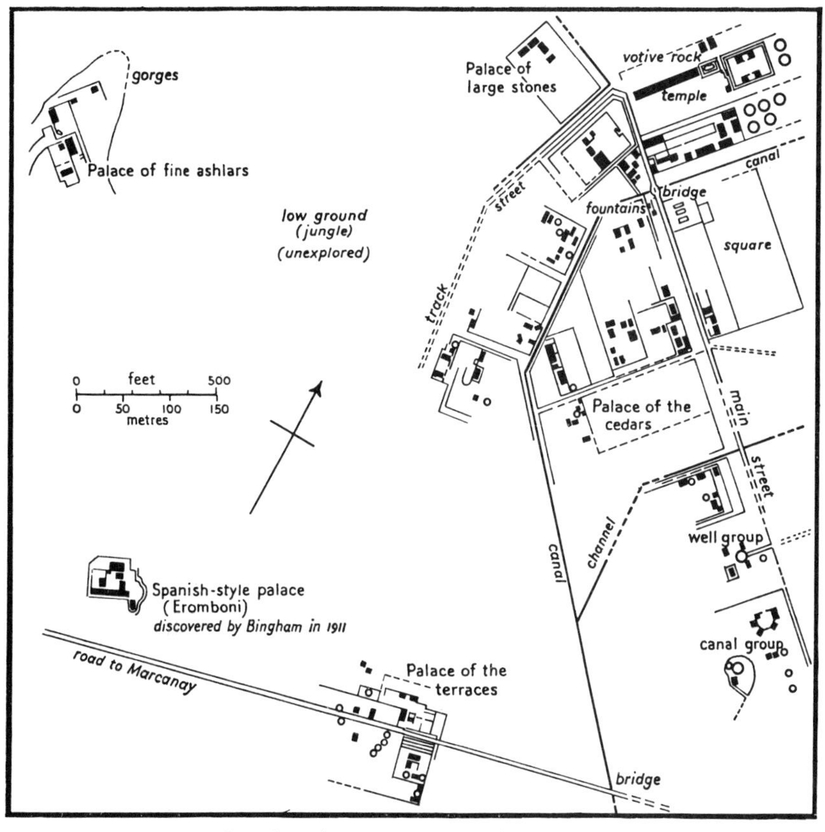

Gene Savoy's excavations at Espíritu Pampa

protection against flooding. The 'Palace of the Terraces' was a group 144 feet long, and the 'House of the Niches' was 99 feet long. Savoy was struck by the use of primitive curved roofing tiles: some of the buildings had been covered with these imitations of Spanish architecture instead of the traditional Inca thatch. The buildings were of fieldstone set in clay – Manco Inca and his sons did not have the manpower or quarries needed for the finest Inca masonry. But many buildings had been faced with a ceramic stucco of high quality, traces of which survived at the bases of the ruined walls.*

Gene Savoy was led to rediscover Espíritu Pampa by consulting the same sources used by Hiram Bingham fifty-three years earlier: Calancha's *Corónica moralizada*, Titu Cusi's *Relación*, and the reports of Diego

Rodríguez and Baltasar de Ocampo. These sources made Bingham debate whether Espíritu Pampa might be Vilcabamba, and made Savoy fairly sure that it was. But both these explorers failed to extract all possible clues from the sources they used. Bingham also suffered from the disadvantage of not knowing three important, detailed and authoritative sources that were discovered after 1911: the first part of Martín de Murúa's *Historia general del Perú*, rediscovered by the Duke of Wellington in 1945 and published by Dr Manuel Ballesteros-Gaibrois in 1962; the second dispatch from General Martín Hurtado de Arbieto to the Viceroy Toledo from Vilcabamba itself, written on 27 June 1572 and first published by Roberto Levillier in 1935; and the chronicle of Toledo's treasurer Antonio Bautista de Salazar.

General Hurtado de Arbieto sent two dispatches to the Viceroy. The first was from Pampaconas, and therefore described in detail the events of the first part of the 1572 campaign. Salazar evidently had a copy of that dispatch. The second dispatch, which described the march beyond Pampaconas and the occupation of Vilcabamba itself, eluded Salazar: he therefore gave no description of the Spanish entry into Vilcabamba. Antonio de la Calancha copied almost verbatim from Salazar and fell into the same error of omission. Bingham, Savoy and others all consulted Calancha and were thus under the impression that the Spanish expedition captured Tupac Amaru without actually occupying Vilcabamba. They had apparently not read Titu Cusi or Pedro Pizarro sufficiently thoroughly to appreciate that Gonzalo Pizarro's expedition had also occupied the same city in 1539.

There are many reasons why I am sure that Espíritu Pampa – and not Machu Picchu – is the correct location of Manco's city of Vilcabamba.

There is the question of altitude. Titu Cusi Yupanqui wrote from Vilcabamba that Manco Inca had made it 'his principal residence, for it has a warm climate' although he frequented Vitcos 'for the cool air, for it is in a cold district'. When Hurtado de Arbieto captured the city he described its tropical produce – coca, cotton and sugar cane – to the Viceroy Toledo. Martín de Murúa frequently commented on the fact that Vilcabamba lay 'in hot country'. He listed a wide variety of tropical plants, trees, birds and animals that left no possible doubt that Vilcabamba lay down in the Amazonian forests. Both authors were obviously surprised to have found the Inca city at such a low altitude. Bingham himself remarked that the climate of Espíritu Pampa at 3,300 feet was as different from that of Vitcos as the climate of Egypt is from Scotland. Vitcos was up at an elevation of some 9,000 feet – and so is Machu Picchu.*

Espíritu Pampa lies close to the navigable stretches of the Cosireni and Urubamba rivers and near the forests of the Pilcosuni. Even Bingham admitted that these were most probably the rivers and forests into which Martín García de Loyola pursued Tupac Amaru after the capture of Vilcabamba in 1572.* It is also probable that the ladies whose 'monastic habits' tormented the Augustinian friars at Vilcabamba were Campas: the Indians who still wear long gowns and live in the forests near Espíritu Pampa.

Another important argument in favour of Espíritu Pampa is the contemporary description of Vilcabamba's topography. Hurtado de Arbieto told Toledo that Vilcabamba lay in a valley that 'has pastures for cattle . . . and is a league in length by half a league wide [four miles by two]'. Murúa gave the same width and said that Vilcabamba lay in a valley rather like that of Cuzco.* No one could possibly describe Machu Picchu's knife-edge as being a broad valley; but the description fits the location of Espíritu Pampa.

More concrete evidence of the location of Vilcabamba is provided by the places encountered along the road towards it. There were six places mentioned as being on the sixteenth-century road between Vitcos and Vilcabamba.* Vitcos is definitely accepted to be the ruin of Rosaspata above the modern village of Puquiura. It is therefore necessary to try to locate the various places passed by the sixteenth-century travellers and expeditions, to see whether they lay to the north-west on the road to modern Espíritu Pampa, or to the south-east towards Machu Picchu. The distance from Vitcos to either ruin is roughly the same.*

There is a modern village called Layancalla or Huarancalla, between Lucma and Pampaconas and to the north of Vitcos. This was probably the place where Diego Rodríguez slept on his way to meet Titu Cusi at Pampaconas in 1565. It was also the place to which Titu Cusi wrote that 'I went out from this town of Vilcabamba to receive baptism' in 1568. And Ortiz passed through Huarancalla on his last journey from Vitcos to martyrdom at Marcanay.* If modern Layancalla was sixteenth-century Huarancalla it would show that Ortiz was travelling northwards, in the direction of Espíritu Pampa.

The most memorable landmark on the friars' journey to Vilcabamba was the swampy lake of Ungacacha. Bingham tried to show that it was the tarn called Yana Cocha on the road to Machu Picchu. It seems more likely that it was Oncoy Cocha, for 'cocha' means lake, and Unga is an alternative spelling of Oncoy: in spelling Indian words Spaniards freely interchanged U and O and G and C. It was at Oncoy, just before reaching Pampaconas, that the expedition of 1572 found herds of cattle and Alvárez Maldonado fell into a swampy bog in his excitement. Hiram Bingham's mule fell into a similarly deceptive swamp at this same place in 1911. Murúa wrote that the silver mines were near here, and Ocampo said that when he moved the Spanish town of Vilcabamba to be near the mines, he chose a site at 'Oncoy, where the Spaniards who first penetrated this land found flocks and herds'. If this Oncoy was Calancha's Unga, the friars were undoubtedly moving towards the Pampaconas valley in the direction of Espíritu Pampa and so was the expedition of 1572.*

Various travellers passed a place called Pampaconas on the road to Vilcabamba. The location of Pampaconas therefore provides a vital clue to the location of Vilcabamba. Titu Cusi dictated his famous narrative in the city of Vilcabamba itself. He said that in 1539 Gonzalo Pizarro's men defeated his father Manco 'three leagues from here' and captured Manco's wife Cura Ocllo. It was at Pampaconas, on the road back from Titu Cusi's Vilcabamba towards Cuzco, that the Spaniards tried to rape Cura Ocllo. The Inca himself thus revealed that Gonzalo Pizarro's men passed through Pampaconas on their way to and from his city of Vilcabamba. In 1565 Diego Rodríguez went as far as Pampaconas to meet Titu Cusi, who had presumably come up from Vilcabamba.

Pampaconas was the rendezvous of the two wings of the Spanish expedition of 1572: Hurtado de Arbieto came from the east, across the bridge of Chuquichaca and past Vitcos; Gaspar Arias de Sotelo marched across the watershed from the Apurímac along an Inca road that has recently been discovered. Murúa wrote that the expedition rested for thirteen days at Pampaconas, 'a very cold place, twelve leagues [fifty miles] from Vilcabamba the Old where the Incas had their capital and court'.

Hurtado de Arbieto wrote to Toledo that he left the horses at Pampaconas before proceeding on foot into the forested valley beyond. Pedro Pizarro explained that the 1539 expedition had had to abandon *its* horses in just the same way. And in 1911 Bingham left his mules below Pampaconas to proceed on foot towards Espíritu Pampa.

If the modern Pampaconas is the same as the sixteenth-century Pampaconas, it is certain that Vilcabamba cannot have been at Machu Picchu. For modern Pampaconas is north-west of Vitcos–Puquiura, in the opposite direction to Machu Picchu and on the direct path to Espíritu Pampa. Bingham said that he had noticed no Inca ruins as he rode through Pampaconas in 1911. He therefore argued, rather wishfully, that the modern village might be in a different location from the old. But a study of the sources shows that Pampaconas was an insignificant village in the sixteenth century, unlikely to have left many ruins. Proof that modern Pampaconas is in the same location as sixteenth-century Pampaconas comes from a careful reading of Ocampo and Murúa. Murúa said that the herds at Oncoy were captured three leagues before Pampaconas; Ocampo said that he founded his mining town of Vilcabamba at this Oncoy; and the mining town is still there, exactly three leagues (twelve miles) before modern Pampaconas.*

The next place that occurs in accounts of both the expeditions of 1539 and of 1572 is the defile of Chuquillusca. The Quipocamayos said that Gonzalo Pizarro's men passed 'in single file to cross a rocky hillside called Chuquillusca'. Murúa said that, after leaving Pampaconas, the 1572 expedition had to go on all fours to pass 'Chuquillusca, which is a cleft outcrop of rock running for a long distance along the course of the turbulent river'.

Beyond Chuquillusca both expeditions found the road blocked by native forts. The 1572 expedition fought at Huayna Pucará, which means New Fort. On the day after capturing this the men 'reached Machu Pucará [Old Fort], where Manco Inca defeated Gonzalo Pizarro'. This was clearly the fort in which Manco ambushed Gonzalo Pizarro's expedition with a cascade of boulders. Titu Cusi wrote in his Relación – *which he dictated to Marcos García in the city of Vilcabamba* – 'My father went out to confront [Gonzalo Pizarro's men] at a fort that he had three leagues *from here*. He fought a fierce battle with them on the banks of a river.' These references leave no doubt that both expeditions were advancing on the Inca's own capital city of Vilcabamba.

The last place before reaching Vilcabamba was Marcanay. It was at Marcanay that Diego Ortiz was martyred in 1571. He had almost reached

Vilcabamba, where the Inca Tupac Amaru was about to be crowned. Calancha wrote that 'the town of Marcanay was two leagues from Vilcabamba the Old'. Murúa said that the 1572 expedition reached Marcanay soon after passing the second fort, Machu Pucará. Both Murúa and Calancha recorded the fact that, after occupying Vilcabamba, the Spaniards exhumed the fragrant body of Ortiz nearby at Marcanay.* The discovery of the body there proves that the city of Vilcabamba conquered by the Spaniards in 1572 was the place where Tupac Amaru was crowned in 1571.

All these clues show inexorably that the Spaniards occupied the Incas' capital city in both 1539 and 1572. On both occasions they were marching north-westwards from Vitcos, past modern Vilcabamba and Pampaconas and evidently into the wild Pampaconas valley beyond. They were moving in the opposite direction to Machu Picchu, a place that was apparently not occupied by Manco and his sons, and was never explored by the Spaniards after their occupation of the area. Machu Picchu was an older city that flourished at the height of the Inca empire. Its buildings of the finest masonry indicate that it was a royal residence: evidently one of the Incas' pleasure houses in the Yucay valley, like Pisac or Ollantaytambo. Its great terraces doubtless served to cultivate coca or tropical luxuries. Manco must have considered that it was too close to Ollantaytambo and Spanish Peru for him to occupy. Its elaborate terracing may not have been suited to the needs of his band of exiles: they wanted easy pasturage and open fields for quick crops.

It remains only to show why the ruins at Espíritu Pampa are certainly those of Vilcabamba. They are in the right location, in the valley beyond Pampaconas; they are at the right altitude, low in the tropical forests; and they are in a broad valley that fits the ancient descriptions. Savoy's expeditions have traced the remains of some three hundred houses at Espíritu Pampa. This tallies exactly with the descriptions by Ocampo, Hurtado de Arbieto and Murúa. It also makes Espíritu Pampa easily the largest ruin in the area; and Calancha described the city of Vilcabamba as the largest in the province.

The search for the lost city of Vilcabamba has been in progress since the mid-eighteenth century. But the best contemporary descriptions of the city – those of Martín de Murúa and Martín Hurtado de Arbieto – seem to have eluded all the explorers. It is therefore all the more convincing that all who have published descriptions of the ruins of Espíritu

Pampa have noticed features that are exactly corroborated in Murúa's description of Vilcabamba.

Gene Savoy noticed that the walls of Espíritu Pampa showed traces of fine stucco instead of the usual ashlars. Murúa wrote that 'the entire palace [was] painted with a great variety of paintings in their style – something well worth seeing'. Howell and Morrison, who visited Espíritu Pampa in 1966, described a layer of ash beneath the carpet of forest leaves. They were convinced that the buildings of Espíritu Pampa had once been burned. They did not know Murúa's passage: 'Next morning, the feast of St John the Baptist, Tuesday, 24 June 1572 . . . at ten o'clock they marched into the city of Vilcabamba all on foot . . . The entire town was found to be sacked . . . [The Indians had] burned the rest of the maize and food that was in the storehouses, so that when the expedition arrived it was still smoking. The temple of the sun, where there was a principal idol, was burned. . . . The Indians had fled, setting fire to all that they could not take.'

All the modern visitors to Espíritu Pampa – Bingham, Savoy, Howell and Morrison – recorded one feature that has never been seen in any other Inca ruin anywhere in Peru. They all noticed piles of ancient curved roofing-tiles, crudely made in imitation of Spanish tiles. But they were unaware that Murúa had noticed this same extraordinary feature in Tupac Amaru's city of Vilcabamba. 'The Incas had a palace on different levels covered in roof-tiles.' Here is final conclusive proof. Unless someone can discover another ruin that so exactly fulfils the geographical and topographical details known about Vilcabamba, and that *also* contains imitation Spanish roofing-tiles, the lost city of Vilcabamba has finally been located at Espíritu Pampa.

Vilcabamba has therefore been rediscovered. Its ruins lie beneath great trees and almost impenetrable undergrowth, far down the turbulent Pampaconas or Concevidayoc river. The city has been stripped by its citizens, ransacked by conquering Spaniards, and suffocated by the Amazonian rain forests. Espíritu Pampa will probably never be excavated. The task of clearing the forests is too great, and the potential rewards for archaeologists or tourists are too meagre. But we now know the location of the last refuge of the Incas, the place that could conceivably have remained the capital of an independent enclave ruled by the descendants of Titu Cusi Yupanqui. And we can recall Murúa's description of the court of Vilcabamba: 'The Incas therefore enjoyed scarcely less of the luxuries, greatness and splendour of Cuzco in that distant or, rather, exiled land. For the Indians brought whatever they could get from outside for their contentment and pleasure. And they enjoyed life there.'

CHRONOLOGY

25 Sept 1513	Panama: Vasco Núñez de Balboa discovers the Pacific
1516–1556	Spain: Reign of King Charles I (Holy Roman Emperor as Charles V, 1519–1558)
1519	Mexico conquered by Hernán Cortés; Panama founded
1522	Pacific coast: Pascual de Andagoya explores to province of Birú
Nov 1524–1525	Pacific coast: Francisco Pizarro's first voyage to river San Juan
10 March 1526	Panama: Contract between Pizarro, Almagro and Luque
1526–late 1527	Pacific coast: Pizarro's second voyage
1527	Pizarro on the Isla del Gallo
26 July 1529	Toledo: Queen grants Pizarro Capitulación to conquer Peru
27 Dec 1530	Panama: Departure of Pizarro's third voyage
16 May 1532	Tumbez: Pizarro marches into Inca empire
16 Nov 1532	Cajamarca: Capture of Atahualpa Inca
5 Jan–25 April 1533	Hernando Pizarro's expedition to Pachacamac
15 Feb–13 June 1533	Reconnaissance by three Spaniards to Cuzco
17 June, 16 July 1533	Cajamarca: Distributions of silver and of gold
26 July 1533	Cajamarca: Execution of Atahualpa Inca
early Aug	Cajamarca: Coronation of Tupac Huallpa; Spaniards march south
Oct 1533	Jauja: Death of Tupac Huallpa
8 Nov 1533	Battle of Vilcaconga between Quisquis and Hernando de Soto
13 Nov 1533	Jaquijahuana: Chalcuchima executed
15 Nov 1533	Cuzco: Entry of Pizarro's expedition
Dec 1533	Cuzco: Coronation of Manco Inca
mid-Feb 1534	Jauja: Riquelme repulses Quisquis's attack
25 Feb 1534	Puerto Viejo: Pedro de Alvarado lands in Ecuador
early March 1534	San Miguel: Sebastián de Benalcázar leaves for Quito

23 March 1534	Cuzco: Foundation of Spanish municipality by Pizarro
c. 3 May 1534	Battle of Teocajas between Benalcázar and Rumiñavi
mid-June 1534	Quito occupied by Benalcázar
26 Aug 1534	Riobamba: Agreement between Diego de Almagro and Pedro de Alvarado
28 Aug 1534	Quito: Foundation of municipality of San Francisco de Quito
6 Jan 1535	Lima: Foundation of Los Reyes by Pizarro
12 June 1535	Cuzco: Agreement between Pizarro and Almagro
3 July 1535	Cuzco: Almagro and Paullu leave for Chile
Oct–Nov 1535	Cuzco: Manco Inca attempts to flee but is imprisoned
6 May 1536	Cuzco: Manco's forces attack and set fire to the city
late May 1536	Cuzco: Juan Pizarro killed during recapture of Sacsahuaman
May–July 1536	Central Andes: Quizo Yupanqui defeats relief expeditions
Aug 1536	Lima besieged; death of Quizo Yupanqui
Nov 1536	Lima: Departure of Alonso de Alvarado's relief expedition
18 April 1537	Cuzco: Almagro seizes Cuzco from Hernando Pizarro
12 July 1537	Battle of Abancay: Rodrigo Orgóñez defeats Alonso de Alvarado
mid-July 1537	Vilcabamba: Rodrigo Orgóñez pursues Manco to Vitcos
July 1537	Cuzco: Almagro crowns Paullu as puppet Inca
26 April 1538	Battle of Las Salinas: Hernando Pizarro defeats Almagro
8 July 1538	Cuzco: Hernando Pizarro executes Almagro
Nov 1538	Battle of Oncoy: Manco defeats Villadiego
Dec 1538	Cochabamba: Gonzalo Pizarro besieged by Tiso and the Chichas
9 Jan 1539	Huamanga: Foundation of municipality by Pizarro
Feb 1539	Charcas: Surrender of Tiso Yupanqui, collapse of rebellion
April–July 1539	Vilcabamba invaded by Gonzalo Pizarro; battle of Chuquillusca
Oct 1539	Condesuyo: Surrender of Villac Umu
Nov 1539	Yucay: Execution of Villac Umu, Cura Ocllo, Tiso and other commanders
1539–1542	Amazon: Francisco Orellana makes first descent of the Amazon
Jan 1540	Cuzco: Pedro de Valdivia leaves for conquest of Chile

1540	Madrid: Hernando Pizarro imprisoned at Medina del Campo
26 July 1541	Lima: Francisco Pizarro murdered by Almagrist supporters
Aug 1541	Puná: Vicente de Valverde killed by natives
16 Sept 1542	Battle of Chupas: Cristóbal Vaca de Castro defeats Diego de Almagro the Younger
20 Nov 1542	Barcelona: King Charles issues New Laws
1543	Lima: Jerónimo de Loayza arrives as Bishop
May 1544	Lima: Arrival of Blasco Núñez Vela as 1st Viceroy of Peru
mid-1544	Vitcos: Manco Inca murdered by Diego Méndez and renegades
Oct 1544	Lima: Entry of Gonzalo Pizarro, expulsion of Núñez Vela
April 1545	Potosí: Discovery of silver
18 Jan 1546	Battle of Añaquito: Gonzalo Pizarro defeats and kills Viceroy Núñez Vela
21 Oct 1547	Battle of Huarina: Gonzalo Pizarro defeats Alvarado
9 April 1548	Battle of Jaquijahuana: Pedro de la Gasca defeats Gonzalo Pizarro
July–Aug 1548	Vilcabamba: Negotiations between Gasca and Sayri-Tupac's regents
18 Aug 1548	Huainarima: Redistribution of encomiendas by Gasca and Loayza
May 1549	Cuzco: Death of Paullu Inca
Aug 1550	Valladolid: Debate between Las Casas and Sepúlveda
1551	Lima: First Ecclesiastical Council of Lima
Sept 1551–July 1552	Lima: Antonio de Mendoza, 2nd Viceroy
mid-1552	Medina del Campo: Hernando Pizarro marries Francisca Pizarro
13 Nov 1553	Cuzco: Outbreak of rebellion of Francisco Hernández Girón
8 Oct 1554	Battle of Pucará: Oidores defeat Hernández Girón
1554–1555	Peruvian encomenderos petition King Charles for perpetuity
Jan 1556	Toledo: King Charles I abdicates; Philip II crowned
June 1556	Lima: Andrés Hurtado de Mendoza, Marqués de Cañete, arrives as 3rd Viceroy

7 Oct 1557	Vilcabamba: Sayri-Tupac leaves Vilcabamba after months of negotiations
5 Jan 1558	Lima: Meeting between Inca Sayri-Tupac and Viceroy Cañete
1559	La Plata: Foundation of Audiencia of the Charcas
1559	Huancavelica: Discovery of mercury
1559	Cuzco: Juan Polo de Ondegardo discovers mummies of the Incas
1559–1560	Curacas petition King Philip to end encomienda system
1560	Yucay: Death of Sayri-Tupac; Titu Cusi crowned in Vilcabamba
14 Sept 1560	Lima: Death of Viceroy Marqués de Cañete
April 1561	Lima: Arrival of Conde de Nieva as 4th Viceroy
May 1561	Trujillo: Hernando and Francisca Pizarro leave prison and return to Trujillo
2 May 1562	Lima: Report by Commissioners on Perpetuity of encomiendas
1563	Quito: Foundation of Audiencia of Quito
Feb 1564	Lima: Death of Viceroy Conde de Nieva
Sept 1564	Lima: Lope García de Castro arrives as President of Audiencia
Dec 1564	Jauja: Discovery of pikes and plotted rebellion
March–May 1565	Vilcabamba: Meetings between Titu Cusi and García de Melo, Rodríguez de Figueroa and Matienzo
1565–1566	Lima: García de Castro introduces corregidores de indios
24 Aug 1566	Vilcabamba: Treaty of Acobamba between Titu Cusi and Spaniards
late 1566	Cuzco: Cristóbal Maldonado rapes Beatriz Clara Coya
Jan 1567	Cuzco and Lima: García de Castro crushes mestizo plot
March 1567–Jan 1568	Lima: Second Ecclesiastical Council of Lima
9 July 1567	Vilcabamba: Titu Cusi performs act of submission to Spain
20 July 1567	Vilcabamba: Quispe Titu baptised at Carco
Aug 1568	Vilcabamba: Titu Cusi baptised at Huarancalla
Sept 1568	Madrid: Junta Magna discusses government of Peru
30 Nov 1569	Lima: Francisco de Toledo arrives as 5th Viceroy
Jan–Feb 1570	Vilcabamba: Friars García and Ortiz visit Vilcabamba city
March 1570	Vilcabamba: Friars burn Chuquipalta; García expelled

Oct 1570	Lima: Special commission endorses forced labour in mines
23 Oct 1570	Lima: Viceroy Toledo leaves on visita general
Nov 1570–March 1571	Jauja to Cuzco: Toledo's *Informaciones* on Inca history
c. May 1571	Vitcos: Death of Titu Cusi, accession of Tupac Amaru
1571–1573	Viceroy Toledo undertakes reduction of Indians into towns
5 Jan 1572	Cuzco: Baptism of Carlos Inca's son Melchor Carlos
March 1572	Chuquichaca: Inca captains kill Atilano de Anaya
14 April 1572	Cuzco: Toledo proclaims war on Vilcabamba
1 June 1572	Battle of Coyao-chaca between Hurtado de Arbieto and Tupac Amaru's captains
24 June 1572	Vilcabamba city occupied by Hurtado de Arbieto
24 Sept 1572	Cuzco: Execution of Tupac Amaru
Jan 1573	Cuzco: Exile of Carlos Inca and other nobles
Feb 1574	Potosí: Toledo organises mine mita
27 Feb 1575	Madrid: King Philip authorises juez de naturales
1578	Lima: Death of Titu Cusi's heir Quispe Titu
Aug 1578	Trujillo: Death of Hernando Pizarro
23 Sept 1581	Lima: Martín Enríquez de Almansa succeeds Toledo as 6th Viceroy
Feb 1582	Cuzco: Death of Carlos Inca
25 Nov 1585–6 Jan 1590	Lima: Fernando Torres y Portugal, Conde de Villar, 7th Viceroy
1588	Madrid: Death of Alonso Atahualpa
1593	Concepción: Ana María Lorenza born to Beatriz and Martín García de Loyola
1598	Madrid: Death of Philip II; Philip III succeeds
23 Dec 1598	Battle of Curalava: Araucanians kill García de Loyola
21 March 1600	Lima: Death of Beatriz Clara Coya
4 Oct 1610	Alcalá de Henares: Death of Melchor Carlos Inca
1614	Yucay valley formed into fief for Marquesa de Oropesa
7 Dec 1630	Madrid: Death of Ana María Lorenza de Loyola, first Marquesa de Oropesa

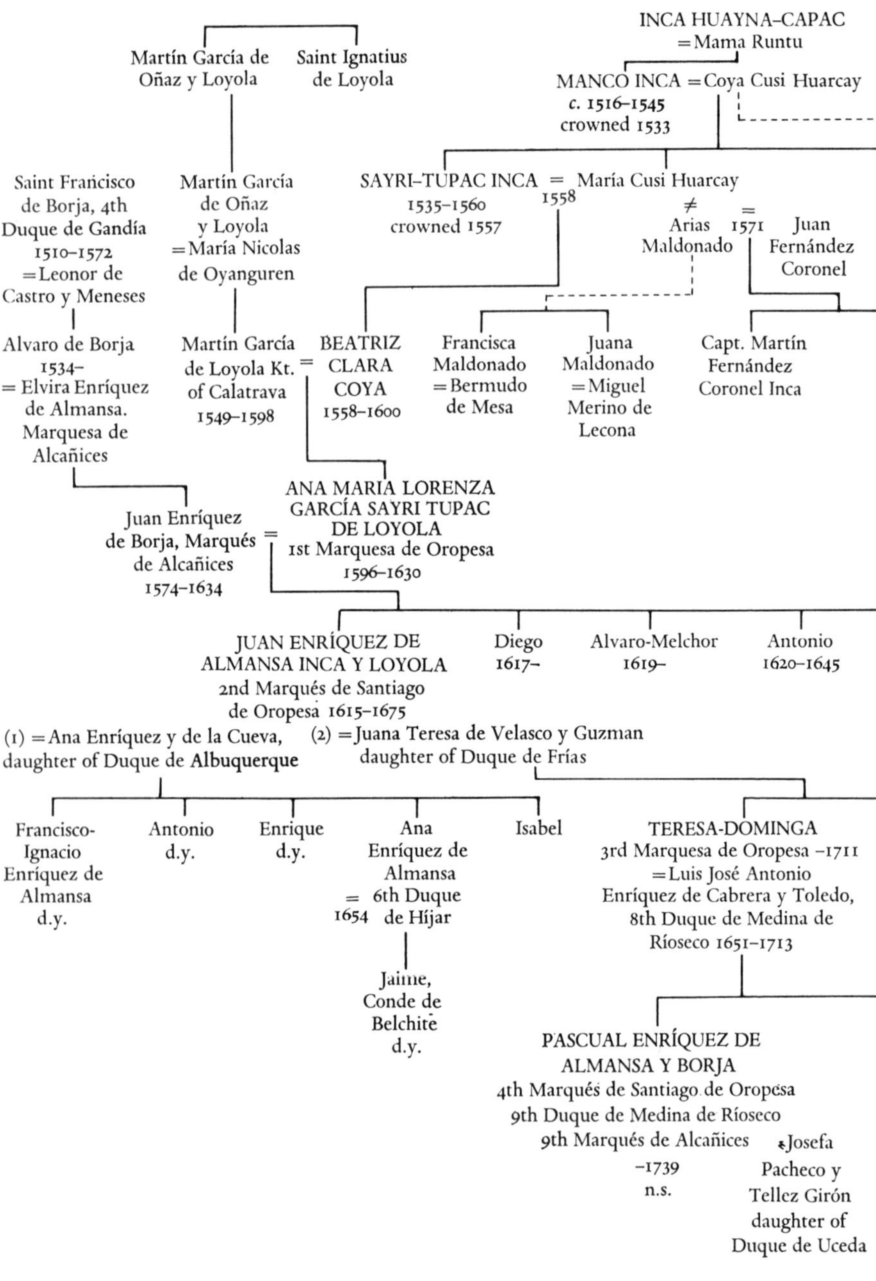
INCA HUAYNA–CAPAC
=Mama Runtu
Martín García de Oñaz y Loyola
Saint Ignatius de Loyola
MANCO INCA =Coya Cusi Huarcay
c. 1516–1545
crowned 1533
Saint Francisco de Borja, 4th Duque de Gandía
1510–1572
=Leonor de Castro y Meneses
Martín García de Oñaz y Loyola
=María Nicolas de Oyanguren
SAYRI–TUPAC INCA = María Cusi Huarcay
1558
1535–1560
crowned 1557
≠
Arias Maldonado
=
1571
Juan Fernández Coronel
Alvaro de Borja
1534–
= Elvira Enríquez de Almansa.
Marquesa de Alcañices
Martín García de Loyola Kt. of Calatrava
1549–1598
=
BEATRIZ CLARA COYA
1558–1600
Francisca Maldonado
=Bermudo de Mesa
Juana Maldonado
=Miguel Merino de Lecona
Capt. Martín Fernández Coronel Inca
Juan Enríquez de Borja, Marqués de Alcañices
1574–1634
=
ANA MARIA LORENZA GARCÍA SAYRI TUPAC DE LOYOLA
1st Marquesa de Oropesa
1596–1630
JUAN ENRÍQUEZ DE ALMANSA INCA Y LOYOLA
2nd Marqués de Santiago de Oropesa 1615–1675
Diego
1617–
Alvaro-Melchor
1619–
Antonio
1620–1645
(1) =Ana Enríquez y de la Cueva, daughter of Duque de Albuquerque
(2) =Juana Teresa de Velasco y Guzman daughter of Duque de Frías
Francisco-Ignacio Enríquez de Almansa
d.y.
Antonio
d.y.
Enrique
d.y.
Ana Enríquez de Almansa
= 6th Duque
1654 de Híjar
Isabel
TERESA-DOMINGA
3rd Marquesa de Oropesa –1711
=Luis José Antonio Enríquez de Cabrera y Toledo,
8th Duque de Medina de Ríoseco 1651–1713
Jaime, Conde de Belchite
d.y.
PASCUAL ENRÍQUEZ DE ALMANSA Y BORJA
4th Marqués de Santiago de Oropesa
9th Duque de Medina de Ríoseco
9th Marqués de Alcañices
–1739
n.s.
=Josefa Pacheco y Tellez Girón daughter of Duque de Uceda

DESCENDANTS OF MANCO INCA

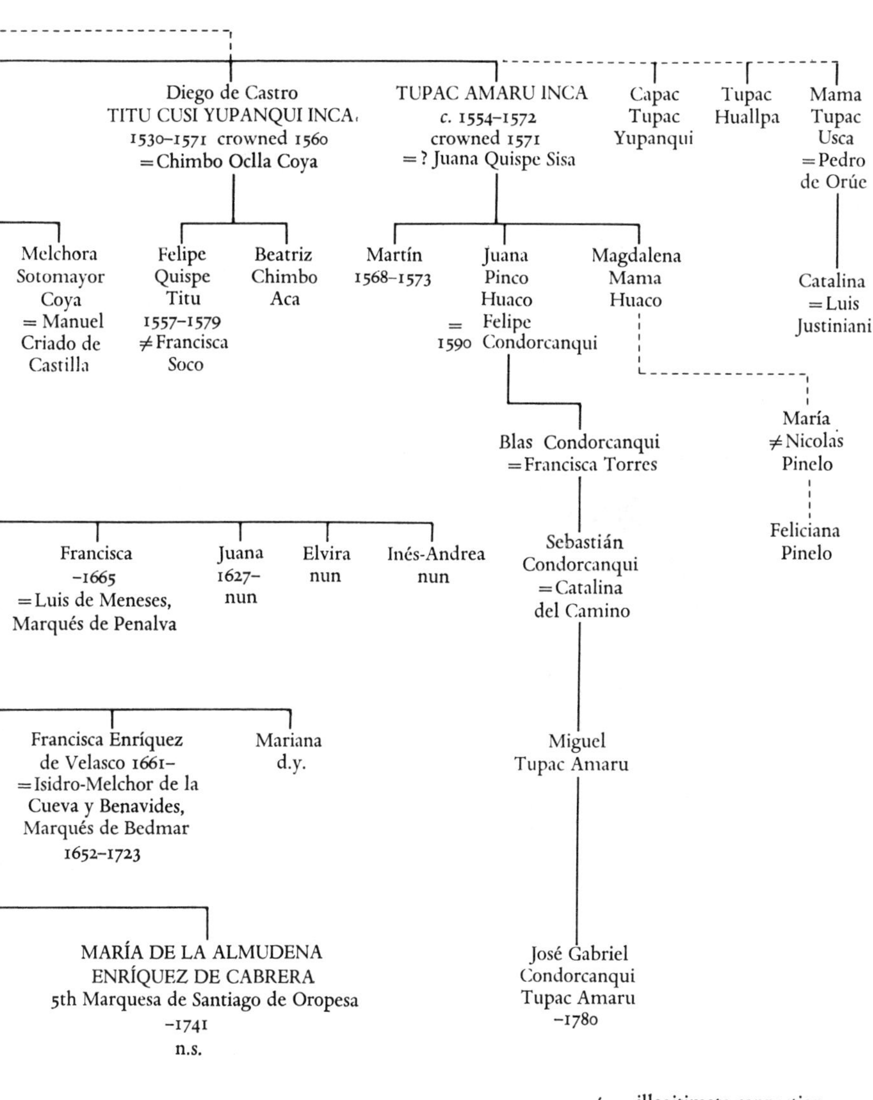

≠ = illegitimate connection
d.y. = died young
n.s. = no succession

DAUGHTERS OF HUAYNA CAPAC

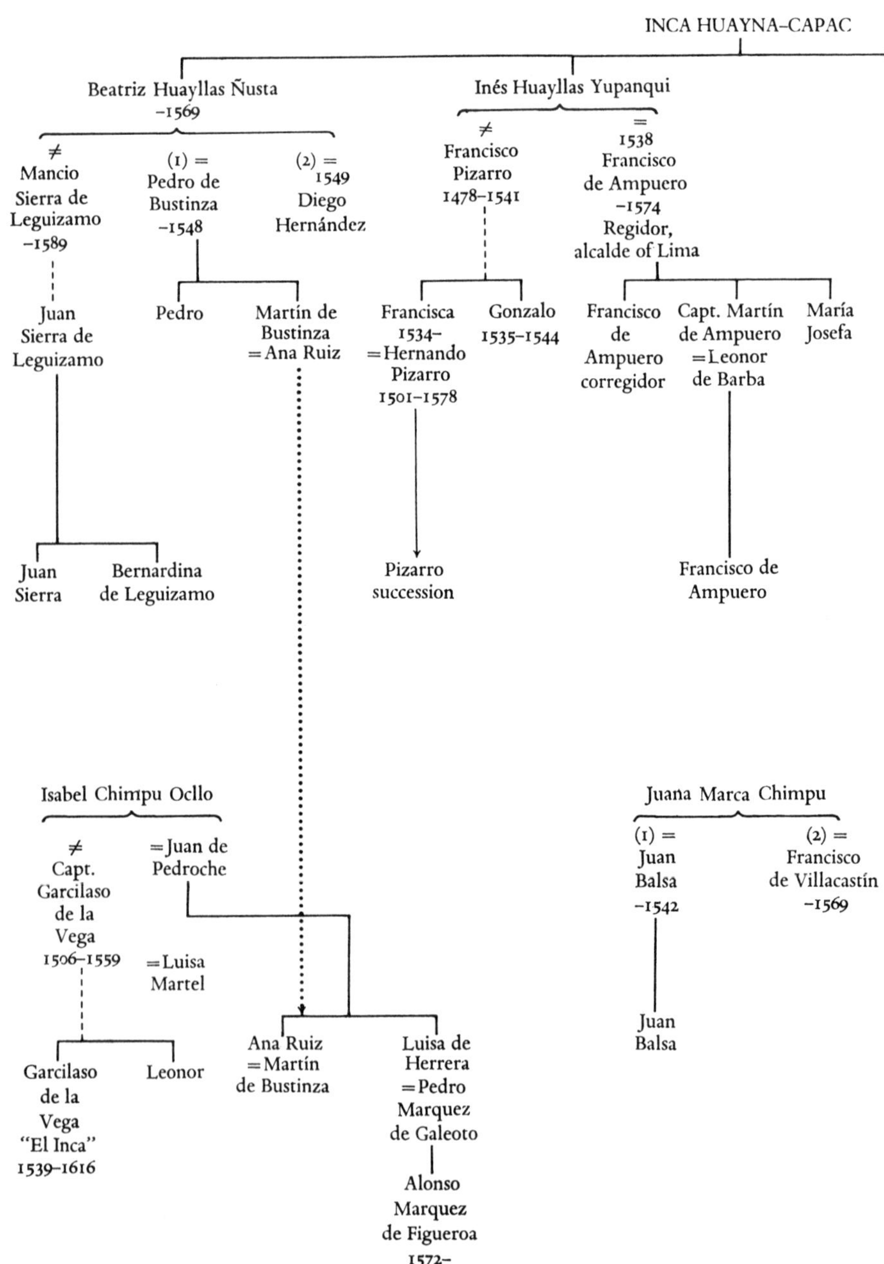

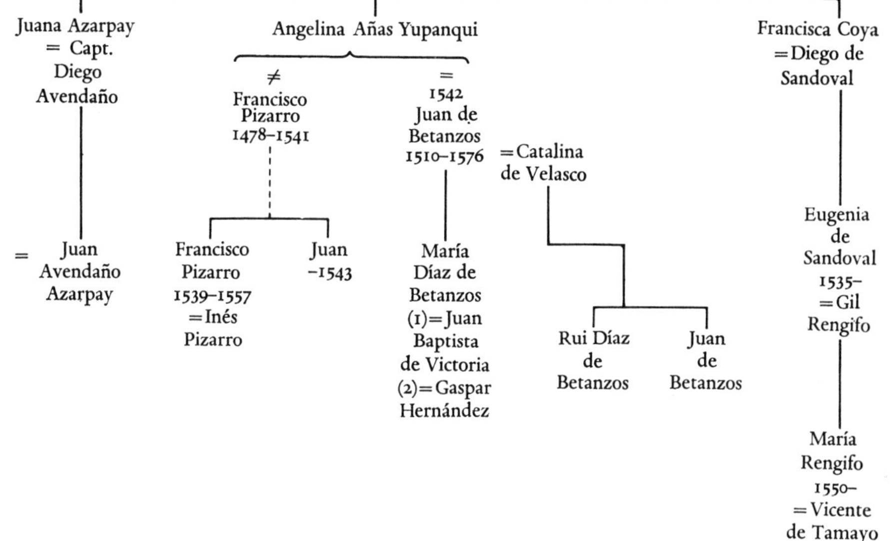

OTHER ROYAL PRINCESSES

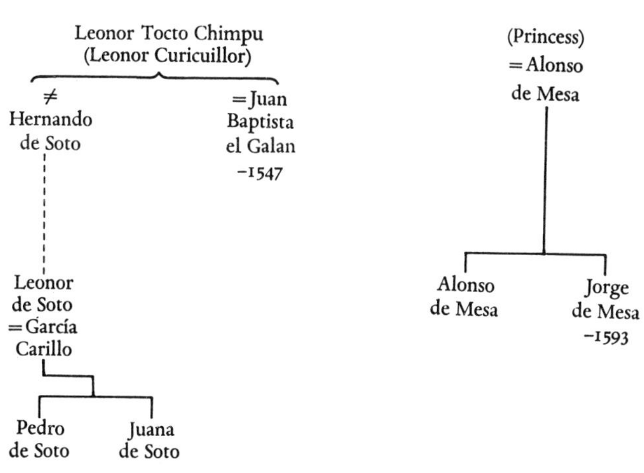

≠ = illegitimate connection
d.y. = died young
n.s. = no succession

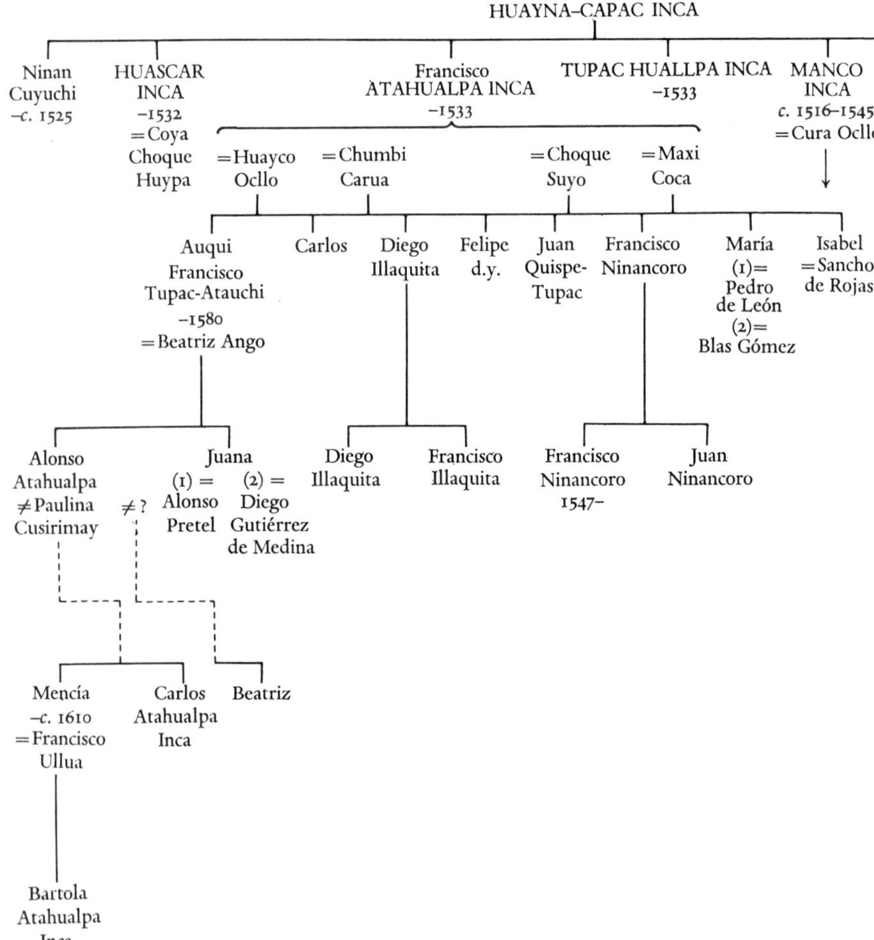

$\neq$ = illegitimate connection
d.y. = died young
n.s. = no succession

THE FAMILIES OF ATAHUALPA AND PAULLU

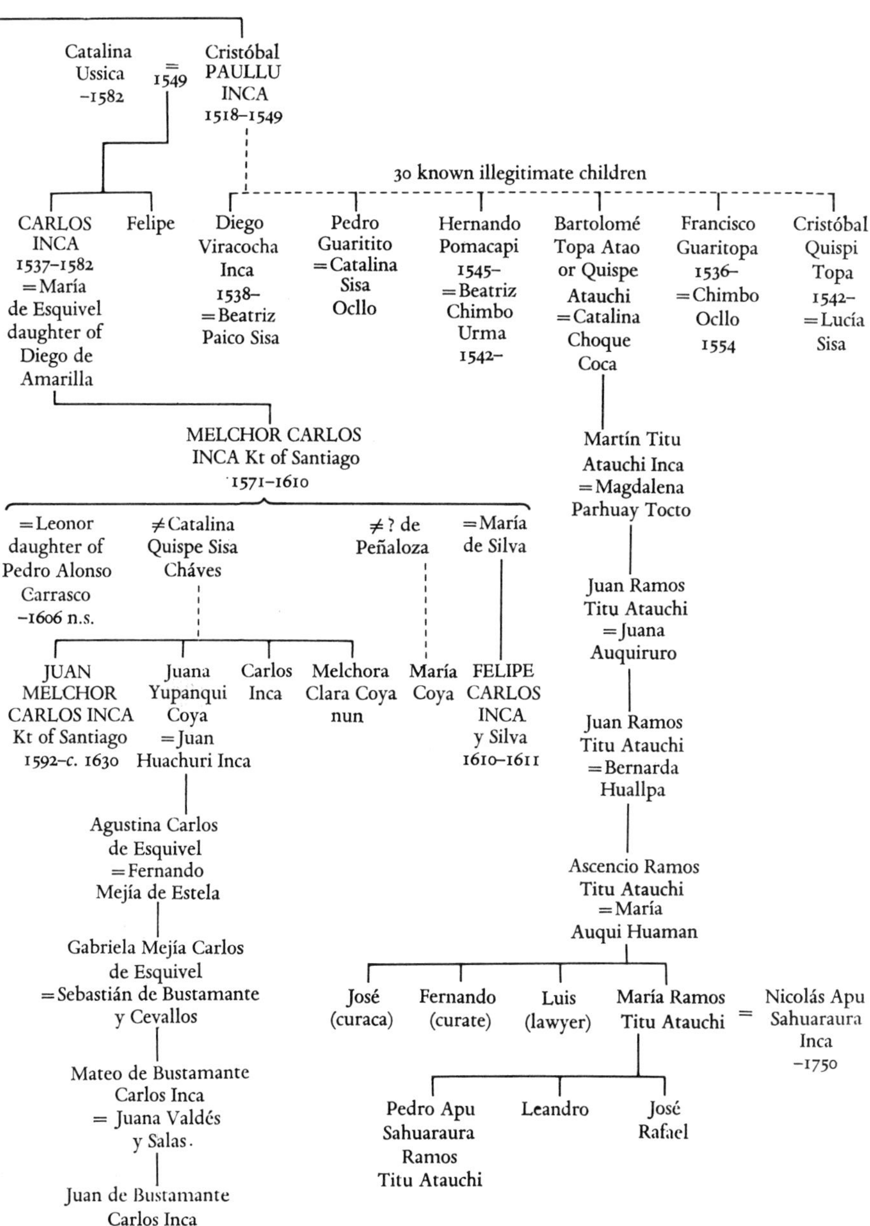

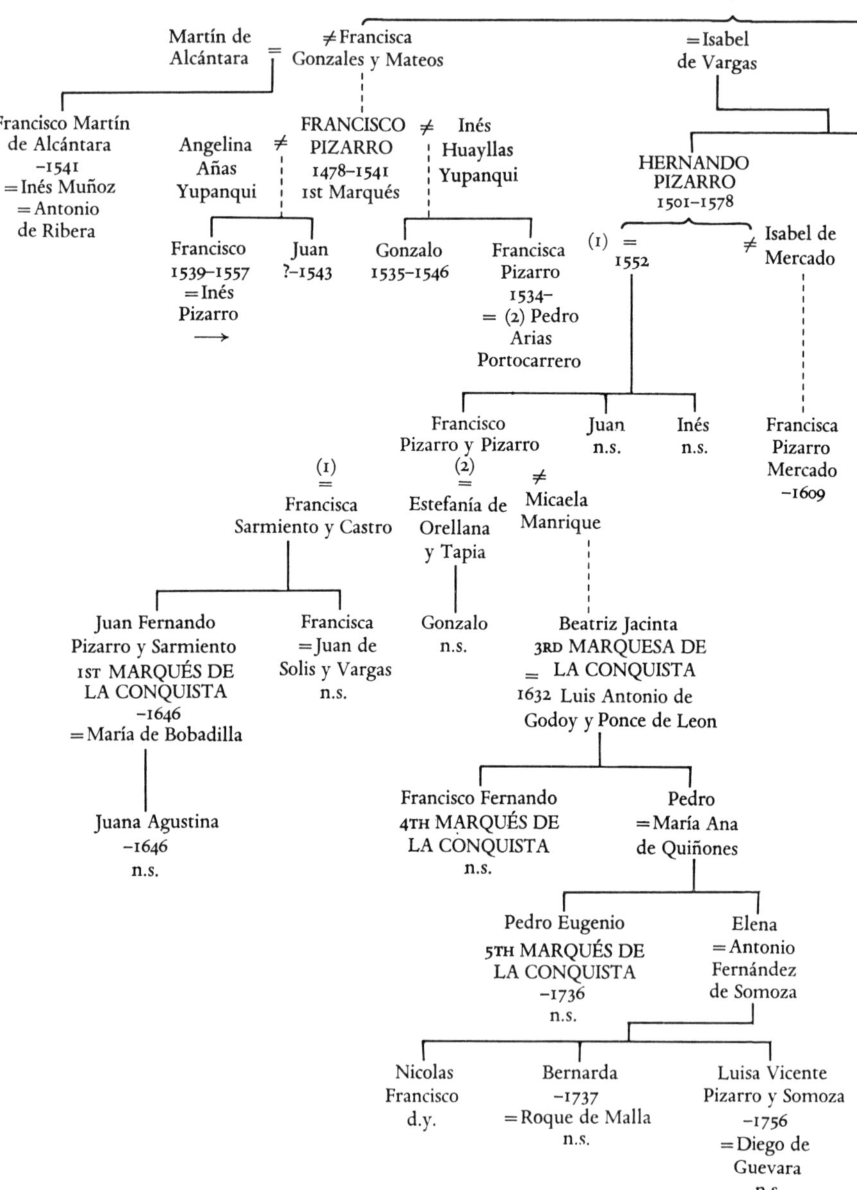

Gonzalo Pizarro 'El Largo'
?–1522
Martín de Alcántara
=
≠Francisca Gonzales y Mateos
=Isabel de Vargas
Francisco Martín de Alcántara
–1541
=Inés Muñoz
=Antonio de Ribera
Angelina Añas Yupanqui
≠
FRANCISCO PIZARRO
1478–1541
1st Marqués
≠
Inés Huayllas Yupanqui
HERNANDO PIZARRO
1501–1578
Francisco
1539–1557
=Inés Pizarro
⟶
Juan
?–1543
Gonzalo
1535–1546
Francisca Pizarro
1534–
= (2) Pedro Arias Portocarrero
(1) =
1552
≠ Isabel de Mercado
Francisco Pizarro y Pizarro
Juan
n.s.
Inés
n.s.
Francisca Pizarro Mercado
–1609
(1)
=
Francisca Sarmiento y Castro
(2)
=
Estefanía de Orellana y Tapia
≠
Micaela Manrique
Juan Fernando Pizarro y Sarmiento
1st MARQUÉS DE LA CONQUISTA
–1646
=María de Bobadilla
Francisca
=Juan de Solis y Vargas
n.s.
Gonzalo
n.s.
Beatriz Jacinta
3rd MARQUESA DE LA CONQUISTA
=
1632 Luis Antonio de Godoy y Ponce de Leon
Juana Agustina
–1646
n.s.
Francisco Fernando
4th MARQUÉS DE LA CONQUISTA
n.s.
Pedro
=María Ana de Quiñones
Pedro Eugenio
5th MARQUÉS DE LA CONQUISTA
–1736
n.s.
Elena
=Antonio Fernández de Somoza
Nicolas Francisco
d.y.
Bernarda
–1737
=Roque de Malla
n.s.
Luisa Vicente Pizarro y Somoza
–1756
=Diego de Guevara
n.s.

THE PIZARRO FAMILY

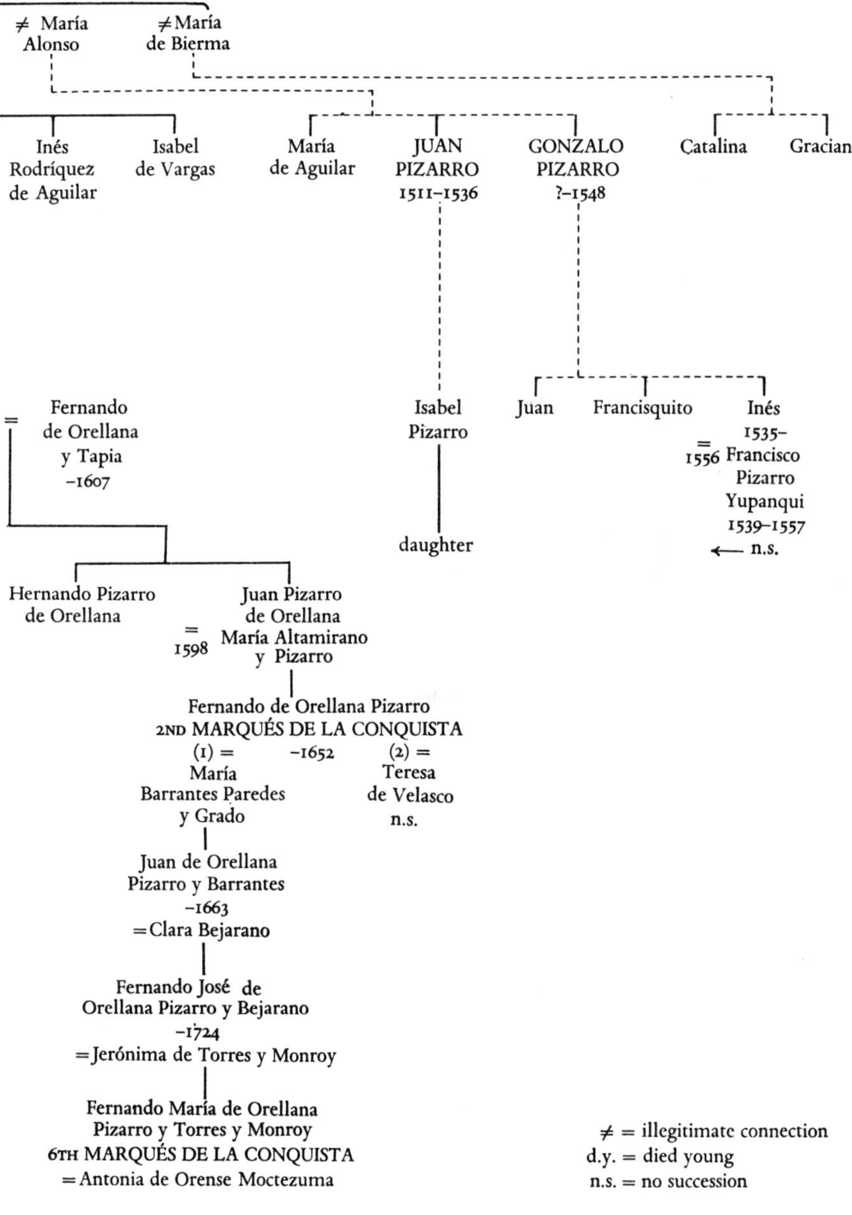

GLOSSARY

(phonetic spellings of Quechua words are in brackets)

ACLLA (*aklya*): girl or woman chosen for royal service

ADELANTADO: marshal, military title for commander of a frontier region

ADOBE: bricks of dried mud and straw

ALCALDE: mayor

ALTIPLANO: plateau around Lake Titicaca and north-western Bolivia

ANTI, ANTISUYU: eastern quarter of the Empire; eastern forest tribes; origin of the name Andes

APACHITA: pile of votive stones and offerings on the top of a pass

APU, APO (*apo*): lord, governor of a suyo or senior military commander

AUDIENCIA: judicial council, area of judicial administration

AUQUA (*awqa*): warrior

AUQUI (*awki*): prince

AYLLU (*aylyo*): kinship group or clan; lineage, through male descent, of an Inca emperor

AYLLU (*aylyo*): bolas, a weapon of twirling balls and tendons

BALSA: something that floats, a raft

BORLA: fringe, particularly the royal Inca fringe

CACIQUE: Caribbean word for chief

CABILDO: municipal council

CAMAYOC (*kamayoq*): specialist, professional

CANIPU: metal disc worn on chest and back by Inca warriors as a sign of distinguished service

CANOPA: household deity

CAPITÁN: an officer, unpaid but rewarded according to status after an expedition; the only title for any sixteenth-century officer

CAPITÁN-GENERAL: strategic commander-in-chief

CAPITULACIÓN: royal licence to undertake a conquest

CHACARA (*cakra*): field

CHACO (*cako*): public hunt in which hundreds of beaters surround game

CHAMPI: battle mace with stone or bronze head

CHASQUI (*caski*): postal runner, courier

CHICHA: fermented drink, generally made from maize

CHINCHAYSUYO (*cinca-soyo*): northern quarter of the Empire

CHUÑO: dehydrated potato

COCA (*koka*): (*Erythroxylon coca*) low tropical bush whose leaves are chewed by Andean Indians

COLLASUYO (*qolya-soyo*): southern quarter of the Empire

CONDESUYO (*konti-soyo*): south-western quarter of the Empire

CORREGIDOR: royal administrator based in Spanish municipality or royal encomienda

COYA (*qoya*): principal wife or sister-queen of the Inca, royal heiress

ÇUMBI (*qompi*): finest-quality woven cloth

CURACA (*koraka*): chief or official

DUHO: low wooden stool, royal throne

ENCOMENDERO: holder of an encomienda

ENCOMIENDA: area allotted to the 'care' of a Spaniard, to whom its inhabitants owe tribute

ESTANCIERO: Spanish peasant supervisor living among encomienda Indians

HATUNRUNA (*hatoñ rona*): adult male, ordinary agricultural tribute-payer

HIDALGO: 'son of somebody', gentleman

HUACA (*wak'a*): shrine, burial mound

HUAMAN (*wamañ*): hawk

HUAMANI (*wamani*): province

HUASI (*wasi*): house

HUAUQUE (*wawqi*): brother, effigy or talisman of an Inca

ICHU: Andean bunch grass

ILLAPA (*ilyapa'*): thunder, thunder god

INCA: Quechua-speaking tribe around Cuzco and the empire it ruled; the Emperor and royal family

INTI: sun, sun god

KERO (*qiro*): wooden beaker

LICENCIADO: the holder of a university degree equivalent to a Master of Arts, above a bachiller (B.A.) and below a doctor.

LLACTA (*lyaqta*): town

LLAUTU (*lyauto*): band or braid wrapped around the head

MACANA (*maqana*): sword-shaped war club

MAESTRE DEL CAMPO: camp master, second-in-command of an expedition, responsible for tactics and supply

MAMACONA (*mama-kona*): mother, senior chosen woman, virgin consecrated to religious service

MANCEBA: concubine or mistress

MASCAPAICHA: Inca's royal headdress, llautu with tassel over forehead and pompom above

MAYORAZGO: entailed estate

MESTIZO: half-caste

MINGA: mine-worker free of the forced mita system

MITA (mit'a): turn or stint of public service, forced labour in Spanish mines

MITAYO: labourer compelled to work in mines or public service

MITIMAES *(mitma-kona)*: settlers transplanted to colonise part of empire

MOCHA: kiss-like obeisance to sun god

MONTAÑA: forested hills, jungle

MORRIÓN: steel helmet with curved brim

ÑUSTA (*nyosta*): princess

OCA: sweet potato

OIDOR: judge, member of an audiencia council

OREJÓN: 'big ears', Spanish term for pakoyoc, Inca nobility who wore disc-plugs in their ear-lobes

PALLA (*palya*): noble lady, generally married

PAMPA: plain (Spanish bamba)

PICOTA: gibbet, symbolic post in each municipality

PROBANZA: legal enquiry or testimonial, proof of service

PROCURADOR: untitled lawyer who practised in lower courts

PUCARÁ (*pokara*): fort

PUMA: mountain lion, jaguar

PUNA: high treeless savannah

PUNCHAO: day, holy image of the sun

QUECHUA: tribe living near Abancay; official language of Incas and modern Andean Indians

QUINUA (*kinowa*): highland grain with spinachy leaves

QUIPU (*khipo*): string and knot device for statistical records

REDUCCIÓN: administrative resettlement of dispersed Indians into villages

REGIDOR: municipal alderman or councillor

REPARTIMIENTO: allocation or share, encomienda holding

RESIDENCIA: critical review of an official's term of office

SEÑOR NATURAL: natural lord, legitimate ruler

SORA: a mash made from maize

SUYO (*soyo*): division of any kind, quarter of the Empire

TAMBO (*tampo*): inn, shelter and storehouse along highway

TAPIA: mud pounded between forms to make a solid wall

TAQUI (*taki*): ritual dance with singing

TERCIO: Spanish infantryman

TOCRICOC (*t'oqrikoq*): governor

USNO (*osno*): raised platform used as imperial judgement seat

VEEDOR: overseer

VILCA (*wil'ka*): (*Piptadenia colubrina*) tree whose pods contain bitter yellow seeds used as purgative

VIRACOCHA (*wiraqoca*): creator god

YANACONA (*yana-cona*): government servants exempt from tribute, Spanish household servants or craftsmen

YUPANQUI (*yopañki*): Inca royal title meaning 'honoured', surname used by Spaniards for members of Inca royal family

TABLES OF MEASUREMENTS AND VALUES

LINEAR MEASUREMENTS	*British*	*metric*
dedo (finger's breadth)	0·71 in	1·8 cm
pulgada (inch)	0·9 in	2·3 cm
jeme (outstretched thumb to forefinger)	5–6 in	12–14 cm
palmo (span) = 12 dedos	8½ in	21·6 cm
pie de Castilla (foot) = 12 pulgadas	11 in	*c.* 28 cm
vara (yard) = 3 pies or 4 palmos	2 ft 7 in	83·6 cm
estado	*c.* 5 ft 2 in	*c.* 1·57 m
braza (fathom)	*c.* 5 ft 6 in	1·67 m
legua (league) = 20,000 pies	*c.* 3½ miles	5·57 km

AREA		
fanega (ground yielding 1 fanega of seed)	*c.* 1·6 acres	0·4 hectare

VOLUME		
azumbre	*c.* 3½ pints	2·02 litres
cántara = 8 azumbres	3·7 gals	16·13 litres
fanega (variable measure)	*c.* 1·6 bushels	*c.* 58 litres

WEIGHT		
peso	0·16 oz	4·55 gm
onza (ounce)	1·01 oz	28·8 gm
libra (pound) = 100 pesos or 16 onzas	1·01 lb	460 gm
arroba = 25 libras	25·3 lb	11·5 kg
quintal = 4 arrobas	101·4 lb	46 kg

PRECIOUS METALS		
gold		
peso de oro, or castellano = worth 450–490 maravedís	0·16 oz	4·55 gm
marco de oro = 50 castellanos		
libra de oro = 100 castellanos or 2 marcos de oro		

silver	*British*	*metric*
onza de plata	1·01 oz	28·8 gm
marco de plata = 8 onzas de plata		
libra de plata = 16 onzas de plata or 2 marcos de plata		

COINAGE

maravedí = basic Spanish monetary unit
real or tomín (= 34 maravedís) a silver coin of 0·6 grammes of silver
real de a ocho, or patacon (piece of eight) = 8 reales or 272 maravedís
ducado (ducat) (= 375 maravedís) a gold coin of 23¾ carats fine
escudo (= (1537–1566) 350 maravedís) a gold coin worth 22 carats fine
escudo = (after 1566) 400 maravedís

BIBLIOGRAPHY

ABBREVIATIONS

Collections of documents and periodicals cited most frequently in the bibliography and notes.

AEA: *Anuario de estudios americanos*, Escuela de Estudios Hispano-Américanos de la Universidad de Sevilla (Seville, 1944–).

AGI: Archivo General de Indias, Seville. This immense archive is estimated to contain fourteen million manuscript pages. It was founded in 1785 and housed in the magnificent sixteenth-century palace of La Lonja, beside Seville cathedral.

The documents are divided into sixteen Sections (ramos or secciones), of which the first six are relevant to this book: 1. Patronato (patronage); 2. Contaduría (paymaster); 3. Contratación (commerce); 4. Justicia (justice); 5. Gobierno (government); 6. Escribanía de Cámara de Justicia (actuarial).

The documents are tied in dockets (legajos) in boxes (cajas) and housed in cabinets (estantes). Each Section is subdivided in different ways, some according to the subject-matter of the documents, others (Sections 4, 5 and 6) according to the Viceroyalties: Lima, Quito, Charcas or La Plata, Chile, etc. References are now given to the Section and subdivision, and to the docket number. Before 1950 references were expressed as three numbers, i.e. 116–1–20.

BAE: *Biblioteca de autores españoles desde la formación del lenguaje hasta nuestros días*, ed. Manuel Rivadeneira, 71 vols (Madrid, 1846–80).

BAE (Cont): *Continuación*, ed. M. Menéndez Pelayo (Madrid, 1905–).

BHA: *Biblioteca Hispano Americana*, 6 vols (Madrid, 1876–82).

CDFS: *Colección de documentos para la historia de la formación social de Hispanoamérica*, ed. Richard Konetzke, 4 vols (Madrid, 1953).

CDH Chile: *Colección de documentos inéditos para la historia de Chile desde el viaje de Magellanes hasta la batalla de Maipo, 1518–1818*, ed. José Toribio Medina, 30 vols (Santiago de Chile, 1888–1902).

CDH Chile (2): Second series (Santiago de Chile, 1956–).

CDHE: *Colección de documentos inéditos para la historia de España*, ed. M. Fernández Navarrete, M. Salvá, P. Sainz de Baranda. *Continued by* Marqués de Pidal y de Miraflores, Marqués de la Fuensanta del Valle, José Sancho Rayon and Francisco de Zabálburu, 112 vols (Madrid, 1842–95).

CDHH-A: *Colección de documentos inéditos para la historia de Hispano-América*, 14 vols (Madrid, 1927–30).

CDIA: *Colección de documentos inéditos relativos al descubrimiento, conquista y colonización de las posesiones españolas en América y Oceanía sacadas en su mayor parte del Real Archivo de Indias*, bajo la dirección de D. Joaquín F. Pacheco, Francisco de Cárdenas, Luis Torres de Mendoza, 42 vols (Madrid 1864–84). *Continued as* CDIU.

CDIU: *Colección de documentos inéditos relativos al descubrimiento, conquista y organización de las antiguas posesiones españoles de Ultramar*, ed. Angel de Altolaguirre y Duvale and Adolfo Bonilla y San Martín, 25 vols (Madrid, 1885–1932).

CL: *Libro primero de cabildos de Lima*, ed. Enrique Torres Saldamando with Pablo Patrón and Nicanor Boloña, 3 vols (Paris, Lima, 1888–1900).

CL (2): Second series: *Libros de cabildos de Lima*, ed. Bertram T. Lee and later Juan Bromley (Lima, 1935–).

CLDRHA: *Colección de libros y documentos referentes a la historia de América*, ed. M. Serrano y Sanz, 21 vols (Madrid, 1904–29).

CLDRHP: *Colección de libros y documentos referentes a la historia del Perú*, ed. Carlos A. Romero and Horacio H. Urteaga, two series, 22 vols (Lima, 1916–35). First series, 12 vols (1916–19). Second series, 10 vols (1920–34).

CLERC: *Colección de libros españoles raros ó curiosos*, 25 vols (Madrid, 1871–96).

CP: *Cartas del Perú, Colección de documentos inéditos para la historia del Perú* **3** (Lima, Edición de la Sociedad de Bibliófilos Peruanos, ed. Raúl Porras Barrenechea, 1959).

DH Arequipa: *Documentos para la historia de Arequipa, Documentos inéditos de los archivos de Arequipa*, ed. Victor M. Barriga (Arequipa, 1939–).

Documenta: *Documenta, Revista de la Sociedad Peruana de Historia* (Lima, 1948–).

Fenix: *Fenix, Revista de la Biblioteca Nacional* (Lima, 1945–).

GP: *Gobernantes del Perú, cartas y papeles, Siglo xvi, Documentos del Archivo de Indias, Colección de publicaciones Históricas de la Biblioteca del Congreso Argentino*, ed. Roberto Levillier, 14 vols (Madrid, 1921–6).

HAHR: *The Hispanic American Historical Review* (Durham, North Carolina).

Hakl Soc: The Hakluyt Society, First series, 100 vols (Cambridge, 1847–98). Second series (Cambridge, 1899–).

Harkness Cal: *The Harkness Collection in the Library of Congress, Calendar of Spanish manuscripts concerning Peru, 1531–1641*, ed. Stella R. Clemence (Washington, 1932).

Harkness Doc: *Documents from Early Peru, The Pizarros and the Almagros, 1531–1578* (Washington, 1936).

IEP: *La iglesia de España en el Perú, Colección de documentos*, ed. E. Lissón-Chaves, 5 vols (Seville, 1943–7).

JLPB: *Juicio de límites entre el Perú y Bolivia, Prueba peruana presentada al Gobierno de la República Argentina*, ed. Victor M. Maúrtua, 12 vols (Barcelona, 1906).

Maggs–Huntington: *From Panama to Peru: the Conquest of Peru by the Pizarros, the rebellion of Gonzalo Pizarro, and the pacification by La Gasca*, ed. Sarah de Laredo (Catalogue of a collection of documents sold to the Henry E. Huntington Library, Berkeley, California: Maggs Brothers, London, 1925).

Mercurio Peruano: *Mercurio Peruano* (Lima, 1918–).

MP: *Manuscritos peruanos*, ed. Rubén Vargas Ugarte, 5 vols (Lima and Buenos Aires, 1935–47), viz. **1** *Manuscritos peruanos en las bibliotecas del extranjero* (Lima, 1935); **2** *Manuscritos peruanos del Archivo de Indias* (Lima, 1938); **3** *Manuscritos peruanos de la Biblioteca Nacional de Lima* (Lima, 1940); **4** *Manuscritos peruanos en las bibliotecas de América* (Buenos Aires, 1945); **5** *Manuscritos peruanos en las bibliotecas y archivos de Europa y América* (Buenos Aires, 1947).

NCDHE: *Nueva colección de documentos inéditos para la historia de España y de sus Indias*, ed. Francisco de Zabálburu and José Sancho Rayon, 6 vols (Madrid, 1892–6).

RAN: *Revista del Archivo Nacional* (Lima, 1920–).

REA: *Revista de la Escuela de Estudios Hispano-Americanos* (Universidad de Sevilla, Seville, 1948–).

RGI: *Relaciones geográficas de Indias*, ed. M. Jiménez de la Espada, 4 vols (Madrid, 1881); 3 vols, BAE (Cont), **183–5**, 1965.

RH: *Revista Histórica, Organo del Instituto Histórico del Perú* (Lima, 1906–).

RHA: *Revista de Historia de América* (Instituto Panamericano de Geografia e Historia, Mexico, D.F., 1938–).

RI: *Revista de Indias* (Consejo Superior de Investigaciones Científicas, Madrid, 1940–).

RMN: *Revista del Museo Nacional de Lima* (Lima, 1932–).

TIFEA: *Travaux de l'Institut Français d'Études Andines* (Paris–Lima, 1949–).

VP: *Virreinato peruano, Documentos para su historia*, ed. Moreyra-Céspedes, 3 vols (Lima, 1954–5).

EARLY SOURCES

The date of completion of a manuscript is given in brackets after its title if it was not published at the time. Where there are several editions an asterisk indicates the one which has been cited in the References without further identification.

Acosta, José de: *De procuranda indorum salute* (1580), ed. Francisco Mateos (Madrid, 1942).

— *Historia natural y moral de las Indias* (Seville, 1590; Mexico, 1940; BAE (Cont) **73**, 1954).* Trans. C. R. Markham (Hakl Soc, 1 ser., **60–1**, 1880).

AGIA, MIGUEL: *Seruidumbres personales de indios, Tratado . . . sobre una cédula real . . . que trata del seruicio personal y repartimientos de indios* (Lima, 1604); ed. F. Javier de Ayala (Seville, 1946).

ALBENINO, NICOLAO DE: *Verdadera relación de lo sucedido en los reynos e prouincias del Perú* (Seville, 1549); ed. José Toribio Medina (Paris, 1930).

ALTAMIRANO, P. DIEGO FRANCISCO: *Historia de la Compañía de Jesus en el Perú* (1705), excerpt in *El suplicio del primer Tupac Amaru* (RH **15**, pts. 1–2, 1942).

ANDAGOYA, PASCUAL DE: *Relación de los sucesos de Pedrarias Dávila en las provincias de Tierra Firme o Castilla del Oro* (1541–2), ed. M. Fernández de Navarrete, *Colección de los Viajes y Descubrimientos* **3** (Madrid, 1829). Trans. C. R. Markham (Hakl Soc, 1 ser., **34**, 1865).

ANON: *Carta de los comisarios: see* COMISARIOS DE LA PERPETUIDAD.

— *Dictamen sobre el dominio de los Ingas y el de los reyes de España en los reynos del Perú* (1571); CDHE **13** 425–69, 1848; CLDRHP, 1 ser., **4** 95–138.

— *Discurso de la sucesión y gobierno de los Incas* (c. 1565), JLPB **8** 149–65; and ed. Julio Luna (Lima, 1962).

— *Memorial para el buen asiento y gobierno del Perú* (1563); CDHE **94** 164–222, 1889.

— *Nouvelles certaines des Isles du Peru* (Lyon, 1534).

— *Parecer acerca de la perpetuidad y buen gobierno de los indios del Perú* (1564); CLDRHP, 2 ser., **3** 145–64, 1920.

— *Relación de las cosas del Perú desde 1543 hasta la muerte de Gonzalo Pizarro* (c. 1550, attributed to Juan Polo de Ondegardo, Rodrigo Lozano or Agustín de Zárate); BAE (Cont) **168** 243–332, 1965.

— *Relación del sitio del Cuzco y principio de las guerras civiles del Perú hasta la muerte de Diego de Almagro* (1539, often attributed to Vicente de Valverde, but more probably by Diego de Silva); CLERC **13** 1–195, 1879*; CLDRHP, 2 ser., **10**, 1934.

— *Relación de los sucesos del Perú con motivo de las luchas de los Pizarros y los Almagros . . .* (1548); CDH Chile **4** 197–212, 1889; GP **2** 389–419, 1921.*

ANTONIO, FRAY (possibly Fray Antonio Vázquez de Espinosa): *Discurso sobre la descendencia y gobierno de los Incas* (c. 1585); CLDRHP, 2 ser., **3**, 1920 (includes Quipocamayos de Vaca de Castro; *see below*).

ARRIAGA, PABLO JOSÉ DE: *La extirpación de la idolatría en el Perú* (1621); CLDRHP, 2 ser., **1**, 1920*; BAE **209** 191–277, 1968.

ARZÁNS DE ORSÚA Y VELA, BARTOLOMÉ: *Historia de la villa imperial de Potosí*, (Buenos Aires, n.d.),* ed. Lewis Hanke and Gunnar Mendoza, 3 vols (Providence, Rhode Island, 1965).

ATIENZA, LOPE DE: *Compendio historial del estado de los indios del Perú, con mucha doctrina i cosas notables de ritos, costumbres e inclinaciones que tienen* (*c.* 1583), in Jacinto Jijón y Caamaño, *La religión del Imperio de los Incas* (Quito, 1931) **1**, app. 1, 1–235.

AUGUSTINIANS ('Primeros Agustinos', also attributed to Juan de San Pedro): *Relación de la religión y ritos del Perú hecha por los primeros religiosos agustinos* (1560), CDIA **3** 5–58, 1865*; CLDRHP 1 ser., **11** 3–56, 1919.

ÁVILA, P. FRANCISCO DE: *Tratado y relación de los errores, falsos dioses y otras supersticiones y ritos diabólicos en que vivían los indios de Huarochirí, Mama y Chaclla* (1608). Trans. C. R. Markham in *Narratives of the Rites and Laws of the Incas*, Hakl Soc 1 ser., **48** 121–47, 1872.

AYALA, MANUEL JOSEF DE: *Diccionario de gobierno*, CDHH-A, **4**, **8**, 1929–30.

— *Notas a la Recopilación de Indias* (1787), ed. Juan Manzano Manzano, 2 vols (Madrid, 1945).

BALLESTEROS, THOMÁS DE: *Ordenanzas del Perú* **1** (Lima, 1685).

BANDERA, DAMIÁN DE LA: *Relación general de la disposición y calidad de la Provincia de Guamanga* (1557), RGI **1** 98–103, 1881; also, BAE (Cont) **183** 176–180, 1965.*

BENZONI, GIROLAMO: *La historia del Mondo Nuovo* (Venice, 1565). Trans. W. H. Smyth, *History of the New World*, Hakl Soc, 1 ser., **21**, 1857.

BERLANGA, TOMÁS DE: *Carta del obispo de Tierra-Firme al Rey* (3 February 1536), CDH Chile, 1 ser., **4** 328–41, 1892.

BERTONIO, LUDOVICO: *Arte breve de la lengua Aymara* (Rome, 1603).

BETANZOS, JUAN DE: *Suma y narración de los Incas* (1551), in M. Jiménez de la Espada, Biblioteca Hispano-Ultramarina **5** (Madrid, 1890); CLDRHP, 2 ser., **8**, 1924*; BAE (Cont) **209** 1–55, 1968.

BUENO, COSMÉ: *Descripción del reyno del Perú* (Lima, 1763).

— *Descripción del obispado del Cuzco* (Lima, 1768).

CABELLO DE BALBOA, MIGUEL: *Miscelánea Austral*, pt 3: *Historia del Perú* (1586), Paris, 1840; CLDRHP, 2 ser., **2**, 1920*; Quito, 1945; Instituto de Etnología, Lima, 1951.

CALANCHA, ANTONIO DE LA: *Corónica moralizada del orden de San Agustín en el Perú* (Barcelona, 1639).

CALVETE DE ESTRELLA, JUAN CRISTÓBAL: *Rebelión de Pizarro en el Perú y vida de D. Pedro Gasca* (1565–7), BAE (Cont) **167–8**, 1964–5.

CASTELLANOS, JUAN DE: *Elegías de varones ilustres de Indias* (Madrid, 1589); ed. Caracciolo Parra, 2 vols (Caracas, 1930–2)*; ed. Miguel A. Caro, 4 vols (Bogotá, 1955).

CASTRO, CRISTÓBAL DE, and DIEGO ORTEGA MOREJÓN: *Relación y declaración del modo que – este valle de Chincha y sus comarcanos se governavan antes que hobiese Ingas* (1558), CDHE **50** 206–20; CLDRHP, 2 ser., **10** 134–49, 1934.

CATAÑO, PEDRO: [Report on the conquest, in Probanza for Gil González Dávila, 5 March 1562], quoted in José Antonio del Busto Duthurburu, 'Una relación y un estudio sobre la Conquista', RH **27** 281–303, 1964.

CIEZA DE LEÓN, PEDRO DE: *Parte primera de la chrónica del Perú* (Seville, 1553); many Spanish editions. Trans. C. R. Markham, *The Travels of Pedro Cieza de Leon*, Hakl Soc, 1 ser., **33**, 1864.

— *Segunda parte de la crónica del Perú, que trata del señorío de los Incas Yupanqui* (1554), ed. Manuel González de la Rosa (London, 1873); many subsequent editions. Trans. C. R. Markham, Hakl Soc, 1 ser., **68**, 1883.

— *The Incas of Pedro de Cieza de Leon* (Combination of parts 1 and 2), trans. Harriet de Onis, ed. Victor W. von Hagen (Norman, Oklahoma, 1959).

— *Tercera parte* (*c.* 1554), *Descubrimiento y conquista*, ed. Rafael Loredo, *Mercurio Peruano*, **27**, **32**, **34**, **36**, **37**, **38**, **39**, 1946, 51, 53, 55–58.

— *La crónica del Perú*, pt. 4: *La guerra de Las Salinas, La guerra de Chupas, La guerra de Quito*, many Spanish editions. Trans. C. R. Markham, *The War of Las Salinas*, Hakl Soc, 2 ser., **54**, 1923; *The War of Chupas*, Hakl Soc, 2 ser., **42**, 1918; *The War of Quito*, Hakl Soc, 2 ser., **31**, 1913.

COBO, BERNABÉ: *Historia de la fundación de Lima* (1639), BAE (Cont) **92**, 1956.

— *Historia del Nuevo Mundo* (1653), ed. Luis A. Pardo, 4 vols (Cuzco, 1956)*; BAE (Cont) **91–2**, 1956.

COMISARIOS DE LA PERPETUIDAD: *Carta de los comisarios á S. M. sobre la perpetuidad y otras cosas* (1562), NCDHE **6** 46–105, 1896.

CORDOVA SALINAS, DIEGO DE: *Crónica franciscana de las provincias del Perú*, ed. Lino G. Canedo (Washington, 1957).

CUEVA, DIEGO DE LA: *Carta escrita a Su Magestad sobre los negocios de las Indias* (*c.* 1562), NCDHE **6** 218–59, 1896.

CUZCO: *Actos de los libros de cabildos de Cuzco, años 1545 a 1548, Revista del Archivo Histórico del Cuzco*, **9** no. 9, 5–13, 37–305, 1958.

— *Acta de fundación del Cuzco* (23 March 1534), in R. Porras Barrenechea, *Dos documentos esenciales sobre Francisco Pizarro y la Conquista del Perú . . . el acta perdida de la fundación del Cuzco*, RH **17** 86–95, 1948; CDHE **26** 221–32.

ENCINAS, DIEGO DE: *Provisiones, cédulas, capítulos de ordenanzas, instrucciones y cartas . . . tocantes al buen gobierno de las Indias*, 4 vols (Madrid, 1596).

— *Cedulario indiano*, 4 vols (Madrid, 1945).

ENRÍQUEZ DE GUZMAN, DON ALONSO: *Libro de la vida y costumbres de don Alonso Enríquez de Guzman* (1543), CDHE **85**, 1889; ed. Hayward Keniston (Madrid, 1960). Trans. C. R. Markham, Hakl Soc, 1 ser., **29**, 1862.

ESCALONA AGÜERO, GASPAR DE: *Gazofilacio regio perubico* (Madrid, 1647).

— (attributed) *Tratado de las apelaciones del Gobierno del Perú*, RAN **2** 79–130, 1921.

ESPINAL, MANUEL DE: *Relación hecho . . . al Emperador de lo sucedido entre Pizarro y Almagro* (15 June 1539), CDIA **3** 92–136, 1865; CL **3** 189–216.

ESTETE, MIGUEL DE: [Narrative of journey to Pachacamac], included in Francisco de Xerez, *Verdadera relación de la conquista del Perú* (Seville, 1534); BAE (cont) **26**, 1947*. Trans. C. R. Markham, *Reports on the Discovery of Peru*, Hakl Soc, 1 ser., **47** 74–94, 1872.

ESTETE, MIGUEL DE: (attributed) *Noticia del Perú*, or *El descubrimiento y conquista del Perú* (c. 1540), *Boletín de la Sociedad Ecuatoriana de Estudios Históricos* (Quito, 1919); CLDRHP, 2 ser., **8** 3–56, 1924.*

FALCÓN, FRANCISCO: *Representación sobre los daños y molestias que se hacen a los indios* (c. 1583), CDIA **7** 451–95, 1867.

FERNÁNDEZ, DIEGO ('El Palentino'): *Primera y segunda parte de la historia del Perú* (Seville, 1571); BAE (Cont) **164–5**, 1963.

FERNÁNDEZ DE OVIEDO, *see* OVIEDO Y VALDÉS, GONZALO FERNÁNDEZ DE.

FERNÁNDEZ DE SANTILLÁN, FELIPE: *Memorial escrito por el año de 1601 sobre las minas de Potosí*, CDIA **52** 445–55, 1868.

GALLEGO, PEDRO: *Querella e información ... contra Francisco Pizarro y Diego de Almagro y oficiales reales del Perú, por fraude a la real hacienda* (1537), GP **2** 83–90.

GARCILASO DE LA VEGA ('EL INCA'): *Primera parte de los comentarios reales de los Incas* (Lisbon, 1609); *Segunda parte de los comentarios reales de los Incas: Historia general del Perú* (Cordoba, 1617); Buenos Aires, 1943–4; BAE (Cont) **134–5**, 1960.* First part trans. C. R. Markham, Hakl Soc, **41**, **45**, 1869, 1871; both parts trans. Harold V. Livermore (London and Austin, 1966).

GUAMAN POMA DE AYALA: *see* POMA DE AYALA.

GUTIÉRREZ DE SANTA CLARA, PEDRO: *Historia de las guerras civiles del Perú y de otros sucesos de las Indias* (?1600), BAE (Cont) **165–7**, 1963–4.

HERRERA TORDESILLAS, ANTONIO DE: *Historia general de los hechos de los castellanos en las islas y tierrafirme del Mar Océano* (Madrid, 1610–15); ed. Antonio Ballesteros Beretta and Miguel Gómez del Campillo, 17 vols (Madrid, 1934–).*

HURTADO DE ARBIETO, MARTÍN: [Report to Viceroy Francisco de Toledo, Vilcabamba, 27 June 1572] in Levillier, *Don Francisco de Toledo* **1** 328–30.

JAUJA, CABILDO OF: *Carta a Su Magestad ... con varias noticias de gobierno e fazienda* (20 July 1534), CL **3**, *documentos justificativos* 1–9; J. Jijón y Caamaño, *Sebastián de Benalcázar* **1** (Quito, 1936) CDIA **42** 114–31.

JEREZ, FRANCISCO DE, *see* XEREZ.

LAS CASAS, BARTOLOMÉ DE: *Apologética historia sumaria* (c. 1550), BAE **13**, 1909. 27 chapters of this were published as *De las antiguas gentes del Perú*, CLERC **21**, 1892; CLDRHP, 2 ser., **11**, 1939.

— *Brevissima relación de la destrucción de las Indias* (Seville, 1552)· CDHE, **71**, 1842.

— *Historia de las Indias* (1552–61), CDHE **62–6**, 1876; Madrid, 1927; 3 vols, Mexico, 1951.

— *De thesauris in Peru*. Trans. Angel Losada, *Los tesoros del Perú* (Madrid, 1958).

LIZÁRRAGA, REGINALDO DE: *Descripción y población de las Indias* (1605), *Revista del Instituto Histórico del Perú*, 1908; *Los pequeños grandes libros de historia americana*, Lima, **12**, 1946.

Loaisa, Rodrigo de: *Memorial de las cosas del Perú tocantes a los indios* (1586), CDHE **44** 554–605, 1889.

López de Caravantes, Francisco: *Noticia general del Perú, Tierra-firme y Chile* (*c.* 1610).

López de Gómara, Francisco: *Hispania Victriz, La historia general de las Indias y conquista de México* (Saragossa, 1552); 2 vols (Madrid, 1922).

López de Velasco, Juan: *Geografía y descripción universal de las Indias* (1571–4), ed. Justo Zaragoza (Madrid, 1894).

Matienzo, Juan de: *Gobierno del Perú* (1567), ed. J. N. Matienzo (Buenos Aires, 1910); ed. G. Lohmann Villena (Paris and Lima, 1967).*

— *Carta a S. M. del Oidor de Los Charcas* (1566), RGI **2**, app. 3, 1885.

Meléndez, Juan: *Tesoros verdaderos de las Yndias: en la historia de la gran Provincia ... del Perú*, 3 vols (Rome, 1681–2).

Mena, Cristóbal de: (attributed to) *La conquista del Perú, llamada la Nueva Castilla* (Seville, April 1534); soon translated into Italian and French. Trans. Joseph H. Sinclair, *The Conquest of Peru, as recorded by a member of the Pizarro expedition*, with a facsimile of the 1534 edition (New York, 1929). Alexander Pogo, 'The anonymous *La Conquista del Peru*', *Proc. Amer. Acad. Arts and Sci.*, Boston, **64**, no. 8 (Jul 1930) gives Spanish and Italian texts.*

Messia, Alfonso: *Memorial sobre las cédulas del servicio personal de los indios* (*c.* 1600), CDIA **6** 118–65, 1886.

Molina, Cristóbal de (el Almagrista or of Santiago): *Relación de muchas cosas acaesidas en el Perú ... en la conquista y población destos reinos* (*c.* 1553), Paris, 1840; CLDRHP, 1 ser., **1** 111–90, 1916*; ed. F. A. Loayza (Lima, 1943); BAE (Cont) **209** 56–96, 1968.

Molina, Cristóbal de (of Cuzco): *Relación de las fábulas y ritos de los Incas* (1573), CLDRHP, 1 ser., **1** 1—103, 1916*; Buenos Aires, 1959. Trans. C. R. Markham, *The Fables and Rites of the Incas*, Hakl Soc **48**, 1873.

Montesinos, Fernando de: *Memorias antiguas historiales y políticas del Perú* (1630), CLERC **16**, 1882. Trans. P. A. Means, Hakl Soc, 2 ser., **48**, 1920.

— *Los anales del Perú*, ed. Victor M. Maúrtua, 2 vols (Madrid, 1906).

Monzón, Luis de: *Relación de la provincia de Rucanas y Soras* (1586), RGI **1** 169–216, 1881; BAE (Cont) **183** 237–48, 1965.

Morales, Francisco de: [Letter to His Majesty, 1561], IEP **1** 179–87, 1943. *Informe sobre lo que es necesario al seruicio de Dios y de Su Magestad en el Perú* (1567), in R. Vargas Ugarte, *Historia de la Iglesia en el Perú* (Burgos, 1959) **2** 529–38.

Morales, Luis de (Provisor del Cuzco): *Relación sobre las causas que convenían proveerse en el Perú* (*c.* 1550), IEP **1** 48–98, 1943.

Morales Figueroa, Luis de: *Relación de los indios tributarios en estos reynos y provincias del Perú, fecha por mando del Señor Marqués de Cañete* (*c.* 1595), CDIA **6** 41–63, 1866.

MURÚA, MARTÍN DE: *Historia general del Perú, Origen y descendencia de los Incas* (1590–1611), ed. Manuel Ballesteros-Gaibrois, 2 vols (Madrid, 1962, 1964)* (Wellington MS.); CLDRHP, 2 ser., **4**, **5**, 1922–5; ed. Constantino Bayle, 2 vols (Madrid, 1946) (Loyola MS.).

NAHARRO, PEDRO RUIZ: *Relación de los hechos de los españoles en el Perú desde su descubrimiento hasta la muerte del Marqués Francisco Pizarro*, CDHE **26** 232–56, 1855*; CLDRHP, 1 ser., **6** 189–213.

[New Laws]: *Leyes y ordenanças nuevamente hechas por su Magestad por la governación de las Indias y buen tratamiento y conservación de los indios* (Alcalá de Henares, 1543). Trans. Henry Stevens and F. W. Lucas, *The New Laws of the Indies* (London, 1893; Amsterdam, 1967). *See also* MURO OREJÓN, *page 540 below*.

NIEVA, CONDE DE (and COMISARIOS DE LA PERPETUIDAD): *Carta a S.M. sobre la perpetuidad y otras cosas* (1562), NCDHE **6** 46–105, 1896.

OCAMPO, BALTASAR DE: *Descripción de la provincia de Sant Francisco de la Vitoria de Vilcapampa* (1610). Trans. C. R. Markham, *Account of the Province of Vilcapampa and a Narrative of the Execution of the Inca Tupac Amaru*, Hakl Soc, 2 ser., **22** 203–47, 1907.*

OLIVA, JUAN ANELLO: *Historia del reino y provincias del Perú y varones insignes en santidad* (1613) (Lima, 1895).

OVIEDO, GABRIEL DE: *Relación de lo que subcedio en la ciudad del Cuzco cerca de los conciertos y horden que su Magestad mandó asentar con el Ynca Titu Cuxi Yopanqui* (1573), in C. A. Romero, *Inédito sobre el primer Túpac Amaru*, RH **2** 66–73, 1907. Trans. C. R. Markham as suppl. to Hakl Soc, 2 ser., **22** 401–12, 1908.

OVIEDO Y VALDÉS, GONZALO FERNÁNDEZ DE: *La historia general y natural de las Indias* (Seville, 1535); including section on Peru: (Salamanca, 1547); (Valladolid, 1557); ed. Juan Pérez de Tudela Bueso, BAE (Cont) **117–21**, 1959.

PALACIOS RUBIOS, JUAN LÓPEZ DE: *De las islas del mar Océano*, ed. Silvio Závala and A. Millares Carlo (Mexico, 1954).

PAULLU INCA, CRISTÓBAL: *Probanza hecha a pedimento de Paulo Inga sobre los servicios prestados a S.M.* (Cuzco, 1540), CDH Chile **5** 341–60, 1889.

— [Will, Cuzco, 1549], in E. Dunbar Temple, 'Los testamentos inéditos de Paullu Inca, don Carlos y don Melchor Carlos Inca', *Documenta* **2** no. 1, 1949–50.

PAZ, MATÍAS DE: *Del dominio de los reyes de España sobre los indios* ed. Silvio Závala and A. Millares Carlo (Mexico, 1954).

PIZARRO, FRANCISCO: *Capitulación de Francisco Pizarro con la corona*, Toledo, 26 July 1529, in R. Porras Barrenechea, *Cedulario del Perú*, **1** 17–24, 1944.

— *Acta de fundación del Cuzco*, 23 March 1534, in R. Porras Barrenechea, 'Dos documentos esenciales sobre Francisco Pizarro y la Conquista del Perú', RH **17** 86–95, 1948.

— [Letter to Cabildo of Panama, Jauja, 25 May 1534], CDIA **10** 134–44; CL **3** 42–9; CP 112–17.

PIZARRO, FRANCISCO: [Despatch to King, Cuzco, 27 February 1539], in R. Vargas Ugarte, 'Dos cartas inéditas de D. Francisco Pizarro y de D. Fray Vicente de Valverde', RHA **47** 152–62, 1959.

— [Will, 1539] in R. Porras Barrenechea, 'El testamento de Pizarro de 1539', RI **2** no. 3, 39–70, 1941.

Información de méritos del Marqués don Francisco Pizarro, Cuzco, 1572, GP **2** 91–203.

PIZARRO, HERNANDO: [Letter to Oidores of Santo Domingo, Panama, 23 November 1533], in Gonzalo Fernández de Oviedo y Valdés, *La historia general y natural de las Indias*, bk 46, ch 16, BAE (Cont) **121** 84–90, 1959. Trans. C. R. Markham in *Reports on the Discovery of Peru*, Hakl Soc **47** 113–27, 1872.

— *Confesión de Hernando Pizarro* (Madrid, 15 May 1540), CDH Chile **5** 405–43, 1889.

— [Wills, Trujillo, June–August 1578] in M. Muñoz de San Pedro, 'Las últimas disposiciones del último Pizarro de la Conquista' and in L. Cuesta, 'Una documentación interestante sobre la familia del conquistador del Perú', RI **8** no. 30, 879–92, 1947.

PIZARRO, JUAN: [Will, Cuzco, 16 May 1536] in L. Cuesta, 'Una documentación interesante sobre la familia del Conquistador del Perú', RI **8** no. 30, 872–78.

PIZARRO, JUAN FERNANDO: *Discurso en que se muestra la obligación que Su Magestad tiene . . . a cumplir . . . la merced que la Magestad Imperial hizo a don F. Pizarro del titulo de Marqués* (Madrid, *c.* 1640).

PIZARRO, PEDRO: *Relación del descubrimiento y conquista de los reinos del Perú* (1571), CDHE **5** 201–388, 1844; BAE (Cont) **168** 159–242, 1965. Trans. P. A. Means as *Relation of the Discovery and Conquest of the Kingdoms of Peru* (Cortes Society, New York, 1921).

PIZARRO Y ORELLANA, FERNANDO: *Varones ilustres del Nuevo Mundo* (Madrid, 1639).

POLO DE ONDEGARDO, JUAN: *Instrucción contra las cirimonias y ritos que usan los indios conforme al tiempo de su infidelidad* (*c.* 1560); *Tratado y auerigación sobre los errores y svpersticiones de los indios* (*c.* 1560) (Lima, 1585); CLDRHP, 1 ser., **3** 3–43, 189–203, 1917.

— *Informe al Licenciado Briviesca de Muñatones sobre la perpetuidad de las encomiendas del Perú* (1561), RH **13** 125–96, 1940.

— *Relación acerca del linaje de los Incas y como conquistaron y acerca del notable daño que resulta de no guardar a estos yndios sus fueros* (1571), CDIA **17** 1–177, 1872; CLDRHP, 1 ser., **3** 45–188, 1917*. Partly trans. C. R. Markham, *Of the Lineage of the Incas and How They Extended their Conquests* in *Rites and Laws of the Incas*, Hakl Soc, 1 ser., **48**, 1872.

POMA DE AYALA, FELIPE GUAMAN: *Nueva corónica y buen gobierno* (?1580–1620), Institut d'Ethnologie, Paris; *Travaux et mémoires*, **23**, 1936; also ed. L. F. Bustios Galvez, 3 vols (Lima, 1956–66).*

QUIPOCAMAYOS of VACA DE CASTRO: *Discurso sobre la descendencia y gobierno de los Incas* (1542–4), CLDRHP, 2 ser., **3**, 1920 (included as part of Fray Antonio).

QUIROGA, PEDRO DE: *Libro intitulado coloquios de la verdad, trata de las causas e inconvenientes que impiden la doctrina y conversión de los indios del Perú, y los daños, malos y agravios que padecen* (1555), ed. Julián Zarco Cuevas (Biblioteca Colonial Americana: Seville, **7**, 1922).

QUITO: *Libro primero, segundo de cabildos de Quito*, ed. José Rumazo González, 4 vols (Quito, 1934).

— *Oficios o cartas al cabildo de Quito por el Rey de España o el Virrey de Indias, 1552–1568*, ed. Jorge A. Garcés G. (Quito, 1934).

RAMÍREZ, BALTASAR: *Descripción del Reyno del Pirú, del sitio, temple, prouincias, obispados y ciudades, de los naturales de sus lenguas y trage* (1597), JLPB **1**; ed. Hermann Trimborn in 'Quellen zur Kulturgeschichte des präkolumbischen Amerika,' *Studien zur Kulturkunde* **3** 10–68 (Stuttgart, 1936).*

RAMOS GAVILÁN, ALONSO: *Historia del celebre santuario de Nuestra Señora de Copacabana* (Lima, 1621).

RIBERA, PEDRO DE and ANTONIO DE CHAVES Y DE GUEVARA: *Relación de la ciudad de Guamanga y sus terminos* (1586), RGI, BAE (Cont) **183** 181–201, 1965.

ROBLES CORNEJO, DIEGO DE: *Memorial sobre el asiento del Pirú*, CDIA **11** 20–9, 1869.

— *Proveimientos generales y particulares del Pirú*, ibid. 29–43.

— *Apuntamientos para el acierto del Pirú y buen gobierno de los naturales* (1570), ibid. 48–55, 97–102.

— *Parecer sobre la perpetuidad de los indios*, ibid. 181–6.

— *Contra el cuaderno que los Comisarios escribierno de la división de la perpetuidad*, NCDHE **6**, 1896.

RODRÍGUEZ DE FIGUEROA, DIEGO: *Relación del camino e viage que . . . hizo desde la ciudad del Cuzco a la tierra de guerra de Manco Inga* (1565), ed. R. Pietschmann, *Nachrichten der K. Gesellschaft der Wissenchaften zu Göttingen, Philologisch historische Klasse* (Berlin, 1910). Trans. C. R. Markham in *The War of Quito*, Hakl Soc, 2 ser., **31** 170–99, 1913.*

ROMÁN Y ZAMORA, JERÓNIMO: *Repúblicas de Indias* (1575), CLERC **14**, **15**, 1897.

RUIZ DE ARCE, JUAN: *Relación de servicios; Advertencias que hizo el fundador del vinculo y mayorazgo a los sucesores en el* (*c.* 1545), *Boletín de la Real Academia de Historia* (Madrid) **102** 327–84, 1933.

SALAZAR, ANTONIO BAUTISTA DE: *Relación sobre el periodo de gobierno de los Virreyes Don Francisco de Toledo y Don García Hurtado de Mendoza* (1596), CDIA **8** 212–93, 1867 (formerly attributed to Tristán Sánchez).

SALINAS Y CÓRDOBA, BUENAVENTURA DE: *Memorial de las historias del nuevo mundo Perú* (Lima, 1631); ed. Luis E. Valcarcel and Warren L. Cook (Lima, 1957).

SÁMANO, JUAN DE: *Relación de los primeros descubrimientos de Pizarro y Almagro*, Wiener Nationalbibliotek, codex cxx, CDHE **5** 193–201, 1842. (Probably by Francisco de Xerez, but forwarded to Emperor Charles V by Juan de Sámano.)

SANCHO, PEDRO: *Testimonio de la acta de repartición del rescate de Atahualpa* (1533). Trans. C. R. Markham, Hakl Soc, 1 ser., **97** 131–43, 1872.

— *Relación para S. M. de lo sucedido en la conquista y pacificación de estas provincias de la Nueva Castilla y de la calidad de la tierra* (1543), CLDRHP, 1 ser., **5** 122–202, 1917.* Trans. P. A. Means, (Cortes Society: New York, 1917). (Original MS. is lost but was translated in Ramusio, *Navigationi e viaggi* **3**, 1550, from which subsequent texts are derived.)

SANTACRUZ PACHACUTI YAMQUI, JOAN DE: *Relación de antigüedades deste reyno del Pirú* (*c.* 1615), ed. M. Jiménez de la Espada in *Tres relaciones de antigüedades peruanas* (Madrid, 1879); CLDRHP, 2 ser., **9**, 1927*; BAE (Cont) **209** 279–319, 1968. Trans. C. R. Markham, Hakl Soc, 1 ser., **48** 67–120, 1873.

SANTILLÁN, HERNANDO DE: *Relación del origen, descendencia, política y gobierno de los Incas* (*c.* 1563), ed. M. Jiménez de la Espada in *Tres relaciones de antigüedades peruanas* (Madrid, 1879), 1–133; CLDRHP, 2 ser., **9** 1–117, 1927*; BAE (Cont) **209** 97–150, 1968.

SANTO TOMÁS, DOMINGO DE: *Relación al Obispo Fr. Bartolomé de Las Casas* (1555), CDIA **7** 371–87, 1867.

— *Léxicon o vocabulario de la lengua general del Perú*, Valladolid, 1560.

— *Grammática o arte de la lengua general de los indios de los reynos del Perú* (Valladolid, 1560). Facsimile edition of both works (Lima, 1951).

SANTOYO, MARTEL DE and ALONSO PÉREZ: *Relación sobre lo que se debe proveer y remediar en el Perú* (1542), IEP **1**, no. 3, 99–120, 1943.

SARMIENTO DE GAMBOA, PEDRO: *Historia indica* (1572), ed. Richard Peitschmann (Berlin, 1906); BAE (Cont) **135** 189–279, 1960*. Trans. C. R. Markham, *History of the Incas*, Hakl Soc, 2 ser., **22**, 1907.

SEPÚLVEDA, JUAN GINÉS DE: *Demócrates Segundo, o De las justas causas de la guerra contra los indios* (Madrid, 1951).

SERVIDORES DE LOS ÚLTIMOS INCAS: *Relación del origen y gobierno que los Ingas tuvieron y del que había antes que ellos señoreasen a los indios deste reino y de que tiempo y de otras cosas que al gobierno convenía* (?1583), CLDRHP, 2 ser., **3**, 1920 (attributed to Cristóbal de Castro).

SILVA Y GUZMAN, DIEGO DE: *Conquista de la Nueva Castilla, 'La Cronica Rimada'* (1538), ed. J. A. Sprecher de Bernegg (Paris and Leon, 1848); also ed. F. Rand Morton (Mexico, 1963).

SOLÓRZANO PEREYRA, JUAN DE: *Política Indiana* (Madrid, 1648); ed. Francisco Ramiro de Valenzuela, 5 vols (Madrid, 1930); ed. Ricardo Levene, 2 vols (Buenos Aires, 1945).

TITU CUSI YUPANQUI, INCA DIEGO DE CASTRO: *Relación de la conquista del Perú y hechos del Inca Manco II; Instrucción para el muy Ille. Señor Ldo. Lope García de Castro, Gouernador que fue destos rreynos del Pirú* (1570), CLDRHP, 1 ser., **2**, 1916.

TOLEDO, VICEROY FRANCISCO DE: *Ordenanzas de Don Francisco de Toledo, virrey del Perú,* (1569–81) ed. Roberto Levillier (Madrid, 1929).

— *Libro de la visita general del Virrey Toledo* (1570–5), RH **7**, pt. 2, 114–216, 1924.

— *Informaciones que mandó levantar el Virrey Toledo sobre los Incas* (1571–2), ed. Roberto Levillier, *Don Francisco de Toledo, supremo organizador del Perú* **2** (Buenos Aires, 1940).

— *Relación sumaria de lo que el Virrey don Francisco de Toledo escribió en lo tocante al gobierno espiritual y temporal y guerra y hacienda* (1571), CDIHE **94** 255–309, 1889 (a contemporary summary of a report, the full text of which is in NCDHE **6** 306–78, 1896).

— [Reports to King of March 1571 and March 1572], GP **3** 450–623, **4** 1–287, 1921.

— *Memorial que dió al Rey Nuestro Señor, del estado en que dejó las cosas del Perú* (1596), in Sebastián Lorente, *Relaciones de los Virreyes* **1** 3–33, 1867; CDIA **6** 516–54, 1866; Ricardo Beltrán y Rózpide, *Biblioteca de Historia Hispano-Americana* **1** 71–106 (Madrid, 1921).*

TORO, DIEGO DE: *Comentario del Perú, Relación del provecho que en las almas se hace en las Indias* (Rome, 1603; Antwerp, 1603).

TRUJILLO, DIEGO DE: *Relación del descubrimiento del reyno del Perú* (1571), ed. R. Porras Barrenechea (Seville, 1948).

ULLOA MOGOLLÓN, JUAN DE: *Relación de la provincia de los Collaguas* (1586), RGI **2** 38–50, 1885.

VACA DE CASTRO, CRISTÓBAL: *Ordenanzas de tambos, distancias de unos a otros, modo de cargar los indios y obligaciones de las justicias respectivas* (1543), ed. C. A. Romero, RH **3**, no. 4, 427–92, 1908.

VALERA, BLAS: *Relación de las costumbres antiguas de los naturales del Perú* (*c.* 1585–9), ed. M. Jiménez de la Espada, *Tres relaciones de antigüedades peruanas* (Madrid, 1879) 135–227; ed. Francisco A. Loayza (Lima, 1945). (Excerpts from a lost MS., *Historia imperii peruani* (*c.* 1590).)

VALVERDE, VICENTE DE: *Carta del Obispo del Cuzco al Emperador sobre asuntos de su iglesia y otros de la gobernación general de aquel país* (20 March 1539), CDIA, **3** 92–137, 1865; CL **3** 89–115; IEP **1** 99–133, 1943.

— [Letter to Audiencia of Panama] (Tumbez, 15 November 1541), CDIA **3** 221–8, 1865.

(Bishop Valverde did *not* write the anonymous *Relación del sitio del Cuzco* which is often attributed to him.)

VAZQUEZ DE ESPINOSA, ANTONIO: *Compendium and Description of the West Indies* (1628). Trans. Charles Upson Clark, Smithsonian Miscellaneous Collections: Washington, D.C., **102**, no. 3646, 1942.

VEGA, BARTOLOMÉ DE: *Memorial al Real Consejo de Indias sobre los agravios que reciben los indios del Perú* (1562), NCDHE **6** 105–31, 1896.

VEGA LOAIZA, ANTONIO DE: *Historia del Colegio y Universidad de San Ignacio de Loyola de la ciudad del Cuzco* (*c.* 1590), ed. Rubén Vargas Ugarte, Biblioteca Histórica Peruana **6** (Lima, 1948).

Velasco, Juan de: *Historia del reino de Quito en la América Meridional*, ed. A. Yerovi, 3 vols (Quito, 1841–4).

Vera, Diego de: *De cosas del Perú en favor de los indios*, NCDHE **6** 131–4, 1896.

Villasante, Salazar de: *Relación general de las poblaciones españoles del Perú*, RGI **1**; BAE (Cont) **183** 121–46, 1965.*

Vitoria, Francisco de: *Relecciones de indios* (Madrid, 1928).

Xerez, Francisco de: *Verdadera relación de la conquista del Perú y provincia del Cuzco* (Seville, 1534); many editions include: CLDRHP, 1 ser., **5** 1–121, 1917; BAE (Cont) **26** 320–46, 1947. Trans. C. R. Markham in *Reports on the Discovery of Peru*, Hakl Soc, 1 ser., **47** 1–109, 1872.

Zárate, Agustín de: *Historia del descubrimiento y conquista del Perú* (Antwerp, 1555); (Seville, 1577); BAE (Cont) **26** 459–574, 1947.* Trans. T. Nicholas (London, 1581), J. M. Cohen (Harmondsworth, Middx., 1968).

Zúñiga, Antonio de: *Carta al Rey Don Felipe II* (15 July 1579), CDHE **26** 87–121, 1855.

MODERN AUTHORS

Armas Medina, Fernando de: 'El clero en las guerras civiles del Perú', AEA **7** 1–46, 1950.

— 'Evolución histórica de las doctrinas de indios', AEA **9** 101–29, 1952.

— *Cristianización del Perú (1532–1600)* (Seville, 1953).

— 'Los oficiales de la Real Hacienda en las Indias', *Revista de Historia* (Caracas, 1963) **16** 11–34.

Ballesteros-Gaibrois, Manuel: *Francisco Pizarro* (Madrid, 1940).

Barra, F. de la: *El indio peruano en las etapas de la conquista* (Lima, 1948).

Barriga, Victor M.: (ed.): *Documentos para la historia de Arequipa* (= DH Arequipa), 3 vols (Arequipa, 1939–55).

— *Los Mercedarios en el Perú en el siglo xvi*, 5 vols (**1** Rome 1933, **2–5** Arequipa 1939–54).

Basadre, Jorge: 'El régimen de la mita', *Letras, Organo de . . . la Universidad Mayor de San Marcos* (Lima, 1937) **3**.

Baudin, Louis: *L'empire socialiste des Inka* (Paris, 1928). Trans. Katherine Woods, *A Socialist Empire: the Incas of Peru* (Princeton, N.J., 1961).

— *La Vie de François Pizarre* (Paris, 1930).

— *La Vie quotidienne au temps des derniers Incas* (Paris, 1955). Trans. Winifred Bradford, *Daily Life in Peru under the Last Incas* (London, 1961).

— *Les Incas* (Paris, 1964).

Bayle, Constantino: *El protector de indios* (Seville, 1945).

— 'Los municipios y los indios', *Missionalia Hispánica* (Madrid, 1950) **7**, no. 21, 409–42.

BAYLE, CONSTANTINO: *El clero secular y la evangelización de América* (Madrid, 1950).
— 'Cabildos de indios en la América Española,' *Missionalia Hispánica* (Madrid, 1951) **8**, no. 22, 5–35.
— *Los cabildos seculares en la América Española* (Madrid, 1952).
BELAÚNDE GUINASSI, MANUEL: *La encomienda en el Perú* (Lima, 1945).
BENNETT, WENDELL C.: 'Excavations at Tiahuanaco', *Anthropological Papers of Amer. Mus. nat. Hist.* **34**, 1934.
— 'Machu Picchu, the most famous Inca ruin', *Natural History* (Amer. Mus. nat. Hist., New York, 1935) **35** pt 1.
— *Andean Culture History* (New York, 1949).
— *Excavations at Wari, Ayacucho* (New Haven, 1953).
— *Ancient Arts of the Andes* (New York, 1954).
Bermejo, Vladimiry: Vida y hechos del Conquistador del Perú, don Francisco Pizarro (Arequipa, 1942).
BINGHAM, HIRAM: *Across South America* (New York, 1911).
— 'The Ruins of Choqquequirau', *American Anthropologist*, n.s., **12** 505–25, 1911.
— 'Vitcos, the Last Inca Capital', *Proc. Amer. Antiq. Soc.*, n.s., **22**, no. 5, 1–64, 1912.
— 'In the Wonderland of Peru', *National Geographic Magazine* **24**, no. 4, 1913.
— 'The Ruins of Espiritu Pampa, Peru', *Amer. Anthropologist*, n.s., **16** 185–199, 1914.
— 'Along the Uncharted Pampaconas', *Harper's Magazine* **129**, August 1914.
— 'The Story of Machu Picchu', *National Geographic Magazine* **26**, February 1915.
— 'Further Explorations in the Land of the Incas', ibid. **27**, May 1916.
— *Inca Land* (Boston, 1922).
— *Macchu Picchu, a Citadel of the Incas* (New Haven, 1930).
— *Lost City of the Incas* (New York, 1948; London, 1951).
BOURNE, EDWARD GAYLORD: *Spain in America, 1450–1580* (New York, 1962).
BOWMAN, ISAIAH: *The Andes of Southern Peru* (London, 1920).
BUSHNELL, G. H. S.: 'Ancient Peoples of the Andes', *Science News* **13** (Harmondsworth, 1949).
— *Peru* (London, 1956).
BUSTO DUTHURBURU, JOSÉ ANTONIO DEL: Trujillo de Extremadura, patria de conquistadores', *Mecurio Peruano* **393**, 1960.
— 'Maldonado el Rico, señor de los Andahuaylas', RH **26** 113–32, 1962–3.
— 'La marcha de Francisco Pizarro de Cajamarca al Cusco', RH **26**, 1962–3.
— *El Conde de Nieva virrey del Perú* (Lima, 1963).
— 'El conquistador Martín Pizarro', *Mecurio Peruano* **432**, 1963.
— *Diego de Almagro* (Lima, 1964).
— 'Ruy Hernández Briceño, el guardián de Atahualpa', *Cuadernos del Seminario de Historia del Instituto Riva Agüero*, Lima, no. 7, 1964.
— 'Una relación y un estudio sobre la Conquista', RH **27** 281–303, 1964.

— 'Los caídos en Vilcaconga', *Historia y Cultura, Organo del Museo Nacional de Historia* **1** 115–25 (Lima, 1965).

— 'Tres conversos en la captura del Inca', *Anales del III Congreso Nacional de Historia del Perú*, Lima, 1965.

— *Francisco Pizarro, el Marqués Gobernador* (Madrid, 1966).

BUSTOS LOSADA, CARLOTA: 'Las hijas de Huainacapac', *Museo Histórico, Organo del Museo de Historia de la Ciudad de Quito*, **3** (1951–2), no. 9, 19–36, nos. 10–11; **4** (1952–3), nos. 12–13, 19–21, nos. 14–15, 42–56; **5–6** (1953–4), no. 16, 50–66, no. 17, 16–28.

CABRERO, MARCOS A.: 'El corregimiento de Saña y el problema histórica de la fundación de Trujillo', RH **1** 151–91, 336–73, 486–514, 1906.

CARRIÓN, BENJAMÍN: *Atahualpa* (Mexico, 1934).

CASTAÑEDA, CARLOS E.: 'The Corregidor in Spanish Colonial Administration', HAHR **9** 446–70, 1929.

CASTELNAU, FRANCIS L. DE (Comte de Laporte): *Expédition dans les parties centrales de l'Amérique du Sud*, Troisième partie: Antiquités des Incas et autres peuples anciens, 15 vols (6 parts, Paris, 1850–9).

CHAMBERLAIN, ROBERT S.: 'The Concept of the Señor Natural as Revealed by Castilian Law and Administrative Documents', HAHR, **19** 130–37, 1939.

COBB, GWENDOLIN BALLANTINE: 'Potosí, a South American Mining Frontier', in *Greater America, Essays in Honor of Herbert Eugene Bolton* (Berkeley and Los Angeles, 1945).

— 'Supply and Transportation for the Potosí Mines, 1545–1640', HAHR, **30**, 1950.

CRESPO RODAS, ALBERTO: 'La "mita" de Potosí', RH **22** 169–82, 1955–6.

CROSBY, ALFRED W.: 'Conquistador y pestelencia: the First New World Pandemic and the Fall of the Great Indian Empires', HAHR **47** 321–38, 1967.

CUESTA, LUISA: 'Una documentación interesante sobre la familia del conquistador del Perú', RI **8**, no. 30, 865–92, 1947.

CÚNEO-VIDAL, RÓMULO: *Vida del conquistador del Perú, don Francisco Pizarro, y de sus hermanos* (Barcelona, 1925).

— *Historia de las guerras de los últimos Incas peruanos* (Barcelona, 1925).

— 'Los hijos americanos de los Pizarros de la Conquista', *Boletín de la Real Academia de la Historia, Madrid*, **87** 78–87, 1925.

— 'El Capitán don Gonzalo Pizarro, padre de Francisco, Hernando, Juan y Gonzalo Pizarro, conquistadores del Perú', ibid. **89**, 1926.

CUNNINGHAME GRAHAM, R. B.: *The Horses of the Conquest* (London, 1930).

DOBYNS, HENRY F.: 'An Outline of Andean Epidemic History to 1720', *Bull. Hist. Medicine*, **37**, no. 3, 493–515 (Baltimore, Md, Nov.–Dec. 1963).

EATON, GEORGE F.: 'The Collection of Osteological Material from Macchu Picchu', *Mem. Connecticut Acad. Arts Sci.* **5**, 1916.

Eguiguren y Escudero, Victor: 'Fundación y traslaciones de San Miguel de Piura', *Boletín de la Sociedad Geográfica de Lima* **4** 260–8, 1895.

Espinosa Soriano, Waldemar: 'El alcalde mayor indígena en el virreinato del Perú', AEA **17** 183–300, 1960.

— 'La guaranga y la reducción de Huancayo: tres documentos inéditos de 1571 para la etnohistoria del Perú', RMN **32** 8–80, 1963.

Fejos, Paul: *Archeological Explorations in the Cordillera Vilcabamba, Southeastern Peru*, Viking Fund Publications in Anthropology, no. 3 (New York, 1944).

Figueroa Marroquín, Horacio: *Enfermedades de los conquistadores* (San Salvador, 1957).

Friede, Juan: 'Las Casas y el movimiento indigenista en España y América en la primer mitad del siglo xvi', RHA **34** 339–411, 1952.

— 'Los Indios y la historia', *América Indígena* **20**, no. 1, 1960.

— *Vida y luchas de Don Juan del Valle, primer obispo de Popayán y protector de indios* (Popayán, 1961).

Gagliano, Joseph A.: 'The Coca Debate in Colonial Peru', *The Americas* **20**, no. 1, 43–63 (Washington, D.C., July 1963).

Gangotena y Jijón, C.: 'La descendencia de Atahualpa', *Boletín de la Academia Nacional de Historia, Quito* **38**, no. 91, 107–24, 1958.

García Soriano, Manuel: *El conquistador español del siglo xvi* (Tucumán, 1954).

Gibson, Charles: *The Inca Concept of Sovereignty and the Spanish Administration in Peru*, University of Texas, Latin American Studies **4** (Austin, 1948).

— *Spain in America* (New York, 1966).

Goldwert, Marvin: 'La lucha por la perpetuidad de las encomiendas en el Perú virreinal (1550–1600)', RH **22–3**, 1955–8.

Gómez Hoyos, Rafael: *Las leyes de Indias y el derecho eclesidistco en la América Española y las Islas Filipinas* (Medellín, Colombia, 1945).

— *La iglesia de América en las Leyes de Indias* (Madrid, 1961).

Hagen, Victor Wolfgang von: *Guide to the Ruins of Sacsahuaman* (New York, 1949).

— 'The Bridge of the Great Speaker', *Geographical Magazine* (London, December 1955).

— *Highway of the Sun* (New York, 1957).

— *The Realm of the Incas* (New York, 1957).

— *Guide to the Ruins of Ollantaytambo* (Lima, 1958).

Hanke, Lewis Ulysses: *The Spanish Struggle for Justice in the Conquest of America* (Philadelphia, 1949).

— 'Pope Paul and the American Indians', *Harvard Theol. Rev.* **30** 65–102, 1950.

— *Bartolomé de las Casas, An interpretation of his life and writings* (The Hague, 1951).

— *Bartolomé de las Casas, Bookman, Scholar and Propagandist* (Philadelphia, 1952).

HANKE, LEWIS ULYSSES: *Bartolomé de las Casas, Historian* (Gainesville, Fla, 1952).

— *The Imperial City of Potosí: an Unwritten Chapter in the History of Spanish America* (The Hague, 1956).

— *Aristotle and the American Indians* (London and Chicago, 1959).

— 'The Other Treasure from the Indies During the Epoch of Emperor Charles V' in P. Rasow and F. Schalk, *Karl V der Kaiser und seine Zeit* (Cologne, 1960).

— 'More Heat and Some Light on the Spanish Struggle for Justice in the Conquest of America', HAHR **44**, no. 3, 293–340, 1964.

HARING, C. H.: *The Spanish Empire in America* (New York, 1947).

HARTH-TERRÉ, EMILIO: 'Fundación de la ciudad incaica', RH **16**, pts 1–2, 98–123, 1943.

— *El indígena peruano en las bellas artes virreinales* (Cuzco, 1960).

HELMER, MARIE: 'La Vie économique au xvie siècle sur le haut-plateau andin', TIFEA **3** 115–50, 1950.

— 'La encomienda à Potosí en 1550 d'après un document inédit', *Proceedings of Congress of Americanists* (Cambridge, England, 1954).

— 'La visitación de los Yndios Chupachos, Inka et encomendero, 1549', TIFEA **5** 3–50, 1955–6.

— 'Notas sobre la encomienda peruana en el siglo xvi', *Revista del Instituto de Historia y Derecho* **10** 124–43 (Buenos Aires, 1959).

— 'Notes sur les esclaves indiens au Pérou (xvie siècle)', *Travaux de l'Institut d'Études Latino-américaines de l'Université de Strasbourg* **5**, 1965.

HELPS, ARTHUR: *The Spanish Conquest in America*, 4 vols (London, 1855–61).

HEREDIA, F. DANIEL: 'Vilcabamba–Apurímac', *Revista del Museo e Instituto Arqueológico, Cuzco*, no. 15, 1953.

HORKHEIMER, HANS: *Guía bibliográfica de los principales sitios arqueológicos del Perú* (Lima, 1950).

JIJÓN Y CAAMAÑO, JACINTO: *La religión del imperio de los Incas* (Quito, 1919).

— *Sebastián de Benalcázar*, 2 vols (Quito, 1936–8).

JIMÉNEZ DE LA ESPADA, MARCOS: *Tres relaciones de antigüedades peruanas* (Madrid, 1879).

— (ed.) *Relaciones geográficas de Indias* (= RGI), 4 vols (Madrid, 1881); 3 vols, BAE (Cont) **183–185**, 1965.

KIRKPATRICK, F. A.: *The Spanish Conquistadores* (London, 1934).

— 'Repartimiento–Encomienda', HAHR **19**, 1939.

— 'The Landless Encomienda', HAHR **22**, 1942.

KONETZKE, RICHARD: *El imperio español, orígenes y fundamentos* (Madrid, 1946).

— 'La esclavitud de los indios como elemento en la estructuración social de Hispanoamérica', *Estudios de historia social* **1** 441–80, 1949.

— 'Estado y sociedad en las Indias', REA **3**, no. 8, 33–58, 1951.

KONETZKE, RICHARD: *Colección de documentos para la historia de la formación social de Hispanoamérica* (= CDFS), 4 vols (Madrid, 1953).

KUBLER, GEORGE: 'A Peruvian Chief of State: Manco Inca (1515–1545)', HAHR **24** 253–76, 1944.

— 'The Behaviour of Atahualpa, 1531–1533', HAHR **25** 413–27, 1945.

— 'The Quechua in the Colonial World', in *Handbook of South American Indians* (Smithsonian Institution, Bureau of American Ethnology: Washington, 1946) bulletin 143, **2** 331–410.

— 'The Neo-Inca State (1537–1572)', HAHR **27**, no. 2, May 1947.

— *Cuzco, Reconstruction of the Town and Restoration of its Monuments* (UNESCO: Paris, 1952).

— 'Machu Picchu', *Perspecta*, Yale University **6** 48–55, 1960.

— *The Art and Architecture of Ancient America* (Pelican History of Art: Harmondsworth, 1962).

LAREDO, SARAH DE: (ed.) *From Panama to Peru: the Conquest of Peru by the Pizarros, the Rebellion of Gonzalo Pizarro, and the Pacification by La Gasca*, catalogue of a collection of documents sold to the Henry E. Huntington Library, Berkeley, California. (= Maggs–Huntington) (Maggs Brothers: London, 1925).

LASTRES, JUAN B.: *Historia de la medicina peruana*, 3 vols: **1** La medicina incaica, **2** La medicina en el virreinato (Lima, 1951).

— *Médicos y cirujanos de Pizarro y Almagro* (Lima, 1958).

LAVANDAIS, M. E. (COMTE DE SARTIGES): 'Voyage dans les républiques de l'Amérique du Sud', *Revue des Deux Mondes*, Paris, **10** 1019–41, June 1851.

LEHMANN-NITSCHE, ROBERT: 'Coricancha, el Templo del Sol en el Cuzco y las imágenes de su altar mayor', *Revista del Museo de La Plata* **31** (La Plata, Argentina, 1928).

LE RIVEREND BRUSONE, JULIO: *Crónicas de la conquista del Perú* (Mexico, 1948).

LEVILLIER, ROBERTO: (ed.) *Audiencia de Charcas, Correspondencia de presidentes y oidores*, 3 vols (Madrid, 1918–22).

— (ed.) *Audiencia de Lima, Correspondencia de presidentes y oidores* (Madrid, 1922).

— *Organización de la iglesia y órdenes religiosas en el virreinato del Perú en el siglo XVI*, 2 vols (Madrid, 1919).

— (ed.) *Gobernantes del Perú, cartas y papeles, siglo XVI; documentos del Archivo de Indias* (= GP), 14 vols (Madrid, 1921–6).

— *Don Francisco de Toledo, supremo organizador del Peru; su vida, su obra (1515–1582)* **1** (Madrid, 1935); **2**, **3** (Buenos Aires, 1940, 1942).

— *Los Incas* (Seville, 1956).

LIPSCHUTZ, ALEJANDRO: 'La despoblación de las Indias después de la conquista', *América Indígena* **26** 229–47, 1966.

LISSÓN CHAVES, EMILIO: (ed.) *La iglesia de España en el Perú, colección de documentos* (= IEP), 5 vols (Seville, 1943–47).

LOCKHART, JAMES: *Spanish Peru, 1532–1560* (Madison, Wisconsin, 1968).

LOHMANN VILLENA, GUILLERMO: 'El Inga Titu Cussi Yupangui y su entrevista con el oidor Matienzo, 1565', *Mercurio Peruano*, año 16, **23**, no. 167, 3–18, 1941.

— *Los Americanos en las órdenes nobiliarias (1529–1900)* (Madrid, 1947).

— 'El señorío de los Marqueses de Santiago de Oropesa en el Perú', *Anuario de Historia del Derecho Español*, Madrid, **19** 347–458, 1948–9.

— *Las minas de Huancavelica en los siglos XVI y XVII* (Seville, 1949).

— *El corregidor de indios en el Perú, bajo los Austrias* (Madrid, 1957).

— *Las relaciones de los virreyes del Perú* (Seville, 1959).

— 'Notes on Prescott's interpretation of the Conquest of Peru', HAHR **39**, no. 1, 46–80, February 1959.

— 'Juan de Matienzo, autor del *Gobierno del Perú* (su personalidad y su obra)', AEA **22** 767–886, 1965.

— 'El testamento inédito del Inca Sayri Tupac', *Historia y Cultura*, Organo del Museo Nacional de Historia, Lima, **1** 13–18, 1965.

LÓPEZ MARTÍNEZ, HECTOR: 'Un motín de mestizos en el Perú (1567)', RI, año 24, nos. 95–6, 367–81, 1964.

LOREDO, RAFAEL: 'Nuevos capitulos de la tercera parte de la crónica del Perú de Pedro Cieza de León', *Mercurio Peruano*, año 21, **27**, no. 233, 1946–.

— *Los repartos; bocetos para la nueva historia del Perú* (Lima, 1958).

LORENTE, SEBASTIÁN: (ed.) *Relaciones de los Virreyes y Audiencias que han gobernado el Perú* **1** (Lima, 1867).

— *Historia antigua del Perú*, 5 vols (Lima, 1860–80).

LOTHROP, SAMUEL K.: *Inca Treasure as Depicted by Spanish Historians* (Frederick Webb Hodge Anniversary Publication Fund **2**: Los Angeles, 1938).

LYNCH, JOHN: *Spain under the Hapsburgs:* **1** *Empire and Absolutism 1516–1598* (Oxford, 1964).

MACKEHENIE, CARLOS A.: 'Apuntes sobre Don Diego de Castro Titu Cusi Yupanqui Inca', RH **3** 371—90, 1909; **5** 5–13, 1913.

MACLEAN Y ESTENÓS, ROBERTO: 'Indios en el virreinato del Perú', *Perú Indígena*, Lima, **3**, nos. 7–8, 136–50, 1952.

— 'La educación en el imperio de los Incas', *Educación*, Lima, **7**, no. 16, 5–75, 1952.

MADARIAGA, SALVADOR DE: *Rise of the Spanish American Empire* (London, 1947).

MARKHAM, SIR CLEMENTS ROBERT: *A History of Peru* (Chicago, 1892).

— *The Incas of Peru* (London and New York, 1910).

MARTÍNEZ CARDÓS, JOSÉ: 'Las Indias y las Cortes de Castilla durante los siglos XVI y XVII', RI **16** 357–76, 1956.

MASON, J. ALDEN: *The Ancient Civilizations of Peru* (Harmondsworth, 1957).

MATEOS, FRANCISCO: 'Los dos concilios limenses de Jerónimo de Loaysa', *Missionalia Hispánica*, Madrid, **4** 479–524, 1947.

— 'Segundo concilio provincial limense, 1567', ibid. **7**, no. 20, 209–92, no. 21, 525–617, 1950.

— 'Constituciones para indios del primer concilio limense, 1552', ibid. **7**, no. 19, 5–54, 1950.

— 'Los Loyola en América', *Razón y Fe* **154**, nos. 702–5, 1956.

MEANS, PHILIP AINSWORTH: 'Indian Legislation in Peru', HAHR **3** 509–34, 1920.

— 'Biblioteca Andina, Part One, the Chroniclers', *Trans. Connecticut Acad. Arts Sci.* **29** 271–525, 1928.

— *Ancient Civilizations of the Andes* (New York, 1931).

— *Fall of the Inca Empire and the Spanish Rule in Peru, 1530–1780* (New York, 1932).

MELLAFE, ROLANDO and SERGIO VILLALOBOS: *Diego de Almagro* (Santiago de Chile, 1954).

MENDIBURU, MANUEL DE: *Diccionario histórico-biográfico del Perú* (Lima, 1874–90), 11 vols (Lima, 1931–4).

MERRIMAN, ROGER BIGELOW: *The Rise of the Spanish Empire in the Old World and the New*, 3 vols (New York, 1918–25).

MÉTRAUX, ALFRED: *Les Incas* (Paris, 1962). Trans. *The Incas* (London, 1965).

MINNAERT, P.: 'La Morale au Pérou', *Bulletin de la Société des Americanistes de Belgique* **16** 25–48, 1935.

MOORE, JOHN P.: *The Cabildo in Peru under the Hapsburgs* (Durham, N.C., 1954).

MOORE, SALLY FALK: *Power and Property in Inca Peru* (New York, 1958).

MORALES, AMBROSIO: 'Documentos para la historia del Cuzco. Las tumbas de los Incas Sairi Tupac', *Revista del Instituto Americano de Arte* (Cuzco, June 1944).

MÖRNER, MAGNUS: *El mestizaje en la historia de Ibero-América* (Mexico, 1961).

— *Race Mixture in Latin America: A History* (Boston, 1967).

MUÑOZ DE SAN PEDRO, MIGUEL: 'Las últimas disposiciones del último Pizarro de la Conquista', *Boletín de la Real Academia de la Historia*, Madrid, **126** 387–425, 1950 and **127** 203–52, 1950.

— *Tres testigos de la conquista del Perú* (Buenos Aires, 1953).

MURO OREJÓN, ANTONIO: *Las Leyes Nuevas de 1542–1543, Ordenanzas para la gobernación de las Indias y buen tratamiento y conservación de los indios* (Seville, 1959).

MURPHY, ROBERT CUSHMAN: 'The Earliest Spanish Advances Southward from Panama along the West Coast of South America', HAHR, February 1941, 3–28.

NAVARRETE, MARTÍN FERNÁNDEZ DE: (ed.) *Colección de documentos inéditos para la historia de España* (= CDHE), 112 vols (Madrid, 1842–95).

NAVARRO, JOSÉ GABRIEL: 'La descendencia de Atahualpa', *Boletín de la Real Academia de la Historia, Madrid,* **97** 817–29, 1930; *Boletín de la Academia Nacional de Historia, Quito,* **56** 216–23, 1940.

NORDENSKIÖLD, BARON NILS ERLAND HERBERT: *Peru under the Incas and After* (London, 1925–).

OTS CAPDEQUÍ, JOSÉ MARÍA: *Instituciones sociales de la América española en el período colonial* (La Plata, 1934).

— *Estudios de la historia del derecho español en las Indias* (Bogotá, 1940).

— *Manual de historia del derecho español en las Indias* (Buenos Aires, 1945).

OUTWATER, J. OGDEN, JR: 'Building the Fortress of Ollantaytambo', *Archaeology* **12**, no. 1, 1959.

PARDO, LUIS A.: *Ruinas precolombinas del Cuzco* (Cuzco, 1937).

PARRY, JOHN HORACE: *The Spanish Theory of Empire in the Sixteenth Century* (Cambridge, 1940).

— *The Sale of Public Offices in the Spanish Indies under the Hapsburgs* (Berkeley and Los Angeles, 1953).

— *The Establishment of the European Hegemony (1415–1715)* (New York, 1961).

— *The Age of Reconnaissance, Discovery, Exploration and Settlement, 1450–1650* (Cleveland, 1963).

— *The Spanish Seaborne Empire* (London, 1966).

PAZ SOLDÁN, MARIANO FELIPE: *Geografía del Perú* (Lima, 1862).

PÉREZ DE TUDELA, JUAN: *Documentos relativos a don Pedro de la Gasca y a Gonzalo Pizarro*, 2 vols (Madrid, 1964).

PETERSON, HAROLD L.: *Arms and Armor in Colonial America, 1526–1783* (Harrisburg, Pa, 1956).

PORRAS BARRENECHEA, RAÚL: 'Hernando Pizarro en Pachacamac', *La Prensa*, Lima, 18 January 1935.

— 'La caída del Imperio Incaico', *Revista de la Universidad Católica del Perú*, no. 13, 142–8, May 1935.

— *Las relaciones primitivas de la Conquista del Perú*, Cuadernos de Historia del Perú **2** (Paris, 1937).

— *El testamento de Pizarro, texto inédito* (Paris, 1936).

— *Una relación inédita de la Conquista del Perú* (Madrid, 1940).

— 'El testamento de Mancio Sierra', RI **1**, no. 1, 63–72, 1940.

— 'El testamento de Pizarro de 1539', RI **2**, no. 3, 39–70, 1941.

— *Pizarro, el Fundador* (Lima, 1941); also RI **3**, no. 7, 5–39, January–March 1942.

— 'Los cronistas de la Conquista del Perú', *Cuadernos de estudio*, Instituto de Investigaciones Históricas de la Universidad Católica, Lima, **1**, no. 3, September, 1941.

— 'Deformación histórica sobre Francisco Pizarro', RI **3**, no. 7, 5–39, 1942.

PORRAS BARRENECHEA, RAÚL: 'Tres cronistas del Incario: Betanzos, Titu Cusi Yupanqui y Santa Cruz Pachacutic', *La Prensa*, Lima, 1 January 1942.

— *Cedulario del Perú* (Ministerio de Relaciones Exteriores, Departamento de Relaciones Culturales, Lima) 2 vols: **1** (1529–1534) 1944; **2** (1534–1538) 1948.

— 'Atahualpa no murió el 29 de agosto de 1533', *La Prensa*, Lima, 31 October 1945.

— 'El pensamiento de Vitoria en el Perú', *Mercurio Peruano*, **27**, no. 234, 465–90, September 1946.

— 'Las conferencias sobre el Conquistador del Perú', *Documenta* **1** 159–74, 1948.

— 'Dos documentos esenciales sobre Francisco Pizarro y la Conquista del Perú: La información sobre el linaje y juventud de Francisco Pizarro hecho en Trujillo de Extremadura en 1529; El acta perdida de la fundación del Cuzco', RH **17** 9–95, 1948.

— 'Jauja, capital mítica', RH **18**, pt 2, 117–48, 1950.

— 'Las primeras crónicas de la Conquista del Perú', *Revista Escorial*, Madrid, no. 56, 1949.

— 'Crónicas perdidas, presuntas y olvidadas sobre la Conquista del Perú', *Documenta* **2** 179–243, 1949–50.

— 'Doña Inés Huaylas ñusta, amante india de Pizarro', *A.B.C.* Madrid, 10 December 1949; *El Comercio*, Lima, 5 April 1953.

— *Fuentes históricas peruanas* (Lima, 1954).

— *Cartas del Perú, Colección de documentos inéditos para la historia del Perú* (= CP) **3** (Lima, Edición de la Sociedad de Bibliófilos Peruanos, 1959).

— 'El testamento de Francisco Pizarro', *Cuadernos Hispanoamericanos*, Madrid, no. 131, 200–84, 1960.

— 'Oro y leyenda del Perú', *Mercurio Peruano* **42**, no. 406, 236–63, February 1961.

— 'Antología de Raúl Porras Barrenechea', *Mercurio Peruano*, **42**, no. 406, February 1961.

— *Los cronistas del Perú* (Lima, 1962).

PRADO, JAVIER: *Estado social del Perú durante la dominación española* (Lima, 1941).

PRESCOTT, WILLIAM HICKLING: *History of the Conquest of Peru*, 2 vols (New York, 1847).

PUIGGROS, R.: *Bartolomé de las Casas* (Mexico, 1953).

QUINTANA, MANUEL JOSÉ DE LA: *Vidas de españoles célebres*, 3 vols (Madrid, 1807–33).

— *La vida de Francisco Pizarro*, BAE **19**, 1861 (London, 1923).

RAIMONDI, ANTONIO: *El Perú*, 6 vols (Lima, 1874–1913).

REGAL, ALBERTO: *Los caminos del Inca en el antiguo Perú* (Lima, 1936).

RIVERA SERNA, PAUL: 'Libro primero de cabildos del Cuzco', *Documenta* **4** 440–80, 1965.

ROMERO, CARLOS A.: 'Inédito sobre el primer Tupac Amaru', RH **2**, 1907.

— 'Informe sobre las ruinas de Choqquequirau', RH **4** 87–103, 1909.

— 'La despoblación del Perú', *Inca* **1**, no. 3, 668–75 (Lima, 1923).

— 'La fundación española del Cuzco', RH **14**, pt 1, 1941.

— 'Un tesoro famoso', RH **16**, pts 1–2, 1943.

ROMERO, EMILIANO: *Historia económica y financiera del Perú: antiguo Perú y virreinato* (Lima, 1937).

ROSENBLAT, ANGEL: *La población indígena de América desde 1492 hasta la actualidad* (Buenos Aires, 1945).

— *La población indígena y el mestizaje en América*, 2 vols (Buenos Aires, 1954): **1** La población indígena 1492–1950; **2** El mestizaje y las castas coloniales.

ROWE, JOHN HOWLAND: 'An introduction to the Archaeology of Cuzco', *Harvard University, Papers of the Peabody Museum of American Archaeology and Ethnology* **27**, no. 2, 1944.

— 'Absolute Chronology in the Andean Area', *American Antiquity* **10** 265–84, 1945.

— 'Inca Culture at the Time of the Spanish Conquest', *Handbook of South American Indians*, Smithsonian Institution, Bureau of American Ethnology, bulletin 143, **2** 183–330, 1946.

— 'Colonial Portraits of Inca Nobles' in *The Civilizations of Ancient America, Selected papers of the XIXth International Congress of Americanists*, ed. Sol Tax (Chicago, 1951), 258–68.

— 'The Incas under Spanish Colonial Institutions', HAHR **37**, no. 2, 155–99, 1957.

— 'The Age Grades of the Inca Census', *Miscellanea Paul Rivet octogenario dicata*, XXIst International Congress of Americanists, Mexico, 1958, Publicaciones del Instituto de Historia, no. 50, **2** 499–522.

SÁENZ, MOISÉS: *Sobre el indio peruano y su incorporación al medio nacional* (Secretaría de Educación Pública, Mexico, 1933).

SALAS, ALBERTO M.: *Las armas de la Conquista* (Buenos Aires, 1950).

SÁNCHEZ BELLA, ISMAEL: 'El gobierno del Perú, 1550–1564', AEA **17** 407–524, 1960.

SANTISTEBAN OCHOA, JULIÁN: 'Fray Vicente Valverde, protector de los indios, y su obra', *Revista de Letras*, Cuzco, año 1, no. 2, 117–82.

SAVOY, GENE: 'Discovery of the ruins of Vilcabamba', *Peruvian Times*, Lima, 14 August 1964.

— 'Vilcabamba, last refuge of the Incas – the Savoy expeditions of 1964–65', *Peruvian Times*, 9 April 1965.

— *Antisuyo – The Search for the Lost Cities of the Amazon* (New York, 1970); as *Vilcabamba – The Last City of the Incas* (London, 1971).

SCHÄFER, ERNST: 'Felipe II, el Consejo de Indias y el Virrey Don Francisco de Toledo', *Investigación y Progreso*, Madrid, July 1931, **5** 103–7.

— *El Consejo Real y Supremo de las Indias*, 2 vols, Seville, 1935.

SPAIN: *Acción de España en el Perú, 1509–1554* (Ministerio del Ejército, Servicio Histórico Militar, Madrid, 1949).

SQUIER, EPHRAIM GEORGE: *Peru, Incidents of Travel and Exploration in the Land of the Incas* (New York, 1877).

TELLO, JULIO C.: *Origen y desarrollo de las civilizaciones prehistóricas andinas* (Lima, 1942).

TEMPLE, ELLA DUNBAR: 'La descendencia de Huayna Cápac', RH **11**, pts 1–2, 1937.

— 'Paullu Inca', RH **11**, pt. 3, 284–323, 1937; **12** 204–45, 1939; **13** 31–7, 1940.

— 'Los caciques Apoalaya', RMN **11**, no. 2, 148–78, 1942.

— 'Los Bustamante Carlos Inca, La familia del autor del Lazarillo de Ciegos Caminantes', *Mercurio Peruano*, año 22, **28**, no. 243, 283–97, June 1947.

— 'Don Carlos Inca', RH **17** 134–77, 1948.

— 'Azarosa existencia de un mestizo de sangre imperial incaica', *Documenta*, año 1, no. 1, 112–56, 1948.

— 'Un linaje incaico durante la dominación española, Los Sahuaraura', RH **18**, pt 1, 45–77, 1949.

— 'Notas sobre el Virrey Toledo y los Incas de Vilcabamba; una carta de Titu Cusi Yupanqui y el testamento inédito de su hijo don Felipe Quispe Titu', *Documenta*, año 2, no. 1, 1949–50.

— 'Los testamentos inéditos de Paullu Inca, don Carlos Inca y don Melchor Carlos Inca', *Documenta*, año 2, no. 1, 630–51, 1949–50.

— 'El testamento inédito de doña Beatriz Clara Coya de Loyola, hija del Inca Sayri Túpac', *Fenix*, **3**, nos. 7–8, 1951–2.

TORRES SALDAMANDO, ENRIQUE: 'Reparto y composición de tierras en el Perú', *Revista Peruana* **3** 28–34, 1879.

— 'Apuntes históricos sobre las encomiendas del Perú', ibid. **3** 99–111, 177–91, 241–56, 329–39, 428–41, 1879; **4** 199–204, 1880.

— 'El marquesado de Pizarro', ibid. **4** 41–6, 1880.

— (ed.) *Libro primero de cabildos de Lima* (= CL), 3 vols (Paris, Lima, 1888–1900).

TRIMBORN, HEINRICH: 'Las clases sociales en el Imperio Incaico', *Revista de la Universidad Católica del Perú*, **3**, 1934.

— 'Quellen zur Kulturgeschichte des präkolumbischen Amerika', *Studien zur Kulturkunde*, Stuttgart, **3**, 1936.

UBIDIA RUBIO, LUIS E.: 'Aporte documental a la captura de Rumiñahui', *Boletín del Archivo Nacional de Historia, Quito*, **2**, nos. 3–4, 5–42, 1951.

UHLE, MAX: 'Fortalezas incaicas', *Revista Chilena de Historia y Geografía*, Santiago de Chile, 1917.

ULLOA, LUIS: 'Visita general de los yndios del Cuzco, año de 1571', RH **3** 332–47, 1908.

URTEAGA, HORACIO H.: (with Carlos A. Romero) *Fundación española del Cuzco y ordenanzas para su gobierno* (Lima, 1926).

— *El imperio incaico* (Lima, 1931).

— *El fin de un imperio* (Lima, 1933).

— 'El virrey Don Francisco de Toledo', *Monografías históricas sobre la ciudad de Lima*, Lima, **2** 250–330, 1935.

VALCÁRCEL, DANIEL: 'Índice de documentos referentes al juicio sobre legítima descendencia del último Inca, Túpac Amaru', *Letras*, Lima, no. 42, 48–110, 1949.

— 'La educación en el Perú autóctono y virreinal', REA **12**, 1956, 305–26, no. 62.

— *Documentos de la Audiencia del Cuzco en el A.G.I.* (Lima, 1957).

— *Historia de la educación incaica* (Lima, 1961).

VALCÁRCEL, LUIS E.: 'Final de Tawantisuyu', RMN **2**, no. 2, 1933.

— 'Sajsawaman redescubierto', RMN **3**, nos. 1–2, 3–36, no. 3, 211–23, 1934; **4**, no. 1, 1–24, no. 2, 161–203, 1935.

— *Ruta cultural del Perú* (Mexico, 1945).

— 'Cuzco Archeology', *Handbook of South American Indians*, Smithsonian Institution, Bureau of American Ethnology, bulletin 143, **2** 177–82, 1946.

— *Historia de la cultura antigua del Perú* (Lima, 1959).

VALDEZ DE LA TORRE, CARLOS: *Evolución de las comunidades de indígenas* (Lima, 1921).

VARGAS, JOSÉ MARÍA: *Fray Domingo de Santo Tomás* (Quito, 1937).

— 'Los hijos de Atahualpa y los padres domenicanos', *Boletín de la Academia Nacional de la Historia, Quito*, **15** 59–64, 1937.

VARGAS UGARTE, RUBÉN: *Manuscritos peruanos* (= MP), 5 vols (1935–47).

— *Historia del Perú, fuentes* (Lima, 1939).

— *Historia del Perú, Virreinato (1551–1600)* (Lima, 1949).

— (ed.): *Pareceres jurídicos en asuntos de Indias* (Lima, 1951).

— *Concilios limenses, 1551–1772*, 3 vols (Lima, 1951–4).

— *Historia de la iglesia en el Perú*: **1** *1511–1568* (Lima, 1953).

— 'Dos cartas inéditas de D. Francisco Pizarro y de D. Fray Vicente de Valverde', RHA 47, 152–62, June 1959.

— *Historia general del Perú*: **1**, *El descubrimiento y la conquista, 1524–1550* (Lima, 1966).

VELLARD, JEHAN: 'Causas biológicas de la desaparición de los indios americanos', *Boletín del Instituto Riva-Agüero*, Universidad Católica del Perú, no. 2, 1956.

VILLALOBOS R., SERGIO: 'Almagro y los Incas', *Revista Chilena de Historia y Geografía*, no. 130, 38–46, 1962.

WHITTAKER, ARTHUR PRESTON: *The Huancavelica Mercury Mine* (Cambridge, Mass., 1941).

WIEDNER, DONALD L.: 'Forced Labor in Colonial Peru', *The Americas*, Washington, D.C., **16**, no. 4, 357–83, April 1960.

WIENER, CHARLES: *Pérou et Bolivie: récit de voyage* (Paris, 1880).

ZÁVALA, SILVIO ARTURO: *La encomienda indiana* (Madrid, 1935).

— *Las instituciones jurídicas en la Conquista de América* (Madrid, 1935).

— *De la encomienda y propriedad territorial en algunas regiones de la América española* (Mexico, 1940).

— *New Viewpoints on the Spanish Colonization of America* (Philadelphia and London, 1943).

ZIMMERMANN, ARTHUR F.: *Francisco de Toledo, Fifth Viceroy of Peru, 1569–1581* (Caldwell, Idaho, 1938).

ZUIDEMA, R. T.: 'Observaciones sobre el Taqui Onqoy', *Historia y Cultura*, Lima, **1** 137–40, 1965.

ZURKALOWSKI, ERICH: 'El establecimiento de las encomiendas en el Perú y sus antecedentes', RH **6** 254–69, 1919.

NOTES AND REFERENCES

In these references, those works that are listed in the bibliography have been cited as briefly as possible. Unless otherwise indicated, page references are to modern editions; and where there are several editions, the reference is generally to that marked in the bibliography with an asterisk.

CHAPTER 1 CAJAMARCA

24 *10 March 1526.* The text of the contract of 10 March 1526 is in CDH Chile **4** 1. There have been doubts about this document, but it is generally accepted to be genuine. In it the partners agreed 'to have equal shares: all three will in a fraternal spirit discover, win, conquer and occupy, without any one having any advantage over any other'. Vargas Ugarte, *Historia general del Perú* **1** 5–7.

25 *scarlet and white.'* The raft was described in a five-page manuscript, Codex cxx in the Imperial Library, Vienna. The manuscript was forwarded to an Austrian prince by Charles I's secretary Juan de Sámano, but the author is thought to have been Francisco de Xerez, who later became Pizarro's secretary (Porras Barrenechea, *Cronistas del Perú* 54).

Throughout this book, I have preferred to translate passages from the original Spanish rather than to accept other English versions when these do sometimes exist. When translating, it is of course impossible to reproduce the quaint and charming spelling of much sixteenth-century Spanish. I have therefore always translated into reasonably fluent modern English, breaking down tortuous sentences and completely revising the original punctuation. I have naturally preserved the original words and meaning, and have adopted the most familiar Spanish transcriptions of native words rather than more correct phonetic spellings. When using published English translations, I have sometimes made small changes to make the texts more easily readable – and conform with the style used elsewhere.

26 *Bishop of Tumbez.* The text of the Capitulación has been published frequently, including that in CL **2** 136 ff.

28 *people also died.'* Murúa **1** 103–4. Many other chroniclers also described the epidemic, including: Cieza de León, *Crónica*, pt. 1, bk. 12, ch. 16; Cabello de Balboa **2** 128; Hernando Pablos, *Relación de Cuenca*, RGI **3** 158; Lizárraga 85; Sarmiento de Gamboa, Hakl Soc, 2 ser., **22** 169, 1907; Santa Cruz Pachacuti in Jiménez de la Espada, *Tres relaciones* 300–7; Fray Antonio 26; Poma de Ayala 85–6; Cobo, bk. 12, ch. 16.

Among modern authors, Lastres favoured malaria as the fatal epidemic, but almost all others were sure that it was smallpox. Lastres (1951) **2** 75; José Toribio Polo, 'Apuntes sobre las epidemias en el Perú', RH **5**, 1913; E. Wagner Stearn, *The Effect of Smallpox on the Destiny of the Amerindian* (Boston 1945); P. M. Ashburn, *The Ranks of Death, a Medical History of the Conquest of America* (New York 1947); Dobyns; Crosby.

Smallpox first reached the Caribbean in 1519 and devastated its native populations. The disease has an incubation period of only twelve days and is transmitted from man to man, not carried by an insect or other parasite. It must therefore have travelled south overland from one Colombian tribe to the next; it cannot have been carried in Pizarro's ships which were at sea for longer than the incubation period, and on which none of the men suffered from the disease.

Various chroniclers named Ninan Cuyuchi as Huayna-Capac's heir, including Cabello de Balboa, Santa Cruz Pachacuti, Fray Antonio, Sarmiento de Gamboa, and Garcilaso de la Vega (**135** 351–2).

29 *remarkable rulers.* Charles Gibson (1948) made the point that struggles for the succession were not new. Chroniclers who said that Huayna-Capac did divide the empire included Gonzalo Fernández de Oviedo y Valdés, Cieza de León, Garcilaso de la Vega, Bernabé Cobo and Fray Antonio. Some chroniclers with strong Quitan sympathies said that Huascar was less legitimate: Santa Cruz Pachacuti, Cabello de Balboa. But most chroniclers regarded Huascar as legitimate and Atahualpa as an usurper, and this attitude has been adopted by modern historians up to Luis E. Valcárcel (1933) and Kubler (1945) 415.

30 *they paid tribute.'* This quotation (Sinclair 27) is from *La Conquista del Perú* published in Seville in April 1534. It was the first published account of the fantastic discoveries and was, for its day, a best-seller, being rapidly translated into other languages. It was regarded as anonymous by Prescott and by two American editors, Sinclair and Pogo, who published texts in 1929 and 1930. In 1935, Raúl Porras Barrenechea showed that the account was almost certainly by Captain Cristóbal de Mena. The Governors of Panama wrote that Mena had arrived there in August 1533 with a narrative account 'that he is going to give to his Majesty', and Mena reached Seville in the first ship containing Peruvian conquerors in November 1533, just in time to prepare his work for publication the following spring. I accept the attribution of this anonymous chronicle to Mena.

In the first months of the conquest, the conquerors thought that 'Cuzco' was the title of the Inca ruler, and they referred to Huayna-Capac and Huascar by this name. It may be that the natives were using this term to describe the faction of the civil war identified with Cuzco – and also to avoid offending Atahualpa by referring to Huascar as Inca.

30 *in his presence.'* Mena (Sinclair) 27.

30 *days among us.'* Ibid.; Estete, *Noticia* 23. This chronicle, also called, *El descubrimiento y conquista del Perú*, is one of the most vivid and charming of the eyewitness accounts of the Conquest. The manuscript in the Archivo General de Indias, Seville, ends abruptly while describing events at Jauja in 1534, but was evidently written in about 1539. Prescott used it extensively and considered it anonymous. The Ecuadorean historian C. M. Larrea showed that it was almost certainly by the Veedor Miguel de Estete, who also wrote a report on the journey to Pachacamac with Hernando Pizarro which was inserted in Xerez's *Verdadera relación*. Although the two accounts differ slightly in some details, they both describe the suspension bridge over the Santa river as looking like a cart with high sideboards; and both authors were on the twenty-man Pachacamac expedition, which was unlikely to have included another person with Estete's education and literary ability. The manuscript also has 'Estete' written at the top of its first page.

30 *fortresses lay ahead.* Mena and Trujillo took this suspicious view of the presents. The more sophisticated writers, Xerez and Estete, said that such presents were common between Peruvian chiefs: the ducks, which were stuffed with wool, were supposed to be ground up to make aromatic powder, while the vessels were for ceremonial drinking.

31 *whatever he did.'* Pedro Pizarro (1844) 223. Pedro Pizarro was a young cousin of the Governor. He was almost too young to fight during the early conquest, but observed everything with adolescent wonder. Years later, in 1571, when he was a highly-respected citizen of Arequipa, he sat down to record his many experiences during the conquest and settlement of Peru. Age had not dimmed his remarkable memory for people and entertaining details.

31 *Venetian glass.* Mena, Xerez, Trujillo. Diego de Trujillo also wrote his experiences as an old man in 1571 and added a few details, such as the fact that the goblet was of Venetian glass.

31 *nor foot-soldiers.'* Hernando Pizarro, Letter to Oidores 85. This letter of 23 November 1533 to the Oidores of the Audiencia Real of Santo Domingo by Francisco Pizarro's younger brother, who was virtual second-in-command of the expedition, was a brief but efficient account of the events up to Hernando's own departure from Cajamarca in mid-1533. It is the most important writing by this man, and ironically, it has survived only as part of the history of Hernando Pizarro's great enemy Gonzalo Fernández de Oviedo, who was one of the Oidores to whom the letter was addressed.

It forms chapter 15 of part 3, book 8 (also confusingly numbered as book 46) of his *Historia general de las Indias*.

32 *town of Cajamarca.'* Ruiz de Arce 359; Estete, *Noticia* 25. Juan Ruiz de Arce was a tough soldier who was one of the first conquistadores to return to Spain and settle with his wealth from the Indies. He was not particularly literate, but wrote a vigorous account of his experiences for the benefit of his heirs, during the 1540s.

33 *will remain alive!"'* Trujillo 54.

33 *I saw fit.'* Hernando Pizarro, Letter to Oidores 85.

33 *with his women:* Ruiz de Arce 360.

33 *so many things.'* Estete, *Noticia* 26.

34 *accustomed to sit,'* Ibid 27.

34 *according to his rank.'* Hernando Pizarro, Letter to Oidores 86.

34 *his entire forehead.'* There are innumerable descriptions of the royal tassel. This one is from Pedro Pizarro (1844) 249.

34 *sign of appreciation.'* Mena 236.

34 *and one horse."'* Xerez 331.

34 *men we were.'* Hernando Pizarro, Letter to Oidores 85.

35 *much of us.'* Ibid. Xerez's and Pizarro's accounts of this conversation are remarkably similar. The savage Indians were presumably the Chachapoyas, who lay some four days' march across the Marañon to the east.

35 *had shown fear.'* Estete, *Noticia* 28. This is a favourite anecdote, repeated by almost every chronicler. Most of the eyewitnesses said that they saw the bodies of the dead soldiers next day, and were told why they had been killed.

36 *star-studded sky.'* Ibid. 28–9.

36 *all were knights.'* Mena 238. The Spanish soldier evoked a scene strangely similar to Shakespeare's description of the eve of Agincourt, written sixty years later.

36 *over eighty thousand.'* Ibid.

36 *spirit they had.'* Pedro Pizarro (1844) 226.

36 *citizens of Panama.* There is a curious tradition that Pizarro was an impoverished and broken old man at the time of the Conquest. It is also said that he was a foundling brought up among pigs. The legend was started by Francisco López de Gómara, a personal enemy of the Pizarros, and was repeated by Prescott and other modern historians. It is entirely disproved by various documents that have come to light in recent years. The most important of these is an investigation commissioned by the Emperor Charles V in August 1529 (before the Conquest of Peru) to see whether Francisco Pizarro was eligible for the habit of a Knight of Santiago. A friar interviewed many citizens of all social classes in Pizarro's birthplace, Trujillo. He established that Pizarro's father was Gonzalo Pizarro 'El Largo', a distinguished captain who fought many campaigns against the Moors and in Navarre. His mother was Francisca Gonzales y Mateos, who later married the father of Francisco Martín de Alcántara, who was Pizarro's companion throughout the Conquest and died at his side. Although he was illegitimate and poorly educated, there was nothing otherwise discreditable about Pizarro's upbringing. He made his name in the Indies, rising to become lieutenant-governor and captain, alderman and mayor of Panama. He possessed large encomiendas and a fortune of 30,000 castellanos deposited with Nicolás de Ribera when he sailed for Peru. Xerez, Estete, Pedro Pizarro, Zárate, Garcilaso and Oviedo y Valdés all said how rich he was. Porras Barrenechea has done much to explode the myths in: 'El testamento de Pizarro de 1539' (1941); 'Deformación histórica sobre Francisco Pizarro' (1942); 'Dos documentos esenciales sobre Francisco Pizarro' – this contains the Información of 1529 – (1948); 'Las conferencias sobre el Conquistador del Perú' (1948).

37 *were so few.'* Trujillo 58.

37 *for that purpose'.* Estete, *Noticia* 32.

37 *on their horses.'* Xerez 332.

38 *that he would'.* Ibid.

38 *pieces of artillery'.* Mena 240.

38 *friend and brother."'* Xerez 331; also Mena 238; Hernando Pizarro, Letter to Oidores 86.

38 *in his hand'.* Estete, *Noticia* 29.

38 *from the camp'.* Xerez 332.

38 *of pure terror.'* Pedro Pizarro (1844) 227.

38 *their ceremonial clothes.'* Mena 238.

38 *swept the roadway.'* Xerez 332.

39 *who heard it.'* Estete, *Noticia* 29–30. Almost all the eyewitnesses mentioned the advance guard wearing chequered tunics and clearing the roadway; others also mentioned the brilliant polished discs and the chanting.

39 *the Governor was.'* Xerez 332.

39 *sign of fear.*' Mena 240. The three earliest eyewitness accounts, Mena, Xerez and Hernando Pizarro, all mentioned this delay outside Cajamarca and the sending of the Spaniard. Diego de Trujillo revealed that the Spaniard was Hernando de Aldana.

39 *underneath their tunics*'. Hernando Pizarro, Letter to Oidores 86. Other chroniclers, Mena, Xerez, Estete, Diego de Silva y Guzman and Pedro Cataño, stressed the weapons under the tunics as though these constituted some breach of faith. The author of the anonymous *Nouvelles certaines des Isles du Peru*, a slim book that appeared in French in 1534, claimed that Atahualpa's men carried battle-axes and silver halberds, and that 'thick clubs hung at their belts . . . and a great many men were carrying javelins, bows and arrows' (*Nouvelles certaines* 6).

39 *his body exposed.*' Estete, *Noticia* 30.

39 *contained the artillery*' Xerez 332.

40 *on a lance*'. Hernando Pizarro, Letter to Oidores 86. Mena also mentioned the native officer with the long lance. He concluded that the raising of the lance was a signal for the advance of men outside the square who were carrying weapons for those in it. This banner was evidently the Inca's royal standard, 'a small, square pennant, ten or twelve palmos in circumference, made of a cotton or wool canvas. It was placed at the end of a long lance, stretched out and stiff so that it did not wave in the breeze. Each king painted his arms and devices on it. They each chose different arms, although the normal symbols of the Inca lineage were the rainbow and two extended snakes.' Cobo, *Historia del Nuevo Mundo*, bk. 12, ch. 36.

40 *the interpreter Martin*', Estete, *Noticia* 30–1.

40 *where Atahualpa was.*' Xerez 332.

40 *from his men*'. Ruiz de Arce 362.

41 *resorting to bloodshed.* It seems clear that Valverde did not attempt to read the exact wording of the Requirement proclamation, as Vargas Ugarte claimed he did in his *Historia de la iglesia en el Perú* **1** 136. Only Pedro Pizarro (1844) 228, who wrote long after the event, said that 'he required of him in the name of God and the King that he should submit to the law of Our Lord Jesus Christ and the service of His Majesty'. Hernando Pizarro, Pedro Cataño and Diego de Trujillo wrote that Valverde mentioned that he had been sent by the Emperor. But all the other eyewitnesses – Mena, Xerez, Estete, Ruiz de Arce, Silva y Guzman, the author of *Nouvelles certaines des Isles du Peru* – simply said that he discussed the Christian religion with no mention of the Spanish crown.

41 *and did so.*' Xerez 332.

41 *a deep crimson.*' Estete, *Noticia* 31.

41 *to the priest.*' Mena 240.

41 *to the ground!*' Ibid.

41 *I absolve you!*' Estete, *Noticia* 31.

41 *become a Lucifer!*' Trujillo 58.

41 *on the ground!*' Murúa 176.

41 *calling on God*'. Ruiz de Arce 363. On the strength of Ruiz de Arce's version, Vargas Ugarte tried to absolve Valverde of instigating the attack, and argued that Prescott and others had treated the friar too harshly (*Historia de la iglesia en el Perú*, bk. 2, ch. 4). But Hernando Pizarro, Diego de Silva y Guzman, the author of *Nouvelles certaines des Isles du Peru* and Pedro Cataño all wrote that Valverde encouraged the attack, largely because the hour was too late for further delay. Francisco de Xerez and Pedro Pizarro simply recorded that Valverde told Pizarro what had happened. In any case, the argument in academic: the Spaniards would certainly have attacked in the circumstances, whatever Atahualpa had done or Valverde said.

41 *not fire more.*' Mena 242. Virtually every eyewitness agreed that the firing of the cannon was the prearranged signal for the attack.

42 *began to kill.*' Pedro Pizarro (1844) 227, 229.

42 *suffocated one another.*' Ruiz de Arce 363.

42 *home the attack.*' Xerez 333.

42 *to any Christian.*' Hernando Pizarro, Letter to Oidores 86.

42 *litters and hammocks.*' Xerez 333.

42 *were all killed.*' Mena 244.

42 *to his lodging.*' Pedro Pizarro (1844) 229–30.

42 *died around him.*' Hernando Pizarro, Letter to Oidores 86. Francisco Pizarro's wound cannot have been serious. It was mentioned only by his brother Hernando, his cousin Pedro, his secretary Francisco de Xerez and by Diego de Silva y Guzman, who wrote an epic poem in his honour. Pedro Pizarro often referred to Francisco Pizarro as the 'Marquis', but since this title was not conferred on him until some

years later, I have translated it as 'Governor' to avoid confusion.

42 *to make war.*' Xerez 333.

43 *were no buildings.*' Estete, *Noticia* 32.

43 *fell on this.*' Mena 244. Mena, who wrote shortly after the event, said that fifteen feet of wall were knocked down. Ruiz de Arce, writing ten years later, said ninety feet. And Pedro Pizarro, writing in 1571, remembered it as 'over two thousand feet'.

43 *against a Spaniard.*' Xerez 333.

43 *made a move.*' Cataño 282. Pedro Cataño said that he himself, his leader Hernando de Soto and some twenty-two other horsemen rode out again and continued the slaughter until midnight.

43 *filled with men.*' Estete, *Noticia* 32.

43 *on the victory.*' Ruiz de Arce 363.

43 *and other wounds.*' Mena 244.

43 *in that battle.*' Ruiz de Arce 363.

43 *with such might.*' Xerez 333.

43 *terrible two hours.* Only three eyewitnesses attempted an estimate of the total of native dead. As usual, their numbers tended to increase with time. Mena was first with 6,000 to 7,000; then Ruiz de Arce with 7,000; and lastly Diego de Trujillo, writing in 1571, with 8,000. Francisco de Xerez (333) said that 'over two thousand dead remained in the square without counting the wounded'.

44 *Governor himself slept.*' Xerez 333.

44 *to kill him.*' Ruiz de Arce 363.

44 *captured by us.*' Mena 245–6.

44 *but not afterwards.*' Trujillo 59.

44 *told him everything.*' Mena 246.

45 *had deceived him.*' Hernando Pizarro, Letter to Oidores 87.

45 *was so pensive.*' Mena 246.

45 *guarding his women.*' Estete, *Noticia* 33.

45 *which is great.*' Mena 244.

CHAPTER 2 ATAHUALPA CAPTIVE

46 *ever been missed.*' Hernando Pizarro, Letter to Oidores 87.

46 *sign of the cross.* Only Cristóbal de Mena mentioned this fascinating signal (Mena 246–8).

47 *act of cruelty.*' Xerez 334.

47 *also ordered this.*' Hernando Pizarro, Letter to Oidores 87.

47 *Indians remained there.*' Cataño 283.

47 *from the square.*' Xerez 334.

47 *great quantity more.*' Ibid. Licenciate Gaspar de Espinosa wrote to the Emperor Charles V from Panama, 21 July 1533, that they took 50,000 pesos of fine gold and over 20,000 marks of silver (CP 59).

47 *and their Emperor.*' Mena 248.

48 *within two months.*' Xerez 335. Pizarro's secretary was very precise about figures, particularly when money was involved. Other writers agreed roughly with his measurements of 22 and 17 feet. Mena said that the room measured 25 by 15 feet; Hernando Pizarro said 30 to 35 by 17 to 18 feet; Ruiz de Arce said 20 by 15 feet; the author of *Nouvelles certaines des Isles de Peru* wrote 20 by 18 feet. Most agreed on the height to be filled.

48 *a formal pledge.*' Pedro Pizarro (1844) 230.

48 *Atahualpa's prison chamber.* Atahualpa's chambers and the ransom room were owned, during the years after the Conquest by a curaca called Carguaguatay and by his son Don Pedro Angasnapon. They consisted of seven large and small thatched buildings. They were confirmed in the possession of Don Pedro's heirs in 1667, but later came into the possession of the Bethlehemite order, which used the site for a hospital. Rubén Vargas Ugarte found documents about them in the archive of the extinct Bethlehemite Order (*Historia general del Perú* **1** 56–7). The buildings were later destroyed to make way for a prison and a town hall.

Various writers visited the famous chamber. Montesinos measured it in the seventeenth century and gave the relatively small measurements of 7⅓ by 5¼ varas. Vázquez de Espinosa said the room was 40 feet long, and José de la Rosa gave 12 by 8 varas, or 35 by 23 feet. Murúa said that Atahualpa's *prison* 'is standing to this day' (1610, **1**, ch. 59, 176). The surviving room was probably the Inca's prison rather than the ransom storehouse. Vázquez de Espinosa 401; José de la Rosa, *Descripción de América*, a manuscript of 1789 in AGI, 112.

48 *would kill him.*' Xerez (335), Pedro Pizarro (1844, 238) and Trujillo (59) all agreed that the Inca feared immediate death.

48 *did no treason'* Mena 250.
48 *had left him.'* Trujillo 59; also Murúa, ch. 59, **1** 176.
49 *showed no pleasure.'* Xerez 336. Pedro Pizarro gave a similar description.
49 *land is calm.'* Licenciate Espinosa to King, Panama, 1 August 1533, CDIA **42** 70, 1884.
49 *they damaged it.'* Pedro Pizarro (1844) 248.
50 *excess involving them.'* Inés Yupanqui's testimony to Probanza on behalf of Don Diego Illaquita, Lima, 28 April 1555, in C. Gangotena y Jijón, 'La descendencia de Atahualpa' 120.
50 *bite the natives.'* Pedro Pizarro (1844) 249–50.
50 *to touch it.'* Ibid. 250–1.
50 *ferocity or authority.'* Ibid. 251.
51 *avoid being bewitched.'* Ruiz de Arce 361. There was another instance of Atahualpa's superstition. He once observed a comet and told the Spaniards that this meant he would soon die (Cieza de León, *Crónica* **1**, ch. 65).
51 *all accorded him.'* Xerez 336.
51 *that befell him.'* Estete, *Noticia* 34.
51 *sign to leave.'* Pedro Pizarro (1844) 247–8.
51 *government and administration.* The surprisingly late date for the start of Inca expansion was established by John Howland Rowe in his 'Absolute Chronology in the Andean Area'. He made an analysis of all the chroniclers who described pre-Conquest Inca history from the natives' oral records. Rowe asked the test question: 'Had the Incas conquered territory more than fifty miles from Cuzco at the beginning of Pachacuti's [the ninth Inca's] reign?' He found that thirteen chroniclers, including *all* chroniclers who published their works in the sixteenth century, gave the answer 'no'. Only eight chroniclers, all of the seventeenth century, said that Inca conquests had been in progress long before that reign. The only important chronicler who maintained that the Incas had long been conquerors was Garcilaso de la Vega, whose mother was an Inca princess. Rowe concluded: 'I agree as to Garcilaso's greatness, and have myself used him as much as anyone, but the weight of evidence is so heavily against his history of the Incas that I feel we must reject it unconditionally.' The short duration of Inca rule has been confirmed by archaeology, where Inca artefacts appear only in the uppermost strata.
54 *Inca's presence.* Pedro Pizarro (1844, 232) claimed that Atahualpa tested Pizarro's reaction to the news of Huascar's death *before* sending orders for the killing. Mena (250) and Estete (*Noticia* 35) both agreed that Atahualpa gave the order. Xerez appeared to accept the Inca's excuse. Murúa said that Atahualpa ordered Huascar's execution, and also that of his brothers Titu Atauchi, Tupac Atau and Huanca Auqui, his mother Rahua Ocllo, wife Chiqui Huipa and high priest Chalco Yupanqui (ch. 60, **1** 180).
54 *brother's [Huascar's] faction.'* Estete, *Noticia* 9.
54 *which he drank.'* Mena 250–2; also Estete, *Noticia* 36; Cieza de León, *Crónica*, pt. **2**, ch. 72.
54 *on the road.* Pedro Pizarro (1844) 233.
54 *sleep elsewhere'* Mena (Sinclair) 41.
54 *his other brothers.'* Pedro Pizarro (1844) 233.
55 *of this country.'* Ibid. 236–7.
55 *increased still further.* The story of the match was told in the grant of an encomienda by Pedro de la Gasca to Alonso Díaz (Vargas Ugarte, *Historia general del Perú* **1** 56).
55 *satiated with it!"'* Licenciate Espinosa to the Emperor, Panama, 21 July 1533, CP 60. The first caravel had sailed north from Peru with news of the capture and ransom, reaching Nicaragua and then Panama in late April 1533.
56 *reach this conclusion.* Pedro Pizarro (1844) 242; Xerez 337; Hernando Pizarro, Letter to Oidores 88; Estete, *Noticia* 36; Mena 252. Most early chroniclers referred to Pachacamac as a 'mosque' and Hernando Pizarro called its priest 'bishop'. The first Augustinian missionaries reported that Atahualpa had been similarly disillusioned by the idol of Catequil, at Porcon near Huamachuco. He sent a captain to smash the idol, set fire to its hill, and bring its gold to Cajamarca for his ransom (*Relación de la religión* 24–6).
56 *coming against him.'* Hernando Pizarro, Letter to Oidores 87.
56 *not dare approach.'* Mena 252.
56 *foot-soldiers to investigate.* Hernando Pizarro, Xerez and Estete agree on the figure

of twenty horse; Trujillo says seventeen, but three men were sent back to Cajamarca from Huamachuco. Estete wrote one report of the trip which was included in Xerez's account published in 1534. He is also assumed to be the author of a report known as the *Noticia del Perú*, written in 1539 or 1540. In the latter he wrote that Hernando Pizarro had fifteen horsemen and ten arquebusiers.

57 *amount of gold.*' Mena 254.

57 *with 55,000 men*'. Estete in Xerez, 338.

61 *to their ankles.* The names of these simple garments were: breechclout, huara; men's tunic, uncu; cloak, vacolla; women's tunic, anacu; cloak, lliclla; decorative pin, tupu; sandals, usuta. In describing Inca customs I have been as brief as possible. There are many admirable books on the Incas. The best in English are probably John Rowe's 'Inca Culture at the Time of the Spanish Conquest', or Louis Baudin's *Daily Life in Peru.* G. H. S. Bushnell and J. Alden Mason have written fine general surveys of all pre-Columbian Peruvian civilisations which are easily obtainable. The best chronicle description of Inca life was that of Bernabé Cobo, a seventeenth-century Jesuit historian. Cobo instinctively adopted the methods of modern anthropologists, interrogating and observing with admirable objectivity. He classified each aspect of Inca life and society, and described all details of their living conditions. His only weakness was to be born too late, too long after the Conquest.

61 *or recording categories.* A quipu consisted of a cluster of strings, some containing coloured threads, which were knotted in different ways as a mnemonic device to record numbers. Many quipus have survived, in varying states of preservation and complexity – some have subsidiary strings attached to the main ones, or special methods of twisting, or complex colour-schemes and knot-arrangements. Leland L. Locke showed that some quipus had different strings for progressive decimal units (*The Ancient Quipu, a Peruvian Knot-Record*, New York, 1923). Erland Nordenskiöld showed that the absence of a knot indicated the concept of zero ('Calculations with Years and Months in the Peruvian Quipus', and 'The Secret of the Peruvian Quipus', *Comparative Ethnographical Studies*, Göteborg Museum, **6**, pts. 1 and 2, 1925. Henry Wassén continued this line of study in vol. **9** of that periodical, 1931. The best summary of investigations into the quipu is Carlos Radicati di Primeglio, 'Introducción al estudio de los quipus', *Documenta*, Lima, **2**, no. 1, 244–339, 1949–50.

62 *after the Conquest.* Murúa's manuscript has recently been rediscovered after a series of extraordinary adventures. It was seen in the Colegio Menor at Cuenca in 1785 by Juan Bautista Muñoz, a remarkable eighteenth-century Peruvianist who formed a collection of documents that was the chief source of William Prescott's classic *Conquest of Peru.* The manuscript apparently passed into the library of the King of Spain. It was found there by Napoleon's brother Joseph Bonaparte, and he removed it, along with other documents and paintings, when he fled Madrid before the army of the Duke of Wellington. Bonaparte had Murúa's manuscript in his private carriage when he rode off towards France after being defeated at the battle of Vitoria, 21 June 1813. He was overtaken by British dragoons, and only escaped by leaping from his carriage on to a fresh horse. Wellington took the captured booty back to England. He tried to return it to the Spanish King and wrote various letters to his brother Sir Henry Wellesley, British Ambassador in Madrid, and to the Spanish Ambassador in London. But the replies were unhelpful: the Spanish Ambassador wrote flowery answers suggesting that this loot should be the noble Duke's spoils of war. So Wellington retained the collection. It was not long since a Scottish historian, Professor Robertson, had produced an early attempt at a history of the conquest of Peru (William Robertson, *The History of America*, 2 vols, London, 1777). The Duke of Wellington wrote to Sir Walter Scott in 1824, suggesting that Robertson would have found Murúa's manuscript invaluable. He sent Scott the manuscript; but Scott, daunted by the pages of early Spanish writing, returned it unread after a few weeks. It languished in the library of Apsley House and was rediscovered by the present Duke of Wellington only in 1945. This 'Wellington' manuscript of Murúa's *Historia general del Perú, Origen y descendencia de los Incas* was edited and published

by Manuel Ballesteros-Gaibrois in two volumes, Madrid, 1962, 1964. The second part of Murúa's work was known by a poor copy, the 'Loyola' copy, and was published on various occasions, notably by Horacio Urteaga and by Constantino Bayle. The author's name was then thought to be Morúa, and few of the important details of sixteenth-century history were included in that second part.

63 *dirty and ugly.'* Estete, *Noticia* 39.

63 *had been living.'* Hernando Pizarro, Letter to Oidores 89. Hernando Pizarro also boasted about this episode in his *Confesión* of 15 May 1540, 406–7.

63 *for its treasure.* Generations of subsequent treasure-hunters have burrowed into the adobe mound, causing it to crumble into pits and sandslides. The first modern excavation of Pachacamac was by Max Uhle in 1903. It revealed an ancient temple of rough stone that had later been covered with débris and used as part of the foundation of the existing step pyramid. Graves found at the base of the older temple appeared on stylistic grounds to be of the Tiahuanaco horizon of about the eleventh century. Later graves produced a debased style that Uhle called 'Epigonal', typical of the central coast in the pre-Inca period, and the sequences end, as do virtually all in Peru, with Inca ware. Uhle's report was translated by C. Grosse and published by the University of Pennsylvania Department of Archaeology. Other archaeological work at Pachacamac is described in Jorge C. Muelle and J. Robert Wells, 'Las pinturas del templo de Pachacamac', *Revista del Museo Nacional, Lima*, **8**, no. 2, 1939; Julio C. Tello, 'Descubrimientos realizados en las ruinas de Pachacamac', *El Comercio*, Lima, 18 July 1940, and his *Memoria sucinta sobre los trabajos arqueológicos realizados en las ruinas de Pachacamac durante los años 1940 y 1941*, Lima, 1943; W. D. Strong and J. M. Corbett, 'A ceramic sequence at Pachacamac', *Archaeological Studies in Peru, 1941–2*, Columbia Studies in Archaeology and Ethnology **1** (New York) 1943.

63 *be no fraud.'* Xerez 336–7. A peso de oro was not a coin but a weight of gold roughly a sixth of an ounce or 4·18 grammes. A thousand pesos de oro was thus about ten pounds or 4·2 kilos. At the time of the Conquest, 450 maravedís made one peso de oro, and 100 pesos de oro made one libra de oro. Later, a peso de oro became known as a castellano, worth 490 maravedís. It is extremely difficult to place a value on a peso de oro in terms of modern money, because of the changes in the relative values of different commodities and services. Prices in the early days of the Conquest of Peru were naturally very erratic because of the scarcity of European goods and the relative abundance of precious metals. Markham (writing before the First World War) calculated that a peso de oro was worth about £2 12*s* 6*d* in the Indies and £5 to £6 in Spain in the early sixteenth century.

64 *in Pizarro's army.* Cieza de León named these three. Martín Bueno returned to Spain in 1535 with a vast fortune. Raúl Porras Barrenechea found a declaration by him in the Archivo General de Indias in Seville, in which he gave a detailed description of his journey, saying that he made it with 'two foot-soldiers and an interpreter'. Pedro Pizarro (1844, 243) also named Martín Bueno and Pedro Martín de Moguer, but none of the other eyewitnesses – Hernando Pizarro, Francisco de Xerez, Pedro Sancho, Miguel de Estete, etc. – bothered to record the names of these relatively unimportant men. Only later chroniclers followed Agustín de Zárate in saying that the envoys had been the more famous Hernando de Soto and Pedro del Barco.

64 *very well served.'* Mena 254. Zárate wrote that Atahualpa reassured Pizarro: 'although it was about two hundred very long leagues of bad road from Cajamarca to [Cuzco], they should not regard that as a long delay since they were to be carried on the shoulders of Indians' (bk. 2, ch. 6, 477).

64 *to rescue him.'* Mena 254.

64 *they themselves related.* Mena (Sinclair) 36–7.

64 *had been nailed.'* Xerez 343.

65 *the other Christians.'* Mena 263.

65 *of His Majesty'.* Xerez 343.

65 *guard of Indians'.* Mena 263.

65 *objects from them.'* Ibid.

65 *army of 30,000* Xerez 343.

65 *Cajamarca, with 35,000.* Hernando Pizarro, Letter to Oidores 89.

66 *the royal road'* Estete in Xerez 340.

66 *experienced great difficulty.'* Hernando Pizarro, Letter to Oidores 89.

66 *tired and unshod.'* Estete, *Noticia* 40.

66 *it was Indians.'* Hernando Pizarro, Letter to Oidores 89–90.

66 *warriors or townspeople.'* Estete in Xerez 340–1.

66 *us in peace'* Hernando Pizarro, Letter to Oidores 90.

66 *for a festival'*. Estete in Xerez 341.

66 *where Atahualpa was'*. Estete, *Noticia* 40.

66 *and seize it'*. Estete in Xerez 340.

66 *stay at Jauja.* This must have been the celebration of the Paucar-hauray, the earth-ripening festival, one of the less important events in the Inca ceremonial calendar.

67 *those on horses.'* Estete in Xerez 341.

67 *own free will'*. Hernando Pizarro, Letter to Oidores 90; also his *Confesión* 407.

68 *in the country.'* Estete in Xerez 341–2.

68 *found 35,000 Indians.'* Hernando Pizarro, Letter to Oidores 89.

69 *swim the river.* Estete in Xerez 342. Estete gave a fine, detailed account of each day's progress. The river at which Chalcuchima had had his engagement was the modern Puccha. This stream flows down to the Marañon past Peru's oldest major ruin, the great rectangular temple of Chavín de Huantar, built well before the Christian era.

69 *of three months.* There is confusion about the date of Hernando Pizarro's entry into Cajamarca. Xerez said it was on 25 March, and Estete said 25 May. Both these dates must be wrong. Estete wrote earlier that Hernando Pizarro had been at Bombón from 11 to 14 March, and Hernando Pizarro said that in that town he talked to a Negro left there by the Spaniards going to Cuzco. Xerez wrote that this Negro returned to Cajamarca on 28 March with 107 loads of gold and 7 of silver, and the news that Hernando Pizarro had gone to Jauja. Estete said that Hernando Pizarro was at Jauja from 16 to 20 March, and Hernando Pizarro also said that they stayed there for five days. Estete described the journey from Jauja to Cajamarca in great detail. The party left Huánuco, half-way to Cajamarca, on 30 March and took nine or ten days to Conchucos, from whence they proceeded to Andamarca. From there to Cajamarca took six days or a week. There may, however, have been some delays, so that the entry was not made until 25 April.

Xerez wrote that Almagro entered Cajamarca on the eve of Easter, 14 April. Estete, in the *Noticia del Perú*, and Diego de Trujillo both said that Almagro was already in Cajamarca when Hernando Pizarro returned; Pedro Pizarro said he came later. Hernando Pizarro himself said that while marching northwards he had 'received news by letters that Don Diego de Almagro had reached Cajamarca' (*Confesión*, 407).

69 *happiness and rejoicing.'* Estete, *Noticia* 41.

69 *loved as dearly.'* Estete in Xerez 343; also *Noticia* 41.

69 *it pleased him.'* Mena 264.

70 *brought it all.'* Ibid.

70 *was tied up.'* Ibid. Soto is shown to have been as brutal as any other conquistador. He had also been in charge of the force that raped the mamaconas of Cajas during the march towards Cajamarca. His reputation among some modern writers of being more humane than his companions is undeserved.

70 *were his servants.'* Ibid, 264, 266.

70 *in my lodging.'* Hernando Pizarro, *Confesión* 408. Hernando Pizarro accused Almagro and the treasurer Riquelme of torturing Chalcuchima.

CHAPTER 3 EXPEDIENCY

71 *conversation with him.'* Herrera, Dec. v, bk. 3, ch. 2, **10** 171.

71 *the Pacific coast.* Letters of Licenciate Espinosa to the King of 15 August and 20 October 1532 and of Licenciate Antonio de la Gama of 20 July, CDHA **42** 69–92, GP **2** 11–31. Xerez said Almagro had 150 men and 84 horses. Espinosa said that he embarked with 200 men and 100 horses, despatch of 21 July 1533, CP 60; Oviedo y Valdés, bk. 46, ch. 10; Herrera, Dec. v, bk. 3, ch. 1, **10** 159.

72 *of his own.* Almagro had already forestalled an attempt by a shipload from Nicaragua, under Francisco de Godoy, to launch out on its own. Other rumours claimed that Pizarro intended to cheat or kill Almagro. One of his secretaries,

Rodrigo Pérez, was tried and hanged by Almagro for spreading this dissension between the two leaders. But Almagro had in fact written to the King requesting permission to conquer beyond the limits of Pizarro's jurisdiction. His letter is lost, but the King replied on 11 July 1532 refusing permission on the grounds that this might jeopardise the entire conquest (CDH Chile **4** 165). Licenciate La Gama reported the rumours of Almagro's threatened rupture to the King on 19 July 1532, and Espinosa mentioned 'the passions that have arisen between the Governor Francisco Pizarro and Captain Diego de Almagro' in his despatch of 20 October 1532 cited in the previous note. Hernando Pizarro claimed that Almagro proceeded to Cajamarca only after abandoning a plan to colonise Puerto Viejo on the Ecuadorean coast (*Confesión* 407).

72 *going to die.'* Pedro Pizarro (1844) 244.

72 *his being killed.'* Ibid.

73 *until 16 July.* Rafael Loredo discovered the two decrees relating to the gold, and published them in *El Comercio*, Lima, on 11 June 1947 and 29 January 1950, and also fragments in his *Los repartos* 72–4 and 117–18. Before this only the document relating to the silver had been known. Francisco López de Caravantes included it in his unpublished *Noticia general del Perú*, written about 1610, and listed most of the awards of gold alongside those of silver, to save time. Manuel José de la Quintana copied this silver decree in 1861, and many later writers copied it from him; Clements Markham translated it (Hakl Soc, 1 ser., **47** 131–43, 1872). Pedro Cieza de León mentioned in chapter 51 of the newly-discovered Third Part of his *Crónica del Perú* that he had seen the distribution of the gold among the papers of secretary Jerónimo de Aliaga in Lima in 1550.

73 *of precious metals.* Important collections of Peruvian gold, such as the Mujico Gallo or Larco collections in Lima or the Bruening museum in Lambayeque, rely mainly on pre-Inca gold excavated from tombs.

73 *Pizarro was leaving.'* Pedro Pizarro (1844) 245. Hernando Pizarro cited this episode in his unsuccessful attempt to escape imprisonment by Charles V (*Confesión* 408). He claimed that Atahualpa had begged him, in front of many witnesses, to take him to Spain to see the King, 'for if he was left there "this fat man (meaning the treasurer [Riquelme]) and this one-eyed man (meaning Don Diego de Almagro) will kill me when you leave".'

74 *total of £2,854,350.* Assuming 1 peso de oro weighed 0·162 ounces avoirdupois, the 1,326,539 pesos weighed 13,431·2 pounds. One troy ounce of 22-carat gold is worth £13 ($31) today. Two marks of silver weighed one pound. This was reasonably good quality fine silver, for which the current price is about £11 ($26) per pound. Rafael Loredo discovered a document signed by Pizarro which gave the dates of the meltings and the total of 1,326,539 pesos de oro of gold and 51,610 marks of silver (*Los repartos* 72–4). Francisco de Xerez and Cieza de Leòn both gave this exact figure, and Sancho gave it in round figures.

74 *they owed them.'* Xerez 344. The dates and quantities of the treasure are found in the accounts of Xerez and of Pedro Sancho, the two official secretaries of the expedition, and also in a report by Sancho on the distribution of the ransom, which was quoted in Francisco López de Caravantes and translated by Markham (Hakl Soc **47** 131–43, 1872). It is interesting to note that Xerez and Sancho were each given an extra 1,110 pesos for their writing, in addition to their quota of 8,880 pesos as horsemen. All the other chroniclers also mentioned the quantity of gold, and Cristóbal de Mena complained that he did not receive enough of it.

74 *also a prisoner.'* Sancho, *Relación* (Means) 13.

75 *was all true.'* Xerez 344.

75 *torture or coercion.'* Sancho, *Relación* 126; also Estete, *Noticia* 41–2, Mena 272, Xerez 344, *Nouvelles certaines* 10.

75 *horses remained saddled.'* Xerez 344. The 150 horsemen included Almagro's new arrivals.

75 *in a barbarian.'* Ibid. Xerez was habitually suspicious of the Inca.

75 *can execute me!" '* Cataño 285.

76 *find him dead'.* Xerez 344.

76 *to rescue him.'* Estete, *Noticia* 41.

76 *would be calmed.'* Ibid. 42.

77 *the enemy army.'* Oviedo y Valdés (pt. 3, bk. 5, ch. 22) **121** 122.

77 *in this way'.* Cataño 284.

77 *love of Cataño?'* Ibid. 285.

77 *the Spaniards' service'*. Xerez 344, Mena 273.
77 *at great speed'*. Xerez 344.
77 *he should die.'* Estete, *Noticia* 42.
77 *evidence was sufficient.'* Sancho, *Relación* 127.
77 *kill the Christians.'* Estete, *Noticia* 42.
77 *cacique Atahualpa immediately'* Mena 274.
78 *converted to Christianity.'* Xerez 344.
78 *he were released.'* Pedro Pizarro (1844) 247.
78 *26 July 1533*. It is an extraordinary fact that the exact date of Atahualpa's execution was not known until very recently. Garcilaso wrote incorrectly that the Inca was christened Don Juan Atahualpa (pt. 2, bk. 1, ch. 36). Because of this Juan de Velasco assumed, in his *Historia del reino de Quito* **1** 372, that the baptism and execution took place on 29 August, the day of St John the Baptist. Prescott accepted this date (ch. 7), and it has been repeated by innumerable later writers to this day.

The Spaniards had in fact marched out of Cajamarca long before 29 August. The legal protocols of the scrivener Jerónimo de Aliaga ceased abruptly at Cajamarca on Sunday, 10 August; he witnessed two documents at Andamarca, far to the south, on 25 and 26 August. Gaspar de Espinosa wrote from Panama, on 10 October, that Pizarro and his men left Cajamarca 'at the beginning of the month of August', some days after having executed Atahualpa (CDIA **42** 73, 1884).

The great nineteenth-century Peruvianist, Marcos Jiménez de la Espada, first queried Velasco's 29 August date, since Atahualpa was christened Francisco (after Pizarro) and not Juan. Jímenez de la Espada wrote in the prologue to Las Casas's *De las antiguas gentes del Perú*: 'I do not trust that date.' Also in his own *Tres relaciones de antigüedades peruanas* 326. Another great Peruvian historian, Raúl Porras Barrenechea, confirmed that the date was wrong in his 'Atahualpa no murió el 29 de agosto de 1533', *La Prensa*, Lima, 31 October 1945; so did Rafael Loredo in *Los repartos* 71, and José Antonio del Busto Duthurburu in 'La marcha de Francisco Pizarro de Cajamarca al Cuzco'. Vargas Ugarte followed Loredo, but wrote (by a misprint?) that the execution was on 16 July – which was not a Saturday (*Historia general* **1** 67). Loredo claimed to possess a document that showed that Atahualpa was executed on Saturday 26 July, after having given Pizarro on that same morning two large golden vases and an ornamented golden dish. The only other Saturday on which the execution could have taken place was the following one, 2 August. The earlier Saturday is confirmed by the fact that Pizarro wrote to the Emperor Charles V describing the execution on 29 July 1533: Pizarro's letter is lost, but the King referred to it in a decree dated 21 May 1534 (CP 64).
78 *around his neck.'* Sancho, *Relación* 127.
79 *he was baptised . . .'* Gangotena y Jijón, 'La descendencia de Atahualpa' 118–19.
79 *sons to Pizarro*. Probanza on behalf of Atahualpa's sons Diego Illaquita, Francisco Ninancoro and Juan Quispe-Tupac, Lima, 28 April 1555, in Gangotena y Jijón, 'La descendencia de Atahualpa' 113, 115, 118–19, 121. Witnesses to this Probanza included Lucas Martínez Vegaso, Juan Delgado, Pedro de Alconchel, Inés Huayllas Yupanqui and Fray Domingo de Santo Tomás, who was not actually present at Cajamarca. Pedro Sancho (*Relación* 127) and Xerez (345) mentioned this bequest. So did Oviedo y Valdés (**121** 104) and Garcilaso (pt. 2, bk. 2, ch. 3). The children, boys and girls, were under Rumiñavi's control at Quito.
79 *done to him.'* Sancho, *Relación* 127–8. The other eyewitness accounts of Atahualpa's death agree exactly with this. All mention the last-minute conversion that earned the Inca his garrotting. Pedro Cataño (285–6) recorded a conversation in which Atahualpa ascertained that only Christians were buried in church or ascended to heaven after death. 'Atahualpa said that the Christians' way was better than theirs and that he wished to be a Christian and be buried in the church. And so he became a Christian.' Xerez (345) confirmed that he died 'with much spirit, revealing no emotion, but saying that he commended his children to the Governor. . . . He died on Saturday, at the same hour on which he was defeated and captured.' Estete (*Noticia* 42) observed that 'for him it was not death but life, for he died a Christian and it is believed that he went to heaven'. Atahualpa's christening as Don Francisco was reported by Santacruz Pachacuti (223) and confirmed in the Probanza made on behalf of Atahualpa's sons in Lima in April 1555.

79 *like drunken men.'* Pedro Pizarro (1844) 248.

79 *to be buried.'* Estete, *Noticia* 42–3. Cristóbal de Mena and Pedro Pizarro also recorded this suicide wish by Atahualpa's wives. One of Atahualpa's despairing sisters, later baptised as Lucía Clara Coya, was given by Pizarro to the conquistador Diego Maldonado to prevent her committing suicide. This was the parentage claimed by Maldonado's son Juan Arias Maldonado in his petition for entry into the Order of Santiago (CL **1** 386). Another full sister was later known as Francisca Coya; she returned to Quito amid the veneration of its native populace, and lived with the conquistador Diego de Sandoval.

80 *do not return.'* Pedro Pizarro (1844) 251.

80 *there was nothing.'* Ibid. 247.

80 *wet with tears.'* Oviedo y Valdés pt. 3, bk. 5, (bk. 46), ch. 22, **121** 122. Oviedo disliked Pizarro.

80 *in that country.'* Pedro Pizarro (1844) 247.

80 *at his capture.'* Pizarro's letter to the Emperor has regrettably been lost. It was referred to and quoted in a royal reply from Toledo, 21 May 1534, CP 64.

Communications between Peru and Spain were very slow: the journey in either direction could take six to eight months, or even more. Ships sailed from San Lucar de Barrameda at the mouth of the Guadalquivir south-westwards to the Canary Islands and thence to the Windward Islands. This crossing took twenty-five to thirty days. From there to Nombre de Dios on the Isthmus took a further fifteen to twenty days. Cargoes were then lightered up the Chagres River and transported by mule train to Panama. On the Pacific side the 1,300-mile voyage to Lima could take many weeks: the prevailing winds are southerly, and the Humboldt Current flows northwards along the coast. There were calms and storms in these waters.

On the return journey, ships had a good run from Nombre de Dios to Havana with the wind on the starboard beam, but then had to beat out against headwinds to clear the Florida Channel. They stood to the north along the Florida coast until they could pick up a westerly trade wind for the crossing to Europe.

Ships making these voyages suffered from shipwreck and attacks by privateers. After the 1560s almost all traffic sailed in convoys protected by armed galleons. The outward annual fleet left Spain in August. The return fleet left the Isthmus in January, joined the Mexican fleet in Havana, and sailed for Spain in the early summer before the hurricane season.

80 *they executed him.'* Hernando Pizarro, Letter to Oidores 90.

80 *from a dream.'* Espinosa, despatch, Panama, 10 October 1533, CP 66.

81 *or other person.'* Ibid.

81 *this with me.'* Ibid.

81 *noble of Castile.'* Ibid.

81 *whom you captured.'* Royal decree, Toledo, 29 July 1533, CP 64.

81 *what is necessary.'* Ibid.

81 *and outstanding evil.'* Oviedo y Valdés, bk. 46, ch. 22, **121** 122.

82 *before Soto's return.'* Andagoya, Hakl Soc **34** 52, 1865.

82 *on the conquistadores.'* Ruiz de Arce 364.

82 *against the Spaniards.'* Felipillo's death in Chile and treacheries there were recorded at length by Gonzalo Fernández de Oviedo y Valdés, **121** 140–1. But it is important that Oviedo – originator of the notion of a trial of Atahualpa – in no way implicated Felipillo in the Inca's death.

82 *to this day.* Zárate, bk. 2, ch. 7; López de Gómara, ch. 118; Benzoni, Hakl Soc, 1 ser. **21**, 1857; Cabello de Balboa, pt. 3, ch. 33; Gutiérrez de Santa Clara, bk. 3, ch. 55, **166** 228; Garcilaso, pt. 2, bk. 1, ch. 36. Pedro Pizarro was the only eyewitness who mentioned Felipillo's false testimony, but he did so in connection with Chalcuchima, not Atahualpa, and he wrote in 1571, by which date Felipillo was well established as the villain. Modern historians who repeated the story include: Prescott **1** 476–82, 1847; Markham, note on 102–3 of his translation of Xerez, Hakl Soc, 1 ser., **162**, 1872; Means, *Fall of the Inca Empire* 42–3; Valcárcel, 'Final de Tawantisuyu' 86; and many others.

83 *the same stamp.'* Oviedo y Valdés, bk. 46, ch. 22, **121** 122.

83 *sixteenth-century Spanish mind.* López de Gómara, ch. 118.

83 *of Atahualpa's execution.* Garcilaso, pt. 2, bk. 1, ch. 37. Herrera, the official historian who like Garcilaso published his history in the seventeenth century, described the summary interrogation and hurried council decision, but no formal trial.

83 *probably be taken'.* Prescott, bk. 3, ch. 7.

83 *of correct forms.'* Edward Hyams and George Ordish, *The Last of the Incas* (London, 1963) 239.

83 *worthy of remembrance'.* Hakl Soc **47** 102–3, 1872.

83 *justice and mercy'.* Hyams and Ordish 240.

83 *at that time.* Of the 'valiant eleven', Fuentes and Mendoza were the only ones definitely in Cajamarca; Mora and Herrada came with Pedro de Alvarado in 1534, and Avila reached Peru in 1535; Atienza was in San Miguel; neither of the 'Chaves brothers' was then in Cajamarca; and there is no record in any document, distribution, passenger list or list of deceased of Moscos, Haro, Ayala or Cuellar. Porras Barrenechea concluded that the story of the brave eleven was invented by Blas Valera as 'the purest popular invention to convert Juan de Herrada, the future assassin of Pizarro, into the protector of the Inca and his posthumous avenger' ('Crónicas perdidas' 210). Herrada was a close friend of Blas Valera's father Alonso Valera, as was Francisco de Chaves, a notoriously bloodthirsty killer of natives.

84 *attack in 1533.* Ruiz de Arce 364, Oviedo y Valdés **121** 123, and to some extent Zárate, bk. 2, ch. 8. Raúl Porras Barrenechea regarded this report as proof that Rumiñavi's army had been outside Cajamarca when Atahualpa was killed.

84 *any native force.* Pedro Pizarro (1844) 246. Diego de Trujillo (59) said that Pizarro sent Soto to the 'river of Levanto' – the Marañon or the Utcubamba, since Chachapoyas was known to the conquistadores as Levanto – and that he himself went on the reconnaissance which found nothing there. The country between Cajamarca and the Marañon is very rugged (modern Peruvians have had to abandon the attempt to connect Cajamarca and Chachapoyas by a road along this direct route) and it is highly unlikely that Rumiñavi would have approached from that direction. But Trujillo may have meant Cajas, which was on a tributary of the Marañon. Prescott, who never went to Peru and whose geography was weak, said that Soto went to Huamachuco – in diametrically the wrong direction – but he clearly confused this mission with Hernando Pizarro's earlier reconnaissance to Huamachuco in search of Chalcuchima's army. Later writers, with less excuse, blithely followed Prescott in despatching Soto southwards to Huamachuco.

85 *had no claim.* Hernando Pizarro (*Confesión* 408), Pedro Pizarro (1844, 245) and Agustín de Zárate all made this valid point about Almagro's motives. Cristóbal de Mena, one of Pizarro's original men, tried to argue that his participation in the kidnapping of the Inca entitled him to a share of all further treasure ever found in Peru.

CHAPTER 4 TUPAC HUALLPA

86 *arrived in Cajamarca.* Tupac Huallpa was given various names by the Spaniards. Pedro Sancho and Francisco de Xerez both called him Atabalipa, others called him Toparpa, Tobalipa or (Herrera) Topaipa.

87 *of His Majesty.'* Sancho, *Relación* 128.

87 *should observe it.'* Ibid. 129.

87 *person of quality.'* Ibid. 130.

87 *crown among them.'* Xerez 343.

88 *and accepted it.'* Sancho, *Relación* 130–1.

88 *and some hidalgos* Hidalgo is a contraction of 'hijo de algo': the son of somebody important.

88 *of the ceremony.'* Sancho, *Relación* 131.

88 *do the same.'* Ibid.

88 *the Governor's house.'* Ibid.; also Cabildo de Jauja, Letter of 20 July 1534, in CL **3** 2. Pizarro referred to the choice of Tupac Huallpa in his despatch of 29 July 1533, CP 64; it was made on Sunday 27 July.

89 *two small drums'.* Xerez 345–6.

89 *are worth seeing'.* The letters to and from the King were: Hernando Pizarro, San Lucar de Barrameda, 14 January 1534; King Charles to Casa de Contración, Calatayud, 21 January, and Sigüenza, 26 January. The objects taken to Toledo were recorded in a letter from the King to the Casa, Toledo, 7 March. The series of letters is in CL **3** 127–30. Further details of the objects taken to Toledo are in a document called *Relación del oro del Perú que recibieron de Hernando Pizarro que trujo la nao de que era maestre Pedro Bernal en febrero de 1534* quoted in Loredo, *Los repartos* 42–3.

89 *he is bringing.*' Council of the Indies to the Emperor, Madrid, 18 January 1534, CP 96.

89 *brought from Cajamarca*'. Cieza de León *Crónica*, pt. 1, ch. 94 (Onis) 255.

89 *left in Seville.* The officials of Seville wrote to the Emperor in 1535 that they had melted down the objects brought by Hernando Pizarro in February 1534. They listed them as: '34 large urns, 2 kettledrums, one figure of the bust of an Indian, another similar of an Indian woman, a small retable, two platters, an idol in the form of a man, gold in powder, golds of 21 to 9 carats. All of this weighed 3,000 marks, 3 ounces, 2 och., 3 grs; or 150,070 pesos, 2 tomines, 3 grs' CP 185. The most complete list of objects brought back to the King of Spain fills pages 12 to 14 of the *Nouvelles certaines des Isles de Peru*, an anonymous French book published at Lyon in 1534.

90 *11 August 1533.* The march has been studied in great detail by Busto Duthurburu in his 'La marcha de Francisco Pizarro de Cajamarca al Cusco'. He compared the narrative accounts, of which the most important are those of Pedro Sancho, the Cabildo of Jauja, Miguel de Estete, Juan Ruiz de Arce, Diego de Trujillo and Pedro Pizarro (written in that order), and related these to the various legal documents issued during the course of the journey. I have followed his conclusions regarding dates.

90 *Inca army camp;* This ruin is baffling because there are almost no potsherds and no streets or passages between the enclosures. It is reminiscent of the ruin Pikillacta at the southern end of the Cuzco valley, and of Incahuasi, an army camp occupied during military operations in the coastal Cañete valley south of Lima.

90 *into city states.* Marca Huamachuco looks remarkably like a crusader castle such as Margat, which was built at approximately the same period. The entire ridge was surrounded by a curtain wall. The many ruins inside this are dominated by two circular corrals, with entrances barely large enough for one man. Huamachuco's ruins were investigated by Theodore H. McCown and described in his *Pre-Incaic Huamachuco: survey and investigations in the northern Sierra of Peru* (University of California Publications in American Archaeology and Ethnology **39**, no. 4, 1945).

91 *this first bridge.*' Estete, *Noticia* 37 and in Xerez 338. Both passages referred to the first crossing of this bridge by Hernando Pizarro's Pachacamac expedition seven months earlier. Hernando Pizarro also described this bridge in his letter of 23 November 1533, and mentioned that there were two bridges, one of which was kept for the exclusive use of imperial officials and armies.

91 *unaccustomed to it.*' Sancho, *Relación* (Means) 60.

91 *Huascarán.* Huascarán at 22,200 feet is 6,400 feet higher than Mont Blanc.

91 *Cajatambo and Oyón.* The army probably reached Cajatambo on 1 October. It is curious to note that Simon Bolívar followed this same route on the campaign that ended Spain's rule in Peru almost 300 years later.

92 *that they needed*'. Cabildo of Jauja, Letter of 20 July 1534, CL **3** 2.

92 *vanguard and rearguard.*' Sancho, *Relación* 135.

92 *to the Christians.*' Ibid.

92 *for the killing.* Herrera, Dec. v, bk. 4, ch. 10.

92 *of the expedition.* According to Pedro Pizarro (1844, 254), the Governor told Chalcuchima to do whatever he wished to improve the commissariat situation. He used this as an excuse to assemble various chiefs of Huamachuco who had displeased him. He lined them up in the square with their heads resting on large stones. 'Taking in his arms the largest rock he could lift, he smashed the first on the head and . . . made a tortilla of it.' Pizarro managed to stop him from continuing this carnage, according to this improbable story.

93 *to govern it.*' Sancho, *Relación* (Means) 34.

93 *to join it.*' Ibid. 33.

93 *the chained Chalcuchima.* Sancho, *Relación* 138; Cabildo of Jauja, Letter of 20 July 1534; Trujillo 60; Herrera, Dec. v, bk. 4, ch. 10.

93 *precipitous, difficult mountainside.*' Sancho, *Relación* 138. Garcilaso (pt. 2, **134** 86) told a story of an attack on the Spanish column by the Inca general Titu Atauchi. This attack was supposed to have resulted in the capture of Sánchez de Cuéllar and various other Spaniards. The natives were then said to have executed those implicated in killing Atahualpa and to have released the rest. The incident cannot have happened.

It would certainly have been mentioned by the eyewitnesses who recorded the march in such detail, and who said that the countryside was entirely peaceful between Cajamarca and Jauja. Moreover Titu Atauchi was one of Huascar's generals, killed with him at Andamarca on Atahualpa's orders some months before Pizarro left Cajamarca.

94 *were all soaked.*' Sancho, *Relación* 138.

94 *Quitan professional soldiers.* Herrera, Dec. v, bk. 4, ch. 10.

94 *its northern end.* The ruins are still there in great quantities, consisting largely of the fieldstone foundations of thousands of rough cottages and burial towers. Hans Horkheimer described them in 'En la región de los Huancas', *Boletín de la Sociedad Geográfica de Lima*, 1951.

94 *Yucra-Hualpa.* Sancho named this commander as well as 'Yguaparro, Mortay and another captain'. The Quipocamayos – the Inca officials responsible for maintaining oral history – interviewed in 1542 named Yucra-Hualpa as one of Atahualpa's leading officers along with Quisquis. Sancho wrote his name as 'Yncorabaliba' and the Quipocamayos (27) as 'Yucra Vallpa'. Herrera, presumably following Cieza de León, said that the commander at Jauja was called Curambayo.

94 *that foreign army,*' Sancho, *Relación* 139.

94 *gold in Jauja*'. Martín de Paredes to the Treasurer Martel de la Puente, San Miguel, 14 February 1534, CP 99; Toribio Montañés to Pascual de Andagoya, San Miguel, 13 March 1534, CP 105. Ruiz de Arce 365; Pedro Pizarro (1844) 255; Cabildo of Jauja, CL **3** 3; Sancho, *Relación* 139–40.

95 *not fifty had escaped.*' Cabildo of Jauja, CL **3** 3; Sancho, *Relación* 140; Herrera, Dec. v, bk. 4, ch. 10; Trujillo 60; Ruiz de Arce 365; Estete, *Noticia* 43.

95 *between the squadrons*'. Ruiz de Arce 365.

95 *gold and silver.*' Ibid.; also Sancho, *Relación* 141–3; Cabildo of Jauja, CL **3** 3.

95 *daughters of Huayna-Capac*'. Herrera, Dec. v, bk. 5, ch. 2.

95 *into unconquered Peru.* Many of these documents survive in the Harkness Collection in the United States Library of Congress. Francisco Pizarro himself issued an authorisation to Juan de Valdivieso and Pedro Navarro to receive merchandise on his behalf, to collect money due to him, and to execute instruments of payment and quittance. Diego de Almagro issued a similar document and, like Pizarro, was unable to sign it because he could not write. Toribio Montañés wrote, correctly, that Riquelme was left in Jauja with forty horsemen and forty foot (Montañés to Pascual de Andagoya, San Miguel, 13 March 1534, CP 105). Francisco de Barrionuevo wrongly told the Emperor that 80 horse and 100 foot were left (Barrionuevo despatch, Panama, 8 April 1534, CP 106).

96 *certainty about this.*' Cabildo of Jauja, CL **3** 3. Sancho, *Relación* 144; Quipocamayos 32; Pedro Pizarro (1844) 252; all repeat the accusation against Chalcuchima.

96 *the Quitan cause.* Garcilaso, pt. 2, bk. 1, ch. 39; Temple, 'La descendencia de Huayna Cápac' 307; Gibson, *The Inca Concept of Sovereignty* 66–7.

96 *came to nothing.* Pizarro's secretary Pedro Sancho was at pains to assure his readers that Pizarro was simply trying to dupe Chalcuchima into arranging a cease-fire and bringing the prince into their custody. Chalcuchima was probably not deceived. He used the offer to obtain release from the chains the Spaniards had placed on him.

97 *Manco Capac.* Bernabé Cobo listed all the huacas (shrines) of Cuzco, which were arranged in 42 ceques or lines radiating from the central square (*Historia del Nuevo Mundo*, bk. 13, chs. 13–16). Cobo's list almost certainly derived from two earlier experts in the Inca religion: Juan Polo de Ondegardo, a mid-sixteenth century corregidor of the city who drew a diagram of the ceques and Cristóbal de Molina (of Cuzco), a priest who wrote a penetrating study of Inca religion in the 1570s and who first recorded the huacas and ceques in a written list.

97 *golden hand-plough.* Inca religion is a vast subject that received considerable attention from the Christian authorities trying to suppress it in the sixteenth century. All the chronicles contain much information on the subject. The best accounts are by Cobo, *Historia del Nuevo Mundo*, bk. 13; Cristóbal de Molina (of Cuzco), *Relación de las fábulas y ritos de los Incas*; Cieza de Leòn, pt. 2, chs. 27–30; Polo de Ondegardo in many of his writings; Pablo de Arriaga, *La extirpación de la idolatría en el Perú*; Francisco de Ávila; Antonio de la Calancha,

bk. 2, chs. 10–12, bk. 3, chs. 1–19. One of the best summaries by a modern author is, as usual, John Rowe, *Inca Culture at the Time of the Spanish Conquest* 293–314; also Jacinto Jijón y Caamaño, *La religión del imperio de los Incas*.

98 *in their hands.'* Betanzos, ch. 2.

98 *foam of the sea".'* Cieza de León, pt. 2, ch. 5 (Onis) 27, 28–9. Both Cieza and Betanzos obtained their information from the natives of Cacha, where there are still extensive remains of a great temple dedicated to Viracocha. The ruins, some seventy miles south-east of Cuzco, are unique in being the only Inca remains of a building that must have resembled an early church with two aisles. The row of central piers still stands, with ashlar bases and adobe tops, and Cacha also boasts a rare three-storey building and a round pillar. The first Augustinians, who reached Huamachuco (not far south of Cajamarca) in 1551, reported a similar legend: 'The Indians say that Viracocha wanted to make them Christians but they drove him out of the land.' *Relación de la religión y ritos del Perú*, CDIA **3** 11.

98 *the universal creator.'* Titu Cusi Yupanqui 8. Sarmiento de Gamboa, writing shortly after Titu Cusi, was sure that Atahualpa identified Pizarro with the returning god (ch. 68, Hakl Soc, 2 ser., **22** 187, 1907). Garcilaso gave additional explanations for the identification of Spaniards with Viracocha (pt. 1, bk. 5, ch. 21). Cobo said that use of this name was still current in the seventeenth century (bk. 12, ch. 19); and it is sometimes used today.

CHAPTER 5 THE ROAD TO CUZCO

101 *them are paved.'* Hernando Pizarro, Letter to Oidores 88.

101 *by their bridles.'* Sancho *Relación* 67. A force of Spaniards was annihilated by the natives at this same ascent three years later.

103 *to the Spaniards.* Ibid. 139.

104 *they then withdrew.'* Trujillo 60–1.

104 *25,000 Indian warriors.'* Ruiz de Arce 366.

105 *over his court.* The pyramid is some 150 feet square at its base, and rises in five tiers to the platform top. The sides are faced in fine coursed masonry, but wherever these retaining walls were buried into the solid fabric of the pyramid, the rectangular coursing gives way to polygonal stonework. Cieza de León left a fond description of the place, even counting the thirty-two steps to the top of the pyramid (*Crónica*, pt. 1, chs. 81, 89, pt. 2, ch. 15). There are other descriptions written by Pedro de Carbajal in 1586, in RGI **1** 167, and by Vázquez de Espinosa, pt. 1, bk. 4, chs. 78–9. Nineteenth-century descriptions include: Raimondi **2** 30, 53; Wiener, *Pérou et Bolivie* 265–70, 1880; Middendorf, *Peru* **3** 551–4. Means described the ruins in *Ancient Civilizations of the Andes* 109–10, and so did Victor von Hagen in his *Highway of the Sun*. These authors call it a sun temple, but I think that it more closely resembles an usno platform. See also: Alberto Arca Parro, *Ante las culturas del antiguo Perú: Vilcas Waman, cuidad y campo*, Lima, 1927; Pedro E. Villar Córdova, 'Las ruinas de Willcas Waman' *El Comercio*, Lima, 29 December 1929. I myself spent two days cutting away the vegetation that was growing between the stones of the pyramid.

105 *were coming behind'.* Trujillo 61.

105 *all almost lost'.* Pedro Pizarro (1844) 256.

106 *with the Apurímac.'* Herrera, Dec. v, bk. 5, **10** 349–50.

106 *groups of four.'* Ruiz de Arce 367.

107 *with greater fury.* Sancho, *Relación* 159.

108 *plenty of fear'.* Ruiz de Arce 367.

108 *Soray and Salcantay.* There were various good descriptions of the important battle of Vilcaconga: Francisco Pizarro to the Cabildo of Panama, Jauja, 25 May 1534, CP 115; Cabildo of Jauja, despatch of 20 July 1534, CL **3** 4; Sancho, *Relación*, ch. 9, 158–61 (Means 80–8); Estete, *Noticia* 44; Ruiz de Arce 367; López de Gomara, ch. 123, **2** 35; Zárate, bk. 2, ch. 8, 480; Pedro Pizarro (1844) 256–8; Trujillo 62; Herrera, Dec. v, bk. 5, ch. 3, **10** 350–2. A useful study on the dead Spaniards is Busto Duthurburu, 'Los caidos en Vilcaconga'.

The slope of Vilcaconga is the western side of the 12,000-foot pass now called

Casacancha, but still known as Huillcaconga when visited by Middendorf in the late nineteenth century. It was roughly midway between Cuzco and the Apurímac, and lay above Limatambo, the Inca post-house of Rimac-tampu. Luis Valcárcel excavated a superb Inca terrace platform (perron) on the hacienda Tarahuasi at Limatambo in 1934. Its three faces are lined with magnificent stonework and niches (José Maria Franco Inojosa and Alejjandro Gonzales, 'Los trabajos arqueológicos en el Departamento de Cusco. Informe sobre las ruinas incaicas de Tarawasi (Limatambo)', RMN **6** 66–80, 1937).

108 *as many horses'.* Cabildo of Jauja, despatch of 20 July 1534, CL **3** 4.

109 *you burned alive.'* Sancho, *Relación* 164.

109 *him on fire'.* Ibid.

110 *to the city.'* Ruiz de Arce 368.

110 *and much determination'.* Estete, *Noticia* 44.

110 *wounded many Christians'.* Ruiz de Arce 368.

110 *on to the plain.'* Sancho, *Relación* 169.

110 *full-scale encounter'.* Francisco Pizarro despatch, Jauja, 25 May 1534, CP 115.

111 *thunder from heaven.'* Titu Cusi Yupanqui 8.

112 *the modern system.'* R. B. Cunninghame Graham, *The Horses of the Conquest* 10–11.

112 *at this price.'* Xerez 344.

112 *period confirm them.* Some deeds were published by Stella R. Clemence, in Harkness Cal. On 21 May 1534 Cristóbal de Sosa bought a horse for 1,500 pesos de oro. On 15 June Melchor Verdugo paid 2,000 pesos for a saddle horse, an Indian slave from Nicaragua, an Indian woman branded on the face, and two dozen hens. On the eve of the Spaniards' departure from Cajamarca, Francisco Solares bought a half-interest in a chestnut horse from Gonzalo Pizarro: he agreed to share any gold, silver or precious stones that he conquered on this horse during the ensuing year.

113 *sign of victory.'* López de Gómara, ch. 125, 40.

113 *the 1914–18 War.* Harold L. Peterson, *Arms and Armor in Colonial America* 111–13. The high crown and peak of the cabasset was later merged to the curved brim of the morion, to produce the morion-cabasset.

113 *the face itself.* This visored helmet was known as a burgonet or 'celada con sobre-vista' (ibid. 119 ff).

114 *the soldier's dream'.* Ibid. 70 ff; Alberto M. Salas, *Las armas de la Conquista*; A Fontecilla Larraín, 'Las espadas de los siglos xvi y xvii', *Revista Chilena de Historia*, 1941.

117 *15 November 1533* Sancho, *Relación* 169–70. Sancho actually wrote Friday, 15 November, but it was in fact a Saturday: Busto Duthúrburu, 'La marcha de Francisco Pizarro de Cajamarca al Cusco' 170; Francisco Pizarro and royal officials to the Cabildo of Panama, Jauja, 25 May 1534, CP 115; Estete, *Noticia* 44; Trujillo 63; Ruiz de Arce 368. All agreed that this last battle was very fierce.

CHAPTER 6 CUZCO

118 *as their lord'.* Sancho, *Relación* (Means) 99.

118 *yellow cotton'* Trujillo 63.

118 *a constant fugitive',* Sancho, *Relación* (Means) 100.

118 *a common Indian.'* Molina of Santiago 156; Trujillo 63; Pedro Pizarro (1844) 191; Cabildo of Jauja, despatch of 20 July 1534, CL 4; Almagro despatch of 8 May 1534, Piura, CDIA **42** 106, 1884; Titu Cusi Yupanqui, 5; Herrera, Dec. v, bk. 6, ch. 3, 23; Zárate 480; López de Gómara, ch. 123, 36. Sancho said that the meeting took place on the morning *after* Chalcuchima's execution at Jaquijahuana, but almost all other chroniclers stated that Manco came to the Spaniards on the slope between Vilcaconga and Jaquijahuana. Trujillo also mentioned that the Cañari chief Chilche had joined the Spaniards a few hours earlier, and had served them loyally 'up to the present day' in 1571.

119 *made a confederation.'* Titu Cusi Yupanqui 23.

119 *from this tyranny.'* Sancho, *Relación* (Means) 100.

119 *come with him.'* Ibid.

119 *of the channel.'* Sancho, *Relación* 192.

120 *dropped while bathing.'* Cieza de León, *Crónica*, pt. 2, ch. 35 (Onis) 203.

120 *in fine gravel.* Both the earliest eyewitness authors mentioned the gravel on the square: Cabildo of Jauja, despatch of 20 July 1534, CL **3** 4; Pedro Sancho, *Relación* (finished on 15 June 1534) 192. Juan Polo de Ondegardo, Corregidor of Cuzco from 1558 to 1561, believed that this gravel or sand had been transported from the coast, and that 'the entire country must have co-operated in the work, for it is a large square and must have absorbed innumerable loads of sand' (*Relación* 107). Polo found that the gravel surfacing was two feet deep in places. The two squares were split by Spanish colonial arcades, and Cusipata has now been occupied by a tourist hotel.
120 *even in Spain.'* Cabildo of Jauja, CL **3** 4.
120 *periods of hunger.'* Sancho, *Relación* 201.
121 *us with goodwill.'* Estete, *Noticia* 43.
121 *four thousand people.'* Garcilaso, pt. 1, bk. 7, ch. 10, Hakl Soc **45** 249.
121 *to earlier Incas.* Pedro Pizarro (1965) 192; Acta de fundación del Cuzco, para. 6, in: Porras Barrenechea, 'Dos documentos esenciales' RH **17** 86–95, 1948; Garcilaso, pt. 1, bk. 7, ch. 10. Juan Pizarro later chose to reside on the terraces on the hillside between the square and Colcampata.
122 *younger Pizarro brothers.* Acta de fundación del Cuzco, para. 4; Pedro Pizarro (1965) 192.
122 *are very remarkable.'* Sancho, *Relación* 192.
122 *and other metals'* Estete, *Noticia* 45. Estete referred to Amaru Cancha as Atahualpa's palace, presumably because Atahualpa had ordered that anything belonging to his father must be protected.
122 *60 feet in diameter.'* Garcilaso, pt. 1, bk. 7, ch. 10; pt. 2, ch. 32. Garcilaso noted that the great tower was torn down during the 1550s to open out the square and regretted its loss. He called it Sunturhuasi, which was the name of Viracocha's palace at the east end of the square.
122 *of the square.* Acta de fundación del Cuzco, para. 9. Hernando Pizarro took possession of Amaru Cancha when he was Governor of Cuzco, after Soto had left for Spain in 1535. The palace was burned during the siege of 1536, and Hernando sold the site to the Jesuits for 14,000 pesos when he was an old man living in Spain. Soto's daughter and son-in-law, a notary called García Carrillo, attempted to sue Hernando for possession of the site in 1572.
123 *festivals and celebrations.'* Cobo, *Historia del Nuevo Mundo*, bk. 14, ch. 3 **4** 197.
123 *all fours to enter.'* Ibid. 196.
124 *appreciate their excellence.'* Ibid. ch. 12, **4** 240.
124 *of their stones'* Cieza de León, *Crónica*, pt. 1, ch. 92.
125 *have been involved.'* Cobo, *Historia del Nuevo Mundo*, bk. 14, ch. 12, **4** 240.
125 *late fifteenth century.'* Nineteenth-century authorities such as Marcos Jiménez de la Espada and Sir Clements Markham thought that the polygonal stones might have come from an earlier 'megalithic' culture somehow linked with Tiahuanaco. They were comparing Peru with Greece, where Schliemann had been excavating Mycenae, whose megalithic walls are built in a polygonal style. These nineteenth-century authorities were impressed by a fanciful list of centuries of Inca rulers that was produced by the seventeenth-century chronicler Fernando Montesinos. The excavations of Max Uhle, Luis Valcárcel and John H. Rowe showed that the two styles were contemporary and dated from the fifteenth century A.D.
125 *on Inca masonry.* This theory for the origin of polygonal and coursed masonry was propounded by John H. Rowe, in 'An Introduction to the Archaeology of Cuzco'.
126 *revere and obey.'* Sancho, *Relación* 170.
126 *all their weapons.'* Ibid. 171.
126 *had ever seen'.* Ibid.
126 *migration towards Quito.* Cabildo of Jauja, CL **3** 4; Sancho, *Relación* 171–2; Ruiz de Arce 374; Titu Cusi Yupanqui 26; Pedro Pizarro (1965) 198.
128 *days in succession.'* Estete, *Noticia* 54–6.
128 *nose is missing'.* Sancho, *Relación* 200, Pedro Pizarro (1965) 192; Cieza de León, *Crónica*, pt. 2, ch. 11; Cobo, *Historia del Nuevo Mundo*, bk. 13, ch. 9.
128 *that they had.'* Sancho, *Relación* 173.
128 *company of Spaniards....'* Estete, *Noticia* 54.
129 *abominate our faith.'* Bartolomé de las Casas, quoted in Hanke, *The Spanish Struggle for Justice in the Conquest of America* 7.
129 *their own land?'* Sermon reported in Las Casas, *Historia de las Indias*, bk. 3, ch. 4.
129 *in peaceful isolation.* Matías de Paz's work was *De dominio Regum Hispaniae super Indios.* Two important supporters of the

conquest were Juan López de Palacios Rubios and Martín Fernández de Enciso. The entire debate is dealt with in Lewis Hanke's brilliant *The Spanish Struggle for Justice in the Conquest of America* and in Silvio Závala's *New Viewpoints on the Spanish Colonisation of America*.

130 *are your fault...'* A translation of the Requirement can be found in Hanke's 'The Development of Regulations for Conquistadores' in *Homenaje al Dr. Emilio Ravignani* (Buenos Aires, 1941) 73–5, and in his *History of Latin American Civilisation* **1** 123–5.

130 *in his force.'* Sancho, *Relación* 200–1.

131 *could be found'*. Loredo, *Los repartos* 131–2.

131 *slightly greater*. Loredo described the legal documents in *Los repartos* 95–107, and quoted some of them (123–30). His comparison of the values of the two meltings was:

		maravedis
Cajamarca:	1,326,539 pesos of gold at 450 maravedis:	596,942,550
	51,610 marks of silver at 1,958 maravedis:	101,052,380
		697,994,930
Cuzco:	588,266 pesos of gold at 450 maravedis:	264,719,700
	164,558 marks of good silver at 2,210 maravedis:	363,673,180
	63,752 marks of poor silver at 1,125 maravedis:	71,721,000
		700,113,880

132 *Pedro Pizarro'*. Ibid. 101.

132 *128 gold marks'* Ibid. 102.

132 *gold and silver.'* Ruiz de Arce 372.

132 *had less publicity.'* López de Gómara, ch. 123, 37; Garcilaso, pt. 2, bk. 2, ch. 7.

132 *such delicate objects.'* Pedro Pizarro (1965) 195–6. The cave may have been the eerie sanctuary of Qenco, to which visitors are taken on a hill above Cuzco. The legendary first Inca was Manco Capac. The discovery of his golden effigy in this find may mean that it was located at the cave of Tambotoco at Pacaritambo, since this cave was where Manco and his three brothers were supposed to have surfaced after an underground journey from Lake Titicaca. Pacaritambo is about eighteen miles south-east of Cuzco.

133 *at roof level.'* Ruiz de Arce 372.

133 *and went in.'* Trujillo 64.

133 *height and strength*. John Howland Rowe obtained permission from the monks to make a thorough investigation of Coricancha. He published his findings in 'An Introduction to the Archaeology of Cuzco'. Many of Rowe's discoveries were laid bare by the earthquake of 1950.

134 *and giving thanks.'* Ruiz de Arce 372.

134 *a quarter deep.'* Lizárraga, ch. 63 (1908) 80.

134 *many precious stones.'* Cieza de León, *Crónica*, pt. 1, ch. 92 (Onis) 146; also Cobo, *Historia del Nuevo Mundo*, bk. 13, ch. 12; an anonymous Jesuit in Marcos Jiménez de la Espada, *Tres relaciones de antigüedades Peruanas* 148; Acosta, *Historia natural*, bk. 5, ch. 12, 153; Poma de Ayala (1936) 258, 264 ff; Santacruz Pachacuti 159–61.

134 *the present day'*. Molina of Santiago 145. Lizárraga thought that Mancio Sierra's 'sun' might have been the golden cover to the stone font that still existed in the late sixteenth century – Lizárraga, ch. 63 (1908) 80. Gonzalo Fernández de Oviedo said that Rodrigo Orgóñez captured the effigy from Manco in 1537 and gave it to his half-brother Paullu (bk. 4, ch. 15), but this was probably a lesser Punchao.

135 *enjoyed and possessed.'* Molina of Santiago 118.

135 *the bare hillside*. I have noticed more accessible, but more ruined, lines of storehouses on the hill above Jauja and in the sands of the coastal desert below the fort of Chancaillo in the Casma valley.

135 *so many items.'* Sancho, *Relación* 195. Estete gave a similar description, *Noticia* 47.

136 *in many colours.'* Pedro Pizarro (1844) 271–2.

CHAPTER 7 JAUJA

138 *a tiny garrison.*' López de Gómara, ch. 127, **2** 42–3.

138 *one of his brothers.* Cabildo of Jauja, letter of 20 July 1534, CL **3** 5; Sancho (*Relación* 174) reported that there had been a rumour that Manco intended to side with the Quitans, but Pizarro satisfied himself that their enmity was too intense for this; Titu Cusi Yupanqui 27; Herrera, Dec. v, bk. 6, ch. 7.

138 *great deal every day*'. Sancho, *Relación* 174.

139 *his Quitan cousins.* Titu Cusi Yupanqui (27) said that Pizarro wrote that he missed Manco and wanted him back; Sancho (*Relación* 176) quoted verbatim the hurried report from Riquelme brought back by Manco.

139 *be kept secret*' Herrera, Dec. v, bk. 6, ch. 7.

139 *their selfish interests*'. Ibid. Although Herrera was the official historian of the Conquest, he wrote almost with regret of Quisquis's lost opportunities at Jauja.

139 *of the city.* Cabildo of Jauja, letter of 20 July 1534, CL **3** 5–6; Despatch from the Governor [Pizarro] and royal officials, Jauja, 25 May 1534, CP 115; Sancho, *Relación* 176. All these contemporary sources reported that Quisquis's main force was by now only 6,000 men.

140 *and fine arms*'. Herrera, Dec. v, bk. 6, ch. 7; The battle is described in Cabildo of Jauja, letter of 20 July 1534, CL **3** 6; Sancho, *Relación* 176; also mentioned briefly by López de Gómara, ch. 127, Pedro Pizarro (1965) 199, Zárate, bk. 2, ch. 8.

140 *returned to Jauja.* Sancho (Means) 133.

141 *a Spanish municipality.* Porras Barrenechea, 'Dos documentos esenciales' 89; Loredo, *Los repartos* 102.

141 *for the natives.* Herrera, Dec. v, bk. 6, ch. 11; Titu Cusi Yupanqui 27. Manco also left Tiso and other generals at Cuzco.

141 *led 203 men.* The figures are from testimony given in 1560 and 1561 by cacique Gerónimo Guacra Paucar and Francisco Cusichaca. They testified before the Audiencia of Lima, and their quipu records were admitted as evidence. The testimony was forwarded to the King with a petition for royal favours (Porras Barrenechea, 'Jauja, capital mitica' 128).

141 *in mid-May.* Jerónimo de Aliaga was appointed as the expedition's overseer on 10 May, Harkness Cal 17. On 25 May, Pizarro wrote that Soto had left eight days previously, CP 116.

141 *part of Peru.* The Cabildo of Jauja, in the letter of 20 July (7), said that Soto's force had advanced for 70 leagues, and Pedro Sancho mentioned that it was authorised to proceed as far as Huánuco. The fortified pass near Bombón was only 40 leagues from Jauja. Titu Cusi Yupanqui (27–8) mentioned many skirmishes, but did not think that his father Manco was on the expedition; the contemporary letters of the Cabildo of Jauja and of Pedro Sancho made it clear that he was. Pedro Pizarro said that Soto pursued Quisquis to beyond the Atavillos.

142 *could join hands.*' Cieza de León, *Crónica*, pt. 2, ch. 16 (Onis) 104.

142 *released the remainder*' Cobo, *Historia del Nuevo Mundo*, bk. 14, ch. 16, (1956) **4** 258. Almost all the chroniclers described these spectacular hunts, including Zárate, ch. 8; Pedro Pizarro (1844) 280; Garcilaso, pt. 1, bk. 6, ch. 6; Gutiérrez de Santa Clara, bk. 3, ch. 57, **166** 235–6; Acosta, *Historia natural*, Hakl Soc **60** 273, 287; Ramírez, JLPB **1** 18.

142 *be on us.*' Estete, *Noticia* 56.

142 *Spaniards and Peruvians.* Cobo, bk. 14, ch. 16; Herrera, Dec. v, bk. 4, ch. 8; Zárate, ch. 8; Pedro Pizarro (1965) 198.

143 *city of Cuzco.*' This document was discovered by Porras Barrenechea, who published its text in 'Dos documentos esenciales' 86–95. The scene was exactly described by the scrivener Sancho Ortiz de Orúe, who was asked by the Corregidor Polo de Ondegardo to record it in 1560 (Harth-Terré, 'Fundación de la ciudad incaica'). The object cut by Pizarro was a picota, a symbolic post that the Spaniards erected in all their towns; it acted as a gibbet, pillory or decorative column.

143 *of this width.* Ibid.

143 *remain as citizens.* Pizarro had invited his men to settle as citizens of Jauja immediately after his arrival on 20 April. The Act of Foundation five days later supplemented an earlier ceremony performed when the Governor passed through Jauja the previous October (Porras Barrenechea, 'Jauja, capital mitica' 121).

143 *the native inhabitants'*. Porras Barrenechea, 'Dos documentos esenciales' 89.
143 *our first ancestor'*. Ibid. 86.
143 *any way whatsoever'*. Meeting of 1 April 1534, Paul Rivera Serna, 'Libro primero de cabildos del Cuzco', *Documenta* **4** 462 1965.
143 *of this order*. 27 June 1534, CL **3** 11.
143 *are more Spaniards.'* CL **3** 13.
143 *from ordinary people*. Ibid.
144 *in his wars'*. Pizarro's second decree of 23 July and the reply of the Cabildo of Jauja formed part of the evidence in a legal action between Pizarro's daughter Francisca and the Crown in 1561 (MS. in AGI, Escribanía de Camara, 496). The decree of 4 August releasing the treasure is in CDIA **10** 248, 1868; Porras Barrenechea, 'Jauja, capital mitica' 132–3.
144 *charge of them.'* Instrucción de Francisco Pizarro a Hernando de Soto, Jauja, 27 July 1534, CDIA **42** 134, 1884.
144 *to the contrary.'* Harth-Terré, 'Fundación de la ciudad incaica' 122.
144 *are tied down'*. Officials of Puerto Rico to Emperor, San Juan, 26 February 1534, CP 100.
144 *greed of Peru'* García de Lerma despatch, Santa Marta, 25 January 1534, CP 98.
144 *go to Peru*. Audiencia of Española despatch, Santo Domingo, 30 January 1534, CP 98–9. Gonzalo de Guzman despatch, Santiago de la Fernandina, 31 October 1534, CP 139–40; Manuel de Rojas despatch, Santiago de la Fernandina, 10 November 1534, CP 141. Many of these despatches exaggerated the scale of the exodus to Peru: few men who were known to be citizens of Santo Domingo in 1528 later appeared in Peru (Lockhart, *Spanish Peru* 135).
145 *feet cut off.'* Francisco Manuel de Lando despatch, Puerto Rico, 2 July 1534, CP 118,
147 *whatever anyone requested.'* Pedro Pizarro (1965) 198.
147 *than 5,000 vassals.'* Ruiz de Arce 372.
147 *for His Majesty.'* Sancho, *Relación* 182.
147 *he reached Jauja*. He awarded an encomienda to Tomás Vásquez at Cuzco on 26 March, probably the day before his departure. The Harkness Collection in the U.S. Library of Congress contains an award to Gonzalo de los Nidos of the territories of two caciques, to the east and south-west of Cuzco (Harkness Cal 14).
148 *protecting the Indians*. Despatch from Pizarro and the royal officials, Jauja, 25 May 1534, CP 112–17. Also Zurkalowski, 'El establecimiento de las encomiendas en el Perú' Kirkpatrick, 'The Landless Encomienda'; Belaúnde Guinassi, *La encomienda en el Perú*; Závala, *La encomienda indiana*.
148 *for their benefit.'* Jauja, 11 August 1534, RAN, pt. 15, **1** 10–11, 1942. This issue of RAN also contains the grant of the natives of Ica to Juan de Barrios, and other awards. An encomienda grant made to Nicolás de Ribera on 1 September is in RAN **4** 12, 1936.
150 *to see them.'* Ruiz de Arce 374.

CHAPTER 8 THE QUITAN CAMPAIGN

152 *the northern army*. Ruiz de Arce 364; Zárate, bk. 2, ch. 8, 480. The Cabildo of Jauja, in its letter of 20 July 1534, mentioned a general called Cocul-Capita seated alongside Chalcuchima and Tiso at the coronation of Tupac Huallpa at Cajamarca in early August 1533 (CL **3** 2). This could have been the officer transcribed as Zope-Zopahua by other chroniclers. He later played a prominent role in the defence of Quito. Velasco described him as Governor of Mocha, to the north of Riobamba (**2** 120). It is not clear whether he was a local Quitan chief or an Inca provincial governor of royal blood. Garcilaso de la Vega argued that his name was Zumac Yupanqui, meaning 'Handsome Yupanqui' (the family name of the Inca royal house), and that he was of royal descent and had been a loyal captain of Atahualpa (pt. 2, bk. 2, ch. 9, 93). But all other sixteenth-century chroniclers transcribed his name with variations of the spelling I have adopted, and they were unlikely to have confused the familiar name Yupanqui.
152 *into a kettle-drum.'* Oviedo y Valdés, bk. 46, ch. 17, 104; also Zárate, bk. 2, ch. 8, 480; López de Gómara, ch. 125, 39; Naharro 251; Gutiérrez de Santa Clara 230; Garcilaso, pt. 2, bk. 2, ch. 3, 83–5; Murúa, pt. 2,

bk. 2, ch. 14 (1946) **2** 139–40; Poma de Ayala **1** 118–19, 377–8; Velasco, bk. 4, ch. 1, **2** 112. Most chroniclers called Atahualpa's brother by the common Spanish surname Illescas; Garcilaso was probably correct in using Quilliscacha. All chroniclers concurred in accusing Rumiñavi of treachery, including the pro-native Garcilaso and Poma de Ayala. But modern Ecuador regards him as a heroic patriot.

152 *be very great'.* Martín de Paredes to Treasurer Martel de la Puente, San Miguel, 14 February 1534, CP 99–100.

152 *to the north.* Porras Barrenechea, *Cedulario del Perú* **1** 17–24.

153 *into the interior.'* Pedro de Alvarado to Francisco de Barrionuevo, Puerto Viejo, 10 March 1534, CP 102–3. He had already written to the King boasting of the size of his expedition, 8 January 1534, CDH Chile **4** 172.

153 *left very depopulated.'* Barrionuevo despatch, Panama, 19 January 1534, CP 97.

153 *San Miguel de Piura.* Espinosa's despatch of 11 September 1533, CDIA **42** 93, 1884; Cabildo of Jauja, letter of 20 July 1534, CL **3** 2; Zárate, bk. 2, ch. 9, 480.

153 *conquest of Quito.* Benalcázar despatch of 11 November 1533, CDIA **42** 93–5, 1884; Francisco Barrionuevo's interrogation of the crew of the ship *Concepción*, which reached Panama from San Miguel, 7 April 1534, CDIA **10** 146, 1868.

153 *sixty-two horses.* Letter of Pizarro and royal officials to Cabildo of Panama, Jauja, 25 May 1534, CDIA **10** 139, 1868. Martín Paredes to Treasurer Martel de la Puente, San Miguel, 14 February 1534, said that Benalcázar was about to leave for Quito, CP 99–100. Francisco de Barrionuevo told the Emperor that Benalcázar had taken '200 men including 140 horsemen', described him as 'a man of the base sort' and recommended that he be punished for his disobedience, despatch of 8 April 1534, Panama, CP 106.

Similar figures were given Zárate, bk. by 2, ch. 9 (200 men, 80 horses), by Nabarro, 251, (200 men, 40 horses), and by Juan de Castellanos **2** 122 (175 men, 64 horses). López de Gómara said 280 men and 80 horses; Herrera 140 men in all. The crew of the ship *Concepción* reported that Benalcázar had 200 to 230 men and 150 horses at San Miguel.

154 *the Quito road.* Letter of 25 May 1534, CDIA **10** 142; Sancho, *Relación* 185–6; Cabildo of Jauja, CL **3** 6; Almagro despatch from San Miguel, 8 May 1534, CDH Chile **4** 220, 1889.

155 *Inca Huayna-Capac.* Herrera, Dec. v, bk. 4, ch. 11 **10** 329; Oviedo y Valdés (bk. 46, ch. 19) described a battle at a town that he called Churnabalta. This may have been a corruption of Zoro-palta. The ruins of Paquishapa are close to the town of Saraguro.

156 *great they were.'* Cieza de León, *Crónica*, ch. 44 (Onis) 69–71.

156 *Lake of Blood.* John Murra, 'The Historic Tribes of Ecuador' in J. H. Steward, ed., *Handbook of South American Indians*, Washington D.C., 1946, **2** 808.

156 *beg for mercy.* The observant Cieza noted that fifteen years later the ratio of women to men was fifteen to one. He marched through here on the way to join the army of President Pedro de la Gasca against the rebel Gonzalo Pizarro. The Spaniards, as usual, ordered the local Indians to carry their baggage, and were surprised when the Cañari sent women and girls to perform this corvée. The surviving men of the tribe were considered too precious to work (*Crónica*, pt. 1, ch. 44 (Onis) 72).

157 *force nor spirit.'* Herrera, Dec. v, ch. 11, **10** 332.

157 *of beaten gold.'* Oviedo y Valdés, bk. 46 (pt. 3, bk. 6), ch. 19, **121** 111.

157 *many horses wounded.'* Ibid.

158 *3 May 1534.* This is the estimate of the eminent Ecuadorean historian Jacinto Jijón y Caamaño, *Sebastián de Benalcázar* **1** 30–2, 42.

158 *50,000 natives deployed.* Oviedo y Valdés, bk. 46, ch. 19, **121** 111; Castellanos **2** 122 estimated 55,000, but López de Gómara said only 12,000 (46). Juan de Velasco said there were 4,000 dead, bk. 4, ch. 2, **2** 117.

158 *throughout the countryside.* The horses belonged to the Spaniards Girón and Albarrán (López de Gómara, ch. 125, **2** 40; Herrera, Dec. v, bk. 4, ch. 11, **10** 332).

158 *preparing their food.'* Castellanos **2** 121.

158 *Colta and Riobamba.* Herrera said that they followed the road of Chimo and of the Puruas. Jijón y Caamaño felt that Chimo referred to the Chimbo river on the Pacific watershed (*Sebastián de Benalcázar* **1** 33). The Spaniards would then have traversed

modern Bolívar province to approach Lake Colta on the road between San Jose de Chimbo and Cajabamba.

158 *of the hills*. Probably the hills of Leompug (Castellanos **2** 121).

159 *to be executed*. López de Gómara, ch. 125, 40; also Zárate, bk. 2, ch. 9, 481.

161 *in these people*.' Cieza de León, *Crónica*, pt. 1, ch. 39 (Onis) 23.

161 *unworthy of a Castilian*' Herrera, Dec. v, bk. 6, ch. 5, **11** 35.

161 *Benalcázar to Quito*. Almagro's despatch from San Miguel, 8 May 1534, CDIA **4** 219–20; *Relación de los sucesos*, CDH Chile **4** 202.

161 *cold Andean nights*. Jijón y Caamaño, *Sebastián de Benalcázar* **1** 67, suggested that Alvarado chose this difficult route instead of the normal road via Angamarca. This would explain the unusually heavy snow and cold.

162 *chains and ropes*.' Información of 12 October 1534 held at San Miguel, CDH Chile **4** 244 ff. Chonanan may have been the village of Chongón, west of Guayaquil.

162 *they were carrying*.' Ibid.

162 *many of them*.' Ibid.

162 *told the route*' Ibid.

163 *on the enemy*.' Oviedo y Valdés, bk. 46, ch. 20, **121** 114, also for other events in this paragraph: Zárate, bk. 2, ch. 9, 482; López de Gómara, ch. 127, 43.

164 *he did so*. The foundation was made on 15 August 1534. From Monday 17 August to Thursday 20 August, 68 out of some 300 members of Almagro's army registered as would-be citizens of the new city. The agreement of 26 August is in the Library of Congress in Washington, D.C. (Harkness Cal 17).

164 *and 20,000 effectives* Zárate (bk. 2, ch. 12, **26** 482) said he had 12,000, López de Gómara (ch. 128, **2** 45) said 5,000 and Oviedo 20,000 (**121** 115).

165 *of Quisquis's camp*.' Zárate, bk. 2, ch. 12, **26** 483. A captain called Martín Monje took credit for capturing Sotaurco. *Información de servicios de Martín Monje*, La Plata, 8 January 1563, CDH Chile **7** 338, 1895.

165 *of San Juan*.' Ibid.; also López de Gómara, ch. 128, **2** 45.

165 *beheaded fourteen Spaniards*.' Ibid.

166 *male and female porters*. Ibid. Oviedo y Valdés (bk. 46, ch. 20 **121** 115, said that 20,000 llamas and 20,000 prisoners were taken at Chaparra.

166 *they were invincible*.' López de Gómara, ch. 128, 45.

166 *among the orejones*.' Ibid.; other authorities for this passage are Zárate, bk. 2, ch. 12, 483; Pedro Pizarro (1965) 199; Titu Cusi Yupanqui 28.

166 *in their grasp*'. Pedro Pizarro (1965) 199.

167 *lances or battle-axes*' Castellanos **2** 128. Francisco de Rodas also claimed, in a Probanza held at Cali on 2 March 1545, that he had participated in this attack (*Boletín de la Real Academia de Ciencias, Bellas Letras y Nobles Artes*, Córdoba, **8**, no. 24, 330–1, 1928). No other sources described it in detail.

167 *a small hut*.' Herrera, Dec. v, bk. 7, ch. 14,. **11** 172.

167 *I captured him*.' Miguel de la Chica claimed to have captured Rumiñavi in an Información of his services held in 1555. The document was found in the central archive of Cauca by Luis E. Ubidia Rubio, who published it in 'Aporte documental a la captura de Rumiñahui', *Boletín del Archivo Nacional de Historia*, Quito, **2**, nos. 3–4, 5–42, 1951. Various witnesses confirmed his claim; and Castellanos described the capture (**2** 128). But other chroniclers made no mention of Rumiñavi's end, except for Oviedo, who said that he was killed by his own troops (bk. 46, ch. 20, **121** 114), but Oviedo may have confused Rumiñavi and Quisquis.

167 *in the crag*'. Probanza of Francisco de Rodas, Cali, 2 March 1545, cited in first note to this page; also Castellanos **2** 128.

Herrera, on the other hand, wrote that Juan de Ampudia called on Zope-Zopahua to surrender, and that the native general did so because of the total collapse of his native supporters, who 'no longer obeyed any law or authority, but thought only of preserving their lives with the conquerors' (Dec. v, bk. 7, ch. 14, **11** 171–2).

167 *the buried treasure*.' Marcos de Niza wrote an 'Información á la corte y al Obispo Zumarraga de Méjico'. Bartoloméde las Casas inserted it into his *Brevissima relación de la destrucción de las Indias* in 1552 (ed. Bouret, Paris, 118–21). Juan Lopez de Velasco quoted it in his *Historia del reino de Quito*, bk. 4, ch. 6, **2** 132. But even Velasco – a rather hysterical eighteenth-century author – warned that not everything Niza wrote should be believed. Niza had in fact

left Quito with Alvarado before these executions, which he claimed to have observed, and he went on to confuse them with the burning of Chalcuchima and Atahualpa. Velasco frequently cited a chronicle by Niza, but Raúl Porras Barrenechea concluded that no such work ever existed ('Crónicas perdidas') 197–202.

167 *his first impression'* Herrera, Dec. v, bk. 7, ch. 14, **11** 172–3.

168 *square of Quito.* Velasco **2** 121. Velasco said in a footnote that Dr José Fernández Salvador, Director of Studies at Quito, told him he had seen an act of the municipality of Quito confirming Rumiñavi's execution in the square.

CHAPTER 9 PROVOCATION

169 *because of him'.* Sancho, *Relación* 200.

169 *and Imperial Majesty'* Almagro's despatch of 8 May 1534, San Miguel, CDH Chile **4** 219, 1889.

169 *of this cacique'.* Cabildo of Jauja, despatch of 20 July 1534, CL **3** 4.

170 *it with admiration.* Falcón 463; Cieza de León, *Crónica*, pt. 2, bk. 2, chs. 20, 23; Sarmiento de Gamboa, ch. 50; Acosta, *Historia natural*, bk. 6, ch. 13; Cobo, *Historia del Nuevo Mundo*, bk. 12, ch. 25; Castro and Ortega Morejón in H. Trimborn, 'Quellen zur Kulturgeschichte' 237; Santillán in Jiménez de la Espada, *Tres relaciones* 17–18; Rowe, *Inca Culture* 263; Espinosa Soriano, 'El alcalde mayor' 187–9; Gibson, *The Inca Concept of Sovereignty* 42 ff.; Means, *Ancient Civilisations of the Andes* 292; Levillier, *Don Francisco de Toldeo* **1** 96–8; Trimborn, 'Las clases sociales'.

172 *crowned by Sacsahuaman.* One of the entries in the distribution of plots granted to two conquistadores 'one lot each in the corral upstream [on the Huatanay], alongside the Governor's plot on one side, and on the other side the new houses which the cacique [Manco] is building'.

173 *broke it first.'* Molina of Santiago 160.

174 *royal Inca blood.* Añas Collque was probably the Juana Tocto mentioned by Sarmiento de Gamboa and Bernabé Cobo as having survived the massacres by Atahualpa's generals (Sarmiento, ch. 67, 185; Cobo, *Historia del Nuevo Mundo*, bk. 12, ch. 20; See also Ella Dunbar Temple's 'Paullu Inca').

174 *as far as Chile.'* Quipocamayos 34. Pedro Pizarro (1965) 196. Cobo, *Historia del Nuevo Mundo*, bk. 12, ch. 20. Sarmiento de Gamboa said that Paullu was in prison in Cuzco when Quisquis arrived and succeeded in persuading the Quitans that his sympathies lay with Atahualpa (ch. 67, Hakl Soc, 2 ser., **22** 185).

175 *a great altercation'.* Herrera, Dec. v, bk. 7, ch. 6.

175 *those who survived.'* The Treasurer Antonio Tellez de Guzman had moved to Cuzco and claimed to have prevented bloodshed there, Tellez de Guzman to King, Seville, 5 May 1536, CDH Chile **4** 60. Also Pedro Pizarro (1965) 200; Zárate, bk. 2, ch. 13; López de Gómara, ch. 130.

175 *Rodrigo Orgóñez.* The text of the agreement of 12 June 1535 was given by Prescott as appendix XI and is in CDH Chile **4**. Rafael Loredo published a different version which he claimed came from chapter 81 of Cieza de León's missing Third Part, in *Los repartos* 381–7.

175 *fifth of everything.* The ovens were functioning from 20 May to the end of July. The exact quantities of treasure and the owners and dates of melting were scrupulously recorded in a Relación published in CDIA **9** 503–82, 1867.

176 *befits your rank.'* Molina of Santiago 157. The subsequent references in this chapter are all to this Cristóbal de Molina.

176 *and his relatives.'* Ibid.

176 *obeisance to him.* Titu Cusi said that Manco's Spanish servant discovered the plot and killed Pascac before he could strike. But Pascac was alive until 1537; the person killed here must have been his agent. (Titu Cusi Yupanqui 59–60.)

176 *in his bed.* Molina (157) described the episode but did not name the brother. Hernando Pizarro named this brother, in 1540, as Atausa and described him as the 'rightful heir' (*Confesión* 443). He may have been Atauchi, whose son of the same name emerged as one of the leading royal claimants in the 1550s. Pedro Pizarro called Almagro's victim Atoc-Sopa (1844, 274),

and Cobo called him Octo-Sopa (*Historia del Nuevo Mundo*, bk. 12, ch. 20). Pedro Pizarro may have been confusing the brother with one Atoc-Sopa, whom Manco appointed regent for his son ten years later. See also, Herrera, Dec. v, bk. 7, ch. 8, **11** 139–40, where the sequence of events told by Molina is transposed.

177 *to the Governor'*. Molina 159.

177 *took no action*. Herrera, Dec. v, bk. 7, ch. 8, **11** 140.

177 *and powerful parentage.'* Cieza de León, *Crónica*, pt. 2, ch. 30, Hakl Soc 2 ser., **68** 95.

178 *dying of hunger.'* Molina 165.

179 *good and fat.'* Ibid.

179 *on the chain.'* Ibid. 171. Hernando Pizarro also accused Almagro of these excesses on the Chilean expedition, *Confesión* 439.

179 *to rob'*. Pedro Pizarro (1965) 201.

179 *was despised.'* Molina 167; Oviedo y Valdés, bk. 47, ch. 3, **121** 134.

179 *pot of water'*. Ibid, ch. 4, 138. Oviedo wrote the most detailed account of the Chilean expedition. He also wrote a deeply moving account of the death of his own son, who was drowned on the return journey.

179 *abuse of them.'* Molina 155.

180 *on the fact.'* Ibid. 114.

180 *their yanacona servants.'* Ibid.

180 *27 May 1536*. Raúl Porras Barrenechea deduced the date of Francisca's birth through meticulous research which he explained in a footnote in 'Jauja, capital mítica' 137–9. He also found the royal decree legitimising Francisca. The three godmothers were Isabel Rodríguez, who called herself La Conquistadora and claimed to be the first Spanish woman in Peru; Francisca wife of Rui Barba; and Beatriz wife of the overseer García de Salcedo.

181 *very good wife.'* Molina 162.

181 *in Quito*. Don Mateo Yupanqui had been the Inca chief of the mitimaes colonists in Quito. The Spaniards gave him the title Chief Constable of the Natives of Quito.

181 *his mistress nearby*. The Acta de Fundación said that a plot was awarded to one Pedro de Ulloa in 'the enclosure in which lives the palla [lady] of the Lieutenant Hernando de Soto, behind Lobillo's lot' (Porras Barrenechea, 'Dos documentos esenciales').

181 *lived in Cuzco*. Cabello de Balboa knew the Carrillos, and Bernabé Cobo knew their son Pedro, who adopted his grandfather's famous name de Soto. Carrillo, with his notary's legal sense, felt that his wife had a claim to Soto's palace of Amaru Cancha, which Hernando Pizarro had later seized and sold to the Jesuits. Carrillo sued the Jesuits for 14,000 pesos in 1572 (Ella Dunbar Temple, 'La descendencia de Huayna Cápac' 161).

181 *jealousy of Inés*. Pedro Pizarro (1965) 225.

182 *who was pregnant*. Lockhart, *Spanish Peru* 211, 215–16.

182 *as my daughter.'* Will of Juan Pizarro, Cuzco, 16 March 1536, RI **8** 873, 1947. The girl was called Isabel, but her mother's name remains unknown.

182 *on their faces'*. Titu Cusi Yupanqui 55.

182 *pieces of you!'* Ibid.

182 *we really want.'* Ibid.

182 *she's the coya."'* Ibid.

183 *any other reason.'* Ibid.

183 *still has her.'* CP (carta 217) 337.

183 *harassment became intolerable*. Manco himself wrote later: 'Diego Maldonado threatened me and demanded gold, claiming that he too was an apo.' CP 337; Titu Cusi Yupanqui 39.

183 *us and abuse.'* Murúa, ch. 66, **1** 196–7.

184 *going that way'*. Herrera, Dec. v, bk. 8, ch. 1.

184 *on his feet.'* Titu Cusi Yupanqui 44. Titu Cusi had made these same charges in the Memorial he handed Juan de Matienzo in June 1565. Matienzo, *Gobierno del Perú*, pt. 2, ch. 18, 301.

184 *on their encomiendas.'* Pedro Pizarro (1844) 287. Herrera (Dec. v, bk. 8, ch. 2) described the native anger, using a source unknown to us, presumably Cieza's missing Third Part.

184 *a lighted candle.'* Despatch from Pedro de Oñate and Juan Gómez Malaver, Cuzco, 31 March 1539, CDH Chile **5** 277–99. This account of a conversation with Manco was written by two men violently opposed to the Pizarro family and faction. Similar charges were made in a 'Causa criminal' issued against the Pizarros by Diego de Almagro the Younger and Diego de Alvaradro, Madrid, 17 April 1540 (CDH Chile **5** 368, 1889). This accusation and a similar one held in December 1541 were resurrected in 1573 and used by the Spanish crown to discredit a petition by Hernando Pizarro

and Francisco's daughter Francisca: evidence of the Royal Fiscal, item 52, CDH Chile **7** 259 ff., 1895.

185 *will be burned."'* Oviedo y Valdés, bk. 47, chs. 7, 8, **121** 153, 155, 157. Oviedo was a strong partisan of Almagro against the Pizarros and therefore repeated the many accusations against the Pizarros and their supporters, possibly with some embellishment.

185 *was deeply distressed.'* Molina 173. Molina was with Almagro on the Chilean expedition at the time.

185 *where people passed.'* Accusation 9 against Hernando Pizarro, 11 December 1541, CDIA **20** 237–8, 387–8, 1873. Hernando had already heard about Manco's arrest when he wrote to the Emperor from Lima, 15 November 1535, CP 176.

185 *others who wished.'* Bishop Tomás de Berlanga to King, Nombre de Dios, 3 February 1536, CP 192. Bishop Berlanga of Panama was sent by the King to investigate the situation in Peru, particularly the growing rift between Pizarro and Almagro, the treatment of the natives, and the position of royal and ecclesiastical authority. He reached Lima in July 1535 and returned to Panama by February 1536 deeply disturbed by what he had seen.

185 *to the east.* Herrera, Dec. v, bk. 8, ch. 3.

185 *the royal mines.* The estimate of thirty dead was made by Alonso Enríquez de Guzman, who arrived in Cuzco early in 1536 (Hakl Soc, 1 ser., **29** 100, 1862). Also Zárate, bk. 3, ch. 3, 486; López de Gómara, ch. 133, 52; *Relación de los sucesos*, 390; Herrera, Dec. v, bk. 8, ch. 2; Pedro Pizarro (1965) 201.

186 *no greater mercy.'* Herrera, Dec. v, bk. 8, ch. 2. The Spanish volunteers included Pedro del Barco and Mancio Sierra. The latter claimed many years later to have been the first Spaniard to penetrate the fortress of Cuzco, but he may have been confusing it with the crag of Ancocagua (Fernando Pizarro y Orellana, *Varones* 199–201). Details of this action were also contained in the Probanza of the Presbyter Rodrigo Bravo held in Cuzco, 23 December 1538, question 4, DH Arequipa **2** 43 ff.

186 *from Juan Pizarro.* This is stated in: *Relación del sitio del Cuzco*, CLERC **13** 5–6, 1879. This important narrative begins with Hernando's arrival. It was once thought to have been written by Bishop Vicente de Valverde, but this was impossible, since it was clearly the work of an eyewitness, and Valverde was in Lima at the time. It might have been by Diego de Silva, a literate soldier, but he was also in Lima. Zárate (bk. 3, ch. 3, 486) confirms this, but other sources said that Hernando released Manco: Pedro Pizarro (Means) 298; *Relación de los sucesos*, 389; Herrera, Dec. v, bk. 8, ch. 4. Hernando may have ordered Manco's release before reaching Cuzco. He promised the King he would do so, despatch of 15 November 1535, CP 176–7.

187 *hoped to wage.'* López de Gómara, ch. 133, 52.

187 *Ronpa Yupanqui'* Titu Cusi Yupanqui 62–3. The tribe of Quito and the northern part of the empire did not respond. They were exhausted by their recent fighting, and evidently refused to obey a summons from an Inca of the Cuzco faction.

188 *interpreter Antonico. Relación de los sucesos* 390; Quipocamayos 38; Zárate, bk. 3, ch. 3; Molina 173; López de Gómara, ch. 133, 52; Herrera, Dec. v, bk. 8, ch. 4.

188 *this sounded plausible. Relación del sitio del Cuzco* 9. A different version of Zamarilla's encounter was given by Diego de Almagro the Younger: 'Alonso García Zamarilla met Manco ten leagues from Cuzco, and the Inca told him to tell Hernando Pizarro that he had rebelled because of the injuries and ill-treatment he had received from him and his servants.' Accusation 11 against Hernando Pizarro, CDIA **20** 389, 1873.

188 *course of action. Relación del sitio del Cuzco* 9–10; *Relación de los sucesos* 390; Zárate, bk. 3, ch. 3; López de Gómara, ch. 133; Molina 173; Garcilaso, pt. 2, bk. 2, chs. 23–4; Pedro Pizarro (1965) 202; Herrera, Dec. v, bk. 8, ch. 4; Titu Cusi Yupanqui 63–4.

CHAPTER 10 THE GREAT REBELLION

190 *filled with stars.'* Pedro Pizarro (1844) 289.

190 *100,000 and 200,000.* Enríquez de Guzman (Hakl Soc **29**, ch. 46, 1862) reckoned between 100,000 and 300,000. Pedro Gallego

said 100,000 in his testimony in 1537 (GP **2** 88). The author of the *Relación del sitio del Cuzco* said 100,000 plus 80,000 porters and auxiliaries 18). Others who estimated 200,000 included Pedro Pizarro, Herrera, Garcilaso and Mancio Sierra de Leguizamo, when he testified in favour of Francisco Pizarro in 1572 (GP **2** 137). Montesinos said over 50,000 (*Anales*, 1906, entry for 1536, **1** 88). Titu Cusi Yupanqui gave the highest figure, 400,000, in his *Relación* (65). Diego de Peralta said 300,000 in his Probanza, Arequipa, 16 February 1566 (DH Arequipa **2** 327).

191 *of the Spaniards.*' Titu Cusi Yupanqui 66.

191 *Paucar Huaman.* Murúa gave the names of these three generals (**1** 198).

191 *debilitated men*'. Pedro Pizarro (Means) 302.

191 *this high altitude.* Ibid. (1844) 294.

192 *a distinct gain.*' *Relación de los sucesos* 391.

192 *they despised them*'. Titu Cusi Yupanqui 67.

192 *with the Spaniards.*' *Relación del sitio del Cuzco* 19.

192 *of the square.* The Acta de Fundación referred to Cora Cora as 'the fortress of Huascar' and said that it was allotted to Diego de Almagro. It had also been the palace of Sinchi Roca, the second Inca.

192 *a hen's egg.* Xerez 334.

193 *thirty yards away.*' Enríquez de Guzman, Hakl Soc **29** 99.

193 *with the enemy.*' *Relación del sitio del Cuzco* 19.

193 *were very intense.*' Molina of Santiago 175.

193 *are hailing furiously.*' *Relación de los sucesos* 392.

193 *to defend themselves.*' *Relación del sitio del Cuzco* 20; Titu Cusi Yupanqui 67.

194 *this extraordinary escape.* Montesinos, entry for 1536 (1906) **1** 88; Garcilaso produced a veritable pantheon of miraculous apparitions, and was sure that the Indians would have won but for these divine interventions, pt. 2, bk. 2, chs. 24, 25, **134** 124–8; Gutiérrez de Santa Clara, bk. 3, ch. 63, 252–3; Murúa, ch. 66, **1** 199; Vázquez de Espinosa 559–60; Poma de Ayala sketched the scene. Declaration by Fray F. de Chaves 1641, Diego de Córdova Salinas, *Crónica, franciscana del Perú*, bk. 3, ch. 8, 536. The miracle was first recorded by Gutiérrez de Santa Clara. It did not appear in any of the early accounts of the siege, nor in López de Gómara, Zárate or Herrera. But for an argument in support of a vision, see Rubén Vargas Ugarte, *Historia del culto de María en América* **2** 232 ff, and his *Historia general del Perú* **1** 114–15. The church of Triunfo abuts against the cathedral on the east side of the main square.

194 *what to do.*' Titu Cusi Yupanqui 67; *Relación del sitio del Cuzco* 20.

194 *few of them.*' *Relación de los sucesos* 393.

194 *save their lives.*' *Relacion del sitio del Cuzco* 21.

195 *some Indian auxiliaries.* Pedro Pizarro (1844) 297.

195 *with these cords.*' *Relación de los sucesos* 392.

197 *65 feet long.* Luis E. Valcárcel, 'Sajsawaman redescubierto', RMN **3**, 1934 and **4**, 1935, and 'Cuzco Archeology' *Handbook of South American Indians* **2** 178–9, 1946. Also: Santiago Astete Chocano, 'Temas arqueológicos sobre la fortaleza de Sajsa-Uma', *Revista del Instituto Arqueológico del Cuzco* **2** (1937) 36–41; J. M. Franco Inojosa y L. A. Llanos, 'Sajsawaman: excavación en el edificio sur de Muyumarca', RMN **9**, no. 1, 22–32, 1940; Luis A. Pardo, 'La ciudadela de Sacsaihuamán, *Revista del Instituto Arqueológico del Cuzco* **2**, no. 3, 3–18; Victor W. von Hagen, *A Guide to Sacsahuaman*, New York, 1949.

197 *visit them all*'. Sancho, *Relación* 193.

197 *its extraordinary construction.* Garcilaso, pt. 1, bk. 7, ch. 29; Vázquez de Espinosa 568–9. The towers and inner buildings were soon destroyed. Garcilaso said that by 1560 'all the buildings above ground were in ruins'. Cieza de León was appalled by the destruction. 'I hate to think of the responsibility of those who have governed here in allowing so extraordinary a monument to be destroyed. The remains of this fortress . . . should be preserved in memory of the greatness of this land.' *Crónica*, pt. 2, ch. 51 (Onis) 155.

197 *at breakneck speed.*' Titu Cusi Yupanqui 68.

198 *terraces of Sacsahuaman.* Pedro Pizarro (1844) 293; Murúa, ch. 66, **1** 198–9; Molina of Santiago 176–7; *Relación del sitio del Cuzco* 26–7.

198 *much was at stake.*' *Relación del sitio del Cuzco* 28.

199 *recognise as daughter.*' The text of the will is contained in Luisa Cuesta, 'Una documentación interesante sobre la familia del

conquistador del Perú', RI **8**, no. 30, 872–8, 1947.

199 *was indeed dead.*' Francisco de Pancorvo's testimony in support of Francisca Pizarro's claim against the Crown, Cuzco, 7 October 1572, GP **2** 153.

199 *200,000 ducats.*' Enríquez de Guzman, Hakl Soc, 1 ser., **29** 100.

199 *in close combat.*' *Relación del sitio del Cuzco* 30.

199 *attacked with determination*'. Ibid.

199 *a thousand times*'. *Relación de los sucesos* 394.

200 *the other tower.* This story is from the anonymous *Relación de los sucesos de la conquista del Perú* written in 1548 (perhaps by the heroic Hernán Sánchez?) 394–5. Herrera repeated it in Decada v, bk. 8, ch. 5. Mancio Sierra de Leguizamo also claimed to have been the first Spaniard to penetrate Sacsahuaman, in front of Juan Pizarro. He made this boast in the enquiry held in Cuzco in 1572, in answer to question 31 (GP **2** 145) and repeated it in his revised will of 18 September 1589. This will was quoted by Calancha, pt. 1, ch. 15 (1639 edition, 98). The full text was published in *Revista Peruana* **2** 1879. See also Ella Dunbar Temple, 'La descendencia de Huayna Cápac', RH **11** 133 ff, and Raúl Porras Barrenechea, footnote in his edition of Diego de Trujillo, 111–13, for information about the life of Mancio Sierra. Pedro Pizarro also said he was with Juan Pizarro in the first entry into Sacsahuaman's defences.

200 *became exhausted.*' *Relación del sitio del Cuzco* 31–2.

200 *stones and arrows.* Pedro Pizarro (1965) 204–5.

200 *climbed to Carmenca.*' Murúa (ch. 67, **1** 200) said that Villac Umu fought his way out of the fortress. Other chroniclers said that he slipped out by night, implying that he deserted his fellow defenders (*Relación del sitio del Cuzco* 31).

200 *top of the tower.*' Pedro Pizarro (1844) 296.

201 *on his head.*' Ibid. 295.

201 *on his arm.*' Ibid. 296.

201 *not been touched.*' *Relación del sitio del Cuzco* 32.

201 *and Cañari Indians.*' Titu Cusi Yupanqui 70.

201 *1,500 of them.*' *Relación del sitio del Cuzco* 33. Alonso Enríquez de Guzman said that three thousand were slaughtered in the capture of Sacsahuaman (Hakl Soc, 1 ser., **29** 98). The valiant orejón is sometimes called Cahuide. This is not a Quechua name and appeared in no contemporary account that I know. Rubén Vargas Ugarte thought that it arose through a mistranscription of the word 'Caribe', meaning an indomitable forest Indian (*Historia general del Perú* **1** 111).

201 *of dead men.*' Titu Cusi Yupanqui 70.

201 *died in it*'. Cedula of Charles V, Madrid, 19 July 1540, CDHH-A **3** 75, 1928.

202 *to do so.*' Titu Cusi Yupanqui 73.

203 *now behind us.*' Pedro Pizarro (1844) 298.

203 *impetuously lancing Indians*'. Ibid.

204 *endure the strain.*' Ibid. 299–300.

204 *they can imagine.*' Enríquez de Guzman, Hakl Soc, 1 ser., **29** 101.

204 *vent to it!*' Cieza de León, *Crónica*, pt. 1, ch. 82 (Onis) 106.

204 *feared to die.*' *Relación del sitio del Cuzco* 43.

205 *to the rest.*' Ibid. 45.

205 *the central highlands.* There is confusion about the name of the native commander of the attack on Lima. The author of the *Relación del sitio del Cuzco* – someone who was in Lima during the attack – referred to him as Tey-Yupanqui. Another resident of Lima, Francisco Martínez, testifying in 1572, called him Hai Yupanqui lord of Curanga (GP **2** 159). Zárate called the general Tizo Yupanqui, and Gómara called him Tizoyo, Garcilaso noted these names, but decided that Titu Yupanqui sounded better (pt. 2, bk. 2, ch. 28). But the three authors who used native sources (Martín de Murúa, Manco Inca's own son Titu Cusi and Poma de Ayala) called the commander Quizo Yupanqui, Queso Yupanqui and Quizu Yupanqui respectively, and made it clear that he was a different man from Manco's other famous general Tiso Yupanqui (Murúa, ch. 68, **1** 203; Titu Cusi Yupanqui 62; Poma de Ayala drawing reproduced on page 221). There seems no doubt that Quizo Yupanqui was the right name.

205 *crossroads of Vilcashuaman.* I deduce the date of Morgovejo's departure from the minutes of the meetings of the town council of Lima. As one of the alcaldes ordinarios of 1536, Morgovejo's presence was always noted. He was present at all meetings up to 5 May, but absent on the meeting of 26 May and thereafter (CL (2) **1** 90;

Cobo, *Historia de la fundación de Lima* 297).

206 *in the land'. Relación del sitio del Cuzco* 56.

206 *grips with them.'* Zárate, bk. 3, ch. 5, 487.

206 *and four horses'*. Titu Cusi Yupanqui 74.

206 *with more authority'*. Murúa, ch. 68, **1** 205.

207 *and Negro servants.'* Ibid. 204–5. Poma de Ayala described the stepped platform of Vilcashuaman as an usno. I am not, however, aware of any such platform at Jauja, where the most conspicuous ruins are a row of Inca storehouses and thousands of Huanca houses and corrals on the surrounding hills.

207 *the bad news'*. López de Gómara, ch. 135, **2** 55; *Relación del sitio del Cuzco* 75; Also Zárate, bk. 3, ch. 5, **26** 488; *Información de méritos del Marqués don Francisco Pizarro*, GP **2** 104, 160, 183; Murúa, ch. 68, **1** 205; Montesinos, *Anales*, 1906, entry for 1536, **1** 90–3. Francisco de Godoy was absent from five meetings of the Cabildo of Lima, from 3 July to 7 August inclusive, but in attendance again from 14 August onwards (CL (2) **1**). The report that Gaete's men were killed by 'the very Indians they were taking with them' came from a letter of October 1536 from Francisco de Barrionuevo, in Panama, to the officials of Seville (CP 224). Francisco de Godoy later did another thing that was considered disgraceful: he sold his encomienda and returned to Spain. An encomienda was a 'trust' to be enjoyed by the recipient, but not sold (Lockhart, *Spanish Peru* 20).

208 *masters were annihilated. Relación del sitio del Cuzco* 65.

208 *to relieve Cuzco*. The detailed account of Morgovejo's expedition was in the anonymous *Relación del sitio del Cuzco* 56–74. Clearly one of the survivors of Morgovejo's contingent gave the eyewitness account to the author of the *Relación*. The most literate survivor was Diego de Silva (Francisco Ponce testimony, Cuzco, 1572, GP **2** 134) and he may well have been the author of the entire *Relación*. See second note to page 186 above.

208 *as many horses.' Relación del sitio del Cuzco* 76

209 *my entire life.'* Francisco Pizarro to Pedro de Alvarado, Los Reyes, 20 July 1536, in Jijón y Caamaño, *Sebastidn de Benalcdzar*, documentos, **1** 173–6.

209 *country has rebelled'*. Pascual de Andagoya to Emperor, Panama, 26 July 1536, CP 218.

209 *possible from here.'* Pascual de Andagoya and Licenciate Espinosa to the Audiencia of Santo Domingo, 27 July 1536, CP 218–19.

209 *who wished to'*. Despatch from Tomás de Berlanga, Nombre de Dios, 3 February 1536, CDH Chile **4** 332–3. Berlanga's mission was to measure the territory awarded to Pizarro, but the aged bishop was insulted and ignored by the rapacious conquistadores.

210 *the Cuzco area*. Despatch from Licenciate Espinosa, Panama, 1 April 1536, CDH Chile **4** 346–50.

210 *gold and silver.'* Despatch from Audiencia of Española, 8 September 1536, CDH Chile **4** 364–8. This first news had been relayed from Panes's ship that reached Panama on 23 June.

210 *Spaniards he found'* Murúa, ch. 68, **1** 205.

210 *join his rebellion*. Ibid. 205–6. Murúa criticised Quizo for this delay: he felt that the Inca general should have attempted an immediate surprise attack on Lima after taking Jauja.

210 *were in it.'* López de Gómara, ch. 135, 55.

210 *his teeth broken'*. Ibid.

210 *was completely finished.' Relación del sitio del Cuzco* 78.

210 *around the city.'* Ibid.

210 *cowardly at night'*. Ibid. 79.

211 *to carry it.'* Ibid. The sugarloaf was named after San Cristóbal to commemorate the siege, in which the Saint was thought to have intervened.

211 *in the city.'* López de Gómara, ch. 135, 56.

211 *sent her into exile.'* Pedro Pizarro (1844) 346–7.

211 *otherwise have endured.' Relación del sitio del Cuzco* 79.

211 *as he said.'* Ibid. 79–80.

212 *race of warriors.'* Ibid. 80.

212 *from Pachacamac*. Murúa, ch. 68, **1** 206.

212 *the Indian redoubt.' Relación del sitio del Cuzco* 80–1. Murúa (**1** 206) said that Quizo Yupanqui was struck on the knee by an arquebus shot and died of his wound in Bombón in the mountains. Poma de Ayala drew a picture of him being lanced; it is reproduced on page 211. Pero Martín de Sicilia claimed that he had aimed his lance at Quizo, who was fighting from his litter, and struck him in the chest. Busto Duthurburu, *Francisco Pizarro* 214.

213 *custody in Lima*. The town council arranged for these coastal chiefs to be given lands near the city from which to support

themselves 'until the kingdom is pacified'. It also imposed heavy fines on any citizen who left the city without permission or who molested the peaceful coastal Indians. Meeting of Cabildo of 11 September 1536, CL (2) **1** 103–4. The various accounts of the siege of Lima are: *Relación del sitio del Cuzco* 81; Montesinos, *Anales* **1** 91–2; Zárate, bk. 3, ch. 6; López de Gómara, ch. 136; Herrera, Dec. v, bk. 8, chs. 4, 5, **11** 207–10; Murúa, ch. 68, **1** 205–6.

213 *by native ambush.* Pedro Pizarro (1844) 303; Enríquez de Guzman, Hakl Soc, 1 ser., **29** 106; Molina of Santiago 175.

213 *each ford was defended.' Relación del sitio del Cuzco* 47.

213 *a horrifying sight.'* Pedro Pizarro (1844) 306.

214 *polygonal Inca masonry.* Few sixteenth- or seventeenth-century writers made the difficult journey to Ollantaytambo. Cieza de León described it briefly (*Crónica*, pt. 1, ch. 94). It caused more excitement in the nineteenth century, and its ruins were described by Francis de Castelnau in 1851 (*Expédition dans les parties centrales de l'Amerique du Sud* **4** 274–9); Clements Markham in 1856 (*Cuzco, a Journey to the Ancient Capital of Peru*, London 1856, 179–85); Ephraim Squier in 1877 (*Peru, Incidents of Travel and Exploration in the Land of the Incas* 493–510); Charles Wiener in 1880 (*Pérou et Bolivie: récit de voyage* 332–43); Ernst Middendorf in 1895 (*Peru, Beobachtungen und Studien über das Land und seine Bewohner*, 3 vols, Berlin, 1893–5, **3** 512–20). Modern studies include: Luis A. Llanos, 'Informe sobre Ollantaytambo', RMN **5**, no. 2, 123–56, 1936; Emilio Harth-Terré, 'Fundación de la ciudad incaica', RH **16** pts. 1–2; 98–123, 1943; Luis A. Pardo, 'Ollantaitampu, una ciudad megalítica', Universidad del Cusco, *Revista de la Sección de Arqueología* **2** 43–73, 1946; Heinrich Ubbelohde-Doering, *Auf den Königstrassen der Inka* (Berlin 1941) 226–51; Victor W. von Hagen, *A Guide to Ollantaytambo* (New York 1949; Lima 1958). Valcárcel, Pardo and many of the nineteenth-century writers thought that Ollantaytambo had originated in the age-old 'megalithic' culture, because the stones of its platform sanctuary looked superficially like those of Tiahuanaco. Most modern historians now accept that it was built by the Inca Pachacuti after he absorbed the local Laris into the empire, in about 1460.

214 *they are dying.' Relación de los sucesos* 397.

214 *well-armed warriors.' Relación del sitio del Cuzco* 47, 48.

214 *hillsides and plains.' Relación de los sucesos* 397

214 *hear the shouting'. Relación del sitio del Cuzco* 48.

214 *have been killed'.* Pedro Pizarro (1966) 209.

215 *by either side.' Relación del sitio del Cuzco* 49; also *Relación de varios sucesos* 396–7.

215 *army under control.'* Herrera, Dec. v, bk. 8, ch. 7, **11** 221.

215 *by Spanish prisoners. Relación de los sucesos* 397, Herrera, Dec. v, bk. 8, ch. 6, **11** 216.

215 *the horses' girths.* All three accounts of the Ollantaytambo campaign mentioned the diverted water. The natives used the same tactic to flood the level ground outside Cuzco, and they diverted the Rimac river at Lima to run beneath the Cerro de San Cristóbal, so that they could harass Spaniards trying to fetch water.

215 *could not skirmish.' Relación del sitio del Cuzco* 49.

215 *a retreat.'* Pedro Pizarro (1844) 307.

215 *horses' tails'. Relación del sitio del Cuzco* 50.

215 *great spirit.'* Pedro Pizarro (1844) 307.

215 *crippled the horses. Relación de los sucesos* 397; Herrera, Dec. v, bk. 8, ch. 5, **11** 211; Titu Cusi Yupanqui 74.

216 *this powerful expedition.* Titu Cusi Yupanqui 75.

216 *in the world.' Relación del sitio del Cuzco* 51.

216 *befallen their compatriots.* Oviedo y Valdés, bk. 46, ch. 17, **121** 104–5; Pedro Pizarro (1844) 300.

216 *Lady the Empress' Relación del sitio del Cuzco* 52.

216 *my old age'.* Enríquez de Guzman, Hakl Soc, 1 ser., **29** 103. Pizarro treated Enríquez de Guzman with deference, because he was one of the first men of aristocratic descent to reach Peru. The Governor lent the impoverished Don Alonso 2,000 pesos when he arrived in Peru, and sent him to Cuzco as nominal second-in-command – although Hernando Pizarro refused to accept him as such.

217 *each new moon.* Pedro Pizarro (1844) 309, 319.

217 *supplies or stores.'* Quipocamayos 40.

217 *intercept the operation.* Pedro Pizarro (1844) 309; *Relación del sitio del Cuzco* 34–5,43.

217 *defence of Cuzco'.* Hernando Pizarro, *Confesión* 443.

217 *and other provisions',* Quipocamayos 40–1. The turncoats were named as Cayo Topa, Don Felipe Cari Topa, Inca Paccac and Huallpa Roca. 'Each came with great contingents of Indians and they gave much advice to the Christians.' The first two men named were both prominent throughout the century among the Cuzco native nobility. Paccac or Pascac was the cousin who contested Manco's claim to be Inca in 1535. According to Hernando Pizarro, Almagro had him secretly murdered by Francisco de Oñate in 1537 because of his Pizarrist sympathies (*Confesión* 443). Huallpa Roca was the commander whom Pedro Pizarro said Manco had sent to attack Lima.

217 *any serious engagement.'* Pedro Pizarro (1966) 208; *Relación del sitio del Cuzco* 44; Herrera, Dec. v, bk. 8, ch. 6, **11** 215.

217 *food and returned.'* Pedro Pizarro (1966) 208; also *Relación del sitio del Cuzco* 44.

218 *was ever seen'. Relación del sitio del Cuzco* 90.

219 *of the Spaniards'.* Herrera, Dec. v, bk. 8, ch. 5.

219 *the Inca capital.* George Kubler analysed the reasons for Manco's failure in his 'The Quechua in the Colonial World' 383–5 and 'A Peruvian Chief of State: Manco Inca'.

219 *effeminates down there!'* Francisco de Barrionuevo to the Council of the Indies, Panama, 21 October 1536, CP 223; Juan de Berrio testimony, Cuzco 1572, GP **2** 166.

219 *arms and horses.' Información de méritos del Marqués don Francisco Pizarro,* GP **2** 103.

219 *Rodrigo de Grijalva.'* López de Gómara, ch. 136, **2** 57.

220 *and 300 horses'.* Despatch from Audiencia of Española, 13 November 1536, CP 227. See also: despatch from this Audiencia of 8 September 1536, CDH Chile **4** 364–8; despatch of the oidores Fuenmayor and Zuazo, 30 May 1537; and a pathetic appeal for compensation by the destitute Diego de Fuenmayor, Lima, 2 October 1537. Fuenmayor's contingent contained the formidable Pedro de Valdivia, future conqueror of Chile.

220 *and fifty crossbowmen.* Decree of 6 November 1536, Valladolid, CDH Chile **4** 381–3.

CHAPTER 11 THE RECONQUEST

221 *with the Indians'.* Letter from Juan de Turuegano to a friend in Seville, Los Reyes, 30 November 1536, quoted in an anonymous letter of 1537, CP 272. The *Relación del sitio del Cuzco* (84) curiously says that Alvarado did not leave Lima until April 1537, but Turuegano, who was on the expedition must be right.

222 *to the knife.'* Ibid. This battle was at the Cuesta de la Sed, now called Lomo de Corvina, at the head of the valley of Lima.

222 *country in rebellion'.* Ibid.

222 *booty from them'.* Ibid.

223 *discovery and Conquest.* Although Francisco Pizarro and Diego de Almagro had originally been equal partners in the plan to explore the Pacific coast of South America, Pizarro had gained the dominant position. He made sure that he or his brother represented the expedition at the Spanish court. Francisco Pizarro went to the Court in Toledo in spring 1528 with news of the exciting find of the Inca raft, and the first landing in Peru at Tumbez. His reward was the Capitulación of 26 July 1529: Almagro was made Commandant of the fortress at Tumbez, but Pizarro was made Governor and Captain-General of Peru; both were to be paid handsome salaries out of the revenues of Peru after they had conquered it.

In the royal grant of 21 May 1534 Pizarro's territory was defined as extending for 200 leagues (some 600 miles) along the coast, from Zemuquella or Santiago (probably the Island of Puna) to 'the village of Chincha' which is well to the south of modern Lima. (The text of the Capitulación is in CDIA **22** 271–85.)

After the melting of Atahualpa's ransom, Pizarro shrewdly sent his brother Hernando to Spain with the first load of treasure. Pizarro's reward was to be given a further 70 leagues beyond Chincha on the coast. This new territory was said to belong to the chiefs named Coli and Chipi, possibly a reference to the Colla and to the pre-Conquest chief Chipana who lived near Lake Titicaca: the first scouts who visited Cuzco would have asked which chiefs ruled to the south, and brought back these garbled names. (Royal grant, Toledo, 4 May 1534, CL **3** 148–54, 154–7.)

Cuzco is in fact on about the same latitude as Chincha, 14° south, and thus fell comfortably within Pizarro's territory after the extension of 70 leagues. But no one knew this at the time, when it seemed to travellers that Cuzco lay far beyond – i.e. south of – Chincha and hence in Almagro's sphere. The King tried to clarify the way in which the 270 leagues were to be measured in a decree of 31 May 1537. Had his system been applied, the boundary between the territories would have run inland from Ica to Vilcanota, to pass midway between Cuzco and Titicaca (CL **3** 167–8; Means, *Fall of the Inca Empire* 25–6, 53–4, 63).

223 *more certain news.* Oviedo unexpectedly interjected a personal tragedy into his history at this point. 'On the road they crossed a river so deep and furious that it was a miracle that the men were not overwhelmed. But the unfortunate Francisco de Valdés, Overseer [Veedor] of Tierra Firme was drowned. [He was] the son of captain Gonzalo Fernández de Oviedo, the chronicler of this History. [His death means that] I can grieve more intimately with the others and can be more closely concerned in these adventures. But my grief was not to be a simple one. There remained a boy and girl, the children of that Veedor, and a few days after I learned of the tragic death of my drowned son, God removed my five-year-old grandson in this city of Santo Domingo of the island of Española. Blessed be God for everything. But although being, as I am, a reasonable man, the loss of these relatives cannot but make me sad. Without doubt the greatest hardship I feel is that God has taken away that young man, at twenty-seven, in the flower of his life, with such a death.' Oviedo y Valdés, bk. 47, ch. 6, **121** 150.

223 *son and brother'*. Ibid. ch. 8, **121** 153.

223 *your beloved wives'*. Ibid. 154.

224 *seven hundred horses'*. Ibid. ch. 7, 151.

224 *two thousand more men'*. Ibid.

224 *it to you.'* Ibid. 152.

224 *felt towards you.'* Ibid.

224 *among the natives. Acción de España en el Perú*, Appendix 14, 506–7; Despatch of Treasurer Manuel de Espinar, 15 June 1539, CL **3** 198; *Relación del sitio del Cuzco* 92–3.

225 *Pizarro's relief expeditions.* Oviedo y Valdés, bk. 47, ch. 8, **121** 154–5; *Relación del sitio del Cuzco* 94–5.

225 *with a candle.* Letter to King from Pedro de Oñate and Juan Gómez Malaver, Cuzco, 31 March 1539, CDH Chile **5** 277–9; Molina of Santiago 177; Oviedo y Valdés, bk. 46, ch. 21 and bk. 47, chs. 8, 13.

226 *at their bases. Relación de los sucesos* 398; Molina of Santiago 178; *Relación del sitio del Cuzco* 95–6; Oviedo y Valdés 155; Diego de Almagro accusation against Francisco Pizarro, 1541, Accusation 44 against Francisco and 32 against Hernando, CDIA **20** 250–1, 401, 1873.

226 *him to Spain. Relación de los sucesos* 399.

227 *the Calca area.* Molina of Santiago 180; Cieza de León, *War of Las Salinas*, chs. 1, 3, 2–4, 15–16; Oviedo y Valdés, bk. 47, ch. 8, **121** 155–6; Murúa, ch. 70, **1** 212.

227 *swollen Yucay river. Relación del sitio del Cuzco* 100; Zárate, bk. 3, ch. 4, 486.

227 *this native aggression. Relación del sitio del Cuzco* 101. Treasurer Manuel de Espinar, despatch of 15 June 1539, CL **3** 198; Oviedo y Valdés, bk. 47, chs. 8, 13; *Relación de los sucesos* 399; Molina of Santiago 180; Cieza de León, *War of Las Salinas* 16–17; Pedro Pizarro (1965) 212; Zárate, bk. 3, ch. 4.

227 *with bare limbs.'* Cieza de León, *War of Las Salinas*, ch. 5, 14; also *Relación de los sucesos* 399.

228 *men obey him'*. Probanza of 15 May 1540, CDH Chile **5** 351.

229 *Spaniards in Cuzco.* Almagro was later accused of having given Paullu too broad authority to round up Spaniards trying to defect to the Pizarran forces. One such called Castañeda tried to escape with property worth 12,000 pesos. Paullu's men caught him, and when he resisted cut off his head and brought this and the money back to Almagro (*Relación del sitio del Cuzco* 85). It was also claimed that over-enthusiastic natives brought back the heads of some of Almagro's own supporters, and the Marshal had to explain the difference between the two factions. Curiously, although it was commonplace for Spaniards to kill one another, there was much indignation over this licensing of natives to do so.

229 *of the field.* Cieza de León, *War of Las Salinas*, ch. 16, 60; Herrera, Dec. v, bk. 2, ch. 9. One Sebastián Rodríguez claimed

that Almagro trained Paullu's Indians 'to wound and kill Christians' and thus won the battle. Rodríguez testimony on behalf of Hernando Pizarro, Madrid, April 1540, CDH Chile **5** 468, 479.

230 *a few days.'* Titu Cusi Yupanqui 77.

231 *the Vilcabamba valley.* Poma de Ayala **2** 35, 299; Murúa, chs. 69, 70 **1** 209, 214. Murúa described the attempt to reach Urocoto. He may well have confused this migration with another towards the Chachapoyas in 1538.

231 *cavalry and foot-soldiers.* Orgóñez's career showed how far a man could rise by bravery alone in that violent period. His parents were poor cobblers and Jews recently converted to Christianity. His mother was persecuted for witchcraft, and the name Orgóñez was taken from a nobleman, Juan de Orgoños, whom he claimed as his father, but who firmly denied the connection. Orgóñez showered his father with gold and silver objects and pleaded: 'Sir, what I beg of you is that it be understood by whatever means that I am legitimate and could thus have the habit of a knight of Santiago' (Letter from Cuzco, 2 July 1535, CP 168). He fled his native Oropesa to avoid the repercussions of a brawl, and made a reputation for outstanding bravery in the Italian wars. He was one of the soldiers who captured Francis I at Pavia, and returned in triumph to Oropesa. After more adventures in Europe, Rodrigo Orgóñez sailed to conquer new worlds in the Americas, with his brother Diego Méndez, a man who was later also to play a crucial role in Manco's life story. After unsuccessful expeditions in the Isthmus and Honduras, Orgóñez arrived at Cajamarca with Almagro, and rose to be the Marshal's most trusted lieutenant in the advance on Cuzco, the Chilean expedition, the capture of Hernando Pizarro, and the battle of Abancay.

232 *of July 1537.* Cieza de León, *War of Las Salinas*, chs. 21, 66, 87–92, 208–9; *Relación del sitio del Cuzco* 131 ff; Molina of Santiago 187; Titu Cusi Yupanqui (82–3) confused Hernando Pizarro's attack on Ollantaytambo with Orgóñez's pursuit and a later expedition under Gonzalo Pizarro; Oviedo y Valdés, bk. 47, ch. 9, 160 and ch. 13, 171; Pedro Pizarro (1844) 323–4. Orgóñez had already returned to Cuzco when the Cabildo wrote to the King on 27 July, MS. in the Henry E. Huntington Library, San Marino, California, f. 352.

232 *security of Vitcos.* Titu Cusi Yupanqui 82.

232 *ill-fated relief expeditions.* Cabildo of Cuzco, despatch of 27 July 1537, MS. in Henry E. Huntington Library, San Marino, California. The other prisoners were Díaz's interpreter Pedro Riquelme, a prisoner called Francisco Martín and one other. Pedro de las Casas, Probanza, Arequipa 1555, DH Arequipa **2** 223. The despatch from the Cabildo of Cuzco of 27 July said that Orgóñez had freed four Spaniards. Diego de Almagro the Younger's Accusation against Francisco Pizarro, 11 December 1541 (Accusations 45 and 64 against Francisco Pizarro and Accusation 33 against Hernando Pizarro), CDIA **20** 252, 264, 402–3, 1873.

232 *the Chilean expedition'.* Molina of Santiago 183.

232 *twenty thousand souls'* Cabildo of Cuzco, despatch of 27 July 1537, MS. in Henry E. Huntington Library, San Marino, California, ff. 352–3; Murúa, ch. 70, **1** 214.

232 *Inca could wage'.* Ibid.

233 *and many coyas.'* Titu Cusi Yupanqui 83.

233 *would repay him.'* Ibid.

234 *due to him.'* Cieza de León, *War of Las Salinas*, ch. 70, 223.

234 *ended with himself.* Ibid.

234 *terms with him.'* Oviedo y Valdés, bk. 47, ch. 9, **121** 160; Molina of Santiago 184; *Relación del sitio del Cuzco* 125; Cieza de León, *War of Las Salinas*, ch. 21, 89; Herrera, Dec. VI, bk. 2, ch. 13. The despatch of the Cabildo of Cuzco of 27 July 1537 said that Almagro crowned Paullu after they returned from the battle of Abancay (12 July) and Orgóñez had gone after Manco, MS. in Henry E. Huntington Library, San Marino, California, f. 352.

234 *to be victorious.'* Cieza de León, *War of Las Salinas*, ch. 21, 88–9.

234 *Lima and Ica. Relación del sitio del Cuzco* 146–7.

234 *as I say.'* This offer was overheard by Dr Sepúlveda who told it to Gonzalo Fernández de Oviedo the following year (bk. 47, ch. 17, 191). Cieza de León, *War of Las Salinas*, ch. 51, 170.

235 *arrested and executed.* Oviedo y Valdés, bk. 47, ch. 17; **121** 193; *Relación del sitio del Cuzco* 157–8; Herrera, Dec. VI, bk. 4, ch. 4.

CHAPTER 12 THE SECOND REBELLION

236 *in the world'*. Oviedo y Valdés, bk. 47, ch. 9, 294–5.

236 *followed their advice.'* Titu Cusi Yupanqui 83.

237 *of his people.'* Cieza de León, *War of Las Salinas*, ch. 66, 209.

237 *exile with him.'* Ibid. Both Cieza and Titu Cusi clearly dated this exodus to the period between the departure of Orgóñez from Vitcos and the battle against Villadiego in late 1538. Cieza somewhat confused the move with the earlier exile to Vitcos, although he mentioned that Manco settled near Huánuco, which would be well on the way to Chachapoyas. Titu Cusi said that Manco and his followers spent a year wandering in the Jauja–Huamanga area, but said that this was after 1539. This was unlikely, since Manco was by then back in Vilcabamba and the Spaniards had founded the city of Huamanga (or Guamanga). I have therefore assumed that the migration mentioned by Titu Cusi – which occurred 'when I was in Cuzco in Oñate's house' (83) and before Gonzalo Pizarro's expedition of 1539 – took place in 1537–8.

237 *south of Chachapoyas*. The chiefs may possibly have referred to the citadel of Rabantu itself, which the Spaniards knew as Levanto and where they founded their first settlement before moving it to the present location of Chachapoyas. Levanto was 'a place strong in natural defences' (Cieza de Leon, *Crónica* pt. 1, ch. 78 (Onis) 99) and its ruins are still extant; but it was in no way comparable to Cuelape in fortifications. Another possibility was that the chiefs intended Manco to occupy the group of cities known as Pajatén or Abiceo, on the eastern side of the Marañon midway between Huánuco and Chachapoyas. The ruins of these cities were discovered by Gene Savoy in 1965.

237 *is towards Quito'*. Titu Cusi Yupanqui 88.

238 *city of Huánuco'* Cieza de León, *War of Las Salinas*, ch. 66, 209.

238 *tribal deity Catequil*. Cieza de León, *War of Las Salinas*, ch. 66. Zárate, bk. 4, ch. 1, 493. These authors called the general Villac Tupac, but he was evidently the same as Murúa's Illatopa or Illa Tupac. There is an excellent description of this local deity, Catequil, by the first Augustinian missionaries at Huamachuco, *Relación de la religión y ritos del Perú* 22–4.

239 *Tiahuanacan ruin Huari*. Titu Cusi said that his father's men built themselves houses and farms 'which are now occupied by the Spaniards who call the place Viñaca because many Castilian vineyards are to be seen there' (88). Cieza de León wrote 'along the Viñaque river vines have been planted ...' He also deduced, with brilliant perception, that the ruins of Huari in that valley 'must have been there for many ages ... long before the Incas reigned', and he later identified them, correctly, as being of the same civilisation as Tiahuanaco (pt. 1, chs. 87, 97 (Onis) 123–4, 284). The magnificent ruins of Huari are on a plateau east of Pacayccasa a few miles north of Ayacucho (Huamanga). Manco's men would not have dared settle in such an exposed location after the foundation of Huamanga at the end of 1538. The migration must therefore have been for a short period during 1538.

239 *well-armed escort.'* Cieza de León, *War of Las Salinas*, ch. 87, 232; Herrera, Dec. VI, bk. 6, ch. 7, **13** 51.

239 *looking for him'*. Titu Cusi Yupanqui 83–4.

239 *upper escape route.'* Francisco Pizarro to the King, Cuzco, 27 February 1539, RHA **47** 156.

239 *a rope torture'* Cieza de León, *War of Las Salinas*, ch. 87, 233.

240 *and great rewards'*. *Cieza de León, Crónica*, pt. 1, ch. 86 (Onis) 122.

240 *his thirty men*. Oncoy lies above a curve in the Pampas river north of Andahuaylas. Polo de Ondegardo described it as a hot repartimiento, *Ordenanzas para las minas de Guamanga*, CLDRHP **4** 148, 1917.

240 *in its honour.'* Titu Cusi Yupanqui 85; also Cieza de León, *War of Las Salinas*, chs. 87–8 and *Crónica*, pt. 1, ch. 86; *Relación del sitio del Cuzco* 190; Valverde, letter of 20 March 1539, CL **3** 107; Pizarro despatch, 27 February 1539, RHA **47** 156; Herrera, Dec. VI, bk. 6, ch. 8. Diego de Almagro the Younger accused Pizarro of causing Villadiego's death by entrusting the campaign to Illán Suárez 'who is not a military man', Accusation 145, CDIA **20** 336.

240 *to help you!"'* Titu Cusi Yupanqui 87.

240 *back to Vilcabamba.* Both Murúa and Titu Cusi, who reported the fighting from the native side, described this victory east of Jauja, and it was also mentioned by Bernabé Cobo, but by no other Spanish source. Paucar was presumably the young commander who had defied Almagro in the Yucay valley in 1537. Murúa did not mention the defeat of Villadiego at Oncoy. But Titu Cusi made it clear that there were two distinct native victories. Murúa, ch. 72, **1** 219; Cobo, bk. 12, ch. 20, **3** 232.

In an earlier Memorial, handed to Juan de Matienzo in June 1565, Titu Cusi mentioned a battle against the Spaniards and their native allies, 'in Pilcosuni' in which 'there was great mortality on both sides'. The Pilcosuni territory lay east of Jauja, between Rupa-Rupa and Vilcabamba. Matienzo, *Gobierno del Perú*, pt. 2, ch. 18, 301.

241 *of subject tribes.* Titu Cusi Yupanqui 86–8. The ruins of the temple of Wari Willca have been excavated some six kilometres south of Huancayo. One 150-foot stretch of its enclosure wall is still standing to a height in places of 16 feet.

241 *cause further damage'.* Francisco Pizarro to the King, Cuzco, 27 February 1539, RHA **47** 156.

241 *killed some Spaniards'.* Información of services of Francisco de Cárdenas, Cuzco, 20 March 1543, quoted by Marcos Jiménez de la Espada in a note on Amador de Cabrera's *Relación de la ciudad de Guamanga* (RGI **1**; BAE (Cont), **183** 181–2, 1965). Alonso de Alvarado first published the foundation of the city when he arrived on 5 September 1538 with a letter from Pizarro (MP **4** 204–5). The settlement flourished and is one of Peru's main cities. Pizarro was clearly a shrewd town-planner, for his other foundations, Lima, La Paz, Trujillo and Arequipa have all grown into modern cities. When Cieza de León visited Huamanga a decade later it had 'the largest and best houses in all Peru, all of stone, brick and tile, with great turrets.' The city was later renamed San Juan de la Vitoria to commemorate the battle of Chupas that took place nearby in 1542, but throughout the colonial period it was known as Huamanga or Guamanga. Its name was changed again to its present Ayacucho ('corner of death') to commemorate Bolívar's final defeat of the last Spanish army in South America outside the city. Cieza de León, *Crónica*, pt. 1, chs. 86, 87 (Onis) 121–4; *War of Las Salinas*, ch. 91, 244; Francisco Pizarro, despatch of 27 February 1539, RHA **47** 156–7; Herrera, Dec. VI, bk. 6, ch. 9, **13** 60.

241 *in this land'* Valverde despatch, Lima, 28 November 1539, RHA **47** 159.

241 *exciting the natives'* Francisco Pizarro, despatch of 27 February 1539, RHA **47** 157.

244 *them all utterly.'* Paullu's Probanza of 6 April 1540, question 16, CDH Chile **5** 343. Alonso de Toro vouched that this was true, Ibid, 347.

244 *suffering little damage'. Relación del sitio del Cuzco*, 183. Also: Cieza de León, *War of Las Salinas*, ch. 86, 229; Oviedo y Valdés, bk. 47, ch. 19, 203; Pedro Pizarro (1844) 335–7; Zárate, bk. 3, ch. 12, 492; Murúa, ch. 71, **1** 215–16; Pizarro y Orellana, *Varones* 332–3.

244 *no harm whatsoever'.* This quotation is from an Información staged in 1597 by two curacas of Copacabana to assert the importance of their grandfather Challco Yupanqui Inca (Villalobos, 'Almagro y los Incas' 42).

245 *was with him.'* Ibid, 44.

had fortified himself'. Pizarro despatch, Cuzco, 27 February 1539, RHA **47** 157.

245 *Garcilaso de la Vega Inca.* Garcilaso de la Vega was born in 1539. His mother was the daughter of Huallpa Tupac, a brother of Huayna-Capac (not to be confused with the young puppet Inca, Tupac Huallpa). She later married a Spaniard, Juan de Pedroche and lived until 1573. Captain Garcilaso de la Vega married a Spanish lady, Luisa Martel, and paid little attention to his illegitimate mestizo son. The young Garcilaso lived with his Inca relatives but was educated with the sons of other conquistadores. Captain Garcilaso died in 1559 aged 53, and his son left for Spain in 1560, never to return to Peru.

245 *use their horses'.* Murúa, ch. 71, **1** 216–17.

245 *with their horses'.* Herrera, Dec. VI, bk. 6, ch. 8, **13** 56.

246 *killed some Indians.'* Paullu's Probanza of April 1540, question 17 CDH Chile **5** 343.

246 *against the enemy.'* Ibid. 352.

246 *would have escaped.'* Ibid. 347.

246 *in this country'.* Murúa, ch. 71, **1** 217.

246 *the Indian auxiliaries'. Relación del sitio del Cuzco* 188.

246 *Christians' greatest enemies'. Relación del sitio del Cuzco* 188–9.

247 *reinforce his brothers*. Francisco Pizarro, despatch of 27 February 1539, RHA **47** 157; Cieza de León, *War of Las Salinas*, chs. 89–90, 237–42; Valverde, despatch of 20 March 1539, CL **3** 107; Illán Suárez de Carvajal, despatch of 25 March 1539, CDIA **42** 162–3; *Relación del sitio del Cuzco* 184–9; Paullu Inca, Probanza, April 1540; Zárate, bk. 3, ch. 12, 492; Pedro Pizarro (1965) 221; Melchor Carlos Inca, *Relación* in R. Cúneo-Vidal, *Historia de las guerras de los últimos Incas* 164; Herrera, Dec. VI, bk. 6, ch. 8, **13** 55–7; Murúa, ch. 71, **1** 216; Pizarro y Orellana, *Varones* 333–6.

247 *this Tiso came.'* Murúa, ch. 71, **1** 217.

247 *the Inca possessed.' Relación del sitio del Cuzco* 193.

247 *greatest good fortune'*. Pizarro, despatch of 27 February 1539, RHA **47** 137. Hernando Pizarro boasted that Tiso had been 'the second person' of the Inca Manco (*Confesión* 414).

247 *friends to us'*. Valverde, despatch of 20 March 1539, CDIA **3** 121; Illán Suárez de Carvajal, despatch of 25 March 1539, CDIA **42** 162–3; *Relación del sitio del Cuzco* 193; Zárate, bk. 3, ch. 12, 492.

247 *conquered on horses'*. Diego de Peralta, Probanza, 16 February 1566, question 8, DH Arequipa **2** 328; Pizarro, despatch of 27 February 1539, RHA **47** 157; Valverde, despatches of 20 March and 28 November 1539, CDIA **3** 121, 139; Grant of encomienda to Juan de Pancorvo, in Ulloa, 'Documentos del Virrey Toledo' RH **3** 320, 1908.

248 *against the Incas*. Cieza de León, *War of Las Salinas*, ch. 65, 206, ch. 79, 284, chs. 84–5, 288–91; Herrera, Dec. VI, bk. 6, chs. 3, 6, **13** 23–4, 43–5. When Pizarro ordered Alvarado to resume his conquest of Chachapoyas, he admitted that the natives 'had caused the deaths of Christians' (Instrucción of 28 June 1538, Jauja, Harkness Doc 109). But when Alonso Mercadillo marched towards the Huallaga he found that the natives had revolted against the 'tyrant' Illa Tupac (Marcos Jiménez de la Espada), 'La jornada del Capitán Alonso Mercadillo a los indios Chupachos e Iscaicingas', *Boletín de la Sociedad Geográfica de Madrid* **37** 197–230, 1895.

248 *gold and silver'*. Meeting of Cabildo of Lima, 7 January 1539, CL (2) **1** 280.

248 *through the Chinchaysuyo*. Orihuela's expedition was mentioned in the *Probanza de méritos* of Pedro de las Casas, Arequipa, 4 February 1555, item 7, DH Arequipa **2** 223.

248 *to his rescue.'* Zárate, bk. 4, ch. 1, 493.

249 *of the Amazon. Probanza de méritos* of Diego de Peralta, Arequipa, 16 February 1566, DH Arequipa **2** 328.

249 *punish its leaders*. Meetings of Cabildo of Lima, 29 June and 2 July 1539, CL (2) **1** 349, 350. The murdered encomenderos were Francisco de Vargas and Sebastian de Torres.

249 *prayed for peace.'* Cieza de León, *War of Chupas*, ch. 17, 50.

249 *a decade earlier*. Royal decree, Innsbruck, 25 December 1551, CDIA **18** 480–1; Cieza de León, *War of Chupas*, ch. 17, 50; *Crónica*, pt. 1, ch. 82 (Onis) 106; Zárate, bk. 4, ch. 1; López de Gómara, ch. 143; Garcilaso de la Vega, pt. 2, bk. 3, ch. 2; Porras Barrenechea, *El testamento de Pizarro* 68, and 'Crónicas perdidas' 204.

249 *is his relative.'* Vaca de Castro despatch, Cuzco, 24 November 1542, CP 506.

249 *after much difficulty'* Cieza de León, *War of Chupas*, ch. 82, 293.

249 *war in 1544*. Zárate, bk. 5, ch. 10, 514.

249 *province of Huánuco*. As far as I know, no other modern historian has acknowledged the role played by Illa Tupac.

249 *number of Indians'*. Cieza de León, *War of Las Salinas*, ch. 91, 244.

250 *revolt once more'*. Pizarro, despatch of 27 February 1539, RHA **47** 156, 157.

250 *in their villages.'* Valverde, despatch of 20 March 1539, CDIA 120.

250 *or a prisoner.'* Pizarro, despatch of 27 February 1539, RHA **47** 157.

250 *always take flight.'* Valverde, despatch of 20 March 1539, CDIA **3** 120.

250 *a warm climate.'* Titu Cusi Yupanqui 88.

250 *will come later*. The name Vilcabamba is used in various ways. The entire area between the Urubamba and Apurímac northwest of Cuzco was the Inca province of Vilcabamba and is still known by that name, as is the range of mountains that runs along it. Manco's capital city was known in the sixteenth century as Vilcabamba Viejo, Vilcabamba la Grande, or, to Titu Cusi, San Salvador de Vilcabamba. Vilcabamba is the modern name of the river that joins the Urubamba at the bridge of Chuquichaca; it used to be called Vitcos river. The

Spaniards founded a city called San Francisco de Vilcabamba which still exists, on the watershed between the Vilcabamba and Pampaconas–Concevidayoc rivers.

250 *resisted Manco single-handed.* The *Relación del sitio del Cuzco* was dated 2 April 1539, the day before Hernando Pizarro's departure. It was an excellent work, although heavily slanted in favour of Hernando. The author remains anonymous. It used to be thought that Bishop Vicente de Valverde wrote it, but it seems to have been written by a soldier who was in Cuzco, and the Bishop was in Lima during the siege. A soldier sufficiently literate to have written it was Diego de Silva, who was on the relief expedition of Morgovejo – an episode described in great detail. Silva was something of a poet, and was a rich citizen of Cuzco, where he could have gathered material for a book. Raúl Porras Barrenechea thought that Silva wrote it ('Crónicas perdidas' 215).

251 *reached a decision.* The grant of a marquisate was the culmination of a series of honours for Pizarro. He was made a knight of Santiago on 26 July 1529 and granted his father's coat of arms on 14 November of that year. He received his own coat of arms from the King from Valladolid, 19 January 1536. The marquisate came when King Charles heard that Manco's rebellion was over, 10 October 1537 (CL **2** 159–88); also E. Torres Saldamando, 'El marquesado de Pizarro'; E. Larrabure y Unánue, *Monografías histórico-americanas* (Lima, 1893) 325–40; Cieza de León, *War of Las Salinas* 245; Zárate, bk. 3, ch. 5; López de Gómara, ch. 133; Garcilaso de la Vega, pt. 2, bk. 2, ch. 22 and bk. 5, ch. 36; Herrera, Dec. v, bk. 6, chs. 9, 13. Pizarro thanked the King for the marquisate in his despatch of 27 February 1539, RHA **47** 154.

251 *of their treasure.* A. Paz y Mélia, *Nobiliario de conquistadores de Indias* (Madrid, 1892) 44–9; Means, *Fall of the Inca Empire* 65–6. This coat of arms can be seen on the palace built by Hernando Pizarro in Trujillo, Spain.

251 *state of suspense.' Relación del sitio del Cuzco*, 194.

251 *and fighting men'* Pedro Pizarro (1844) 341.

251 *to count them'.* Titu Cusi Yupanqui 88.

251 *friend Paullu Inca.* Quipocamayos 42; Montesinos, *Anales* **1** 107.

252 *jungle and undergrowth.'* Quipocamayos 42.

252 *of huge boulders'* Ibid. 43.

252 *had grown cowardly'*, Pedro Pizarro (1965) 223–4.

252 *killed at Chuquillusca.* Quipocamayos 44; Pedro Pizarro (1844) 342; Titu Cusi Yupanqui 89; Murúa (ch. 70, **1** 214) said that they lost thirteen Spaniards and six horses.

252 *depart with mine.'* Titu Cusi Yupanqui 89.

253 *had been executed'.* Ibid. Murúa named these same brothers, but said that Inquill fell from a cliff; Manco killed only Huaspar in front of his sister (ch. 70, **1** 214). In the Memorial he handed to Juan de Matienzo in June 1565, Titu Cusi said that these brothers were killed in the battle at the fort (pucará) and that they were the only surviving sons of Huayna-Capac, apart from Manco and Paullu. Matienzo, *Gobierno del Perú*, pt. 2, ch. 18, 301.

253 *from Vilcabamba itself.* Titu Cusi Yupanqui (88) said that the fort was three leagues from the city of Vilcabamba. Murúa located it in roughly the same place, ch. 80, **1** 253.

253 *when it emerged.'* Pedro Pizarro (1965) 224.

253 *had his residence'.* Ibid.

253 *beside this fort.'* Ibid.

253 *he was there.'* Ibid.

253 *Manco Inca!'* Titu Cusi Yupanqui 89.

253 *to his forefathers'. Información de méritos del Marqués don Francisco Pizarro*, testimony of Mancio Sierra de Leguizamo, 7 October 1572, GP **2** 146.

253 *on his trail'*, Pedro Pizarro (1965) 224.

254 *reached [Ollantay]tambo.'* Titu Cusi Yupanqui 90.

254 *be fully satisfied.'* Ibid.

254 *sane Christian man'.* Herrera, Dec. vi, bk. 7, ch. 1, **13** 77. His cousin Pedro Pizarro felt that Francisco Pizarro's murder was divine retribution for this cruelty (1965) 224.

254 *loved her much.'* Murúa, ch. 70, **1** 214; Illán Suárez de Carvajal, despatch of 3 November 1539, CDIA **42** 172; Cieza de León, *War of Chupas*, ch. 1, 3; Pedro Pizarro (1844) 346; Herrera, Dec. vi, bk. 7, ch. 1, **13** 76–7; Montesinos, *Anales*, entry for 1539.

255 *peace more quickly.'* Illán Suárez de Carvajal, despatch of 3 November 1539, CDIA **42** 173.

255 *their rear secure.'* Titu Cusi Yupanqui 91; Murúa, ch. 70, **1** 214. Luis de Morales's *Relación* although a fine and sympathetic

work, has never been fully published. Hernando Pizarro mentioned the executions briefly in a letter to the King from Madrid, 19 March 1541, CDH Chile **6** 180. Diego de Almagro's son accused Francisco Pizarro of 'secretly burning many Indians and native lords, including Villac Umu' (Accusation against Francisco Pizarro, 11 December 1541, accusations 177, 187, CDIA **20** 351, 354–5). A royal decree of 28 October 1541 ordered the restoration of lands confiscated from the widows of caciques 'recently executed in royal justice', CDIA **42** 184–5.

255 *making the attempt*. The second rebellion of 1538–9 has received little attention, largely because no single chronicler recorded it. John H. Rowe remarked that the coherent story of fighting during these years had never been told, although it would be perfectly possible to reconstruct it from published records ('The Incas Under Spanish Colonial Institutions', HAHR **36**, no. 2, 155). I have attempted to do this.

255 *to an end*.' Hernando Pizarro, letter to the King, Madrid, 19 March 1541, CDH Chile **6** 180.

CHAPTER 13 PAULLU AND MANCO

256 *well disposed towards us*.' Valverde letter of 20 March 1539, CL **3** 102.

256 *and well disciplined*', Molina of Santiago 184.

256 *of His Majesty*'. Martín de Gueldo testimony in answer to question 7 of Paullu's Probanza, Cuzco, 6 April 1540, CDH Chile **5** 351.

256 *courageous in battle*' Gómez de Alvarado testimony to question 8 of the Probanza, ibid. 349.

256 *much about warfare*'. Martín de Gueldo testimony to question 8, ibid. 351.

257 *on many occasions*', Alonso de Toro, answers to questions 8, 11 and 14 of Probanza, ibid. 346–7.

257 *friendship towards us*'. Valverde letter of 20 March 1539, CL **3** 102.

257 *than to us*'. Quipocamayos 44.

257 *discovered against him*.' Answer to question 10 of Probanza of 6 April 1540, CDH Chile **5** 354.

257 *for royal favours*. Ibid.

257 *council of Cuzco*. According to Fernando Montesinos the royal factor Illán Suárez de Carvajal made an official complaint about this to the Cabildo of Cuzco on 10 September 1540. He demanded that the Collao raid be stopped and suggested that Paullu be sent instead to Huánuco or Piscobamba to ferret out treasure to finance the struggle against Manco. The Cabildo sent to recall the two treasure-hunters to the city (Montesinos, *Anales*, entry for 1540, **1** 114).

258 *of the war*'. Encomienda award, Cuzco, 22 January 1539, quoted by Cúneo-Vidal, *Historia de las guerras de los últimos Incas* 170.

258 *of 12,000 pesos*. Quipocamayos 47; *Cédula de concesión del repartimiento*, quoted in Cúneo-Vidal, *Historia de las guerras* 170.

258 *just outside Cuzco*. Temple, 'Paullu Inca', RH **13** 47.

258 *Juan de León as tutor*. Calvete de Estrella **168** 69; Gutiérrez de Santa Clara **166** 229.

258 *against your service*'. Royal decree, Seville, 29 November 1541, CDH Chile **6** 198.

259 *to perform henceforth*'. Prince Philip to Paullu Inca, Valladolid, 28 September 1543, CDH Chile **4** 277. The Prince instructed the new Viceroy, Blasco Núñez Vela, to favour Paullu.

259 *legitimised en masse*. Royal decree, Valladolid, 1 April 1544. Temple, 'Paullu Inca', and 'Un linaje incaíco durante la dominación española, Los Sahuaraura' 45. Urteaga gave the text in *El imperio incaíco*, Appendix D, 263–6.

259 *golden Jerusalem crosses*. Granted by Charles V, Valladolid, 9 May 1545; text in Urteaga, *El imperio incaíco* 182–4.

259 *the puppet Inca*. President Cristóbal Vaca de Castro wrote to the King, 24 November 1542, that Paullu was almost ready for baptism. GP **1** 72; CP 508.

259 *hill of Huanacauri*. Cobo, *Historia del Nuevo Mundo* **4**, bk. 13, ch. 15; Molina of Cuzco, ch. 5.

259 *mother and relatives*'. *Relación* of Bachiller Luis de Morales, quoted by Ella Dunbar Temple in her 'Paullu Inca', RH **13** 59. These mummies were not as important as those later ferreted out by Polo de Ondegardo.

259 *by my name*'. Vaca de Castro's will dated

26 January 1571, quoted by Temple in 'Paullu Inca', RH **13** 59. Garcilaso de la Vega said that his father was one of Paullu's godfathers, along with the Inca's own uncle Titu Auqui (pt. 1, bk. 6, ch. 2, 1871, 105).

259 *and many others'*. Quipocamayos 46. Nobiliario of the Sahuaraura, manuscript in the Biblioteca Nacional, Lima, quoted by Temple.

260 *had been baptised'*. Gutiérrez de Santa Clara, bk. 3, ch. 55, **166** 230.

260 *the greatest privations.'* Letter from Pedro de Valdivia to Hernando Pizarro in Spain, La Serena, 4 September 1545, Maggs–Huntington 66. Valdivia's expedition met determined resistance, presumably inspired by Manco, in the Coquimbo valley and during its march south. Michimalongo, chief of the valley of Aconcagua, led the attack on Santiago in September 1541. His Indians killed various groups and settlements of Spaniards and continued the resistance in this southern province of the Inca empire until Michimalongo's 'pacification' in 1549. The most formidable opposition came later from the Araucanians south of the River Maule. This heroic people had resisted the Incas and they fought the Spaniards with equal success. They were by far the most formidable warriors encountered anywhere in the Americas. They killed the resolute Pedro de Valdivia and all his men at Tucapel on 25 December 1553, and continued to block Spanish advance into southern Chile for centuries.

260 *difficulty until 1545*. Ibid.

260 *Christians and natives'*. Letter from Francisco Pizarro to Garcí Manuel de Carvajal, Lieutenant-Governor of Arequipa, Lima, 7 May 1541; first published in Carlos A. Mackehenie, 'Apuntes sobre Don Diego de Castro Titu Cusi Yupanqui, Inca', RH **5** 5–13. Also, CLDRHP, 1 ser., 2 115–17; DH Arequipa **2** 110–11.

261 *done this summer.'* Ibid.

261 *a dozen bolts'*. Council meetings of 11 December 1540 and 20 March and 22 April 1541, mentioned in Montesinos, *Anales* **1** 114, 118.

261 *fortify themselves better'*. Murúa, ch. 72, **1** 220. This proposed migration apparently came after Gonzalo Pizarro's invasion of 1539 and would account for the activity reported near Huamanga. It may well have been the same as the exodus towards Rabantu reported by Titu Cusi, which I have dated in 1537 or 1538, between the Orgóñez and Pizarro invasions of Vilcabamba.

262 *Captain-General of Peru*. The best account of Pizarro's assassination is in Cieza de León, *War of Chupas*, chs. 29–31, 96–110; also, Pedro Pizarro (Means) 414–23; Zárate, bk. 4, chs. 7–8; Oviedo y Valdés, bk. 48, ch. 1, 215–18; Garcilaso, bk. 3, chs. 6–7, **134** 179–83.

262 *a decade earlier*. Letter of Martel de Santoya of 1542, AGI, Patronato 185, no. 31.

262 *of the world'*. López de Gómara, ch. 144, **2** 79.

263 *with lowered lances*. Cieza de León, *War of Chupas*, ch. 78. After the battle the standards of the dead were displayed in the tiny church of San Cristóbal at Huamanga (Ayacucho) – probably Peru's oldest surviving colonial building.

264 *Corregidor of Cuzco*. Busto Duthurburu, 'Maldonado el Rico, señor de los Andahuaylas' 130.

264 *give you pleasure.'* Melchor Verdugo letter of 1543 quoted by Rafael Loredo, *Los repartos* 29.

265 *and long marches.' Vaca de Castro* 428.

265 *law of Mahomet'* Morales in Porras Barrenechea, 'Crónicas perdidas' 233; this *Relación*, long unpublished, gave a moving account of the hardships of the natives in the 1540s.

265 *committed other atrocities'*. Cieza de León, *War of Chupas* 338.

265 *at their peril*. Hanke, *The Spanish Struggle for Justice in the Conquest of America* 85.

265 *from the Spaniards.'* Cieza de León, *War of Chupas* 338.

265 *and his Church.'* Salamanca's sermon quoted by Las Casas in his *Historia de las Indias*, bk. 3, ch. 135, translated by Hanke, *The Spanish Struggle for Justice* 86.

266 *however ineffectively.'* Box in a book-review in *History* **20**, 1935, quoted by Hanke, *The Spanish Struggle for Justice* 89. The literature about the 'leyenda negra' is endless. A recent instance of Las Casas arousing the fury of Spanish historians is Ramon Menéndez Pidal's *El padre Las Casas, su doble personalidad* (Madrid, 1963), criticised by Hanke in 'More Heat and Some Light on the Spanish

Struggle for Justice in the Conquest of America'.

266 *grants in perpetuity.* Law of 26 May 1536: text in CDIA **46** 198–204, 1884, and CDFS **1** 171–4; Marvin Goldwert, 'La lucha por la perpetuidad de las encomiendas en el Perú virreinal, 1550–1600'.

266 *20 November 1542.* The text of the New Laws has been published often, for instance in Antonio Muro Orejón, *Las Leyes Nuevas, 1542–1543* (Seville, 1946); CDFS **1** 216–20; facsimile and translation as *The New Laws of the Indies* by Henry Stevens and F. W. Lucas (London, 1893; Amsterdam, 1967); Cieza de León, *War of Chupas*, ch. 99, 340–60.

266 *have drafted them.* Hanke, *The Spanish Struggle for Justice* 91.

267 *part to another.'* Cieza de León, *War of Chupas*, ch. 100, 362.

267 *the New Laws.* Maggs–Huntington 17–20.

268 *Illán Suárez de Carvajal.* The events leading up to the civil war were amply chronicled: Cieza de León (*War of Quito*), Diego Fernández ('El Palentino'), Calvete de Estrella, Zárate, López de Gómara and Garcilaso de la Vega were the principal historians of this period.

268 *lances and horses.'* Letter to Alonso de Alvarado, then in Spain, justifying the rebellion, Lima, 17 October 1546, Maggs–Huntington 146–7. The proposal to marry Gonzalo to an Inca princess is mentioned in Gutiérrez de Santa Clara, bk. 3, chs. 45, 46; Diego Fernández, bk. 2, ch. 13; Calvete de Estrella, bk. 3, chs. 3 and 6; López de Gómara, ch. 173; Garcilaso de la Vega, pt. 2, bk. 4, chs. 40–2.

269 *for many years.* Inés married Ampuero in 1538, soon after the birth of her son Gonzalo Pizarro. Ampuero signed an agreement on behalf of his wife on 6 July 1538 'because she cannot write', Harkness Cal 88. Francisco de Ampuero was encomendero of Chaclla and one of the original citizens of Lima. As a regidor he took part in the reception of the first Viceroy in 1544 (Cobo, *Historia del Nuevo Mundo*, bk. 12, ch. 20, *Fundación de Lima*, bk. 1, chs. 8, 17).

269 *and much honoured'. Méritos de Vicente de Tamayo, Diego de Sandoval y Gil Rengifo*, 1575, in Carlota Bustos Losada, 'Las hijas de Huainacapac', *Museo Histórico* **3**, no. 9, 29, 31.

269 *their father's request.* Royal decree, Monzón, 12 October 1537, CP **2** 51–61; and Valladolid, 19 January 1544, MP **5** 52. Francisco Martín de Alcántara was the son of Governor Pizarro's mother Francisca González y Mateos, who was never married by the Governor's father, Captain Gonzalo Pizarro. Francisco Martín's widow Inés Muñoz claimed to have been the first married Spanish woman in Peru. She bombarded the authorities with letters about the estates of her husband and the Pizarro children in her care. These won the results due to an importunate widow. Some of the letters are: 30 March 1543, Harkness Cal 128; 5 May 1543, ibid. 137; 8 May 1543, CDIA **42** 197–200, 1884.

269 *war of Chupas.* Cobo, *Historia de la fundación de Lima*, bk. 3, ch. 16, 430.

269 *the best off.'* Hernando Pizarro to Gonzalo Pizarro, La Mota de Medina del Campo, 2 December 1544, Maggs–Huntington 31. Gonzalo had arranged for Francisco de Robles to administer the Marquis's encomienda of Huaylas on behalf of the orphans, on 4 April 1543, at which time Juan was still alive, Harkness Cal 133.

269 *sailors and soldiers'.* Zárate, bk. 5, ch. 11, 517; Calvete de Estrella, bk. 1, ch. 4, **167** 252.

269 *marriage for her'.* Carvajal to Gonzalo Pizarro, Lima, 25 October 1545, Maggs–Huntington 71.

270 *an independent Peru.* Gutiérrez de Santa Clara, bk. 4, ch. 48, **166** 399. The rumour of this possible union reached the King, who wrote to Gasca to prevent it at all costs. But Gasca replied that 'it never crossed Gonzalo's mind, for this marriage would not have been sanctioned by the Spaniards or the natives', despatch of 25 September 1548, CDHE **49** 424.

270 *died in 1546.* Gonzalo Pizarro to Hernando Pizarro, 29 May 1546, Maggs–Huntington 105.

270 *and jump together'.* Garcilaso de la Vega, pt. 1, bk. 9, ch. 38, Hakl Soc **45** 525.

270 *charming little ladies'* Tomás Vásquez to Gonzalo Pizarro, Cuzco, 25 February 1547, Maggs–Huntington 272.

270 *play the virginals.'* Juan de Frías to Gonzalo Pizarro, Cuzco, 28 March 1547, Maggs–Huntington 302; also Juan de Acosta to Francisco de Herrera, Cuzco, 19 December 1546, Maggs–Huntington 205.

270 *in August 1544*. Royal decree, Valladolid, 29 August 1544, MP **5** 52.
270 *from the Indies*. CDFS **1** 234; Hanke, *The Spanish Struggle for Justice*.
271 *in civil strife*. Malines, 20 October 1545, CDFS **1** 236–7.
271 *defend our own*.' Instructions to his lieutenants by Gonzalo Pizarro, Lima, 18 April 1547, Maggs-Huntington 328.
271 *and loyal vassal*.' Gonzalo Pizarro to Emperor Charles V, Lima, 20 July 1547, Maggs–Huntington 406–8.
271 *to be said*.' Pedro Hernández Paniagua de Loaysa to Pedro de la Gasca, San Miguel de Piura, 1 August 1547, Maggs–Huntington 416, 418.
271 *path to power*.' Pedro de la Gasca to Gonzalo Pizarro, Jauja, 16 December 1547, Maggs–Huntington 441.
272 *side at Chupas*. López de Gómara said that Almagro had 'a great mass of Indians, with Paullu, whom his father had made Inca' (ch. 149, **2** 86) – but Paullu was probably not there, as no other more reliable chroniclers mentioned his presence.
272 *to the coast*. Zárate, bk. 5, ch. 4, 510.
272 *rebels occupying Cuzco*. Calvete de Estrella, bk. 4, ch. 1, **167** 393.
273 *respect for them*'. Gasca despatch to Council of the Indies, Cuzco, 27 June 1547, CDHE **49** 309.
273 *of the campaign*. Gutiérrez de Santa Clara, bk. 5, ch. 24, **167** 105; Calvete de Estrella, bk. 4, ch. 3, bk. 6, chs. 1, 4, **168** 2; Gasca despatch of 27 June 1547, CDHE **49** 309.
273 *rebels in Cuzco*'. Gutiérrez de Santa Clara, bk. 5, ch. 24, **167** 105.
273 *throughout the land*'. Quipocamayos 33.
273 *Manco Yupanqui at Vitcos*.' Cieza de León, *War of Chupas*, ch. 80, 287–8.
273 *the Inca Manco*' Ibid. ch. 82, 292.
273 *with great difficulty*'. Cabildo of Cuzco to Vaca de Castro, Cuzco, 23 September 1542, CP 479.
273 *he was able*.' Ibid.
273 *prison of Cuzco*. Ibid; Cabildo of Cuzco to Emperor, Cuzco, 20 January 1543, CP 531. Cieza de León, *War of Chupas*, ch. 82, 292–3.
274 *Almagro the Younger*. Not many years earlier, Rodrigo Orgóñez had written to his father: 'My brother Diego Méndez is well, thank God, and rich. He wishes very much to go and rest in Spain and marry there. I think that I will take him with me when I return.' (Rodrigo Orgóñez to Juan de Orgóños, Jauja, 20 July 1534, CP 132). Like so many conquistadores, neither brother returned to rest in Spain, and both died violent deaths.
274 *shoot an arquebus*. Montesinos, *Anales*, entry for 1545, **1** 163; Zárate, bk. 4, ch. 21, 506.
274 *his own brothers*.' Titu Cusi Yupanqui 94.
274 *would attack it*'. Cabildo of Cuzco to the King, Cuzco, 20 January 1543, CP 526.
274 *three principal captains*'. Vaca de Castro to King, Cuzco, 24 November 1542, CP 504.
275 *became extremely deferential*.' Ibid. 504–5.
275 *these were provided*. Ibid. The five types of land that Manco requested were: an area containing Indians trained as litter-bearers, clearly the Rucana; one containing 'a place for recreation', presumably the Yucay valley; a hot coca plantation; lands to produce maize and to support herds of llamas; and provision for certain of his leading orejones.
275 *of the kingdom*.' Cieza de León, *War of Quito*, ch. 51, 124.
275 *promises of pardon*. Garcilaso, pt. 2, bk. 4, ch. 6, **134** 233–4.
275 *fear of him*'. GP **2** 362; Zárate, bk. 5, ch. 4, 510; *Relación de las cosas del Perú* 254–5. This flattering quotation was in a document written in Lima on 15 September 1546, when Gonzalo Pizarro was controlling the city. It purported to be a copy of an agreement made between the Viceroy Blasco Núñez Vela and the oidore of the Audiencia of Lima on 20 October 1544 appointing Gonzalo as commander-in-chief against the Inca.
276 *they could muster*' Cieza de León, *War of Quito*, ch. 51, 123.
276 *Spaniards on horseback*'. Ibid. 124.
276 *make himself Inca*. Murúa, ch. 72, **1** 221; Cobo, *Historia del Nuevo Mundo*, bk. 12, ch. 20.
276 *would reward them*'. Murúa, ch. 73, **1** 223.
276 *warn the Inca*. Ibid.
277 *to find me*.' Titu Cusi Yupanqui, Hakl Soc, 2 ser., **31** 165, 1913. The boy was wounded on the leg and showed the scar to a Spaniard, Rodríguez de Figueroa, twenty years later, Rodríguez de Figueroa, ibid. 178.
277 *some were burned*.' Titu Cusi Yupanqui, ibid. 165.
278 *was premeditated murder*. Six reliable chroniclers wrote that the assassin was

Diego Méndez and that he planned the murder during a game of quoits at a time when the Inca army was absent: Titu Cusi Yupanqui 92; Pedro Pizarro (1965) 231; Diego Rodríguez de Figueroa, Hakl Soc, 2 ser., **31** 178, 1913; Fernándo Montesinos, *Anales* **1** 163; Martín de Murúa, ch. 73, **1** 223–5 and **2** ch. 13; Bernabé Cobo, *Historia del Nuevo Mundo*, bk. 12, ch. 20, **3** 234. The Quipocamayos (23) said that Méndez was the murderer, but gave no further details.

It was, as usual, the fertile Francisco López de Gómara who embellished the story: he said that the stabbing was not planned but arose from an argument during a game of bowls (not quoits in his version), and named Gómez Pérez as the assassin. His version was followed by Garcilaso, who made some minor changes that he claimed to have heard from his Inca relatives; and Calancha in turn followed Garcilaso. López de Gómara, ch. 156, **2** 102–3; Garcilaso pt. 2, bk. 4, ch. 7, **134** 234; Calancha, bk. 4, ch. 2, (1639) 792. Poma de Ayala (**2** 35) named Diego Méndez but accepted the story about a quarrel over the game. Cieza de León merged the names of the assassins to produce Diego Pérez. He mentioned no game, but said that a dispute arose as the Spaniards were about to depart, and that they killed the Inca in self-defence. Cieza wrote soon after the event, and his informant was Paullu's priest, who heard the story from natives freshly arrived from Vitcos. *War of Quito*, ch. 51, 124–5. Herrera followed Cieza de León – his favourite source – Dec. VII, bk. 8, ch. 6, **15** 144–5.

Of modern writers, Means predictably followed Garcilaso's version (*Fall of the Inca Empire* 109), and so did Kirkpatrick (*The Spanish Conquistadores* 191). Prescott, with his usual admirable prudence, wrote: 'It is impossible to determine on whom the blame for the quarrel should rest, since no one present at the time has recorded it' (bk. 4, ch. 8) – he cited only Pedro Pizarro and Garcilaso and did not know that the boy Titu Cusi was to give an eyewitness account. Kubler saw the motive prompting Méndez to murder, but accepted the quarrel version ('A Peruvian Chief of State').

CHAPTER 14 SAYRI-TUPAC

281 *destroyed two bridges'*. Cieza de León, *War of Chupas*, ch. 1, 1.

281 *treat him well'*: Gasca despatch to Council of the Indies, Lima, 25 September 1548, CDHE **49** (1866) 406–7 and in GP **1** 116.

281 *in those provinces'*. Cobo, *Historia del Nuevo Mundo*, bk. 12, ch. 21, **3** 238.

281 *come by force.'* Letter of 25 September 1548, CDHE **49** (1866) 406, GP **1** 116.

282 *to the King*. The career of Don Martín was described by Lockhart, *Spanish Peru* 213–15.

282 *have in Jaquijahuana'*. Gasca despatch, 25 September 1548, CDHE **49** (1866) 417–18.

282 *is Hernando Pizarro'* Ibid. 418.

282 *whenever they wish'*. Ibid.

282 *off to him'*. Gasca despatch of 2 May 1549 to Council of the Indies, CDHE **50** 61.

282 *road into Vilcabamba*. Cobo, *Historia del Nuevo Mundo*, bk. 12, ch. 21.

283 *of this chief.'* Molina of Santiago 158.

283 *from Don Carlos.'* Gasca despatch of 17 July 1549, CDHE **50** 69, GP **1** 198.

284 *as a father'*. Ibid.

284 *of what nation.'* Cieza de León, pt. 2, ch. 6, (Onis) 31; Documents of 8 October and 9 November 1549, in Barriga, *Los Mercedarios* **2** 161–6. Cayo Topa was described as 'son of Tupac Inca Yupanqui and nephew of Huayna-Capac' and he later described himself as son of 'the second-in-command of Huayna-Capac'. His father fought in Manco's second rebellion.

284 *various royal favours*. Alonso's father was also called Titu Atauchi and was a son of Huayna-Capac. He had fought for Huascar against the Bracamoros and later against Atahualpa's generals. Sarmiento de Gamboa said that he was killed with Huascar at Andamarca (ch. 68). Blas Valera claimed that he led an attack on Pizarro's army as it marched through Huamachuco after leaving Cajamarca – an attack totally ignored in all the eyewitness journals of Pizarro's march. Garcilaso and Anello Oliva both repeated Valera's story, in which Titu Atauchi captured various Spanish prisoners, punished those involved in Atahualpa's death

and released the others (Garcilaso, pt. 2, bk. 1, ch. 18). Raúl Porras Barrenechea argued that this utterly improbable story of native vengeance formed part of the legend in favour of Juan de Herrada and Francisco de Chaves that originated with Blas Valera's father Alonso (Luis) Valera in the Chachapoyas campaigns ('Cronicas perdidas' 208–13).

284 *who seems well-inclined'*. Gasca despatch of 25 September 1548, CDHE **49** 418 and Maggs–Huntington 487.

284 *in his place'*. Ibid.

284 *early in 1549*. Gasca despatch of 2 May 1549, CDHE **50** 48; Royal decree of 11 March 1550, Valladolid, CDIA **18** 7–9; MP **5** 212–13.

284 *of their grants*. They were escorted by Baltasar Daza. Gasca provided that if either girl died without children or became a nun, her share of the inheritance would pass to her cousin (Calvete de Estrella, bk. 4, ch. 12, **168** 63–4).

284 *cousins to Spain*. Gutiérrez de Santa Clara, bk. 5, ch. 47, **167** 171.

285 *to this effect*. Petition of 29 December 1550, Harkness Cal 183; Porras Barrenechea, 'El testamento de Pizarro de 1539', and 'Deformación histórica sobre Francisco Pizarro'. The will is quoted in Manuel Tovar, *Apuntes para la historia eclesiástica del Peru* (Lima, 1873) 461.

285 *which was inflamed'*. Oviedo y Valdés bk. 46 (pt. 3, bk. 8), ch. 1, **121** 33.

285 *jewels and horses*.' Ibid. bk. 48, ch. 6, **121** 230.

286 *conquered native rulers*. I am indebted for my information on the Palacio de la Conquista at Trujillo to Mr William Oddie, who very kindly visited the town and took photographs of the statues for me. Prescott (bk. 4, ch. 5) cited Richard Ford, (*Murray's Handbook for Travellers in Spain* **1** 535) for the figures of manacled Indians, 'fit badges of the bloody Conquest'. Rómulo Cúneo-Vidal mentioned a statue of Francisca's mother Inés ('Los hijos americanos de los Pizarros de la Conquista' 81). He also added a few years to the ages of Hernando and Francisca at the time of their marriage, claiming that she was 20 and he was over 70. See also Miguel Muñoz de San Pedro, 'Las últimas disposiciones del último Pizarro de la Conquista' 389; Juan Fernando Pizarro, 4–5

287 *dropping the claim*. The 20,000 vassals were awarded in a decree of 10 October 1537, GP **2** 51–61. Probanza of 1561 by Fiscal Licenciate Gamboa, CDH Chile **7**; probanza by Francisca and Hernando Pizarro, 15 March 1561, CDH Chile **7** 250–64; *Información de meritos del Marques don Francisco Pizarro*, GP **2** 91–203; Gamboa probanza of 18 June 1570, Romero, 'Un tesoro famoso', RH **16** 153–9.

288 *published in Rome*. Sepúlveda's manuscript was called *Democrates alter, sive de justis belli causis apud Indios* (Hanke, *The Spanish Struggle for Justice* 113–14, and *Aristotle and the American Indians*).

288 *be paid for*. Valladolid, 29 April 1549, CDFS, no. 168, **1** 257–8.

288 *security of conscience'*. 3 July 1549, Hanke, *The Spanish Struggle for Justice* 116

288 *they were just*.' Ibid. 117.

290 *fighting the encomenderos*. Las Casas, 'Treatise concerning the imperial sovereignty and universal pre-eminence which the Kings of Castile and Leon enjoy over the Indies', in Hanke, *The Spanish Struggle for Justice* 150–5; Parry, *The Spanish Theory of Empire in the Sixteenth Century*; Chamberlain 'The Concept of the Señor Natural as Revealed by Castilian Law and Administrative Documents' 131; Gibson, *The Inca Concept of Sovereignty* 88; Porras Barrenechea, 'El pensamiento de Vitoria en el Perú'; Zavala, *New Viewpoints on the Spanish Colonization of America* 42.

290 *it reached Peru*. Guillermo Lohmann Villena, 'El señorio de los marqueses de Santiago de Oropesa en el Perú' 350. The King's letter was copied in a memorial in AGI which referred to a lawsuit between Martín García de Loyola and the Fiscal of the Audiencia of Lima, towards the end of the century.

291 *any other person*.' Diego Fernández, pt. 2, bk. 3, ch. 4, **165** 76. This was the best account of the Cañete–Sayri-Tupac negotiations. Garcilaso used it, with acknowledgement (pt. 2, bk. 8, chs. 8–9); so did Antonio de la Calancha (bk. 2, ch. 29).

291 *Mancio Sierra de Leguizamo*, Mancio Sierra de Leguizamo was the man who boasted of having gambled away the sun effigy before dawn, and of having been the first man to scale the walls of Sacsahuaman during the siege (Garcilaso, pt. 1, bk. 3, ch. 20, Hakl Soc **41** 272). He probably did

neither. But he lived to a great age, until 1589. His will showed that the Princess Beatriz had never been his wife as Garcilaso claimed (pt. 1, bk. 9, ch. 38). This document also contained an emotional and often-quoted passage in which he deplored the Conquest and wished to clear his conscience 'as I was a guilty party in it. By our bad example we destroyed a people of good government . . . There were robbers among us and men who incited their women and daughters to sin . . . These natives have . . . changed from that extreme of doing nothing wrong to now doing nothing, or almost nothing, good.' (Calancha **1**, ch. 15, 98). These pious sentiments were rather contradicted by Sierra de Leguizamo's earlier complaining at having to surrender some of his lands to Paullu, and by his testimony in 1572 when he claimed that the Incas had sacrificed children and made drums of the skins of their enemies.

291 *half-brother Paullu.* Garcilaso, pt. 2, bk. 6, ch. 3; 'they married the wife of Martín [sic] de Bustincia, who was a daughter of Huayna-Capac . . . to a good soldier, a fine man called Diego Hernández, of whom it was said (with more falsehood than truth) that in his youth he had been a tailor. When the infanta heard this she refused the marriage, saying that it was not right to marry the daughter of Huayna-Capac Inca to a ciracamayo, which means tailor. The Bishop of Cuzco and Captain Diego Centeno begged and importuned her . . . but to no avail. They then sent to call Don Cristóbal Paullu her brother . . . He came and took his sister aside into a corner of the room, and when they were alone told her that it was not advisable for her to refuse that marriage. By so doing she would make all members of their royal line unpopular with the Spaniards . . . who would never be friendly towards them. She consented to what her brother ordered, although with bad grace. They therefore went before the bishop . . . but when the interpreter asked her . . . if she wished to become the wife of that man . . . she replied, "Perhaps I do, perhaps I do not." With that the marriage went ahead.'

291 *had been cut'*. Diego Fernández, pt. 2, bk. 3, ch. 4.

291 *Melchor de los Reyes.* Juan Sierra was a classmate of Garcilaso de la Vega and at the close of the century was a friend of Bernabé Cobo (*Historia del Nuevo Mundo*, bk. 13, ch. 5). Juan de Betanzos compiled the first Quechua–Spanish dictionary, to which he referred in the dedication of his *Suma y narración de los Incas*, which was commissioned by the first Marquis of Cañete but not published until the nineteenth century. The book was a curious mixture of Spanish and Quechua, but was the only source of many Inca legends, doubtless narrated by his wife and her family.

292 *aunt Doña Beatriz.'* Diego Fernández, pt. 2, bk. 3, ch. 4, **165** 77.

292 *come of age'*. Ibid.; also quoted in Garcilaso, pt. 2, bk. 8, ch. 9.

292 *cloud and mist.'* Garcilaso, pt. 2, bk. 8, ch. 10, **135** 143; Murúa ch. 74, **1** 229.

293 *me my life.'* Garcilaso **135** 143.

293 *about the departure'*. Diego Fernández, bk. 3, ch. 4, **165** 79.

293 *King of Peru'*, Poma de Ayala **2** 61.

294 *of the past.'* Garcilaso, pt. 2, bk. 8, ch. 10, **135** 144; Murúa, ch. 74, **1** 229.

294 *and stirrup-guards.'* Letter of 23 December 1557 from Factor Bernardino de Romaní to the President of the Council of the Indies. This was a secret report, full of venomous accusations against the Viceroy. Romaní claimed that Cañete had grossly over-valued the saddle-cloth when charging it to the royal treasury. The Factor insisted, to Cañete's fury, that it was of local manufacture and worth only 120 pesos (GP **2** 501).

294 *of the oidores'*. Montesinos, *Anales* 1558 entry, **1** 251; Poma de Ayala **2** 61.

294 *was not unreasonable.* Garcilaso, pt. 2, bk. 8, ch. 10.

294 *their country houses.* Francisco Pizarro had settled this repartimento on his Inca princess Angelina, and it was later administered by her husband Juan Betanzos. The Crown confiscated the land with the disgrace of the Pizarro family after the failure of Gonzalo's rebellion. But it awarded pensions from it to the Marquis's and Angelina's son Francisco and to his girl cousins. Betanzos was still administering the property when it was granted to Sayri-Tupac; and its chief beneficiary, the seventeen-year-old Francisco Pizarro conveniently died soon afterwards.

295 *resembled him closely'*. Cobo, *Historia del*

Nuevo Mundo, bk. 12, ch. 21, **3** 240.

295 *people pulling them.'* Garcilaso, pt. 2, bk. 8, ch. 11, **135** 146.

295 *on his face.'* Ibid. 145–6.

296 *of her beauty'*. Ibid. 146.

296 *of pure chastity'*. Murúa, bk. 2, ch. 15, (1946) **2** 145.

297 *Indians adore him.' Colección de las memorias o relaciones que escribieron los virreyes del Perú*, ed. Ricardo Beltran y Rospide (Madrid, 1921) **1** 61.

297 *in that place'*. Garcilaso, pt. 2, bk. 8, ch. 11, **135** 146.

297 *Hinojosa as godfather*. Garcilaso said that his own father, Captain Garcilaso de la Vega, was to have been godfather, but was prevented by illness.

297 *the Yucay valley*. The will was dated 25 October 1558. A copy was found by Guillermo Lohmann Villena among the papers in the Archive of the Indies in Seville concerning the relations of Cristóbal Maldonado with the young Beatriz Clara Coya. Dr Lohmann Villena published it in 'El testamento inédito del Inca Sayri Tupac'. I have seen no other reference to Sayri-Tupac's sister Doña Inés. One of the captains mentioned in the will was Curi Paucar, who later returned to Vilcabamba and became one of its military leaders.

298 *council of Lima*. The two Jesuits Bernabé Cobo and José de Acosta used this report, Cobo, *Historia del Nuevo Mundo*, bk. 11, ch. 2.

298 *off their caps.'* Garcilaso, pt. 1, bk. 5, ch. 29, Hakl Soc **45** 92, 93–4, also Garcilaso, pt. 1, bk. 3, ch. 20.

298 *still perfectly embalmed*. Polo de Ondegardo, *Relación* 97, 118; Polo's testimony to Información of 17 January 1572, CLERC **16** 245–57; Sarmiento de Gamboa, chs. 14–18, 25; Acosta, bk. 5, ch. 6, bk. 6, ch. 22; Cobo, *Historia del Nuevo Mundo*, bk. 12, chs. 4, 11, 13; Benzoni, *History of the New World* 184; Velasco **1** 237. Sancho had seen Huayna-Capac's mummy when he first entered Cuzco.

CHAPTER 15 NEGOTIATIONS

299 *until they die.'* Bartolomé de Vega, NCDHE **6** 130.

300 *the civil wars*. Titu Cusi Yupanqui 99–100; Rodríguez de Figueroa, 188–9; Murúa, pt. 2, bk. 2, ch. 16 (1946) **2** 149; Garcilaso, pt. 2, bk. 8, ch. 10; Cobo, *Historia del Nuevo Mundo*, bk. 12, ch. 21. Garcilaso wrote that Sayri-Tupac left the royal fringe in Vilcabamba at the advice of his captains, in order not to offend the Spaniards by appearing to reclaim the Inca empire. This was a cunning excuse by the captains. The Spaniards of course wanted their puppet Inca to wear all his regalia, and in fact gave him the fringe that had belonged to Atahualpa.

300 *not obey him'*. Rodríguez de Figueroa 189. Titu Cusi later tried to pretend that he had been legitimate all along. He claimed in his *Relación*, dictated in 1570, that he had been ruling Vilcabamba as legitimate successor of Manco ever since the latter's death. He said that, as ruler, he had authorised the departure of his younger brother Sayri-Tupac for Spanish Peru.

301 *body in Vilcabamba.'* Murúa, ch. 74, **1** 230.

301 *him uti* [*impotent*].' Sarmiento de Gamboa, ch. 70, Hakl Soc **22** 193. Other chroniclers such as Poma de Ayala and Garcilaso de la Vega chose to omit all reference to the reign of the 'usurper' Titu Cusi.

301 *proved against him*. Cobo, *Historia del Nuevo Mundo*, bk. 12, ch. 21. Murúa, ch. 74, **1** 230. Chilche survived this slur and was still curaca of Yucay in 1571 when he was a witness to Viceroy Toledo's enquiry of 13 March. Diego de Trujillo described (also in 1571) how Chilche had greeted Francisco Pizarro in 1533 with the words 'I come to serve you and will not oppose the Christians as long as I live', and, added Trujillo, 'he has not done so up to the present' (Trujillo 63).

301 *its native community*. Poma de Ayala **2** 60.

301 *had my father.'* Titu Cusi Yupanqui 100.

302 *of the King'*, Ibid.

302 *could come later*. Ibid. 101; Governor Lope García de Castro despatch of 30 April 1565, GP **3** 82. It was García de Castro who credited García de Melo with the idea that Titu Cusi's son Quispe Titu should marry Sayri-Tupac's daughter Beatriz.

302 *grandson of Huayna-Capac.* Decree of 29 November 1563, signed at Monzón de Aragon, CDIU **15** 270.

303 *pay no rent'* García de Castro despatch of 30 April 1565, GP **3** 83.

303 *monastery of Cuzco'* Ibid. 82.

303 *400 pesos a year'.* Ibid.

303 *my own expense.'* Ibid. 83.

304 *of its corregidor.* The corregidor was Dr Gregorio González de Cuenca, who had been in office from 1561 to 1563, during which time he had had some belligerent correspondence with Titu Cusi. He called the Inca 'a drunken dog and highwayman'. Hernando Bachicao took this letter to the Inca, and testified in Matienzo's secret enquiry into González de Cuenca's conduct. Matienzo, *Gobierno del Perú*, 296.

304 *to enter Vilcabamba.* See Lohmann Villena's 'El Inca Titu Cussi Yupangui' and his introduction to Matienzo, *Gobierno del Perú* (1967) xxxiv.

304 *some three hours.* Matienzo, *Gobierno del Perú*, pt. 2, ch. 18 (1910) 193.

304 *truly desires peace'.* Ibid.

305 *the Huamanga militia.* Montesinos, *Anales* **1** 242.

305 *Castro into action.* GP **3** 98–100.

305 *from various places'.* Titu Cusi Yupanqui 101.

305 *without an escort.'* Cobo, *Historia del Nuevo Mundo*, bk. 12, ch. 21, **3** 240–1.

305 *in many places.'* Rodríguez de Figueroa, 188. Titu Cusi also insisted at his meeting with Juan de Matienzo that he had not harmed Spaniards or churches. Matienzo, *Gobierno del Perú*, pt. 2, ch. 18, 294.

305 *llama with it.'* Ibid.

305 *almost his neighbours'.* García de Castro despatch, Lima, 6 March 1565, GP **3** 59; Matienzo said that Titu Cusi was plotting a federation with the Chiriguanas and Diaguitas, *Gobierno del Perú*, pt. 2, ch. 18, 295.

306 *illuminating one another.'* Letter from Felipe de Segovia Balderábano Briseño to Governor Castro, 3 December 1564, in testimony by his son Felipe, 8 June 1565, Manuel de Odriozola, *Documentos históricos del Perú* **3** 3–9, 1872.

306 *it with horror'.* García de Castro despatch of 6 March 1565, GP **3** 99. The royal factor Bernardino de Romaní also expressed alarm in a despatch of 24 February 1565, AGI, Lima section, 121. Other authorities mocked at Castro's fears, notably Archbishop Loayza who told the King that there was 'no foundation to the story some people are telling that the Indians of Jauja wished to revolt', despatch of 1 March 1566, IEP **2** 313, and Licenciate Monzón and Jerónimo de Silva shared this scepticism: R. Vargas Ugarte, *Historia del Perú, Virreinato (1551–1590)*, Lima, 1942, Appendix 1; Vargas Ugarte, *Manuscriptos peruanos* **2** 86; Vargas Ugarte, *Historia del Perú, Virreinato (1551–1600)* 160–1; Consejo de Indias, CDIU **15** 246; Lohmann Villena, *El corregidor de indios* 41–2.

306 *would have died'.* García de Castro despatch of 6 March 1565, GP **3** 59.

306 *the frightened Governor.* Guacra Pucara was 'cacique principal' of Hatun Jauja, son of the chief who had helped Francisco Pizarro and Hernando de Soto against the Quitans in 1534. Guacra Pucara enterprisingly went all the way to Madrid to plead for favours from Philip II – one of the first Peruvian chiefs to cross the Atlantic. He presented a battery of memoranda of service (Madrid, 3 September 1563, AGI, Lima, 205) and was granted a coat of arms, Barcelona, 18 March 1564, as well as the pension of 600 pesos mentioned by García de Castro. Cúneo-Vidal, *Historia de las guerras de los últimos Incas*; Temple, 'Los caciques Apoalaya'; Espinosa Soriano, 'El alcalde mayor indígena en el virreinato del Perú 207.

307 *pikes were made'.* García de Castro despatch of 6 March 1565, GP **3** 60.

307 *arquebus very well.'* Ibid.

307 *to the owners.* Ibid.

307 *was afoot there.* Cabildo of Cuzco, session of 12 March 1565, (*Libro de Cabildos del Cuzco*, in Biblioteca del Ministerio de Relaciones Exteriores del Perú); Letter of corregidor Juan de Sandoval, 5 May 1565, AGI Lima, 121; Lohmann Villena, *El corregidor de indios* 41.

307 *Captain Francisco Mercado.* García de Castro despatch of 23 September 1565, GP **3** 97; Vargas Ugarte, *Historia del Perú, Virreinato (1551–1600)* 160. The Indians killed seven Spaniards at Valladolid and besieged fifteen others, before being driven off by reinforcements from Loja and Piura.

310 *lose its inspiration.* R. T. Zuidema, 'Observaciones sobre el Taqui Onqoy', *Historia y Cultura*, Lima, **1**, no. 1, 137–40, 1965;

Porras Barrenechea, *Fuentes históricas peruanas* 53; Vargas Ugarte, *Historia de la iglesia en el Perú* **1** 117–19; Armas Medina, *Cristianización del Perú*, 552, 583–4; Kubler, 'The Quechua in the Colonial World' 396–8.

310 *cruellest possible war'*. Titu Cusi Yupanqui 101.

311 *bygones by bygones'*. Ibid. 102. Also, Memorial of June 1565 handed to Juan de Matienzo, in Matienzo, *Gobierno del Perú*, pt. 2, ch. 18, 296, 302.

311 *incursions towards Amaibamba*. Rodríguez de Figueroa 187. Gaspar de Sotelo received permission from García de Castro for a similar expedition in 1565, Matienzo, *Gobierno del Perú*, pt. 2, ch. 18, 298.

311 *the whole kingdom'* Rodríguez de Figueroa 186.

311 *of the Spaniards.'* Ibid.

311 *that he holds.'* Letter of 6 March 1565, GP **3** 59.

311 *transported across it.'* Testimony of Tomé de Villagran, *Información de méritos y servicios del Licenciado Matienzo*, La Plata, 19 January 1580, Levillier, *Audiencia de Charcas* **2** 528.

312 *of Santa Clara*, This convent also contained the daughter of Juan de Betanzos and of his Inca wife Angelina. The girl, María de Betanzos Yupanqui, was later seduced and abducted by one Juan Baptista de Vitoria while a novice in the convent. Her furious father disinherited her in 1573, in a document now in the Archivo Nacional in Lima; but she married Vitoria.

312 *María Cusi Huarcay.* Ella Dunbar Temple, the indefatigable researcher into the descendants of the Incas, discovered a document of 1599 in the Biblioteca Nacional of Peru. This was a division of María Cusi Huarcay's personal estate among her heirs. It revealed that she had two daughters called Doña Francisca Maldonado and Doña Juana Maldonado, who married Bermudo de Mesa and Miguel Merino de Lecona respectively ('El testamento inédito de doña Beatriz Clara Coya de Loyola' *Fenix* **3**, nos. 7–8, 1951–2).

312 *as a seducer.* Cristóbal Maldonado had raped and abducted the daughter of one Gerónimo Zurbano, but had escaped punishment through a legal technicality as to whether Cuzco lay within the jurisdiction of the Audiencia of Lima or of Charcas for such cases.

312 *them in Cuzco.'* García de Castro despatch to the King, 12 January 1566, GP **3** 155.

312 *to his brother'* Letter to King of February 1567, GP **3** 238.

312 *has succeeded him'*. Archbishop Loayza to King, Lima, 1 May 1566, MP **2** 86.

312 *report to us'*. The decree clearly assumed Maldonado's guilt. It said: 'Because this Cristóbal Maldonado had not been punished for ravishing a daughter of Gerónimo Zurbano, with whom he was also said to have been secretly married, he dared to commit this crime.' CDIA **18** 81.

313 *to Cristóbal Maldonado.'* Ibid.

313 *he now is'*. The treaty is in the Archivo General de Indias in Seville, still unpublished in full. Its terms were noted in the annual records of the Council of the Indies for 1569: CDIU **15** 271–3. Also: Temple, 'Notas sobre el Virrey Toledo y los Incas de Vilcabamba', *Documenta* **2** no. 1, 617. The Acobamba river flows into the Apurímac and marked a limit of Spanish-occupied Peru.

313 *without further warning'*. Ibid. 273.

313 *and signed it'*. Ibid.

314 *overcame many difficulties'*. García de Castro despatch of 2 September 1567, GP **3** 263.

314 *Figueroa for ever . . .'* CDIU **15** 275.

314 *Kings of Spain'*. Ibid.

314 *his own hands'*. Ibid. 273.

314 *cuts in it.'* Diego Rodríguez de Figueroa's testimony to an 'Información de méritos y servicios del Licenciado Matienzo', La Plata, 19 January 1580, Levillier, *Audiencia de Charcas* **2** 525.

314 *placed upon it'*. Temple, 'Notas sobre el Virrey Toledo y los Incas de Vilcabamba' 617–18.

314 *His Majesty if summoned'*. CDIU **15** 275.

315 *Figueroa were godfathers.'* This statement was in a report in AGI. The Royal Scrivener Alvaro Ruiz de Navamuel certified that Rodríguez de Figueroa copied the passage from the original (Temple, 'Notas sobre el Virrey Toledo y los Incas de Vilcabamba' 616 and CDIU **15** 275). Carco or Zarco was in the western part of Vilcabamba, on the road across the Apurímac to Huamanga immediately above the mouth of the Acobamba.

315 *the holy baptism'*. Ibid.

315 *to this heir'*. Unpublished Información of 1567 also found by Ella Dunbar Temple in

AGI, 'Notas sobre el Virrey Toledo y los Incas de Vilcabamba' 618.
315 *went against me.'* García de Castro despatch, 2 September 1567, GP **3** 263.
315 *for 40,000 pesos.'* Ibid. 265.
315 *is well done.'* Ibid.
315 *to his service.'* García de Castro despatch, 20 December 1567, GP **3** 270.
315 *to their custom'*. Rubén Vargas Ugarte found the letter, dated Madrid, 12 August 1568, in the archive of the Spanish embassy in Rome. He quoted it in MP **1** 129–30.
315 *Inca Titu Cusi.* The Capitulación was confirmed by the King at Madrid on 2 January 1569, MP **2** 214.
316 *government of Peru.* Titu Cusi's letter was written from San Agustín de Atunzalla – one of the villages in which the Augustinians had evidently established a church – on 24 May 1569. Vargas Ugarte noted the presence of the letter in AGI, MP **2** 367. Ella Dunbar Temple gave its text in 'Notas sobre el Virrey Toledo y los Incas de Vilcabamba'.
316 *ratified completely'* Titu Cusi Yupanqui 103. The *Relación* was dictated at San Salvador de Vilcabamba, 6 February 1670.

CHAPTER 16 VILCABAMBA

317 *call spiritual affairs'*, Rodríguez de Figueroa 188.
317 *represents the day.'* Calancha 794.
318 *preserved through it.'* Toledo despatch, 20 March 1572, GP **4** 344–5.
318 *of these medallions.'* Ibid.; also Sarmiento de Gamboa, ch. 29, Hakl Soc, 2 ser., **22** 97.
318 *the Antisuyo.* Rodríguez de Figueroa 179, 180, 186, 189.
318 *in their hands.'* Ibid. 186.
319 *that sun temple.'* Calancha 796.
319 *wrought on them.'* Ibid.
319 *as something divine.'* Ibid. 806. Murúa, ch. 75, **1** 232. Murúa called the shrine Chuquipalta, and Calancha called it Chuquipalpa.
319 *the Inca rule.'* 'Vitcos, the Last Inca Capital' 184. Bingham later repeated the exciting story of his discovery in his books *Inca Land*, 247–51 and *Lost City of the Incas* (1951) 116.
319 *support of man."'* Temple, 'Notas sobre el Virrey Toledo y los Incas de Vilcabamba' 617; CDIU **15** 275.
320 *could see them.'* Temple, 'Notas' 616.
320 *give him any'*. Book of the Contador Miguel Sánchez, entry for 6 December 1569: this mentioned that the payments being made to the then resident missionary were identical to those previously paid to Vera. The entry was discovered by Carlos A. Mackehenie, who quoted it in his two articles 'Apuntes sobre Don Diego de Castro'. The same account book also mentioned 132 pesos being spent by Antonio de Vera and Francisco de las Veredas, of which twelve pesos were for 'two pieces of cloth given to two captains of the Inca'.
320 *live with him.'* Declaration by Antonio de Vera copied by Rodríguez de Figueroa, in Temple, 'Notas sobre el Virrey Toledo' 616.
320 *I had brought.'* Rodríguez de Figueroa 184.
320 *to be killed.'* Ibid.
320 *remember the name.'* Ibid. 192.
321 *to be made'* Ibid.
321 *hear me preach'*. Ibid.
321 *Lord Jesus Christ.'* Ibid. 193–4.
321 *converted a Christian'*, Titu Cusi Yupanqui 104.
321 *an important person.'* Ibid. 105.
322 *himself a Catholic'*, Calancha, bk. 4, ch. 2, 795.
322 *in the land.'* Titu Cusi Yupanqui 106.
322 *their parents' consent'*. Ibid.
322 *the entire region'*. Ibid.
322 *of Our Lord.'* Titu Cusi to Vivero, Pampaconas, 24 November 1568, CLDRHP 1 ser., **2**, Appendix B, 121.
322 *dissolution held sway'*. Calancha 795.
322 *a dozen strokes'*; Ibid. 798.
323 *with some liberty'*. Report on the martyrdom of Diego Ortiz by Doña Angelina Llacsa, one of Titu Cusi's wives. CLDRHP 1 ser., **2**, Appendix E, 133–7.
323 *with apostolic zeal'* Calancha 798.
323 *types of food.'* Ibid. 801.
323 *and clothed them'*. Ibid. 802.
323 *Pilcosuni jungle Indians.* Ibid. 806.
323 *by the dozen'*. Ibid. 803.
324 *of their abominations.'* Ibid. Also Murúa, ch. 75, **1** 231–2.
324 *covered in mud.'* Calancha 804; Murúa (ch. 75, **1** 232) also thought that the flooding was deliberate, and was intended to show

the priests how bad was the road to Vilcabamba.
324 *with the witch-doctors'*. Calancha 804.
324 *little by little.'* Titu Cusi Yupanqui 107.
325 *women returned crestfallen.'* Calancha 804–5; also Murúa ch. 75, **1** 232.
325 *up terrible temptations.'* Calancha 805.
325 *of their chastity.'* Ibid.
325 *of tender age . . .'* Raimondi, *El Perú*, segundo fasciculo, cuaderno XLVII, año 1866, 64, 71. Hiram Bingham also saw the Campas wearing their long robes when he visited the Pampaconas valley in 1911, and the Indians still wear them.
325 *to their fury.'* Calancha 806.
327 *would kill me.'* Rodríguez de Figueroa 174.
327 *be much afraid.'* Ibid. 174–5.
327 *we made friends.'* Ibid. 176.
328 *of different colours'*. Ibid. 180.
328 *permission to kill'*. Ibid. 183.
328 *with his lance.'* Ibid. 180.
328 *to his seat.'* Ibid. 186.
329 *me until morning.'* Ibid. 189–90.
329 *only their fun.'* Ibid. 190.
329 *with the Inca'*. Ibid. 179–80.
329 *least among us.'* Oviedo y Valdés, bk. 47, ch. 13, **121** 171.
329 *Titu Cusi's close friends'* Calancha 812.
329 *the Inca's Relación*. Titu Cusi Yupanqui 107.
330 *very old cloak'*. Rodríguez de Figueroa 179.
330 *2,000 other Indians'*. Ibid. 194.
330 *behind these mountains'* Ibid. 189.
330 *who eat human flesh'* Ibid. 180.
330 *term for cannibals*. Garcilaso, pt. 2, bk. 8, ch. 9, **135** 141.
330 *they eat human flesh'*. Titu Cusi to Vivero, Pampaconas, 24 November 1568, CLDRHP, 1 ser., **2**, Appendix B, 121.
330 *tribute to him.'* Matienzo, *Gobierno del Perú*, pt. 2, ch. 18, 294. This same list was given in Antonio Vázquez de Espinosa, *Compendium and Description of the West Indies* 552. Antonio Bautista de Salazar's *Relación sobre el periodo de gobierno de los Virreyes Don Francisco de Toledo y Don García Hurtado de Mendoza* (formerly attributed to Tristán Sánchez) said that Titu Cusi's territories were bounded to the west: by 'Curamba, Pingos, Marcahuasi, Mollepata' 270. These are towns in a line north of Andahuaylas.
331 *has in revolt'*. Toledo report of 1 March 1572, GP **4** 89. Matienzo in fact referred to this tribe as Pilcomu and Toledo as Pilioconi; in Titu Cusi's letter they were Pellcosuni, but in his *Relación* (88) Pillco suni; Calancha and Ocampo both called them Pilcosones.
331 *grains and sugar-cane'*. Ocampo, Hakl Soc, 2 ser., **22** 238. Ocampo led an expedition into Pilcosuni territory in 1610.
331 *come to me.'* Toledo, report of 1 March 1572, GP **4** 94.
331 *province of Vilcabamba'*. Ocampo, Hakl Soc, 2 ser., **22** 216.
332 *and almost impregnable'*. García de Loyola, encomienda grant, Potosí, 10 February 1573, CDH Chile (2) **4** 206; Probanza de servicios, Cuzco, 2 October 1572, JLPB **7** 22–3; Petition to King, 26 August 1576, CDH Chile (2) **4** 215.

CHAPTER 17 TITU CUSI AND CARLOS INCA

334 *than other Indians'* Matienzo, *Gobierno del Perú*, pt. 2, ch. 18, 294.
334 *severe and manly'*. Rodríguez de Figueroa 182.
334 *kill fifty Spaniards'* Ibid. 189.
334 *that I desired'*. Ibid. 181.
334 *know that liquor.'* Ibid.
335 *bracelets of silver'* Ibid.
335 *a paper book'* Ibid. 183.
335 *with moving tears'*. Matienzo, *Gobierno del Perú*, pt. 2, ch. 18, 301.
335 *had murdered him.'* Rodríguez de Figueroa 198. The passage named Juan Pizarro instead of Gonzalo as leader of the invasion of Vilcabamba. Hernando's conquest was presumably the attack on Ollantaytambo in 1536.
336 *propagation of Christianity*. Titu Cusi Yupanqui 100.
336 *by political convenience'*. Ella Dunbar Temple, 'Notas sobre el Virrey Toledo y los Incas de Vilcabamba' 614–5; Rubén Vargas Ugarte, *Historia del Perú, Virreinato (1551–1600)* 167; Kubler, 'The Neo-Inca State'.
336 *to the present.'* Titu Cusi Yupanqui 104.
336 *ravine of Purumata*. Ocampo 231–2; Calancha, bk. 4, ch. 2, 794.

336 *that contained silver'*. Calancha 795.
337 *the Spaniard Romero.'* Ibid. bk. 4, ch. 4, 810; Murúa **1** 233.
337 *would be trouble'*. Rodríguez de Figueroa 191.
338 *and his vassals'*. Titu Cusi Yupanqui 104.
338 *I received him.'* Ibid.
338 *my father possessed.'* Ibid. 103–4.
339 *and other finery.'* Memorial of Don Melchor Carlos Inca, quoted in Temple, 'Don Carlos Inca' 142 and 'Azarosa existencia' 121.
339 *and rich men'*. Garcilaso, pt. 1, bk. 2, ch. 28, Hakl Soc **41** 205.
339 *arms and music'*. Quipocamayos 47.
340 *of Paullu Inca.'* Garcilaso, pt. 2, bk. 1, ch. 23, **134** 48.
340 *solemnised their festivals'*. Vázquez de Espinosa 554.
341 *nonsense and bathos.'* Garcilaso, pt. 2, bk. 8, chs. 1, 2, **135** 128–9. There are twelve very dusty paintings of such a Corpus Christi procession in the tiny church of Santa Ana, in the Carmenca district of Cuzco, on the Lima road as it climbs out of the city. The church was begun in 1560 and completed in 1622 and is the oldest surviving church in Cuzco. But the paintings are from the late seventeenth century; they depict Bishop Manuel de Mollinedo, who was in office from 1673 to 1699. Markham thought that they might date from the time of Carlos Inca. Some Inca nobles are shown watching the procession and participating in it. They wear a headdress with a crimson fringe and plumes of feathers. 'Round the neck is a broad collar of several colours with a long yellow fringe. The tunic is of white cotton covered with ornaments, and confined round the waist by a very broad belt of richly worked cloth. On the breast there is a golden sun. Garters confine the pantaloons above the knee, which are of black cloth. The shoes are also of black cloth. Pumas' heads of gold, set with emeralds, on the shoulders, secure a long scarlet mantle with full white sleeves bordered with wide lace.' Markham's note, Hakl Soc, 2 ser., **22** 207–8. Humberto Vidal, *Visión del Cuzco* (Cuzco, 1958) 129–32.
341 *Our Lord God'*. Garcilaso **135** 128.
341 *once been theirs*. Ibid.
341 *Alderman of Cuzco*. Ella Dunbar Temple discovered a reference to Carlos's being a regidor or alderman in an unpublished transaction between Mancio Sierra and Carlos Inca, in the Archivo Nacional del Perú. Her exhaustive research provides almost all that is known about the Inca nobility in Cuzco. Much is contained in her 'Don Carlos Inca' and 'El testamento inédito de doña Beatriz Clara Coya de Loyola, hija del Inca Sayri Túpac'.
341 *of much Christianity'* Quipocamayos 46. Carlos's wife was born in Trujillo in 1542. Her father, Diego de Amarilla, died when she was an infant, and her mother Doña Catalina decided to take her to Peru, where the Pizarros and many other natives of Trujillo had made their fortunes. But her mother died on the journey to America in 1550, and María was adopted by a family called Esquivel. She therefore changed her name from Amarilla to Esquivel. Ella Dunbar Temple discovered this from investigations made in Trujillo: 'Los testamentos inéditos de Paullu Inca, don Carlos Inca y don Melchor Carlos Inca,' 614–29.
341 *and respected him.'* Poma de Ayala **2** 60.
341 *his uncle Huascar*. Sarmiento de Gamboa, ch. 69, Hakl Soc 189; Diego Fernández, bk. 3, ch. 5, **165** 82; Gutiérrez de Santa Clara, bk. 3, ch. 50, **166** 215.
342 *a coat of arms*. Waldemar Espinosa Soriano, 'El alcalde mayor indígena en el virreinato del Perú' 206–7. There is another document, a royal decree issued at Valladolid, 1 October 1544, that legitimised the numerous bastard offspring of one Don Alonso Titu Uchu Inca, which he had fathered 'even though a bachelor'. Ramos Gavilan and Calancha both mentioned this legitimation, and its text has been published by Urteaga, *El imperio incaico*, Appendix C, 249–52; and Konetzke, CDFS **1** 231–4; and mentioned by Vargas Ugarte, MP **2** 368. It has caused considerable confusion. Ella Dunbar Temple felt that 1544 was clearly too early for the young Alonso Titu Atauchi to have fathered any bastards, and assumed that 1554 was the correct date ('Un linaje incaico durante la dominación española, Los Sahuaraura', 45). I think she was wrong here. The date 1544 was written in words in the document; and 1554 would still have been too early for the King to have heard about the exploit at Pucará. The 1544 document in fact mentioned that the recipient, Don Alonso Titu Uchu Inca, was 'the son of Huascar and grandson of

Huayna-Capac'. It surely referred to a *cousin* of the young Alonso Titu Atauchi. The latter quoted the 1544 decree at Cuzco, 16 June 1566, and asked for similar legitimation for his own bastard children as had been granted to his cousin's twenty-two years before.

342 *than the Spaniards'*. The decree against holding public office was issued at Valladolid, 27 February 1549; against loading Indians at Valladolid, 1 June 1549; against carrying arms, Madrid, 19 December 1568. The texts are in CDFS **1** 256, 259, 437. Philip II prohibited mestizos from becoming caciques in Indian communities or from holding the office of protector de indios: decrees of 18 January 1576 and 20 November 1578, CDFS **1** 491, 512.

342 *their great numbers'*. Nieva despatch of 4 May 1562, GP **1** 423. See also: Richard Konetzke, 'El mestizaje y su importancia en el desarrollo de la problación hispano-americana durante la época colonial' – a rather heavy-handed study of the problem (RI **7** 1946 no. 23).

343 *encomienda in Peru*. Juan Arias Maldonado claimed that his mother had been Atahualpa's full sister, later baptised as Lucía Clara Coya. Maldonado said that Pizarro had given her to his father to prevent her committing suicide on the day of Atahualpa's funeral. Busto Duthurburu, 'Maldonado el Rico, señor de los Andahuaylas', 115; CL **1** 386, probanza for entry into the Order of Santiago.

343 *the food supplies'*, Información sent by Corregidor Jerónimo Costilla to García de Castro, López Martínez, 'Un motín de mestizos' 377. The author based this article on a large enquiry into the rebellion held by President Castro in Cuzco: AGI, Justicia, 1086. López Martínez curiously failed to mention that Castro had just prevented the Maldonado brothers' attempt to marry Beatriz, or that they had been mounting a virulent oral and written attack against the President during the months before the rebellion.

343 *a triumphant Castro*. Letter of 19 April 1567 by Licenciate Serrano Vigil quoted by Roberto Levillier in *Don Francisco de Toledo* **1** 420. Diego Maldonado el Rico had refused to recognise his son Juan as his heir, and would not allow him to enter his house in Cuzco when he fled from the authorities in 1567. But he now relented and travelled to Lima to intercede with Castro. He lived near the city with his son until his death in 1570. Busto Duthurburu, 'Maldonado el Rico, señor de los Andahuaylas' 131; López Martínez, 'Un motín de mestizos'.

344 *over Carlos's head*. Levillier, *Don Francisco de Toledo* **1** 367.

344 *with one another*.' Salazar 252–3.

344 *in those days'*. Ocampo 207.

344 *of Your Majesty'*. Undated letter from Don Carlos Inca to the King. Roberto Levillier reproduced the letter, complete with Carlos's flamboyant signature, *Don Francisco de Toledo* **1** 292. It is reproduced here as plate 41.

345 *of this country*.' Ibid.

345 *province of Quito*. During the Quitan campaign, Benálcazar captured eleven royal children from Rumiñavi in the forests of the Yumbos, and Almagro captured three sons held by the curaca of Chillo, RGI **3**, Appendix 4.

345 *Domingo de Santo Tomás*. Navarro, 'La descendencia de Atahualpa' (1930) 821; Vargas, 'Los hijos de Atahualpa y los padres domenicanos'.

345 *Francisco de Ampuero*. This enquiry established a convincing line of argument: it showed that the boys must have been sons of Atahualpa, for they were the children of his wives and no other male dared approach *them*; Atahualpa had commended his sons to the care of Francisco Pizarro; Pizarro accepted this charge as head of state, but had never given any of them any gift or support from his vast fortune; his meanness should therefore be remedied by the Crown. The text of the 1555 Probanza is in Gangotena y Jijón, 'La descendencia de Atahualpa' 109–22. The witnesses included Juan Delgado, who had been in Cajamarca and was godfather to Juan Quispe-Tupac, Pedro de Alconchel, the trumpeter who had rescued Soto at Vilcaconga, Lucas Martínez Vegaso, Inés Yupanqui, and Domingo de Santo Tomás.

345 *and Francisco Ninancoro*. Informaciones of 14 November 1554, Cuzco, and 28 April 1555, Los Reyes. Another son, Juan Quispe-Tupac, was included in these proofs of parentage but for some reason received no pension. Navarro, 'La descendencia de Atahualpa' 822; Jiménez de la Espada note

in his edition of Santacruz Pachacuti, 233.

345 *all consulted them.* Garcilaso pt. 1, bk. 9, ch. 38; Poma de Ayala, said that Juan Ninancoro and Francisco Illaquita Inca were grandsons of the Inca kings who served as chiefs of Lurin-Cuzco, (1936) folio 740; Sarmiento de Gamboa had Francisco Ninancoro testify to his *History*, ch. 71; Cobo said that he knew three grandsons of Atahualpa called Diego Illaquita, Francisco Illaquita and Juan Ninancoro, who were living in Cuzco in 1620, *Historia del Nuevo Mundo*, bk. 12, ch. 19, **3** 202. Atahualpa's daughters married Spaniards. María first married one Pedro de León and then a soldier called Blas Gómez, who unwisely supported the revolt of Francisco Hernández Girón. Isabel married Sancho de Rojas and went to live in her native Quito. She was given a pension of 400 pesos a year to last for two lifetimes. This poor princess soon suffered the death of her husband Rojas and her only son. The tax authorities, in their usual charming way, claimed that this disposed of the 'two lives' of the original grant and refused further payment of her pension. But the Viceroy Toledo decided that the lady should continue to receive 200 pesos a year for the rest of her life (Toledo despatch of 1 March 1572, Cuzco, GP **4** 55). Garcilaso could not recall whether it was María or Isabel who married Blas Gómez from Extremadura. He thought that the same lady had then married a mestizo gentleman called Sancho de Rojas (pt. 1, bk. 9, ch. 38). The testimony of Juan Delgado in 1555 made it clear that María married León and then Gómez.

345 *held such grants.* Jiménez de la Espada found a note about the boys' baptism. Carlos's encomienda was a poor one, with an income of only 170 pesos; it was listed in the *Relación de los vecinos encomenderos que hay en estos reinos del Perú* quoted in Navarro, 'La descendencia de Atahualpa' 823–4.

345 *which to marry'.* Francisco de Morales to King, 22 September 1552. Royal decree, 12 July 1556, mentioned that the Franciscan monastery in Quito contained 'two sons of Atahualpa and one of Huayna-Capac' (RGI **3**, Appendix 4, cxliv). Francisco received 300 pesos from the royal treasury at Trujillo and later a further 700 from the treasury at Quito. This Francisco Tupac-Atauchi living in Quito has occasionally been confused with Atahualpa's other son called Francisco Ninancoro who lived in Cuzco. The third son in Quito, Felipe, presumably died as a child.

346 *himself with it'.* Pedro de Valverde and Juan Rodríguez, *Relación de la provincia de Quito*, 30 December 1576. The authors recommended that Francisco be given an encomienda instead of the pension. Navarro, 'La descendencia de Atahualpa' 825–6. Francisco was not in fact wholly content with his income: he petitioned the King for more and received a further 600 pesos a year. He and his sister Isabel (who married a Spaniard called Pretel) each received pensions of 1,000 pesos, Papers of Council of the Indies for 1573, CDIU **16** 73.

346 *of San Francisco.* Toledo despatch, 1 March 1572, Cuzco, GP **4** 55. Francisco M. Compte, *Varones illustres de la orden seráfica en el Ecuador* (2 vols, Quito, 1885) **1** 67.

346 *master Captain Sandoval.'* Catalina, serving woman of Hernando de Rojas, interrogated in Tunja, 3 December 1575, for Información on behalf of Vicente de Tamayo, in Bustos Losada, 'Las hijas de Huainacapac', *Museo Histórico* **3**, nos. 10–11, 51.

346 *and other conspirators.* Navarro, 'La descendencia de Atahualpa' 829.

CHAPTER 18 OPPRESSION

347 *to great pity.'* Valverde despatch, Cuzco, 20 March 1539, CL **3** 44.

347 *Quito to Cuzco.'* Vaca de Castro, preamble to *Ordenanzas de tambos* 427.

348 *scarcely two thousand.'* Molina of Santiago 126.

348 *by the 1560s.* Cieza de León, pt. 1, ch. 74 (Onis) 346; Vega 105. The royal Comisarios de la Perpetuidad reported in 1562 that Chincha had only 500 tributaries, *Carta de los Comisarios*, NCDHE **6** 95. In 1558 a Spanish friar, Cristóbal de Castro, and an official, Diego de Ortega Morejón, produced a special study of the way in which Chincha had been ruled under the Incas and before their advent; but they did not

mention its present depopulation, CDHE **50** 206–20.

348 *only one hundred. Relación de los señores que sirvieron*, CLDRHP, 2 ser., **3** 66.

348 *from its fertility'*, Cieza de León, pt. 1, chs. 70, 75.

348 *the year 1600.* Garcilaso, pt. 1, bk. 6, ch. 39. Domingo de Santo Tomás described heavy decline in a Colombian coastal valley in a letter to Las Casas in 1555, CDIA **7** 371.

348 *of the kingdom'.* Fernando de Armellones to Council of the Indies, Lima, 10 December 1555, MP **2** 153.

348 *on the increase'. Relación de los señores que sirvieron*, CLDRHP, 2 ser., **3** 66; Santillán, *Relación* 66–7.

348 *cease before long'* Acosta, *Historia natural*, bk. 3, ch. 19, 78–9. Acosta reckoned the rate of decline on the coast to be 30 to 1 by the 1580s.

348 *it will cease'.* Despatch from Domingo de Santo Tomás, Lima, 13 March 1562, IEP **1** 193–4.

348 *of these Indians.'* Robles, Memorial of 5 April 1570, CDIA **3** 48.

348 *situation is remedied.'* Rodrigo de Loaisa, Memorial of 1586, CDHE **94** 586.

349 *1·5 to 1.* These reports were made, at government request, between 1582 and 1586, and published in Marcos Jiménez de la Espada, RGI **1**, BAE (Cont) **183**. Diego Dávila Brizeño estimated 1·5 to 1 decline for the Yauyos (in the hills above Lima); Luis de Monzón estimated 1·6 to 1 for the relatively unmolested Sora, south of Vilcashuaman; Pedro de Ribera gave 2 to 1 for Huamanga; Andrés de Vega gave 3·75 to 1 for Jauja; and Pedro de Carbajal reported very heavy depopulation for Vilcashuaman. Santillán said that in some districts decline was 4 to 1, in others relatively slight (*Relación* 44).

349 *after the Conquest.* Rowe, 'Inca Culture at the Time of the Spanish Conquest' 184–5. Rowe based his estimates on the RGI reports of the 1580s, on Cristóbal de Castro's and Diego Ortega Morejón's penetrating contemporary report on the valley of Chincha, Bernabé Cobo, and Luis de Morales Figueroa, *Relación* 41–61.

349 *a shattering figure.* Kubler also relied on Morales in his discussion of population decline, in which he reckoned that the rate of decline between the Conquest and 1560 was only two to one, 'The Quechua in the Colonial World' 334–40.

349 *some seven million.* Juan Canelas Albarrán, *Discripción de todos los reinos del Perú* (MS. of 1586, no. 3178, Biblioteca Nacional de Madrid) gave a figure for each province, producing a total of 1,796,312 Indians and 55,700 Spaniards, mestizos and mulattos. A comparable figure was given by the anonymous *Relación de los naturales que ay en los repartimientos del Perú* (MS. of 1561, Colección Muñoz, manuscritos de la Academia de la Historia, Madrid, LXV, folio 46): total population, 1,758,563, of which 396,866 were tribute-paying adult males. The anonymous *Memorial para el buen asiento y gobierno del Perú* of 1563 estimated a population of 'almost two million', CDHE **94** 164. Juan de Matienzo estimated 535,000 tributaries and a total population five times as great (2,675,000) in his *Gobierno del Perú* of 1567 (1910,55). These estimates are higher than that of Luis de Morales Figueroa, with 311,257 tributaries and a total population of some 1,500,000. Two other competent contemporary writers also attempted estimates: Vázquez de Espinosa, writing in 1628, reckoned only 287,395 tributaries for 1572; but Juan López de Velasco, writing 1571–4, gave 680,000 tributaries.

The native population of Peru, Ecuador and Bolivia in the late sixteenth century thus lay somewhere between the 1,500,000 of John Rowe, George Kubler and Silvio Závala (*La encomienda indiana* 323–6) and a more recent estimate of 2,600,000 by Angel Rosenblat (*La población indígena y el mestizaje en América* **1** 88, 252).

The population of the Inca empire had been between 3 million (Kubler, 'The Quechua in the Colonial World' 339; Rosenblat, *La población indígena . . . desde 1492*, pt. 1, 88, 93); 6 million (Rowe, 'Inca Culture at the Time of the Spanish Conquest' 185); perhaps 7 million (Vargas Ugarte, *Historia de la iglesia en el Perú* **1** 10); and much higher traditional guesses – 10 million (Sebastián Lorente, *Historia antigua del Perú* 5 vols., Lima 1860–80, **1** 207) or 10 to 12 million (Romero, *Historia económica y financiera del Perú* 92; Baudin, *A Socialist Empire* 24). It all depends on the ratio of decline and the estimate accepted for the end of the sixteenth century. The situation is further complicated by the fact that population statistics were based on the number of tribute-paying males, and it is not certain which age-groups came into that definition under the Incas. A man owed tribute for less of his life under the Incas than under the Spaniards: tribute-payers were probably one ninth of the total population under the Incas, but one fifth under the Spaniards. There is also the problem that certain areas were declining before the Conquest – particularly along the coast – and other districts had been denuded of men in the Inca wars.

The best of all contemporary census reports was made by Garcí Diez de San Miguel in the province of Chucuito, on the south shore of Lake Titicaca, in 1567. This report has served three modern historians in further exploration of the problem of depopulation: Waldemar Espinosa Soriano, 'Visita hecha a la provincia de Chucuito por Garcí Diez de San Miguel en el año 1567' and J. V. Murra, 'Una apreciación etnológica de la visita de Garcí Diez de San Miguel', both in *Documentos regionales para la etnologia y etnohistoria andina*, Lima, **1** 1944; and C. T. Smith, *Depopulation of the Central Andes in the Sixteenth Century*, MS., Cambridge, England, 1968. Personally, I would accept a rate of decline of 3·5 to 1, since this is roughly that of Jauja, one of the most heavily devastated of the sierra provinces. I would also take 1,800,000 as the population in 1580, thus arriving at 6,300,000 for the Inca empire at the time of the Conquest.

349 *before Pizarro's arrival.* Sarmiento (1907) 169. Cobo, Cieza, Garcilaso and others described this attack on Huayna-Capac's court. Lizárraga said that the epidemic consumed 'the greater part of them', 85.

349 *died from it'.* Herrera **4** 385. Also Lastres, *Historia de la medicina peruana* **2** 75. José Toribio Polo, 'Apuntes sobre las epidemias en el Perú', RH **5** 1913. Dobyns 499.

349 *provinces of Peru'. Información del trabajo y tratamiento que se da a los indios en las minas de Potosí*, Potosí, 31 May 1550, in Barriga, *Los Mercedarios* **4** 27. This Información is a very slanted eulogy of Potosí. It tried to show that the climate there was exceptional, since the epidemic of 1549 had not reached Potosí.

349 *like hay fever'.* Ibid. 32. The disease 'like hay fever' sounds like influenza, which commonly kills those who have no immunity to it.

349 *quantity of Indians'* Conde de Villar despatches of 19 April 1589, GP **11** (1925) 207; and 13 June 1589, GP **11** 284.

350 *for five years.* Montesinos, *Anales*, entry for 1590. Polo, 'Apuntes sobre les epidemias en el Perú' 14–16; Lastres, *Historia de la medicina peruana* **2** 75–7; Dobyns 501–8.

350 *in Inca times.'* Servidores de los últimos Incas 67; Santillán said much the same, 66–7.

350 *sensed its importance.* Andrés de Vega, report on Jauja, 1582; Luis de Monzón, report on Ataunsora, 1586, RGI **1**, BAE (Cont) **183** 167, 221. Both officials thought the birth-rate had fallen partly because of enforced monogamy.

351 *meat] was wasted.'* Santillán 45, 53.

351 *were totally consumed'.* Gutiérrez de Santa Clara, bk. 5, ch. 25, **167** 108.

351 *of their ill-treatment.'* Cieza de León, *War of Las Salinas*, ch. 87, 230.

351 *die of hunger.'* Pascual de Andagoya despatch, Panama, 22 July 1539, CP 371. The same story was told in a letter from Dr Robles to the Cardinal of Sigüenza, Panama, 20 September 1539, CP 374.

352 *from their encomenderos'.* Robles, *Apuntamiento para el acierto del Pirú*, 1570, CDIA **3** 184.

352 *tributes on them.'* Morales despatch, 1561, IEP **1** 183.

352 *to such tributes'* Falcón, CDIA **3** 481.

352 *consumed very rapidly.'* Ramírez 26; also Domingo de Santo Tomás to King, 16 March 1562, IEP **1** 197.

352 *outside their homelands.'* Bartolomé de Vega, NCDHE **6** 106. Vicente de Valverde despatch, 20 March 1539, CDIA **3** 105–7.

353 *humble or well-behaved.'* Santillán, *Relación* 72–3.

353 *of each encomendero.'* Ibid. 66.

353 *Gonzalo Pizarro rebelled.* Francisco López de Caravantes, *Noticia general del Perú*, RGI **1**, Appendix 3, BAE (Cont) **183** 258. The royal decrees were: 19 July 1536, 19 June 1540, Annexe to New Laws, 4 June 1543, CDFS **1** 223–4.

354 *Atahualpa's ransom gold.* Calvete de Estrella, bk. 4, ch. 8, **168** 36.

354 *would be obeyed.'* Ibid, ch. 11, 58; also López de Gómara, bk. 12, ch. 188. Various reports from the inspectors have survived. Rafael Loredo reproduced those relating to

the Charcas, Arequipa, Chuquiabo (La Paz), and Huamanga (Ayacucho), in *Los repartos* 149–213. Huánuco is in RAN **19** pt. 1, 14–25; and the nearby Conchucos in Marie Helmer, 'La visitación de los Yndios Chupachos, Inka et encomendero, 1549'.

354 *Holy Catholic Faith ...'* Award of repartimiento of Los Conchucos by Gasca, Lima, 1550, CDIA **25**, 5 ff. The assessment for Ilabaya, near Arequipa, was made by three of the most liberal ecclesiastics in Peru – Archbishop Loayza, the Dominican Provincial Tomás de San Martín, and Domingo de Santo Tomás, a prior of the same order. But even this award was quoted by Bartolomé de Vega in 1562 as being severe (Vega 107–8); DH Arequipa **2** 203 ff. Gasca's assessment for Pachacamac is in CL **2** 152–6.

355 *to provide them ...'* Domingo de Santo Tomás, *Relación* to Bartolomé de las Casas, 1555, CDIA **7** 374, 1867; also Falcón 480; Vega 106, 109; Loaisa 590–2; Rowe, 'The Incas under Spanish Colonial Institutions'.

355 *enough precious metals.* Vega 106; Loaisa 592.

355 *on his land.'* Robles, Memorial of 5 April 1570, CDIA **3** 51.

355 *the Spanish city.'* Vega 112; Robles, *Memorial sobre el asiento del Pirú*, CDIA **11** 24; Falcón 481.

356 *added to them.'* Santillán, *Relación* 53; Gibson, *The Inca Concept of Sovereignty* 89; Lohmann Villena, *El corregidor de indios* 18–19.

356 *and highly paid.* A good account of these Spanish managers is in Lockhart, *Spanish Peru* 23–6.

357 *was the case.* Royal provision, Valladolid, 4 June 1543, CDFS **1** 224. Robles, *Proveimientos generales y particulares del Pirú* 32; *Carta de los comisarios a S.M. sobre la perpetuidad y otras cosas*, NCDHE **6** 87–8.

357 *the Inca era'* Falcón 460–1. The same damaging comparisons were made by Santillán 63–9.

357 *of 1538–9 failed.* Gibson, *The Inca Concept of Sovereignty* 91; Lohmann Villena, *El corregidor de indios*, ch. 1; Trimborn, 'Las clases sociales en el imperio incaico'; Levillier, *Don Francisco de Toledo* **1** 96–8; Baudin, *A Socialist Empire* 135, 159; Cobo, bk. 12, chs. 25, 27.

358 *the entire kingdom.'* Damián de la Bandera, RGI **1**, BAE (Cont) **183** 178.

358 *property of others.'* Cieza de León, pt. 2, ch. 20 (Onis) 167.

358 *in his district.'* Polo de Ondegardo, *Relación*, 26 June 1571, CDIA **17** 89, 163; also Matienzo, *Gobierno del Perú* 17; Pedro de Quiroga 68; Santillán 58.

358 *to by birth.'* Poma de Ayala **2** 36–7.

358 *as independent lords.* Domingo de Santo Tomás, letter of 1 July 1550, IEP **1** 195–7.

358 *treasure from them'*, Diego de Vera 133.

358 *of a flogging.* Antón Quadrado to Gonzalo Pizarro, Trujillo, 11 March 1547, Maggs–Huntington 284.

358 *by Hernando Pizarro.* Gasca despatch, Lima, 25 September 1548, CDHE **49** 405; Almagro accusation of 1541, accusation 67 against Hernando Pizarro, CDIA **20** 162.

359 *this year 1550'.* Cieza de León, *War of Las Salinas*, ch. 87, 230.

359 *burned to death.'* Ibid. 231.

359 *from paying it.* Solórzano Pereyra, bk. 2, ch. 21, para. 21.

359 *towards their Indians.* Valverde despatch of 20 March 1539, CDIA **3** 105, 118.

359 *caciques are concerned.'* Cañete despatch, 15 November 1556, CDIA **4** 108–9 and GP **4** 290.

359 *tribute from them'*, Gasca to Philip II, Villamuriel, 17 October 1554, Maggs–Huntington 541.

359 *fleece the natives'*. Santillán 50.

359 *to their caciques'* Santo Tomás 384.

359 *without being discovered.' Carta donde se trata el verdadero y legitimo dominio de los Reyes de España sobre el Perú*, Yucay, 16 March 1571, CDHE **13** 436.

359 *my own Negroes'*, Matienzo, letter of 20 October 1561, La Plata, Levillier, *Audiencia de Charcas* **1** 58–9.

359 *in the world.'* Agia 86.

360 *bear the burden.'* Antonio de Zúñiga, despatch of 15 July 1579, CDHE **21** 104; also García de Castro despatch of 12 January 1566, GP **3** 135; Santillán 48, Vera 132; *Memorial para el buen asiento*, para. 50, 177.

360 *their main tribute.'* Vega 110; also Diego de la Cueva 247–8.

360 *the payment themselves'*. Damián de la Bandera RGI **1**, BAE (Cont) **183** 180; also Santillán 49; *Memorial para el buen asiento*, para. 90, 188.

360 *not dare complain.'* Loaisa, ch. 47, 587; García de Castro, despatch of 12 January 1566, GP **3** 135; Francisco de Toledo,

memorial of 1572, RGI (1881) **1**, Appendix 3, cliii.

360 *of public prostitution.'* Vera 132.

360 *to pay it'.* Santillán 74.

360 *collection of tribute.* Vega 114–15; Lohmann Villena, *El corregidor de indios* 18. This practice was severely prohibited by royal decree, Valladolid, 12 May 1550, CDFS **1** 265.

360 *than any Spaniard.'* Loaisa, ch. 47, 587; also Vera 131–2.

360 *a thankless task.* Falcón 485.

361 *in the village.'* Zúñiga 105; Vera 131–2; Vega 126.

361 *that was lacking',* Santillán 54.

361 *an additional tribute.* Domingo de Santo Tomás, letter of 10 December 1563, Vargas Ugarte, *Historia de la iglesia en el Perú* **1** 126.

361 *instructing his Indians.'* Vega 109.

361 *to prepare food'.* Mateos, 'Constituciones para indios del primer concilio limense, 1552' para. 30, 43.

361 *to twenty children.'* Poma de Ayala **2** 61. *Carta de los comisarios . . . sobre la perpetuidad,* NCDHE **6** 95.

362 *their hair shaved.* Mateos, 'Constituciones para indios del primer concilio limense', para. 26, 38; *Memorial para el buen asiento* 174–5; García de Castro, despatch of 12 January 1566, GP **3** 136; *Carta de los comisarios . . . sobre la perpetuidad* 95.

362 *their Spanish masters.* Ramírez 22; Matienzo, *Gobierno del Perú* 18–20; Loaisa chs. 60, 61, 603–5; Kubler 'The Quechua in the Colonial World' 377–8; Wiedner, 'Forced Labor in Colonial Peru' 358, 360, 378; Gibson, *The Inca Concept of Sovereignty* 91.

362 *aptitude and ingenuity'.* Loaisa, ch. 60, 604.

362 *of the Indians.'* Vega 110–11.

363 *to clothe him.'* Ibid. 111.

363 *will to anybody . . .'* Robles, *Memorial* of 5 April 1570, CDIA **3** 48; Gasca to King, 17 October 1554, Maggs–Huntington 541; Francisco de Morales, letter to King, 1561, IEP **1** 183.

363 *and not eat'.* Zúñiga, letter to King, 15 July 1579, CDHE **26** 109–12.

363 *men, but slaves.'* Agia 37.

363 *working for nothing'.* Decree of 22 February 1549, repeated 16 August 1563, CDIA 504–10, CDFS **1** 252–6.

364 *the loads themselves.'* Domingo de Santo Tomás to Council, 1 July 1550, in Vargas Ugarte, *Historia del Perú, Virreinato (1550–1600)* 37.

364 *there, all dead.'* Santillán 89. Santillán named some of the worst expeditions: 'that of Diego de Rojas, that of the Chunchos, of Felipe Gutiérrez, of Candía, or of Diego de Almagro to Chile, of Francisco de Villagran . . . and Pedro de Valdivia to the same province . . . that of Juan de Salinas, of Gómez Arias to Ruparupa, of Pedro de Ursúa . . .'

364 *with dead Indians';* Santillán 89.

364 *the utmost cruelty.'* Diego de Almagro, Accusation of 11 December 1541, Accusation 60 against Hernando Pizarro, CDIA **20** 430.

364 *escape being captive.'* Villasante 139.

364 *a Christian town'.* Santillán 89. Ursúa led an expedition to Omagua on the upper Marañon on which he himself was killed by a lieutenant, and the expedition continued down the Amazon under the bloodthirsty Lope de Aguirre. The depopulated Christian town was presumably that of Valladolid which Juan de Salinas Loyola had founded on the Marañon in 1557 and the natives destroyed in 1565.

364 *road to Cuzco'* Cieza de León, *War of Quito,* ch. 31, 81.

364 *artillery towards Lima.* Gutiérrez de Santa Clara **165** 219.

365 *all must fall.'* Gasca despatch to Council of the Indies, 28 January 1549, CDHE **50** 27.

365 *upon the Indians'.* Vega 126; also Santillán 89–90. Diego Dávila Brizeño, Andrés de Vega, Pedro de Ribera, Antonio de Chaves y de Guevara and Luis de Monzón all named porterage during the wars as a reason for the decline of the native population. RGI **1**, BAE (Cont) **183** 156, 167, 184, 221.

365 *compelled to carry.'* Francisco de Morales to King, 1561, IEP **4**, no. 269, 179–87.

365 *into the mountains'.* Cabildo of Jauja, 29 November 1534, CL **1** 1. It was partly to avoid this porterage that the capital was moved to Lima on the coast. Cabildo of Quito, 9 July 1537, repeated on 28 July 1544 and in 1545. The council decreed a fine of ten pesos or a hundred lashes for each Indian removed, and later regulated the weight and distance for native porters. *Libro primero de Cabildos de Quito* **1** 278; *Libro segundo de Cabildos de Quito* **1** 80; Bayle, 'Los municipios y los indios' 418, 420.

365 *burdened by force',* Despatch of Domingo

de Santo Tomás, Lima, 16 March 1562, IEP **1** 197.

365 *for more leagues'*. Despatch of Archbishop Loayza, Lima, 20 April 1567, IEP **2** 365.

365 *of such tasks.'* Gasca to King Philip II, Villamuriel, Spain, 17 October 1554, Maggs–Huntington 541.

365 *shoulders of men'*. Agia 91. Antonio de Zúñiga described how the natives of Quito were forced to carry firewood, fodder and water over long distances, Letter to King, 15 July 1579, CDHE **26** 111.

365 *in flat country*. Francisco de Toledo, report to King, 1 March 1572, GP **4** 103; Diego Fernández, pt. 2, bk. 2, ch. 2, **164** 288; Montesinos, *Anales* entry for 1552; Vargas Ugarte, *Historia del Perú, Virreinato (1551–1600)* 24.

366 *form of labour'* Polo de Ondegardo, *Informe* 165.

366 *any person whatsoever.'* New Laws, Barcelona, 20 November 1542, CDFS **1** 218. Repeated 1 June 1549 and 6 August 1563, CDIA **18** 498–504.

366 *work for others*. Royal decree, Valladolid, 22 February 1549, CDIA **18** 504–10 or CDFS **1** 252–5.

366 *explosive to reveal*. Despatch of 28 January 1549, CDHE **50** 26.

366 *to discuss it.'* Diego Fernández, pt. 2, bk. 2, ch. 2, **164** 288–9; Garcilaso, pt. 2, bk. 6, chs. 17–20. The Viceroy died on 21 July 1552; A. S. Aiton, *Antonio de Mendoza, First Viceroy of New Spain* (Durham, 1927) 191–2; Means, *Fall of the Inca Empire* 99–100; Vargas Ugarte, *Historia del Perú, Virreinato (1551–1600)* 24, 41.

366 *consulted the encomenderos. Carta de los comisarios . . . sobre le perpetuidad* NCDHE **6** 89–90. The commissioners felt that it was most rash of the assessors to ignore the encomenderos when dealing with 'such belligerent people in a new and unsettled territory'.

367 *of the natives.'* Robles, *Memorial* of 5 April 1570, CDIA **3** 51.

367 *in heathen rites*, Polo de Ondegardo, *Tratado* 20–1; Murúa (Loyola MS.) CLD-RHP, 2 ser., **4** 230–1; Vega 127; Augustinians, *Relación de la religión* 15. Gagliano, 'The Coca Debate' 43.

367 *Lima of 1551*. Vargas Ugarte, *Historia de la iglesia en el Perú* **1** 242, Mateos, 'Constituciones para indios del primer concilio limense'.

368 *Majesty's vassals perish.'* Zúñiga, letter to King, 15 July 1579, CDHE **26** 93.

368 *and severe anemia*. The tiny insect is called *Verrucarum*. This disease, *verruga peruana*, killed hundreds of workmen building the railway into the Andes over Lima in the 1870s. D.D.T. is now checking the insect, and the disease responds to antibiotics. Pedro de Ribera and Antonio de Chaves y de Guevara said, in their description of Huamanga of 1586, that coca workers caught 'diseases of incurable sores', RGI **1**, BAE (Cont) **183** 191.

368 *by the cancer'*. Vega 128. Robles called it 'an incurable cancer', CDIA **3** 51. Similar ravaged faces can be seen in the hospitals of eastern Peru to this day, and many Mochica pots showed heads without noses. Rodrigo de Loaisa called the disease 'andeongo' and said that it 'ate away the nose and produced maggots in it', 601.

368 *they suffer there'*. Vega 128. Even Juan de Matienzo, who favoured the trade, reckoned the death rate at forty per cent, *Gobierno del Perú* 94; Francisco Falcón said that it caused 'infinite deaths', CDIA **3** 484; Baltasar Ramírez (38–40), Hernando de Santillán (108), Rodrigo de Loaisa (584), Antonio de Zúñiga (93), Pablo José de Arriaga, (44–5, 139) and Reginaldo de Lizárraga (**1** 207), all condemned the coca trade for similar reasons.

368 *they never recuperate.'* Decree of 18 October 1569, Gagliano, 'The Coca Debate' 50–1.

368 *so highly valued.'* Cieza de León pt. 1, ch. 87 (Onis) 260.

368 *by the cocaine*. Acosta, *Historia natural*, bk. 4, ch. 22, 117. Blas Valera (*Relación de las costumbres antiguas*) wrote: 'Indians who chew it appear stronger and more disposed to work'. *Memorial para el buen asiento* 214–15; Cieza de León pt. 1, ch. 87; Loaisa 584; Cobo, *Historia del Nuevo Mundo*, bk. 13, ch. 22; Gagliano, 'The Coca Debate' 60–1; W. G. Mortimer, *Coca, the Divine Plant of the Incas* (New York, 1901); L. N. Sáenz, *La coca* (Lima, 1938).

369 *of the Spaniards'*. Santillán 108.

369 *soon in progress*. Interrogation of Diego Gualpa made on instructions of Viceroy Francisco de Toledo, Potosí, 31 December 1572, by Rodrigo de la Fuente Sanct Angel, RGI **2**, BAE (Cont) **183** 357–61. Slightly different versions were given by Diego

Fernández, bk. 2, ch. 11, **164** 113; Acosta, *Historia natural*, bk. 4, ch. 6, 95–6; Cieza de León, pt. 1, ch. 99; Nicolao del Benyno, *Relación* of 9 October 1573, RGI **2**, BAE (Cont) **183** 364.

369 *free of disease*. Wiedner, 'Forced Labor in Colonial Peru' 361; Alberto Crespo Rodas, 'La "mita" de Potosí', RH **22** 170. Diego de la Cueva described the skilful native smelting ovens, which funnelled high winds to whip up an intense heat, in an otherwise insufferably boring treatise: *Carta escrita a Su Magestad sobre los negocios de las Indias* 254–5.

370 *provisions and ordinances*.' Domingo de Santo Tomás, despatch of 1 July 1550, Vargas Ugarte, *Historia del Perú, Virreinato (1551–1600)* 36–7. An Información by Fray Bartolomé de Montesinos of 1551 painted a far less harrowing picture, but Ortega de Melgosa was already complaining that the production of silver was suffering from lack of Indians, GP **2** 532.

370 *ambitions in Europe*. Acosta wrote that Potosí had produced 70 million pesos of silver up to 1574, and a further 35 million in the eleven years to 1585. These pesos were worth 13¼ reales or 450 maravedis in contemporary coinage; a ducat was worth 11 reales; Acosta, *Historia natural*, bk. 4, ch. 7, 97–8.

371 *in the world*. Acosta heard from Cabrera how he found the Cerro Rico, *Historia natural*, bk. 4, ch. 11, 103–4; Montesinos, *Anales*, entry for 1571; Lohmann Villena, *Las minas de Huancavelica*, 13, 53–7.

371 *the tunnels' end*. Vázquez de Espinosa 504, 543. The Viceroy Marquis of Montesclaros penetrated the mine in 1608 and was equally appalled by what he saw. So was Miguel Agia, who descended some 600 feet to the lowest workings in 1603.

372 *led to pneumonia*. Lohmann Villena, *Las minas de Huancavelica* 171–4, 212.

372 *them to die*'. Agia 62.

372 *crippled on Saturday*.' Loaisa 601.

372 *the slightest pretext*.' Messia 140.

372 *pounds a week*. Wiedner, 'Forced Labor in Colonial Peru' 371; Cobb, 'Potosí, a South American Mining Frontier' 41–2; Escalona Agüero, *Gazofilacio regio perubico*, bk. 2, pt. 2, ch. 1, 113; Robles, *Apuntamientos* 52; Arzáns de Orsúa y Vela 189–90; Basadre 346–7; Rowe, 'The Incas under Spanish Colonial Institutions'.

373 *hating to serve*.' Agia 56.

CHAPTER 19 EXPERIMENTS IN GOVERNMENT

375 *ruled their country*. James Lockhart has demonstrated how quickly the Spaniards produced a settled society in Peru, in his *Spanish Peru, 1532–1560*.

376 *of native kings*. Guillermo Lohmann Villena produced a fine survey of the political writings of the 1560s in the introduction to his 'Juan de Matienzo' 767 ff. Chamberlain, 'The Concept of the Señor Natural as Revealed by Castilian Law and Administrative Documents' 131; Gibson, *The Inca Concept of Sovereignty* 88.

376 *from paying tribute*. Sarmiento de Gamboa listed many Inca nobles, with their ayllus, ch. 71 and certification, Hakl Soc, 2 ser., **22** 197–8. Viceroy Toledo recorded their land holdings, 11 August 1572, text in Urteaga, *El imperio incaico*, Appendix B, 229–35.

376 *having Spanish friends*.' Santillán 73.

377 *or guarding cattle*.' Luis de Morales, *Relación*, quoted in Porras Barrenechea, 'Cronicas perdidas' 234.

377 *Vilcashuaman with 40,000*. *Ordenanza para el tratamiento de indios*, Valladolid, November 1536 (CDFS **1** 180–1) had ordered this. So did an instruction to Berlanga in 1535, CDIU **10**, 1897, 466–7; Espinosa Soriano, 'El alcalde mayor indígena'.

377 *province of Vilcas*. The Chachapoyas had been auxiliaries brought by Alonso de Alvarado to help the royal army in the battle of Chupas. After his victory Vaca de Castro settled them nearby in a town called Santa Lucía de Chiara and perpetually exempted them from paying tribute.

377 *from among themselves*' Royal decree, Valladolid, 9 October 1549, CDFS **1** 260–1; Bayle, 'Cabildos de indios', Espinosa Soriano, 'El alcalde mayor indígena' 200–1.

377 *each Spanish city*. Gasca to King, Villamuriel, Spain, 17 October 1554, Maggs–Huntington 540. The same suggestion was made by the 1552 ecclesiastical council of Lima.

377 *consider such appointments* Instructions to

Cañete, Brussels, 10 June 1555, article 10, GP **2** 441–2.

377 *to be alcalde.* Montesinos, *Anales* **1** 258. The order to Polo was given in 1559, and Cañete issued ordinances for these alcaldes in 1560. Espinosa Soriano, 'El alcalde mayor indígena' 204; Bayle, 'Cabildos de indios' 17.

377 *and other things'.* *Memorial para el buen asiento* 174.

378 *to the corregidor'.* Bayle, 'Cabildos de indios' 18.

378 *in judicial affairs.* Don Diego de Figueroa y Cajamarca was grandson of Apo Guacal, one of Huayna-Capac's commanders, and son of Carguatanta. He opposed Gonzalo Pizarro, who exiled him to Chile. From here he escaped to Mexico and returned with Gasca's royal army. Although he had plenty of authority, he received no salary and issued two lengthy petitions for some such favours. He was given more titles in 1579 and continued to administer efficiently until the end of the century, but he was never given a salary. When Don Mateo Yupanqui died in 1578, Don Diego asked for his title of 'Alguacil Mayor', but it was passed to Mateo's son Don Antonio Silquigua. Espinosa Soriano, 'El alcalde mayor indígena' 24–5, 34–7; Bayle, 'Cabildos de indios' 18.

378 *Alcalde Mayor of Quito.* Zambiza was appointed at Guayaquil in 1579 and succeeded in Quito at the end of the century. He commissioned a fine portrait of himself: one of the first ever painted in South America. He is wearing golden facial ornaments and a Spanish ruff, and holds two spears. This portrait appears in colour in Hammond Innes, *The Conquistadors* (London, 1969) 254.

378 *their own affairs.* Instructions of Audiencia of Lima to visitadores of Huánuco, Lima, Huamanga and Arequipa, 15 December 1561. AGI, patronato 188, ramo 28, quoted in Espinosa Soriano 'El alcalde mayor indígena', 210.

378 *corregidor and justice.* Matienzo letter from La Plata, 20 October 1561, Levillier, *Audiencia de Charcas* **1** 58–9.

378 *for such officials.* Loayza despatch, 2 August 1564, IEP **2** 273; Lohmann Villena, *El corregidor de indios* 17.

379 *self-governing native communities.* Cañete despatch, 8 April 1561, GP **1** 429–30.

379 *not take root.* Nieva explained his ideas about native judges, 4 May 1562, GP **1** 429, and about his reluctance to give them any criminal jurisdiction, 15 July 1563, GP **1** 524; Lohmann Villena, *El corregidor de indios* 16, 24.

379 *to force them.* King Philip ordered the reduction of natives in a decree of 13 September 1565 to García de Castro, CDFS **1** 416, and 15 January 1567 to the Audiencia of Charcas, CDIA **18** 514–6 (1872).

379 *in encomienda government.* Bayle, 'Cabildos de indios' 18; Espinosa Soriano, 'El alcalde mayor indígena' 30–1.

380 *in April 1538.* Instructions to Valverde of 14 July 1536, CL (2) **1** 194; and decree of 2 April 1538, ibid. 181; Bayle, *El protector de indios* 32, 61–2.

380 *and defend them.'* Valverde despatch, Cuzco, 20 March 1539, CDIA **3** 105, or CL **3** 97.

380 *of cruel conquistadores.* Valverde sentenced one Juan Vegines to a fine of 30 pesos and five days' imprisonment for beating a native woman called Menzia and putting her in a chain at night to prevent her running off to serve another Spaniard (3 February 1539). The same sentence was given to Francisco Gonzales for similar treatment of a woman called Pospocolla, 3 March 1539, AGI, Lima, 305; Bayle, *El protector de indios* 71–6. The protector's powers were defined in *Disposiciones complementarias de Leyes de Indias* (3 vols, Ministerio de Trabajo y Previsión, Madrid, 1930) **1** 142.

380 *for these people.'* King to Valle, Valladolid, 13 August 1557, CDHE, 2 ser., **17** 63.

381 *make to them'.* CDIA **5** 494. Friede, *Vida y luchas de Don Juan de Valle.*

382 *of the Indians'.* Audiencia of Lima, 1550, CL (2) **3** 258.

382 *so did Cañete.* Gasca despatch of 25 September 1548, GP **1** 125; Cañete's instructions to corregidores, 1558, CL (2) **5** 141. Bayle, *El protector de indios* 51, 125–6.

382 *to the Indians'.* Matienzo despatch of 31 January 1562, Levillier, *Audiencia de Charcas* **1** 49. Matienzo in fact favoured the idea of protectors or corregidores to defend the Indians, *Gobierno del Perú*, pt. 1, ch. 20, 73–5.

383 *they were thieves.* Loayza despatch, 1 March 1566, IEP **2** 311–12. Rowe, 'The Incas under Spanish Colonial Institutions'.

383 *dozens of lawsuits'*. Toledo despatch, 1578, GP **6** 45, Bayle, *El protector de indios* 93.

383 *found to register*. García de Castro despatch of 26 April 1565, GP **3** 70–1, and of 15 June 1565, GP **3** 89. The levy of two tomines represented roughly the cost of a cock. There are six tomines to a peso.

383 *of those innocents'*; Morales despatch of 1561, IEP **2** no. 269, 184.

383 *the new officials*. Loaisa 581–3; Lizárraga, bk. 2, ch. 24. Lohmann Villena, *El corregidor de indios*, ch. 4, 62–5.

384 *of the priests*. Loayza letter of 1 March 1566, IEP **2** 310–12; García de Castro despatches of 12 January and 1 October 1566, GP **3** 135, 199–200. Castro heard about the secret collection and stopped it when the total was at 6,000 pesos. He tried to return the money to the natives from whom the curacas had extorted it.

384 *conspiracy of silence*. García de Castro despatch, 12 January 1566, GP **3** 137; Audiencia de la Plata despatch, 10 June, 1566, Levillier, *Audiencia de Charcas* **2** 451–5.

385 *Indians before departing'*. Petition by Alonso de Villanueva and Gonzalo López, quoted by Marvin Goldwert, 'La lucha por la perpetuidad' **22** 343.

387 *with your recommendations'*. Royal decree, Ghent, 23 July 1559, CDFS **1** 370–6. Goldwert, 'La lucha por la perpetuidad' **22** 352–8; Vargas Ugarte, *Historia del Perú, Virreinato (1551–1600)* 136–7; Zurkalowski, 'El establecimiento de las encomiendas'. The three commissioners were: Diego Bribiesca de Muñatones of the Council of Castile, Diego de Vargas Carbajal, Postmaster General of the Indies, and Ortega de Melgosa of the Chamber of Trade of the Indies.

387 *Peru in 1560*. Ribera brought with him two urns full of olive shoots; these were the forerunners of the venerable olive trees that still grow in protected isolation amid the rich gardens of the Lima suburb of San Isidro. The Oidor Salazar de Villasante left Peru in 1560 and wrote that Ribera's olive trees had not yet borne fruit, *Relación* 123. Ribera planted his olives in a famous huerta or kitchen garden started by Pizarro's half-brother Francisco Martín de Alcántara. The garden had been maintained by his widow Doña Inés Muñoz, who later married Ribera. Garcilaso, pt. 1, bk. 9, ch. 17. Cobo, *Historia de la fundación de Lima*, bk. 3, ch. 16, 429–31.

387 *avalanche of treatises*. Many of these were published in NCDHE **6**; Vargas Ugarte mentioned a dossier of documents on perpetuity in the archive of the Biblioteca del Palacio Real de Madrid, and the other Spanish archives are also full of literature on the subject; *Historia del Peru, Virreinato (1551–1600)* 148–9.

387 *advanced by academics.'* Domingo de Santo Tomás, letter from Andahuaylas, 5 April 1562, AGI, Lima, 313, quoted in Vargas Ugarte, *Historia del Perú, Virreinato (1551–1600)* 140.

387 *under royal administration*. Meeting of 21 January 1562. The curacas chose as their delegates: Jerónimo de Loayza, Archbishop of Lima, Francisco de Morales, Franciscan Provincial, Domingo de Santo Tomás, Dominican Provincial, and named as alternate delegates: Pedro de Cepeda, Augustinian Prior in Cuzco, Bartolomé de las Casas, Bishop of Chiapa, Bravo de Saravia, Oidor of the Audiencia of Lima, Gil Ramíres Dávalos, former corregidor of Quito and Cuzco, and Alonso Manuel de Anaya, a liberal citizen of Lima. Goldwert, 'La lucha por la perpetuidad', **23** 214–15.

388 *the rapacious curacas*. Nieva and commissioners, despatch of 4 May 1562, GP **1** 415–16.

389 *equally unwelcome conquerors*. Polo de Ondegardo, *Informe al Licenciado Briviesca de Muñatones*; Matienzo, *Gobierno del Perú*, pt. 1, ch. 30; Santillán 111–17. Lohmann Villena, 'Juan de Matienzo' 103–11; Goldwert, 'La lucha por la perpetuidad' **22** 355–6; **23** 216.

389 *perpetuity with jurisdiction*. *Carta de los comisarios*, 2 May 1562, GP **1** 398–9, NCDHE **6** 50. Eight other cities had not yet responded, and these included Quito, which would presumably have offered about half a million pesos. The total from all the cities would therefore have been about 5,500,000 pesos, or roughly three-quarters of Ribera's original offer to the King. Vargas Ugarte and Goldwert mentioned only the contributions from the first six cities, omitting those of Arequipa and Huamanga.

389 *were given orders.'* Ibid. NCDHE **6** 54.

389 *from the encomenderos'*. Ibid.

389 *so very good'*, Ibid. 81.
389 *the first instance.'* Ibid. 82.
390 *parties who intervene'.* Licenciate Monzón, despatch of 2 January 1563 in Vargas Ugarte, *Historia del Perú, Virreinato (1551–1600)* 149.
390 *to the King.* Loayza despatch of 15 March 1564 quoted in Vargas Ugarte, *Historia del Perú, Virreinato (1551–1600)* 144; Goldwert, 'La lucha por la perpetuidad' 222; Sánchez Bella, *El gobierno del Perú, 1556–1564*, 498–502; Schäfer, *El Consejo Real y Supremo de las Indias* **2** 46.
390 *to a balcony'*, Pedro de Mexia de Ovando, unpublished *Memorial político*, quoted in Vargas Ugarte, *Historia del Perú, Virreinato (1551–1600)* 149.
390 *his private parts'.* Ibid.
391 *enforce royal laws.* Luis Sánchez memorial to Cardinal Espinosa, President of the Council of Castile, 26 August 1566, CDIA **9** 163–70.
391 *the all-important mines.* The Junta Magna consisted of Cardinal Espinosa, Luis Quijada, President of the Council of the Indies and an old servant of Charles V, Ruy Gómez de Silva, a powerful minister of Philip II, the Duke of Feria, Suárez de Figueroa and Gómez Zapata of the Council of the Indies. The Junta Magna first met on 27 July 1568. Levillier, *Don Francisco de Toledo* **1** 78–80; Vargas Ugarte, *Historia del Perú, Virreinato (1551–1600)* 196–7. The Council of the Indies resented Toledo's attempt to bypass their authority; they retaliated by obstructing and misrepresenting his achievements as Viceroy. Ernst Schäfer, 'Felipe II, el Consejo de Indias y el Virrey Don Francisco de Toledo'.

CHAPTER 20 TOLEDO'S SOLUTIONS

393 *would live together'.* Royal instruction, Alcalá de Henares, 20 March 1503, CDFS **1** 9.
393 *Castro in 1565.* Royal instruction, Segovia, 13 September 1565, CDFS **1** 416.
393 *worshipping their idols.'* Ibid.
393 *up to now'.* Toledo, *Libro de la visita general* 160.
394 *of good climate.'* Ibid. 163.
394 *were then destroyed. Instrucción general* for the visita, Lima 1570, with notes added by Toledo, Cuzco 17 June 1571 and 8 September 1571. Texts are in Toledo, *Libro de la visita general.* Toledo wrote Instructions to be followed by those responsible for the reduction of Indians, 6 March 1573. These Instructions are in an unpublished manuscript in the Biblioteca Nacional of Peru (MS. B 511) called Códice de Toledo: Lohmann Villena, 'Juan de Matienzo'.
394 *and good treatment.'* Toledo letter quoted by Levillier, *Don Francisco de Toledo* **1** 247–8.
394 *from their hideaways.'* Toledo, *Memorial que dió al Rey* 88–9; Toledo, *Libro de la visita general* 163.
394 *very Christian manner.'* Bartolomé Hernández to Juan de Ovando, President of the Council of the Indies, quoted by Levillier, *Don Francisco de Toledo* **1** 250.
395 *and even daughters'.* Toledo to King, date missing through fire damage, *Libro de la visita general* 190.
395 *notice of them'.* Ibid. 191. Rowe, 'The Incas under Spanish Colonial Institutions' 156.
395 *live at present'.* Diego Dávila Brizeño, *Descripción y relación de la provincia de los Yauyos*, 6 January 1586, RGI, BAE (Cont) **183** 155.
395 *greatly in this.'* Ibid. 160.
396 *have tile roofs.'* Andrés de Vega, *La descripción que se hizo en la provincia de Jauja*, ch. 31, RGI, BAE (Cont) **183** 171.
396 *and a scrivener.* Matienzo despatch, 21 January 1573, Levillier, *Audiencia de Charcas* **2** 467.
396 *of their Indians.'* Juan Maldonado de Buendía to King, Lima, 25 March 1575, in Levillier, *Don Francisco de Toledo* **1** 270. Vargas Ugarte, *Historia del Perú, Virreinato (1551–1600)* 243; also Ulloa, 'Visita general de los yndios del Cuzco RH **3** 332–47 (1908); Espinosa Soriano, 'La guaranga y la reducción de Huancayo' and 'El alcalde mayor indígena' 212–5; Helmer, 'La Vie économique au xvie siècle'.
396 *and cultivate them.'* Luis de Monzón, RGI, BAE (Cont) **183** 238–9.
396 *of the district.'* Pedro de Rivera and Antonio de Chaves y de Guevara, ibid. 185.
397 *temporal and spiritual'.* Ramírez 65.
397 *the natives' Christianity.'* Toledo, Memorial of 1582, RGI **2**, Appendix 3, BAE (Cont) **183** 259.

397 *of the Indians'*. Vargas Ugarte *Concilios limenses* **1** 254. Paragraphs 95–113 of second part of Council of 1567, 252 ff, summarise the council decisions.

397 *of the devil.'* Ibid. **2** 160.

398 *his missionary career*. Porras Barrenechea, *Fuentes* 55.

398 *their ancient idolatries'*. Quoted in ibid. 56. See also Vargas Ugarte, *Historia de la iglesia* **1** 119.

398 *Indians of Chile*. Betanzos said that he had written a dictionary in the dedication to his *Suma y narración de los Incas*. Santo Tomàs called his work *Léxicon o vocabulario de la lengua general del Perú*. Bertonio's *Gramática o arte de la lengua Aymara* was published in Lima in 1612. González Holguín studied Quechua for twenty-five years before publishing his brilliant *Gramática y arte nueva* in Lima in 1607, and his *Vocabulario* the following year.

398 *in their work.'* Ordinance of 18 October 1572, Toledo, *Ordenanzas* 125.

399 *oppression and cruelty'*. Toledo despatch, Cuzco, 1 March 1572, GP **4** 89; also 101.

400 *families of Peru*. Goldwert 227 ff.

400 *with their wishes.'* Toledo despatch, undated but probably 1571, GP **5** 315–16.

400 *temporal and spiritual'*, Ibid. 315.

401 *what they ought.'* Vivero to King, Cuzco, 1572, AGI, 70–326, quoted in Bayle, *El protector de indios* 87–8.

401 *minors cannot plead'*; Ordinance concerning defensor general de los naturales, Arequipa, 10 September 1575, Toledo, *Ordenanzas* 297.

401 *Spaniards in Peru*. Toledo despatch to King, 30 November 1573, GP **5** 253–4; Royal decree of 27 February 1575, ch. 27; Bayle, *El protector de indios* 139–42; Lohmann Villena, *El corregidor de Indios* 83–93.

401 *dozens of lawsuits'*. Toledo despatch to King, Lima, 6 April 1578, GP **6** 45.

402 *priests and caciques*. The text of much of this important legislation remains unpublished. Some is in CL (2) **9** 196–207; more is in Lohmann Villena, *El corregidor de indios*, Appendices 2 and 3, 519–64.

402 *of the caciques.'* Toledo, memorial of 1583, *Memorial*, ch. 20, 91–2.

402 *rule of Peru*, Royal decree, 15 July 1584, CDIA **18** 180.

403 *by diabolical persuasion'*. *Ordenanza para los indios*, Arequipa, 6 November 1575, Toledo *Ordenanzas* 315.

403 *from their encomenderos'*. Ibid. The text of these ordinances is in Lorente, *Relaciones de los Virreyes* **1** 155–217; Toledo *Ordenanzas* 304–82. See also Bayle, 'Cabildos de indios' 25–7; Lohmann Villena, *El corregidor de indios* 17.

403 *their own workshops.'* Agia 90.

404 *could be ignored.'* Poma de Ayala **2** 62.

404 *to approve it'*. Toledo to King, 1571, GP **5** 319–20.

405 *during daylight hours*. Lohmann Villena, *Las minas de Huancavelica* 97.

405 *and great confusion.'* Vázquez de Espinosa 543.

405 *distress and harm'*. Vargas Ugarte, *Historia del Perú, Virreinato (1551–1600)* 235–6 quoted extracts from the unpublished collective letters, which are in AGI, Lima, 310, 314.

406 *health and lives.'* Ibid.

406 *order their revocation!'* GP **5** 319. Lohmann Villena, *Las minas de Huancavelica* 93–5. Solórzano Pereira, *Política indiana* **1** 125.

407 *in the world'*. Montesinos, *Anales* **1**, entry for 1571.

407 *envy of kings.'* Hanke, *The Imperial City of Potosi* 30.

407 *the whole world.'* Ibid.

407 *flow of silver*. Ibid. 21; William E. Rudolph, 'The Lakes of Potosí', *Geographical Review* **26** 529–54, 1936; Wiedner 372. Arzáns de Orsúa y Vela 466, 475; Ramírez 57. The fine original waterworks started in 1572 were destroyed when a dam broke in 1621.

407 *in February 1574*. *Ordenanzas acerca de los descubridores, registros y estacas de las minas*, La Plata, 7 February 1574, Toledo, *Ordenanzas* 143–240.

408 *the return journey.'* Messia 140–1.

408 *until they returned*. Ibid. Crespo Rodas 175; Kubler, 'The Quechua in the Colonial World', 372–3. Those who went had to pay extra tribute because they had been relieved of paying it at home – 'an extraordinary practice'.

408 *of the earth.'* Messia 142.

409 *from this quota*. Escalona Agüero, *Gazofilacio* bk. 2, ch. 1, 113; Arzáns de Orsúa y Vela 189–90; Cobb, 'Potosí' 41–2; Basadre 346–7; Wiedner 371; Rowe, 'The Incas under Spanish Colonial Institutions'.

409 *wrath about this*. Loaisa 593; Arzáns de Orsúa y Vela, 'Anales de la villa imperial de Potosí', *Biblioteca Boliviana* **3** 20–4, 1939;

Hanke, *The Imperial City of Potosí* 40; Wiedner 371.

409 *of the mita*. Muñoz de Cuéllar to King, La Plata, 1 March 1615, quoted in Crespo Rodas 177. By 1628 another oidor, Gabriel Gómez de Sanabria, estimated that two-thirds of the Indians from round La Paz were missing (Gómez de Sanabria to King, La Plata, 18 January 1628, AGI, Charcas, legajo (bundle) 19).

CHAPTER 21 THE INCA PROBLEM

411 *will be ratified'*. Letter of 8 February 1570, GP **3** 401.

412 *to Titu Cusi*. Las Casas, 'Treatise concerning the imperial sovereignty and universal pre-eminence which the Kings of Castile and Leon enjoy over the Indies', 1553, in *Colección de obras*, (Paris, 1822) **2** 315. An anonymous tract written in Peru in the mid-1560s posed the question: 'Should the King bring this Inca Titu [Cusi] out of there [Vilcabamba] and give him the kingdom of Peru, keeping for himself the supreme and overall sovereignty?' *Duda sobre los tesoros de Caxamarca* 149. Gibson, *The Inca Concept of Sovereignty* 106 ff. Hanke, *The Spanish Struggle for Justice* 160 ff.

412 *protecting the Indians'*. Royal instructions to Francisco de Toledo, 28 January 1568, quoted in Lewis Hanke, 'Was Inca Rule Tyrannical?' in Lewis Hanke (ed.), *History of Latin American Civilization* **1**, *The Colonial Experience*, Boston 1967, 87–8.

412 *rule in America'*. Toledo to King, 26 December 1573, GP **5** 312; Hanke, *The Spanish Struggle for Justice* 163.

413 *his native policy*. There have been some wild accusations against Toledo's *Informaciones* by the sentimentalist school of historians. It was assumed, often without reading the *Informaciones*, that they were intentionally defamatory of the Incas. It was also hinted that a corrupt interpreter had distorted the answers; in fact, three interpreters were employed. The historians who hated to hear a harsh word about the Incas included Mendiburu, Markham, José de la Riva Agüero, Urteaga, Romero and Means.

413 *corner of it*. Toledo dispatch, Cuzco, 25 March 1571, GP **3** 443.

414 *here in Peru.' Carta donde se trata el verdadero y legitimo, dominio de los Reyes de España sobre el Perú*, Yucay, 16 March 1571, CDHE **13** 433. The author may have been Toledo's chaplain, the Franciscan Pedro Gutiérrez, or the Jesuit Jerónimo Ruiz de Portillo, or even Polo de Ondegardo.

414 *you are conducting'*. Ibid.

414 *the Inca heartland*. Fragments of the *Informaciones* were first published by Marcos Jiménez de la Espada, CLERC **16** 1882; and in full by Levillier in *Don Francisco de Toledo* **2** and commented on in **1** 197–221, 273–91.

415 *published until 1906*. Sarmiento's manuscript was sent by Toledo to King Philip in 1572 and found its way into the famous library of Abraham Gronovius, which was sold in 1785. It was acquired by the Library of the University of Göttingen, where it was rediscovered and presented to the librarian, Dr Richard Pietschmann, at the beginning of this century. Pietschmann published it in 1906 after three years of careful, dispassionate study, with excellent notes and introduction. Sir Clements Markham translated Sarmiento and it was published the following year by the Hakluyt Society. Markham casually assumed that 'The Viceroy made some final interpolations to vilify the Incas . . . which are so obvious that I have put them in italics within brackets.' (x, xiii.) He was angrily refuted by Levillier in *Don Francisco de Toledo* **3**.

415 *of each Inca'*. Alvaro Ruiz de Navamuel, quoted in Jiménez de la Espada, *Tres relaciones* xxviii. Molina of Cuzco 10.

415 *had the Incas.'* Letter of 9 April 1572 by a priest called Juan de Vera to the Council of the Indies, MS. in AGI, Lima 270, discovered by Levillier and quoted in *Don Francisco de Toledo* 1, 286–7.

415 *Juan Fernández Coronel*. María Cusi Huarcay had two children by Coronel: Captain Martín Fernández Coronel and Doña Melchora Sotomayor Coya who married General Manuel Criado de Castilla. Temple, 'El testamento inédito de doña Beatriz Clara Coya de Loyola'.

416 *of this kingdom."'* Letter of Juan de Vera to Council of the Indies, Cuzco, 9 April 1572, written shortly after the conversation

took place, Levillier, *Don Francisco de Toledo* **1** 286–7.

416 *to endow them.* Toledo despatch, Lima, 8 February 1570, GP **3** 344, 401.

416 *for the King.* These reports were sent on 1 March 1572. They dealt with ecclesiastical affairs, with war, with his tour of inspection and other affairs. In all, this year's reports occupy four hundred pages of Levillier's *Gobernantes del Perú* (GP **3**, **4**). This is a large book, closely printed; and Toledo's reports are published as they were written, with almost no punctuation or paragraphs. Such was the size of one annual report by this competent but verbose viceroy.

416 *Lima to Cuzco.*' Toledo despatch, Cuzco, 25 March 1571, GP **3** 452.

417 *of the Indians.*' Ibid. 453.

417 *and he expired.*' Calancha, bk. 4, ch. 5 812–13. Martín de Murúa, who wrote some years before Calancha, told the same story but with less picturesque detail. Murúa did not mention the fencing with Pando after Titu Cusi's return from the shrine 'where Diego Méndez killed his father Manco Inca', nor did he mention the lethal medicine. He said that the Inca died within twenty-four hours of the first pain in his side and effusions of blood (ch. 75, **1** 234). Both Calancha and Murúa used interrogations of eyewitnesses made by the Augustinians.

418 [*the Inca's*] *secretary.*' Angelina Llacsa testimony, Appendix E, CLDRHP, 1 ser., **2** 134; also Murúa, ch. 76, **1** 235.

418 *him for this*'. Llacsa testimony 135.

419 *rites and ceremonies.*' Ibid. 136.

419 *struck Ortiz withered.* Ibid. Murúa, chs. 76, 77, **1** 238–41. Calancha, bk. 4, ch. 6, (1639) 820–7. Bishop Antonio de Raya held an enquiry into Ortiz's life and martyrdom, and the Augustinians of Cuzco held another enquiry in 1582. It was hoped that the martyr would be canonised. But some witnesses testified about Ortiz's involvement in the preparation of the fatal medicine. This meant that the friar had not died solely because of his profession of Christianity, and he therefore failed to be sanctified. He nevertheless became the proto-martyr of Peru, and his remains were greatly venerated. They were eventually placed in the main chapel on the gospel side of the Augustinian monastery in Cuzco. Bishop Raya's evidence has been lost, but the Augustinian enquiry, held by their prior Gerónimo Núñez in September 1582, is in AGI, Lima, 316. Parts of it were quoted by Levillier, *Don Francisco de Toledo* **1** 342–4. Calancha used it (and his own fervent imagination) to produce thirteen long pages of minute descriptions of every blow suffered by the martyr.

419 *he remained there*'. Murúa, ch. 74, **1** 230.

419 *heir Quispe Titu.* Tupac Amaru became Inca in 1571, twenty-seven years after the death of his father Manco Inca. There is a curious tradition that Tupac Amaru was a young man, even a minor, at this time. Poma de Ayala said that he was fifteen, and Lizárraga said he was eighteen or twenty. Modern authors make the same error. Cúneo-Vidal claimed that the new Inca was born 'in 1558' in the same paragraph that he said that Manco died in 1544 – Manco's widow must have had a fourteen-year pregnancy! (*Historia de las guerras* 225). Markham referred to Tupac Amaru as 'innocent young prince', 'lad' and 'youth' (Introduction to Sarmiento de Gamboa, Hakl Soc, 2 ser., **22** xx). Means dated Manco's assassination in 1545 (*Fall of the Inca Empire* 109) and then said, six pages later, that Tupac Amaru was 'about twenty one years old' in 1568. This process of artificial rejuvenation was apparently intended to win greater sympathy for the Inca because of his youth and innocence.

419 *him uti* [*impotent*]'. Sarmiento de Gamboa, ch. 70, 193.

420 *removed from power.*' Cúneo-Vidal, *Historia de las guerras* 267–8.

420 *had won power.* Oviedo himself later wrote that it was discovered in June 1572 that Titu Cusi had been dead for almost a year. Oviedo, Hakl Soc **22** 405.

421 *his son's affairs.*' Vivero letter of 24 January 1572, quoted by Levillier, *Don Francisco de Toledo* **1** 322.

421 *at the time.*' Toledo despatch of 1 March 1572, GP **4** 294.

421 *am sending you.*' Ibid. 295.

421 *then be made.*' Letter written by Toledo in the Yucay valley. Quoted by Antonio Bautista de Salazar (266) and said to have been written on 16 October 1571, although this seems too early for the Viceroy to have known of Oviedo's failure. I think that the letter was sent in March 1572.

422 *had business relations*'. Oviedo, Hakl Soc

22 404. Salazar (268), writing a few years later, said that Atilano de Anaya 'had formerly been the Inca's majordomo'.

422 *of the Inca'* Calancha, bk. 4, ch. 8, 831.

422 *of the Indians'* Ocampo 211.

422 *cross with him.'* Salazar 268. Gabriel de Oviedo (Hakl Soc **22** 404) said that the Spaniard was approached by a patrol of some thirty native troops under two officers. He entertained them with some of the thirty loads of food, drink and presents he was bringing with him. Other versions were Ocampo 216–17; Cobo, *Historia del Nuevo Mundo*, bk. 12, ch. 21; Calancha 831; Murúa, ch. 78, **1** 245–6. All except Calancha blamed the local garrison for the outrage. Ocampo named Curi Paucar as the commander responsible; Murúa named Paucar Inca, Curi Paucar and Colla Tupac. Calancha, on the other hand, wrote that Atilano de Anaya was enticed across the river and murdered on the personal orders of the Inca.

423 *fetched and buried.'* Ibid.

423 *of His Majesty.'* Murúa, ch. 78, **1** 246.

423 *in every way.'* Salazar ch. 28, 271.

423 *was in communication.'* Ibid.

424 *fire and blood'* Ibid. ch. 29, 271; Oviedo, Hakl Soc **22** 404. Montesinos, *Anales*, gave the date as 14 April. Sarmiento de Gamboa, testifying in the Probanza of one Captain Valenzuela, simply said that it was in April. Murúa said Quasimodo or Low Sunday, ch. 78, **1** 247.

CHAPTER 22 THE VILCABAMBA CAMPAIGN

425 *Juana Marca Chimpu.* Murúa **1** 247; Calancha, 831. This force also included Pedro de Orúe who later married an Inca princess – said to have been a daughter of Manco Inca – called Mama Tupac Usca.

425 *the royal force.'* Ocampo 220. When Rodríguez de Figueroa first entered Vilcabamba in 1565, he crossed at Chuquichaca in a basket slung on a cable. The bridge was then rebuilt for the meeting between Titu Cusi and Matienzo in June 1565. Matienzo, *Gobierno del Perú* pt. 2, ch. 18, 296.

426 *the greatest vigilance'* Murúa **1** 247.

426 *to enter Vilcabamba'.* Ibid.

426 *they already foresaw'.* Ibid. 248.

426 *to their income.'* Toledo, *Memorial* 81.

426 *and gallant men'*, Murúa **1** 248.

426 *of this kingdom'.* Ibid. 249.

427 *of their tribe.* Ibid. 249–50. Murúa often mentioned this Francisco Cayo Topa, who was presumably one of his informants. His name was surprisingly absent from the witnesses to Sarmiento de Gamboa's *Historia Indica* on 29 February 1572. The well-known Diego Cayo of the ayllu of Pachacutec was present on that occasion; and one of Paullu's illegitimate sons was also called Diego Cayo Topa. Toledo described the Cañari as 'valiant and diligent people' in his report of 1 March 1572 (GP **1** 119), and as a reward for their services in Vilcabamba, he renewed their exemption from paying tribute in his ordinances for the government of Cuzco, issued at Checacupe, 18 October 1572, Toledo *Ordenanzas* 106–8.

427 *and precipitous paths'.* Murúa **1** 249.

427 *in the kingdom'* Ibid.

427 *to the north-west.* Ibid. Ocampo 220; Salazar, ch. 29, 272; Oviedo, Hakl Soc **22** 405. The passage of Cusambi is modern Osmabre, where the road south-westwards from Pampaconas crossed the Apurímac on an Inca bridge. All the Spanish contingents were instructed to allow the Inca to surrender on the conditions made on behalf of the King, should he do so peacefully without knowing of their attack. Cobo, *Historia del Nuevo Mundo*, bk. 12, ch. 21. One Spaniard advised the Viceroy to enter Vilcabamba by the high pass between Soray and Salcantay to avoid crossing either river (Anonymous report in CDIA **24** 166–7, 1875).

427 *no impediment whatsoever'* Murúa **1** 250.

427 *and Coyao-chaca.'* Ibid.

427 *in their favour'.* Ibid.

427 *into the river'* Ocampo 221.

427 *a half-giant'.* Murúa **1** 250.

428 *from the scabbard.'* Ocampo 221.

428 *his captain's life.'* Murúa **1** 250.

428 *called "Loyola's leap" '.* Ocampo 221. This incident is also in Salazar 274, and Calancha 832.

428 *with the Indians.* Martín García de Loyola, petition of 26 August 1576, and grant of encomienda, Potosí, 10 February 1573, CDH Chile, (2) **4** 205, 215.

428 *the far bank.*' Salazar, ch. 30, 274. Calancha copied, and garbled, this account, bk. 4, ch. 8, 832.
428 *come to grips.*' Ibid.
428 *many brave Indians*'. Murúa **1** 251.
428 *more level place*'. Ibid.
429 *other leading Indians*' García de Loyola, Probanza, Cuzco, 2 October 1572, JLPB **7** 22. Probanza of Francisco de Valenzuela, Lima, 27 June 1578, JLPB **7**; MP **2** 76. Toledo's grant of encomienda to Loyola, Potosí, 10 February 1573, CDH Chile, 2 ser., **4** 206.
429 *Titu Cusi Yupanqui died*'. Murúa **1** 252.
429 *sheep and pigs*'. Ibid.; also Ocampo 221.
429 *into a swamp.*' Ibid.
429 *that treacherous bog*'. Bingham, 'Along the Uncharted Pampaconas' *Harper's Magazine* **123**, August 1914, 455; *Inca Land* (London, 1922) 271–2.
429 *very cold place*' Murúa **1** 252.
429 *form of measles.*' Ibid.
429 *on the expedition*'. Ibid.
430 *the other prisoners.* Ibid.
430 *until the Friday.*' Report by Hurtado de Arbieto to the Viceroy Toledo, 27 June 1572, AGI, 70-1-29, unpublished except for an excerpt in Levillier, *Don Francisco de Toledo* **1** 328. None of the searchers for Vilcabamba have come across this extremely important battlefield dispatch, nor the very detailed narrative by Martin de Murúa: the best published accounts of the campaign. Curamba was the place from which Gaspar de Sotelo started on the main road near Abancay, Carco was on the Vilcabamba side of the Apurímac crossing – the place where Titu Cusi's son Quispe Titu was baptised by Antonio de Vero on 20 July 1567.
430 *difficulty and danger,*' Murúa **1** 253.
430 *each difficult place*'. Ibid.
430 *artillery and arquebuses*'. Hurtado de Arbieto, 328.
431 *were more experienced*'. Murúa **1** 253.
431 *Atilano de Anaya*'. Ibid. 254.
431 *resist until death*'. Ibid.
431 *to the Spaniards*'. Ibid. 255.
431 *to be dangerous*'. Ibid.
431 *their own expense.*' Ibid.
431 *four small towers.*' Hurtado de Arbieto, 328.
431 *shoot from slings*'. Murúa **1** 256.
431 *enter the fort*'. Hurtado de Arbieto, 328.
432 *them tumbling down.*' Murúa **1** 256. See also García de Loyola, Probanza, Cuzco, 2 October 1572, JLPB **7** 23; grant of encomienda by Toledo, 10 February 1573; petition to King, 26 August 1576, CDH Chile (2) **4** 206, 215.
432 *to do so*'. García de Loyola, grant of encomienda, CDH Chile (2) **4** 206.
432 *method of fighting*'. Murúa **1** 257.
432 *destroy the Spaniards*'. Ibid.
433 *spirit and bravery*', Ibid.
433 *the Christians there.*' Hurtado de Arbieto 329. All the best contemporary sources gave similar accounts of the battle of Huayna Pucará: Hurtado de Arbieto's dispatch written a few days later, Martin de Murúa, who clearly interrogated some excellent eyewitnesses, Martín García de Loyola, in the wording of his encomienda grant a few months later and petition for royal favours, and Sarmiento de Gamboa testifying on behalf of Francisco de Valenzuela in 1578 (text in JLPB **7** and in Levillier, *Don Francisco de Toledo* **1** 326).

I used to think that the battle of Coyao-chaca was the same as Huayna Pucará. Antonio Bautista de Salazar, who wrote in 1595, described only Coyao-chaca; and Calancha, writing twenty years later copied his account using many identical passages. But Murúa and García de Loyola both made it clear that there were *two separate* battles, and Murúa located Coyao-chaca precisely as being three leagues before Puquiura. The despatch from Hurtado de Arbieto of 27 June 1572 (which was in the Archivo General de Indias in Seville) stated clearly that he had already sent Toledo a report on events before the expedition reached Pampaconas. Salazar, who had travelled to Peru with Toledo as his accountant and was constantly with the Viceroy, evidently had in his possession that *first* report by Hurtado de Arbieto from Pampaconas, which never found its way to Spain. When that report was written the only fighting had been at Coyao-chaca. Salazar's version was apparently also used by Baltasar de Ocampo (writing in 1610) who, as a resident of Vilcabamba province, should have known more about the campaign beyond Pampaconas. Hiram Bingham identified Huayna Pucará as being four days' march beyond Pampaconas, and two or three days before

Vilcabamba (for a cumbersome column moving slowly).

433 *defeated Gonzalo Pizarro'.* Murúa **1** 257.

433 *have been roasted.'* Ibid. 257–8.

433 *fort of Vilcabamba.'* Hurtado de Arbieto, 329.

433 *yucas and guavas'* Murúa **1** 258.

433 *short of provisions'.* Ibid.

433 *suitable for horses.'* Ibid.

433 *acting as witnesses.* All the best sources agreed on the date 24 June: Hurtado de Arbieto, Murúa, García de Loyola's various petitions, and Valenzuela's probanza. Also, Cúneo-Vidal, *Historia de las guerras* 280; Levillier, *Don Francisco de Toledo* **1** 326, 330–1. Gabriel de Oviedo also mentioned the occupation of the city of Vilcabamba (Hakl Soc, 2 ser., **22** 405–6). Salazar remembered the Viceroy being worried in Cuzco at this time, but, for the reasons given in my second note to page 433, he omitted all the details of the latter part of the campaign. He was followed in this by the hagiographer Antonio de la Calancha, who was the chief authority used by most modern investigators. Baltasar de Ocampo, who wrote when he was a confused old man, became hopelessly muddled about dates and places (222).

434 *the Incas burned.'* Hurtado de Arbieto 328.

434 *to settle it.'* Murúa **1** 258–9.

435 *enjoyed life there.'* Ibid. 260. Hurtado de Arbieto also described these tropical foods in his dispatch, and stated that the valley of Vilcabamba was half a league (2¼ miles) wide and a league long.

435 *few days earlier.'* Hurtado de Arbieto, 329–30.

435 *country for them'.* Ibid.

435 *rest had vanished.'* Murúa **1** 259.

435 *is hot country.'* Ibid. 260.

436 *had been held.'* Información of the services of Captain Francisco de Camargo, San Francisco de la Vitoria de Vilcabamba, 16 February 1573, JLPB **7**; MP **2** 25; Cúneo-Vidal, *Historia de las guerras* 283–4.

436 *enjoyed their share.'* Murúa **1** 260–1.

436 *Indians and captains.'* García de Loyola petition, AGI, estante 1, caj. 5, leg. 2, 913, in Cúneo-Vidal, *Historia de las guerras* 280–1. Murúa frequently mentioned Colla Topa as being an Inca commander, and so did Hurtado de Arbieto. Paucar Unia or Paucar Inca was one of the captains also named by Calancha as being involved in the martyrdom of Diego Ortiz. Curi Paucar was regarded by most Spanish authorities as being the leader of the Vilcabamba resistance.

437 *Atilano de Anaya'.* Murúa, ch. 82, **1** 261.

437 *are Chuncho Indians'.* Murúa **1** 262. Salazar (276) said that Loyola had over twenty men. Murúa named the river, which I assume to be the Pampaconas or Concevidayoc.

437 *from Vilcabamba'* Ibid.

437 *about his whereabouts.'* Ibid. 262–3.

437 *of the Inca'.* Ibid.

437 *they had made'.* Ibid.

437 *were with him'.* García de Loyola, grant of encomienda, Potosí, 10 February 1573, CDH Chile (2) **4** 207.

438 *currents and rapids'.* García de Loyola, probanza de servicios, Cuzco, 2 October 1572, quoted in Cúneo-Vidal, *Historia de las guerras* 281.

438 *lives by swimming.'* Ibid.

438 *knew as Simaponte.* The Simaponte was evidently the modern Cosireni, into which the Pampaconas or Concevidayoc flows, some twenty miles downstream of Vilcabamba. The Spaniards had left Vilcabamba on the river now called Pampaconas or Concevidayoc, but twenty miles downstream this joins the larger Cosireni and flows north-eastwards towards the Urubamba.

438 *retreated further inland.'* García de Loyola, probanza de servicios, in Cúneo-Vidal, *Historia de las guerras*, 281.

438 *of Your Majesty.'* Ibid.

438 *to his lord.'* Murúa **1** 264.

438 *in short stages.'* Ibid.

438 *eight of cassava.'* Ibid.

439 *the far side.'* Ibid. 264–5.

439 *jungle with torches.'* García de Loyola, grant of encomienda, CDH Chile (2) **4** 207.

439 *on the river.'* García de Loyola, petition to the King of August 1576, CDH Chile (2) **4** 216.

439 *in the morning.'* Murúa **1** 265. The two mestizos who saw the Inca's fire were both sons of public scriveners of Cuzco, Gómez de Chavez and Benito de la Peña respectively.

439 *to the Spaniards.'* Calancha, bk. 4, ch. 8 832.

440 *wives and children.*' García de Loyola, petition of 1572 in Cúneo-Vidal, *Historia de las guerras* 281.

440 *and its booty*'. Salazar, ch. 30, 277.

440 *to this conquest*'. Ibid.

440 *eloquent and intelligent*'. Murúa **1** 265.

CHAPTER 23 THE ELIMINATION OF THE INCAS

441 *nephew Tupac Amaru.*' Murúa, ch. 84, **1** 269.

442 *of the Inca.*' Ocampo 224.

442 *round his neck*', Salazar, ch. 30, 278.

442 *Silva y Guzman.* The house still stands in a small square called Plaza de Silbaq after its sixteenth-century owner. He died shortly before Toledo's visit, but the Viceroy rewarded his widow Teresa Orgóñez for her hospitality by renewing her husband's rich encomienda for a further lifetime. Diego de Silva was probably the author of the verse chronicle *Conquista de la Nueva Castilla*, and was one of the few survivors of Morgovejo de Quiñones's expedition to relieve Cuzco in 1536. Lizárraga described Silva's house as the only gay one among the sombre houses of Cuzco, because of its fine furnishings and many distinguished visitors, and Baltasar de Ocampo confirmed this (224).

442 *without being seen*', Salazar 278.

442 *for not obeying.* Murúa **1** 269–70. The same story was told by Poma de Ayala. See E. Mendizábal Losack's 'Las dos versiones de Morúa', RMN **32** 153–85.

442 *pleasure of conquest*' Ocampo 224.

442 *and majestic buildings*' Ibid.

442 *with a spoon*'. Ibid. 226.

443 *or by day*', Antonio de Vega Loaiza, *Historia del Colegio y Universidad de San Ignacio de Loyola*, quoted in Vargas Ugarte, *Historia del Perú, Virreinato (1551–1600)* 257.

443 *the native religion.* He had just completed his famous *Relación de las fabulas y ritos de los Incas.*

443 *more each day.*' Oviedo (1908) 406.

443 *Christian name Pedro.* Salazar, Calancha, Oviedo and Garcilaso all gave the name Pedro. Cobo gave Felipe. Other chroniclers did not mention a name.

443 *rested with them*'. Vega Loaiza, in Vargas Ugarte, *Historia del Perú, Virreinato (1551–1600)* 257.

443 *to die Christians*'. Oviedo, Hakl Soc, 2 ser., **22** 406

444 *it before now.*' Memorial of 24 October 1572 sent to the Viceroy at Checacupe by the Cabildo of Cuzco, signed by such notables as Sierra de Leguizamo, Hurtado de Arbieto, Polo de Ondegardo and Pancorvo. This document was discovered by Levillier and published in his *Don Francisco de Toledo* **1** 337–9.

444 *of that war*'. García de Loyola, petition of 26 August 1576, CDH Chile (2) **4** 216.

444 *and were hanged.*' Ocampo, JLPB **7** 308 or Barriga, *Los Mercedarios en el Perú* **5** 182. Sir Clements Markham translated this passage but made casual changes to pretend that the captains had been tortured to death (Hakl Soc, 2 ser., **22** 225). Markham apparently thought that 'la chepetonada' was a form of torture, although any good dictionary would have told him that it was a respiratory disease common among Europeans in Peru. This was evidently the same disease that killed Huallpa Yupanqui before he reached Cuzco. Murúa said that the other two Inca captains, Colla Tupac and Paucar Unia, were punished by having their hands cut off (ch. 84, **1** 270).

445 *province under arms.*' Toledo's appointment of Martín Hurtado de Arbieto as Governor of Vilcabamba, 30 July 1572, in Levillier, *Don Francisco de Toledo* **1** 336. A similar list of accusations was repeated by the city council in its Memorial of 24 October 1572 addressed to the Viceroy (Ibid. 339).

445 *the wretched missionary.* The Augustinian order arranged a probanza in 1582 to establish the facts of the martyrdom of Diego Ortiz. One question put to the witnesses was: 'After the death of Titu Cusi, did the Inca who succeeded in his place martyr one of the friars who was called Diego Ortiz?' Two witnesses, Gómez de Tordoya and Juan Pérez de Prado (both of whom had been on the Vilcabamba expedition), replied that Tupac Amaru had martyred Ortiz. But the former envoy García de Melo said that it was the Indians and not the Inca who had done so.

445 *was clearly established*'. Vega Loaiza, *Historia del Colegio . . . del Cuzco*, quoted in

Vargas Ugarte, *Historia del Perú, Virreinato (1551–1600)* 257. Roberto Levillier, author of a brilliant biography of Toledo, tried hard to establish Tupac Amaru's guilt from contemporary sources. But none of his quotations implicated the Inca in anything but very general terms, and their authors were generally men with an interest in disposing of the Inca. He quoted: Toledo's grant of Vilcabamba to Hurtado de Arbieto; the Memorial of the Cabildo of Cuzco of 24 October 1572; Lope de Atienzo, *Compendio historial del estado de los Indios del Perú*, and letters by Licenciate Juan de Matienzo and Licenciate Pedro Ramírez de Quinoñes of the Audiencia of Charcas.

446 *first informed him.*' Cobo, *Historia del Nuevo Mundo*, bk. 12, ch. 21.

446 *would be executed.*' Garcilaso, pt. 2, bk. 8, ch. 18.

446 *do the same.*' Vega Loaiza quoted in Vargas Ugarte, *Historia del Perú, Virreinato (1551–1600)* 257.

446 *of the Inca.*' Ocampo 227–8; Lizárraga, bk. 2, ch. 23 (1908) 141. Bernabé Cobo, Garcilaso de la Vega and Antonio de la Calancha also mentioned the intercessions. It is quite possible that the list grew in later years, when it became more fashionable to condemn Toledo's action.

447 *to be saintly*'. Vega Loaiza in Vargas Ugarte, *Historia del Perú, Virreinato (1551–1600)* 257.

447 *tears and fervour*' Ibid.

447 *in this case.*' Murúa **1** 271.

447 *crowd of ecclesiastics.*' Vega Loaiza in Vargas Ugarte, *Historia del Perú, Virreinato (1551–1600)* 257–8.

447 *the people packed.*' Ocampo 226.

447 *of the Viceroy*'. Murúa **1** 271.

447 *kinds of people.*' Ocampo 228.

447 *grave and humble.*' Vega Loaiza, in Vargas Ugarte, *Historia del Perú, Virreinato (1551–1600)* 258.

448 *deplore his death.*' Murúa **1** 271.

448 *tears and sobbing.*' Ibid.

448 *to their lords*'. Salazar 279.

448 *his right thigh.*' Garcilaso, pt. 2, bk. 8, ch. 19, **135** 171.

449 *to tell them.*" ' The Inca's speech was first recorded by Antonio de Salazar (279–81) who was admittedly a member of Toledo's official mission. It was later mentioned by another eyewitness, Gabriel de Oviedo, and repeated by later chroniclers such as Antonio de la Calancha, and Bernabé Cobo. Murúa mentioned the Inca's silencing the crowd, but recorded no speech.

449 *of these peoples.*' Letter of 19 October 1572 from Checacupe, GP **4** 343.

449 *then came forward.*' Ocampo 228.

449 *on the dais*' Murúa **1** 271.

449 *eyes of all.*' Ocampo 228.

449 *our Lord God.*' Ibid. 228–9.

450 *his confession heard.*' Letter of 19 October 1572; also Ocampo 229, Oviedo, Hakl Soc, 2 ser., **22** 407–8.

450 *shares of treasure.*' Toledo despatch, Checacupe, 20 October 1572, GP **4** 345.

450 *to His Holiness.*' Ibid.; also letter to Cardinal of Sigüenza, 19 October 1572, GP **4** 501–2; *Relación de los pleitos y pretensiones de justicia del Virrey D. Francisco de Toledo*, Madrid, 2 November 1596, GP **7** 497; Jiménez de la Espada, 'El cumpi uncu,' *Inca*, Lima **1** 904 (1923); Vargas Ugarte, *Historia del Perú, Virreinato (1551–1600)* 266.

450 *in the Vatican.* Jiménez de la Espada quoted a letter from a French Ambassador to Madrid, Father Muret, who wrote in 1667 that he had seen a collection of all the most precious objects from the Indies in the Palace of the Buen Retiro ('El cumpi uncu', *Inca* **1** 928). Robert Lehmann-Nitsche in 'Coricancha' described the various sun images, but concluded that this last was the holiest. See also Cunéo-Vidal, *Historia de las guerras* 283–9; Means, 'Biblioteca Andina' 491–3, *Fall of the Inca Empire* 126, 137.

451 *in this kingdom*'. Toledo despatch, Cuzco, 8 May 1572, GP **4** 366.

451 *memory and ayllos.*' Ibid.

451 *of Your Majesty.*' Carlos Inca to King, Cuzco, 1571, Levillier, *Don Francisco de Toledo* **1** 293, plate 18. See plate 41.

451 *force of artillery*'. Toledo despatch, Cuzco, 8 May 1572, GP **4** 366.

452 *and unjustified attack.* Pedro de la Gasca described Cayo Topa as a grandson of Huayna-Capac, and Cieza de León said that he was 'the one living male descendant of Huayna-Capac' in Cuzco in 1550 (Gasca despatch of 27 June 1547, CDHE **49** 309; Cieza de León, pt. 2, ch. 6). Cayo Topa described himself as 'son of Tupac Inca Yupanqui and nephew of Huayna-Capac' in a donation of some lands to the Mercedarians on 8 October 1549, Barriga, *Los*

Mercedarios en el Perú **2** 161–6. Toledo's *Información* of 6 September 1571, CDIA **21** 205.

452 *coming out peacefully'.* Sentence on Incas of Cuzco, unpublished except for a quotation in Levillier, *Don Francisco de Toledo* **1** 367.

452 *please his master.* The oidor of the Charcas, Licenciate López de Armendáriz, accused Toledo, in a letter of 25 September 1576, of having stopped the interpreter's trial to avoid further scandalous revelations, and of then having had the prisoner garrotted in his cell (Levillier, *Audiencia de Charcas* **1** 337).

453 *on 18 March.* Toledo had wanted to exile the Cuzco Incas to Spain, but the King had said that 'he was unenthusiastic because of the obligations and problems that would arise from this'. Toledo to King, 20 May 1573, in Levillier *Don Francisco de Toledo* **1** 369.

453 *the Viceroy's provision'.* Diego López de Herrera to Council of the Indies, Lima, 16 April 1573, in Levillier, *Don Francisco de Toledo* **1** 371.

454 *crimes of treason.'* Loarte quoted in Levillier, *Don Francisco de Toledo* **1** 383. The case against Loarte is dealt with in detail in pages 377–95 of that book.

454 *than any other'.* Letter of 8 November 1574 to President Ovando, GP **5** (1924) 449.

454 *been better omitted'.* King to Toledo 27 February 1575, Angel de Altolaguirre and Adolfo Bonilla de San Martín, Papeles del Consejo de Indias, CDIU **16** (1924) 76. Garcilaso claimed that the King later became far angrier about Toledo's treatment of the Incas, and that it was because of this that Toledo received no rewards or high position when he returned to Spain after his distinguished viceroyalty. The King was supposed to have said to Toledo that 'he had not sent him to Peru to kill kings, but to serve them' (pt. 2, bk. 8, ch. 20, **135** 172).

455 *a Christian ceremony. Suplicación de los hijos naturales de Paullu Inca*, before Audiencia of Lima, 11 December 1573, MS. in Biblioteca Nacional of Lima. The brothers argued that they had all been legitimised under the royal decree of 1 April 1544. The court eventually found that, although they were all entitled to inherit, Paullu's will had left his entire estate to Carlos and Felipe, sons of Catalina, whom Paullu married just before his death (Temple, 'Don Carlos Inca').

455 *the northern sierra.* Provision of 31 January 1573, Potosí, quoted in Levillier, *Don Francisco de Toledo* **1** 370–1.

455 *Potosí in 1610.* Murúa, bk. 2, ch. 15. Murúa said that Titu Atauchi had been lodged in an obscure prison in Lima.

455 *including two children'* Garcilaso, pt. 2, bk. 8, ch. 18. Calancha also copied this wild charge.

455 *to carry arms'.* Ibid., ch. 17.

455 *the New World'.* Ibid.

455 *of southern Chile'.* Markham, *The Incas of Peru* 298.

455 *Juan Arias Maldonado.* Toledo letter of 1 April 1571, quoted Levillier, *Don Francisco de Toledo* **1** 421–2; and Toledo's Report on temporal affairs, Cuzco, March 1571, para. 50, CDHE **94** 297. Toledo said that he thought the sentences imposed by García de Castro were unduly harsh; he also mentioned the fine service of Juan Arias's father Diego Maldonado 'el Rico'. In view of these interventions, it was most unjust of this Maldonado to misrepresent Toledo's help in his conversations with Garcilaso. Otherwise it is quite possible that Garcilaso invented his sweeping condemnation of Toledo without even this one mestizo as evidence for it.

456 *his native land'.* Garcilaso, pt. 2, bk. 8, ch. 17, **135** 168.

CHAPTER 24 THE INCA SURVIVORS

458 *country in 1585.* Request by Alonso Atahualpa for a repartimiento of 20,000 pesos rent, 1582–6; licence to pass to Spain, Quito, 12 March 1585, MP **2** 215.

458 *for similar favours'.* This quotation and the facts in the ensuing paragraph are taken from these documents: Mateo Vázquez to Hernando de Vega, President of the Council of the Indies, Madrid, 8 December 1586; Vega's reply, and the King's approval, 19 January 1587; letter from Hernando de Vega, Madrid, 26 February 1589, and

accompanying list of 'debts left by Don Alonso Atahualpa, deceased' (RGI **3** 1897, Appendix 4, cxlv–cxlix).

458 *her grandfather's estates.* Navarro 826–7. In a more compassionate mood, the Council of the Indies recommended in 1592 that Alonso's two children receive pensions, CDIU **3** 35–6.

458 *Atahualpa's direct line.* Claim by Doña Bartola Atahualpa Inca, daughter of Doña Mencía Atahualpa; for 1,000 pesos' rent that the King was said to have awarded Don Alonso Atahualpa only twelve days before his death. The King later granted 2,000 pesos to Mencía, Quito, 7 March 1610, discovered by Vargas Ugarte in the Biblioteca y Archivo Nacional, Quito, MP **4** 161. Mencía and her brother Don Carlos petitioned for rent passed to their father by their grandmother Beatriz, widow of the Auqui Don Francisco, Quito, 8 April 1606, MP **2** 213–14. At about this time, one Bartolomé Inca de Orozco claimed that he was son of Joseph Orozco and Doña Ana Azarpay Coya, daughter of Atahualpa Inca; but I know of no other evidence of this daughter's existence. Petition by Bartolomé Inca de Orozco, Lima, 25 January 1607, MP **2** 214.

459 *on any pretext.* María Cusi Huarcay to the Viceroy Count of Villar, Cuzco, 23 December 1586, GP **11** 231–6.

460 *now aged twenty.* Toledo grant of encomienda to García de Loyola, Potosí, 10 February 1573, CDH Chile (2) **4** 208. Toledo also sent Loyola back to Madrid in 1574 and helped him obtain an additional income of 1,000 pesos a year from the King in return for his services in Vilcabamba.

460 *in peaceful possession.'* Petition by Cristóbal de Maldonado to King, 1577, CDH Chile (2) **4** 218–19.

460 *have two husbands.'* Ibid.

460 *with her husband'.* Ibid.

460 *of the Viceroy'.* Ibid. Murúa **1** 272. Maldonado had already survived a royal legal action claiming damages from him for his rape of Beatriz. A document about this was found by Vargas Ugarte in the AGI. It was dated Madrid, 21 June 1572, MP **5** 72.

461 *corsair Thomas Cavendish.* This naval appointment may have inspired García de Loyola's future mother-in-law María Cusi Huarcay to offer her Vilcabamba mines to help against the Lutheran intruders.

461 *ceremonial drinking vessel.* The skull of García de Loyola was recovered in a peace settlement in 1641. Garcilaso felt that his death represented some form of belated retribution for the capture of Tupac Amaru twenty-seven years earlier, and in his excitement wrongly described the Araucanians as 'vassals of the prince whom [García] captured'. But other Spaniards regarded the victor of Vilcabamba and martyr of Curalava as a heroic figure.

461 *[of my funeral]'.* Ella Dunbar Temple discovered Beatriz's will in the Archivo Nacional of Peru, 'El testamento inédito de doña Beatriz Clara Coya de Loyola'.

461 *Domingo in Lima.* Lohmann Villena, 'El señorío de los Marqueses de Santiago'.

462 *in my stables.'* This passage (originally written in the third person) is from a Memorial called *Ascendencia de Don Melchor Carlos Inca*, written in Trujillo de Extremadura in 1603, a 200-page manuscript in the Biblioteca Nacional of Madrid. It was in four parts: the royal lineage of Cristóbal Paullu Inca; the royal descent of Melchor Carlos Inca; the services to the crown of Paullu Inca; and a petition for royal favours by Melchor Carlos. A meticulous study of the descendants of Paullu has been done by Ella Dunbar Temple in her 'Don Carlos Inca', 'Azarosa existencia de un mestizo de sangre imperial incaica' (about Melchor Carlos Inca), 'Los testamentos inéditos de Paullu Inca, don Carlos Inca y don Melchor Carlos Inca'. In addition there have been fine researches on this and related subjects by Lohmann Villena, Porras Barrenechea and Vargas Ugarte.

462 *in this land'.* Lohmann Villena, *Los Americanos en las ordenes* **1** 200.

462 *the royal fringe.* Rowe, 'Colonial Portraits of Inca Nobles.'

463 *and Alonso Pérez.* The older Pedro Alonso Carrasco was married to Leonor Arias de Castillejo. Her sister, Doña María de Arias wife of Martín de Olmos, had acted as Melchor's godmother at the famous christening ceremony in 1571.

463 *of this land.'* Velasco despatch, 15 June 1599, GP **14** 193.

463 *orders and violence.'* Velasco despatch, Lima, 7 December 1600, GP **14** 288.

463 *league with them'* Murúa **1** 273.

464 *assist his passage'.* Ibid. 274.

464 *Lima before sailing.* Temple, 'Azarosa

existencia de un mestizo de sangre imperial incaica' 129.

464 *to this day.* Porras Barrenechea has shown that if the sketch was done from life it must have been made in Cuzco in 1580 or in Lima in 1602, since these were the only occasions on which Melchor and Poma de Ayala were in the same place at the same time. The earlier date seems improbable, since no amount of European blood could have produced such a luxuriant growth on the face of a nine-year-old boy. Porras Barrenechea, *El cronista indio Felipe Huamán Poma de Ayala* (Lima, 1948).

464 *to be greater.'* Garcilaso, pt. 1, bk. 9, ch. 38, Hakl Soc **45** 522, 1871.

465 *painted with figures'.* Dr Lohmann Villena wondered whether these could have been the famous paños sent to Spain in 1572 by Toledo, but it seems improbable.

465 *descendants of kings'.* Garcilaso, pt. 1, bk. 9, ch. 40. Hakl Soc **45** 531.

465 *white China silk'.* Ibid. The authorisation by the Cuzco Incas to Garcilaso and the others has been found in the AGI, dated Cuzco, 20 March 1603, MP **2** 214.

465 *hoped to receive.'* Garcilaso, pt. 2, bk. 8, ch. 21, **135** 174. The white silk with the busts of the Incas, from the first 'Manco Inca to Huayna-Capac and his son Paullu', has disappeared, although a later king asked his secretary to search for it in 1748. This series of busts is not to be confused with the paños sent by Toledo in 1571. The latter probably formed part of the frontispiece to the 1615 edition of Herrera. The royal order to search for the white silk version was dated 18 February 1748. It mentioned Garcilaso de la Vega had given the cloth to Melchor Carlos Inca and Alonso de Mesa at Valladolid, MP **2** 214–15.

465 *his native land.* This Memorial is in the Biblioteca Nacional in Madrid (Temple, 'Azarosa existencia de un mestizo de sangre imperial incaica' MP **2** 367).

466 *journey to Spain.* Lohmann Villena, 'El señorió de los Marqueses de Santiago' 414.

466 *in the honour.* The Información is in Ordenes Militares, Santiago, exp. 4081, in the Archivo Histórico Nacional, Madrid: Lohmann Villena, *Los Americanos en las órdenes nobiliarias* **4** 200; Temple, 'Azarosa existencia de un mestizo de sangre imperial incaica' and 'Los testamentos inéditos'. The only earlier American native to enter the order was Cortés's half-caste son Don Martín Cortés. The Council's report on Melchor's claim, dated 8 November 1607 is in the AGI (MP **2** 216).

466 *died of melancholy.* Garcilaso, pt. 2, bk. 8, ch. 18, **135** 169. Memorial by Doña María de Silva, Madrid, 7 September 1611, MP **2** 216.

467 *his only achievement.* The first petition by Juan Melchor Carlos Inca was made in Madrid, 4 December 1620. The Council of the Indies reported on it, Madrid, 27 August 1626, MP **2** 216.

467 *from paying tribute.* These sons of Paullu were legitimised by royal decree, Valladolid, 1 April 1544, Temple 'Un linaje incaico' 45–6. Although Francisco de Toledo did not consider them worthy of full legal persecution, he decided to reduce all Inca nobles in Cuzco to ordinary tribute-paying Indians. On his orders, García de Loyola sent a visitador to register all the exempt Incas living in Cuzco in 1572, and these included the sons of Paullu. The order for this inspection, Cuzco, 11 August 1572 is in Urteaga, *El imperio incaico* 229–35.

467 *envied half-brother Carlos.* Toledo revoked their exemption from paying tribute. They therefore sent a delegation to Lima to plead for restitution of their earlier immunity. Their powerful arguments won the day. Their privileges were restored with compensation by the Audiencia, as part of its reversal of Toledo's attempted purge. The sentence was reversed on 3 March 1576. On 30 August the brothers claimed and received repayment of the tribute they had paid between 1572 and 1576. Texts in Urteaga, *El imperio incaico*, Appendix B, 236–40.

467 *for his history.* Sarmiento mentioned Diego Viracocha Inca, born 1535–8, who was one of Paullu's most respected sons (*History of the Incas* 199). Bernabé Cobo said that he knew Don Fernando Poma Capi and Don Alonso Tupa Atau, *Historia del Nuevo Mundo*, bk. 12, ch. 20. Clements Markham also said in *The Incas of Peru* that Cobo knew descendants of Paullu.

467 *by his father.* José Rafael Sahuaraura was the curate who spied against Condorcanqui and attacked him savagely in a book called *Estado del Perú*. Pedro Sahuaraura was the

sargento who fought Condorcanqui and helped denounce him to the Spanish authorities. His son Justo was a canon, whose book *Recuerdos de la monarquía peruana* was lavishly published in Paris in 1850 and was full of romantic praise of his family, with not a word about his father's collaboration (Temple, 'Un linaje incaico durante la dominación española, Los Sahuaraura').

468 *Martín de Ampuero*. He was the son of Francisco de Ampuero and his wife Inés Huayllas, Francisca's mother.

468 *was a wastrel*. Letter of 25 May 1578 to Martín de Ampuero, Harkness Cal 252. Cuesta, 'Una documentación interesante' contains the text of the will, the creation of the entail and codicil. Miguel Muñoz de San Pedro, 'Las últimas disposiciones del último Pizarro de la Conquista', examined the settlement, as did Roberto Moreno y Mórrison, 'El centenario de Pizarro', *Revista de Historia y Genealogía Española*, Madrid, 2 ser., **111**, no. 14, 133 ff., 1929.

468 *Pedro Arias Portocarrero*. Hernando and Francisca had five children in all, but two died young and two more, Juan and Inés, died without succession. Francisca and her second husband had no children, but a granddaughter of Portocarrero's by his first marriage married Francisca's heir Francisco. These had children, but their line ended with a childless granddaughter. Francisco married again and produced a son who also died childless. He then had a long liaison with one Micaela Manrique, and one of their daughters started a succession that lasted for many generations. This Francisco Pizarro was a wealthy figure; he befriended Paullu's grandson Melchor Carlos, who named him as a protector of his children in his will of 1610.

468 *de la Conquista*. Juan Fernando's 160-page plea appeared in 1622 as *Discurso en que se muestra la obligación que su Majestad tiene a don Juan Fernando Pizarro, bisnieto y heredero del Marqués don Francisco Pizarro*. The claim by Hernando and Francisca was for tribute from 20,000 Indians awarded to her father, and for 300,000 ducats spent by him in the Conquest and suppression of Manco's rebellion. As the years dragged by, the arrears on these claims had become enormous, and the Crown was finding it difficult to deny their justice.

470 *from the corregidor*. I assume that this Martín Fernández Coronel Inca was descended from the second marriage of Ana María's grandmother María Cusi Huarcay to Juan Fernández Coronel, and was thus her cousin.

471 *Codesal y Ayo*. Ana María had four sons and six daughters. Her son Diego became a Knight of Calatrava, Alvaro-Melchor a Knight of Santiago, and Antonio a Knight of Calatrava. Her daughter Francisca married the Portuguese Marqués de Penalva and was a lady-in-waiting to Queen Mariana of Austria, wife of Philip IV. Three daughters entered the convent of Santa Cruz in Valladolid, and the other two died young.

471 *Charles V to Paullu*. Rowe, 'Colonial Portraits of Inca Nobles' 258. The portrait of Chiguan Topa carries an inscription referring to various people living in Peru between 1740 and 1745. The other paintings in the series are contemporary, although the ladies were shown wearing Inca dress, as was one Don Alonso Chiguan Inca. Francis Comte de Castelnau first reproduced this painting of Marcos Chiguan Topa and said that it represented an Inca prince at the time of the Conquest (Castelnau, pt. 3, 'Antiquités des Incas', plate 59). Later authors repeated his error. Urteaga. and Romero reproduced it in their edition of Titu Cusi's *Relación* and claimed that it was a contemporary portrait of Sayri-Tupac; so did Cúneo-Vidal, *Historia de las guerras* 202.

472 *buttocks like mules'*. John Rowe discovered this order from the fifth Marquis dated 15 March 1738 among some volumes of papers accumulated at the end of the eighteenth century by a Spanish officer called Vicente José García. Rowe, 'Colonial Portraits of Inca Nobles' 265 and Appendix D, 267–8.

472 *of the Conquest*. My information on the holders of the Marquisate of Santiago de Oropesa is taken from a superbly detailed study by Guillermo Lohmann Villena, 'El señorío de los Marqueses de Santiago', and Temple, 'El testamento inédito de doña Beatriz Clara Coya de Loyola'.

472 *until the present*. Temple, 'Notas sobre el Virrey Toledo y los Incas de Vilcabamba'.

472 *south of Cuzco*. Another daughter, Doña Magdalena Mama Huaco Inca, petitioned

the King from Cuzco, 7 March 1610. She said that, 'since Tupac Amaru's male children have died without succession, she alone remains as a legitimate daughter'. She said that she had an illegitimate daughter called Doña María, and that this María had an illegitimate daughter by Nicolás Pinelo, royal treasurer in Cuzco. The granddaughter was called Doña Feliciana Pinelo. MP **2** 214.

CHAPTER 25 THE SEARCH FOR VILCABAMBA

474 *up the valley*' Hurtado de Arbieto despatch of 27 June 1572, 330.

475 *and chief justice*'. Award of Vilcabamba to Hurtado de Arbieto, Cuzco, 30 July 1572, JLPB **7**, or Levillier, *Don Francisco de Toledo* **1** 334–5.

475 *they were conquered*.' Ibid.

475 *entry into Cuzco*. Ocampo 222. Ocampo gave the location in Hoyara but muddled the date of the foundation. Murúa said that the new name was applied to Vilcabamba itself, but Ocampo was more specific about the location. In a testimony on 5 January 1589, Hurtado de Arbieto quoted Toledo's original grant of 30 July 1572, and mentioned that the new city was to be founded 'in the valley of Viticos', GP **11** 259. Salazar 277; Calancha 832, 838; Murúa **1** 267, 268; Oviedo, Hakl Soc **22** 406; despatch of Licentiate Pedro Ramírez de Quiñones, 1575, in Levillier, *Don Francisco de Toledo* **1** 341.

475 *famine and destruction*.' Calancha 835.

476 *to these excesses*. Villar to King, Lima, 25 April 1588, GP **11** 85, 100; 12 May 1589, GP **11** 223–70, including Hurtado de Arbieto's defence, Cuzco, 5 January 1589, GP **11** 256–64.

476 *a second lifetime*. Hurtado de Arbieto's fifteen-year-old son should have succeeded according to Toledo's grant, but the Viceroy García de Mendoza argued that the father had violated it. He therefore made Antonio de Cabrera corregidor of Vilcabamba in 1590 (García de Mendoza despatch of 25 February 1590, GP **11** 95). He later made 'a Mexican gentleman called Antonio de Monrroy Corregidor of Vilcabamba because of his experience of silver mines in Mexico', despatch of 12 April 1594, GP **13** 139.

476 *martyr Diego Ortiz*. Calancha 835.

476 *did not smell*'. Murúa **1** 267.

476 *many leagues away*'. Calancha 838.

476 *the martyr's death*. The Augustinian Información of September 1582 is in the AGI. Levillier published some extracts in *Don Francisco de Toledo* **1** 342–4. Romero found the very mutilated manuscript of Angelina Llacsa's testimony on the martyrdom of Ortiz in the Biblioteca Nacional, Lima, and published it as Appendix E of his edition of Titu Cusi's *Relación* (CLDRHP, 1 ser., **2** 136).

476 *of Titu Cusi*. Calancha drew heavily on these interrogations: his *Corónica* contained pages of details about the sufferings of Ortiz. Murúa also wrote much about the missionary (chs. 75–84). He revealed that the Spaniards totally destroyed the town where the death occurred. He also mentioned that the Franciscans made an enquiry about the martyrdom, at which a leading witness was Juana Guerrero, widow of the mestizo Martín Pando who was killed with Diego Ortiz.

477 *herds of cattle*. Doña María Cusi Huarcay to the Viceroy Conde de Villar, Cuzco, 23 December 1586, GP **11** 231–6; Villar to King, 24 December 1586, 12 December 1587, 25 April 1588, GP **10** 239, 250, 252–3 **11** 54; Ocampo 222.

477 *by the Mercedarians*. Ocampo was proud of his role in moving the town, a change that he regarded as a great service to God and the King (222–3, 241). A plot of land for the Mercedarian convent in San Francisco de la Vitoria de Vilcabamba was granted by Governor Hurtado de Arbieto, 5 March 1587; another in the Villa Argete la Rica de Vilcabamba (the town near the mines) was granted by the Viceroy García de Mendoza, 16 August 1590. Both awards are in the Mercedarian Convent, Cuzco, MP **4** 297.

477 *or no result*' Messia, *Memorial*, CDIA **6** 133.

477 *days were numbered*. Ocampo (231–2) deplored the decision to remove the mitayos from work in Vilcabamba. Vázquez de Espinosa 550. Murúa **1** 252. The Viceroy García de Mendoza first allocated mitayos

to work in Vilcabamba in 1590 (Despatch of 27 April 1590, GP **12** 112–13), but the quota was stopped in 1593. The miners complained, and the Viceroy again allocated two hundred mitayos, after negotiating better conditions for them (Marquis of Cañete despatch, 22 January 1593, 18 May 1593, GP **13** 16–17, 42).

477 *silver each year'*. Anonymous *Descripción del Virreinato del Perú*, ed. Boleslao Lewin (Rosario, 1958) 90. One Miguel Aincildegui Orez wrote to the King from Lima, 18 April 1685, recommending that the corregimiento of Vilcabamba be suppressed and merged with that of Calca and Lares, MP **2** 76.

478 *Vilcabamba cedar wood*. Ocampo 230.

478 *bells and images'*, Ibid. 242. Ocampo also described expeditions made by him and others to the Manarí Indians, who proved friendly, and the Pilcosuni, who did not.

478 *Father Diego Ortiz.'* Bueno, *Descripción del reyno del Perú, Descripción del obispado del Cuzco* (unpaginated).

478 *palaces were recognised.'* Ibid.

478 *the Peruvian Republic.* Pablo José Oricaín, *Compendio de noticias geográficas del Cuzco*, 1790.

479 *somewhere called Vilcabamba*. The Comte de Sartiges described his visit, using the pseudonym M. E. de Lavandais, in 'Voyage dans les républiques de l'Amérique du Sud'. Carlos A. Romero, 'Informe sobre las ruinas de Choqquequirau', RH **4** 90–5, 1909; Bingham, *Inca Land* 199.

479 *this savage asylum'*. Angrand's notes were published by Ernest Desjardins in his *Le Pérou avant la conquête espagnole* (Paris, 1858) 137–45.

480 *of his visit*. Ernest Grandidier, *Voyage dans l'Amérique du Sud* (Paris, 1861) 152 ff. Antonio Raimondi (**3** 401 ff.) repeated Gastelú's claim. Sámanez Ocampo, *Exploración de los rios peruanos Apurímac, Eni, Tambo, Ucayali y Urubamba* (Lima, 1883). Carlos A. Romero, 'Informe sobre las ruinas de Choqquequirau', 100–1.

480 *long days' march'* Calancha 794.

480 *above the Apurímac*. Raimondi, parte preliminar, 1869, bk. 2, ch. 7. Paz Soldán, *Geografía del Perú*. Wiener, *Pérou et Bolivie*. Raimondi and Wiener both travelled down the Urubamba river without finding any of its ruins. So also did another eminent traveller, Francis Comte de Castelnau, *Expedition dans les parties centrales de l'Amérique du Sud* (Paris, 1851) **4** 22 ff. Raimondi entered the area by the pass of Panticalla and emerged by the high pass between Soray and Salcantay.

480 *the legendary ruin*. Bingham, 'The Ruins of Choqquequirau'; *Across South America*; *Lost City of the Incas* 84–99.

481 *totally without foundation'*. Carlos A. Romero, 'Informe sobre las ruinas de Choqquequirau' 99.

481 *the Vilcabamba state*. Ibid. 102.

483 *in South America*. Bingham published the discovery of Machu Picchu in various works: 'The Discovery of Machu Picchu', *Harper's Magazine* **126** 709–19, April 1913; 'In the Wonderland of Peru'; 'The Story of Machu Picchu'; *Inca Land* 315–23; *Lost City of the Incas* (1951) 138–43. Bingham noted that the French traveller Charles Wiener had searched for ruins in the Urubamba valley in 1875 and had been told that there were fine ruins at a place called Huaina-Picchu or Matcho-Picchu. Wiener marked the two peaks of that name on his map but failed to investigate them because the riverside road was not built. Bingham's guide Arteaga referred to the hillside by that name.

483 *up from Vilcabamba*. Bingham, 'Vitcos, the Last Inca Capital'; Ocampo 221, 222.

483 *in May 1565*. Rodríguez de Figueroa 176.

484 *Titu Cusi Yupanqui died'*; Murúa **1** 252.

484 *late sixteenth century*. The ruin of this ore-crushing plant is near the junction of the Tincocacha and Vilcabamba rivers; it has some massive millstones and a primitive mortar – a large boulder in which four Indians could rock a semi-circular pestle. Baltasar de Ocampo described Ortiz's chapel as being 'near my houses and on my own lands, in the mining district of Puquiura, close to the ore-crushing mill of Don Cristóbal de Albornoz, former precentor of the cathedral of Cuzco' (214). It was in this chapel that Ortiz celebrated masses for the Inca Titu Cusi, and tried to revive him after his death.

484 *and elaborately cut.'* Ocampo 216.

484 *die in Puquiura*. Bingham, 'Vitcos, the Last Inca Capital'; 'A Search for the Last Inca Capital' *Harper's Magazine* **125**, 695–706, 1912; *Inca Land* 239–45; *Lost City of the Incas* 113–14. Many sources confirm that this ruin beside modern Puciura must have

been Vitcos. Titu Cusi noted that Vitcos, unlike Vilcabamba, was a place of 'cool air, for it is in a cold district' (91). Diego Rodríguez's narrative revealed that Vitcos was close to Lucma (he travelled between them on 12 May 1565) and that it was high, for he was shown the rocks *down* which the Spaniards fled after murdering Manco Inca (178). Martín de Murúa confirmed that Manco was killed at Vitcos (ch. 73, **1** 223–5), that his son Titu Cusi prayed at this place before dying (ch. 75, **1** 234), and that Titu Cusi's body was brought to be resuscitated by Ortiz at Puquiura which was therefore close to Vitcos (ch. 76, **1** 236; ch. 80, **1** 252). Calancha and Angelina Llacsa also made it clear that Titu Cusi's body was brought to Puquiura after his prayers at the scene of Manco's murder.

484 *spring of water'*. Calancha 796.

485 *of Inca rule'*. Bingham, 'Vitcos, the Last Inca Capital' 184.

485 *and legendary site*. Bingham, *Inca Land* 246–51; *Lost City of the Incas* 115–17. The modern name of the ruin was Ñusta España, but the area was known to Bingham's guides as Chuquipalta. Murúa called it Chuquipalta (ch. 75, **1** 232) but in Calancha it was transcribed as Chuquipalpa.

487 *Yupanqui in 1565'*. Bingham, 'The Ruins of Espiritu Pampa, Peru' 199; 'Along the Uncharted Pampaconas' 461–3; *Inca Land* 293–6; *Lost City of the Incas* 132.

488 *in the area*. Many learned reports resulted from these expeditions, and the Vilcabamba area received its first scientific survey. Hiram Bingham went on to lead a distinguished life. He taught at Princeton and Yale; became a military pilot and served, between the Peruvian expeditions, as a lieutenant-colonel commanding a flying school in France during the 1914–18 War. He entered politics and was a Senator for Connecticut from 1924 to 1933; was a businessman in banking and oil; and from 1951 to 1953 was Chairman of the Loyalty Review Board that heard appeals from civil servants suspected of being 'soft on communism'. Dr Bingham died in 1955 aged eighty.

489 *of their idolatries'*. Calancha 803.

491 *of Machu Picchu*. The expedition cleared Llacta Pata (also called Patallacta) on the Aobamba valley, and located Cedrobamba (later called Sayac Marca), Ccorihuayrachira (later called Phuyu Pata Marca), the round rest-house of Runcu Raccay, and the settlement of Choquesuysuy down beside the Urubamba.

492 *court in Cuzco*. Fejos, *Archeological Explorations in the Cordillera Vilcabamba* 59–60.

492 *the last Incas'*. The expedition was led by G. Brooks Baekeland and Peter R. Gimbel on behalf of the National Geographic Society and the New York Zoological Society. G. Brooks Baekeland, 'By Parachute into Peru's Lost World', *National Geographic Magazine* **126**, August 1964.

493 *the ruined walls*. Savoy's discovery was first reported in the *Peruvian Times*, 18 September 1964, and subsequent reports appeared in the issues of 9 April and 12 November 1965. The discovery was also reported in *Time* Magazine, 28 August 1964 and the *Illustrated London News*, 13 November 1965.

494 *a warm climate'* Titu Cusi Yupanqui 88.

494 *a cold district.'* Ibid. 91.

494 *in hot country'*. Murúa, ch. 82, **1** 260.

495 *is Machu Picchu*. Ibid. Murúa mentioned, among other things, that Vilcabamba produced peanuts and macaws. When Diego Rodríguez was waiting at Pampaconas to meet Titu Cusi, the Inca sent him a present of peanuts and macaws. Bingham himself commented that he was given many peanuts to eat at Espíritu Pampa, and saw macaws there. Rodríguez de Figueroa 177, 181, 182, 193. Bingham, 'The Ruins of Espiritu Pampa, Peru'. In his despatch of 27 June 1572, Hurtado de Arbieto said he would send the natives 'up' towards Vitcos.

495 *Vilcabamba in 1572*. Bingham, *Inca Land* 298, 336. Bingham knew the story of García's pursuit from Calancha, who copied it from Salazar; but neither of those authors made it clear that the Spaniards had already occupied the city of Vilcabamba. Bingham tried to show that when Calancha talked about 'Vilcabamba the Old' he was not describing the same place as the 'Vilcabamba the Old' of Ocampo. Bingham argued that Calancha based his narrative on 'notes' sent out by the Augustinian priests before Vilcabamba was invaded. In fact Calancha wrote in the 1620s, fifty years after Toledo's invasion of Vilcabamba. He used the enquiries about

Ortiz's martyrdom made by the Augustinians in the 1570s and 1580s. These enquiries used the name 'Vilcabamba the Old' to differentiate the Inca capital from the Spanish city of Vilcabamba originally founded in the Hoyara valley and later moved up to the silver mines. Ocampo used the name in the same way.

495 *miles by two*]'. Hurtado de Arbieto despatch, Vilcabamba, 27 June 1572, 330.

495 *that of Cuzco*. Murúa **1** 260.

495 *Vitcos and Vilcabamba*. Gonzalo Pizarro's expedition passed Pampaconas, Chuquillusca and Marcanay and occupied Vilcabamba. Diego Rodríguez went beyond Vitcos to Huarancalla and Pampaconas. The Augustinian friars passed Huarancalla, the swamp of Ungacocha and Marcanay on their visit to Vilcabamba in 1570 and during Ortiz's martyrdom in 1571. The expedition of 1572 found European cattle at Oncoy, rested at Pampaconas, passed Chuquillusca and the old and new forts, and discovered Diego Ortiz's body in its pit at Marcanay.

495 *roughly the same*. Calancha once wrote that it was 'two long days' march' from Puquiura to Vilcabamba (794). It took Bingham five days to go from Puquiura to Espíritu Pampa and rather longer to make the journey from Puquiura to Machu Picchu. But in the sixteenth century the roads were unobstructed, and the Incas and friars were sturdy walkers.

495 *to receive baptism*' Titu Cusi Yupanqui 105.

495 *martyrdom at Marcanay*. Rodríguez called the place Arancalla (177, 196); also Rodríguez testimony, *Información de méritos del Licenciado Matienzo*, Levillier, *Audiencia de Charcas* **2** 524. Matienzo called it Rangalla, *Gobierno del Perú*, pt. 2, ch. 18, 296. Titu Cusi called it Rayangalla (105, 106). Calancha said that Ortiz passed through Huarancalla soon after leaving Vitcos–Puquiura: bk. 4, ch. 3, 800. In another passage Calancha said that Huarancalla, the site of Ortiz's church, was 'two or three days' march' from Puquiura. This statement – apparently a slip by Calancha – led Raimondi and Bingham to conclude that the Huarancalla of Ortiz's church was across the mountains at the modern hacienda of Huarancalque near Carco above the Apurímac (Raimondi **2** 162). Raimondi's assumption seems to be disproved by a passage in which Titu Cusi wrote that García and Martín Pando once left Rayangalla and *crossed the passes* to the country around Carco (106).

496 *flocks and herds*'. Ocampo 222.

496 *expedition of 1572*. Even if this Oncoy was not the friars' swamp, the modern village of Vilcabamba was undoubtedly near the Oncoy passed by the 1572 expedition on its way towards Vilcabamba. Murúa, ch. 80, **1** 252; Ocampo 221, 222. Calancha; 803. An expedition that visited Espíritu Pampa shortly after its rediscovery by Gene Savoy reported several miles of deep mud lower down in the Pampaconas–Concevidayoc valley – but there would be no reason for that stretch of road being called Ungacocha. Mark Howell and Tony Morrison, *Steps to a Fortune* (London, 1967) 28.

496 *leagues from here*' Titu Cusi Yupanqui 88.

496 *capital and court*'. Murúa **1** 249.

497 *before modern Pampaconas*. The insignificance of Pampaconas is shown by the fact that the Indians had to construct special temporary buildings for the meeting between Titu Cusi and Diego Rodríguez in 1565. Rodríguez said that 'the road by which the Inca was to come was very clean and passed over a great plain'. This tallies with Murúa's description of 'a very cold place' on the high treeless puna. An anonymous report to the Viceroy Toledo about routes into Vilcabamba described Pampaconas as 'in puna country', CDIA **24**, 166, 1875. Modern Pampaconas is in bare grasslands at 10,000 feet. Rodríguez de Figueroa 178, 195; Murúa, ch. 79, **1** 249, ch. 80, **1** 252; Hurtado de Arbieto despatch 328; Ocampo 221, 222. Bingham, 'Along the Uncharted Pampaconas' 456, *Inca Land* 277–280, *Lost City of the Incas* 123–4.

497 *hillside called Chuquillusca*'. Quipocamayos 42.

497 *the turbulent river*'. Murúa **1** 253.

497 *defeated Gonzalo Pizarro*'. Murúa **1** 257.

497 *of a river*.' Titu Cusi Yupanqui 88–9. Pedro Pizarro (1844) 342; Hurtado de Arbieto despatch 329. Bingham tried to argue that the Vilcabamba towards which the 1572 expedition marched was not the 'lost city' visited by the Augustinian missionaries. It is therefore important to note Titu Cusi's reference to the Old Fort in the

narrative that he dictated to the friar García at Vilcabamba itself.

498 *Vilcabamba the Old'*. Calancha 820.

498 *nearby at Marcanay*. Murúa **1** 238–9, 258, 266–7; Calancha, bk. 4, ch. 6, 823 ff.; Angelina Llacsa, CLDRHP, 1 ser., **2**, Appendix E, 136; Bingham must have read the passage in which Calancha described the discovery of Ortiz's body at Marcanay near Vilcabamba by the Spaniards in 1572. He chose to ignore this passage when he produced his argument that Calancha's Vilcabamba the Old was a lost city never found by the Spaniards (*Inca Land* 298).

499 *well worth seeing'*. Murúa **1** 260.

499 *could not take.'* Ibid. 258.

499 *covered in roof tiles.'* Ibid. 260. Bingham described finding 'a few red Spanish roofing-tiles of various sizes' 'Along the Uncharted Pampaconas' 462. Savoy noted that the colour and incisions on the ceramic tiles showed that they were of Inca manufacture ('Last Refuge of the Incas', *Peruvian Times*, 9 April 1965). Mark Howell and Tony Morrison noticed that the antiquity of the tiles was proved by the fact that some of them were deeply embedded into a tree (*Steps to a Fortune* 31).

499 *enjoyed life there.'* Murúa **1** 260.

INDEX

Spanish compound surnames are indexed thus 'Núñez de Balboa, Vasco'. Incas, whether baptised or not, are indexed under their first names unless they have taken *Spanish* surnames. Exceptions are cross-referenced. Italic figures (*123*) refer to the Notes and References on pages 547–624, in which only the main topics are indexed; the number given is *that to which the note refers* (printed before each note and at the head of the page) *not* to the page on which the note is printed. Bold figures (**45**) refer to plates by number.